Studies in Systems, Decision and Control

Volume 590

The series "Studies in Systems, Decision and Control" (SSDC) covers both new developments and advances, as well as the state of the art, in the various areas of broadly perceived systems, decision making and control–quickly, up to date and with a high quality. The intent is to cover the theory, applications, and perspectives on the state of the art and future developments relevant to systems, decision making, control, complex processes and related areas, as embedded in the fields of engineering, computer science, physics, economics, social and life sciences, as well as the paradigms and methodologies behind them. The series contains monographs, textbooks, lecture notes and edited volumes in systems, decision making and control spanning the areas of Cyber-Physical Systems, Autonomous Systems, Sensor Networks, Control Systems, Energy Systems, Automotive Systems, Biological Systems, Vehicular Networking and Connected Vehicles, Aerospace Systems, Automation, Manufacturing, Smart Grids, Nonlinear Systems, Power Systems, Robotics, Social Systems, Economic Systems and other. Of particular value to both the contributors and the readership are the short publication timeframe and the world-wide distribution and exposure which enable both a wide and rapid dissemination of research output.

Indexed by SCOPUS, DBLP, WTI Frankfurt eG, zbMATH, SCImago.

All books published in the series are submitted for consideration in Web of Science.

Bahaa Awwad
Editor

AI and IoT: Driving Business Success and Sustainability in the Digital Age

Volume 2

Set 2

Springer

Editor
Bahaa Awwad
Department of Computerized Banking
and Financial Sciences
Palestine Technical University—Kadoorie
Tulkarm, Palestine, State of

ISSN 2198-4182 ISSN 2198-4190 (electronic)
Studies in Systems, Decision and Control
ISBN 978-3-031-88873-1 ISBN 978-3-031-88874-8 (eBook)
https://doi.org/10.1007/978-3-031-88874-8

This Springer imprint is published by the registered company Springer Nature Switzerland AG
The registered company address is: Gewerbestrasse 11, 6330 Cham, Switzerland

Contents

Green Human Resource Management Practices and Their Impact on Employee Environmental Behavior in Oman's Government Sector

Aflah Said Amur Saif Alhinai and Mohamed Elgeddawy

Abstract The study on "The Green Human Management Practices and their Roles in Enhancing the Environmental Behavior of Employees in the Governmental Sector in the Sultanate of Oman". It tackled the impact of Green Human Resource Management (GHRM) practices in boosting the environmental behavior of employees in Omani government institutions. It tried to explore to what extend such practices as environmental recruitment, training and incentives can enhance employees' environmental awareness and urge them to undertake these environmentally friendly based green behaviors that would contribute to the achievement of the sustainable development goals and Oman's (2040) vision. For the analysis of the practices of the (GHRM) in the Omani government sector descriptive deductive approach was used via questionnaires and interviews conducted with the employees and officials. Statistical and qualitative methods were used to collect and analyze the data gathered to draw logical conclusion. The study concluded that, using environmentally inspired practices such training, incentives innovation has a positive effect on employees' environmental behavior, and which will in turn reinforce their commitment to an environmental sustainability. However, some challenges such as lack of environmental understanding and poor supportive policies have been identified to impact the effectiveness of these practices. In conclusion, the researcher recommended the importance of developing comprehensive environmental training programs, promoting green employment policies, and providing incentives which would encourage employees to adopt friendly environmental practices. Eventually, the researcher suggested that environment-based policies should be regularly updated in order to assure that, sustainability objectives are successfully achieved in the Omani government sector.

A. S. A. S. Alhinai (✉) · M. Elgeddawy
Modern College of Business and Science, Muscat, Oman
e-mail: 20246736@mcbs.edu.om

M. Elgeddawy
e-mail: Mohamed.Elgeddawy@mcbs.edu.om

B. Awwad (ed.), *AI and IoT: Driving Business Success and Sustainability in the Digital Age*, Studies in Systems, Decision and Control 590,
https://doi.org/10.1007/978-3-031-88874-8_51

Keywords Green human resource management · Environmental behavior · Sustainable development · Government sector · Sultanate of Oman

1 Introduction

Due to the increasing environmental challenges encountering the world, sustainable development issues and environmental protection has become a top concern of the governments and institutions all over the world. Role of government and institutions in the implementation of environmental policies is increasingly emphasized as these institutions are a key driver in implementing the Sustainable Development Objectives. In this context, it has become necessary to integrate environmental practices into all aspects of the work of organizations, including human resources administration. Green Human Resource Management (GHRM) practices are important, which aim to link environmental considerations with the recruitment, training, motivation, and performance appraisal processes of organizations. These practices are an effective way to increase environmental awareness among employees, thereby promoting sustainable environmental behavior in the work environment. In general, it can be asserted that the introduction of information and communication technologies is a true picture in the department of the change it brings about in the style, effectiveness and performance of administrative work. Hence, an online management means moving from local work to information applications, including computer networks, to join organizational units with each other to facilitate access to data and information to make appropriate decisions, complete work and provide services to beneficiaries efficiently, at the lowest cost and as soon as possible.

As a result of an increasing understanding of the importance of environmental sustainability in public organizations, it has become necessary to undertake green human resource management (GHRM) practices that integrate environmental considerations into all HR policies such as recruitment, training, motivation, and performance appraisal. Green HRM practices intent to improve the environmental behavior of employees via several techniques that include hiring environmentally conversant individuals, training them in sustainable behaviors, and motivating them to reduce the consumption of natural resources. These practices can contribute significantly to building a durable environmental culture within government institutions, which supports employees' commitment to environmental sustainability goals and contributes to improving the environmental performance of organizations.

In fact, Omani government sector is still late in the application of these practices compared to some other countries which are seeing early growing interest in this field. The Sultanate of Oman, as part of the GCC, adopts ambitious environmental strategies in line with Oman Vision 2040 which aims to achieve sustainable development and conserve natural resources. In addition, the Omani government sector still faces some challenges in carrying out green HRM practices, such as the lack of comprehensive training programs in this area, and the lack of integrated environmental policies that support these practices continuously and effectively. Environmental culture still needs more stimulation to promote positive environmental behavior among employees within the government work environment.

Based on the previous studies on the subject, it was found that there is a growing interest in the research literature in green human resource management practices and their impact on environmental behavior in different institutions, whether in the private or public sector. Although these studies vary in their scope of application, most focus on the importance of integrating environmental values into human resources policies and how this affects the environmental behavior of employees. Some studies suggest that practices such as environmental training, motivating employees to reduce the consumption of natural resources, and stimulating environmental innovation within organizations play a large role in achieving the SDGs. For example, a study by Jamaluddin et al. in the UAE showed that employees who are exposed to environmental training programs or incentives that encourage them to adopt environmentally friendly behaviors show higher levels of environmental awareness and apply these behaviors in their working lives. Ali and Sami study [1] in Egypt also confirmed that the integration of environmental standards in performance evaluation and motivation enhances employees' commitment to the values of environmental sustainability, which contributes to meliorating the performance of organizations and achieving their environmental goals. However, some studies indicate that the implementation of green HRM practices in the government sector faces some challenges related to the lack of adequate awareness of the importance of these practices and the lack of clear policies to support them on an ongoing basis. Al-Marzouqi's [2] study in Saudi Arabia confirmed that government institutions need to develop strategies to promote a culture of sustainable environmental action, by linking environmental policies with human resource management systems. Although these studies focusing on the private and public sectors in some Arab countries, the literature on the application of in the government sector in the Sultanate of Oman is still limited, which makes this study contribute to minimizing this knowledge gap by analyzing the impact of these practices on the environmental behavior of employees in Omani government institutions and giving new insights that contribute to promote the environment oriented sustainability in the Oman's public.

2 Literature Review

Jamal Eldin et al. (2021) study on "Green Human Resource Management Practices in Government Institutions in the UAE" [3]. It examines the influence of green human resource management practices in improving the employee's environmental performance in government institutions in the UAE. The highlights the role of incorporating environmental values into human resources policies and its effects on the employees' environmental behavior. The results revealed that environmental training programs and incentives related to environmental behaviors have an essential role in increasing environmental understanding among employees and encourage them to adopt environmentally friendly behaviors. The study recommended the expansion of environmental training programs and executing durable incentive policies to ensure the ongoing of these behaviors within government institutions.

On her study, "The Impact of Green Human Resource Management Practices on Environmental Performance in the Government Sector in Egypt." Ali and Sami [1] aimed to measure the impact of green human resource management practices on the environmental performance of employees in the Egyptian government sector. The necessity of the study is to highlight the impact of incorporating environmental standards in performance appraisal and stimulate employees to take up durable environmental behaviors. The findings revealed that incorporating environmental motivation into evaluations and performance development processes meliorate employees' adherence to environmental practices, resulting in overall improvement of environmental performance of the organization. The study recommended the importance of introducing environmental standards at all levels of management evaluation and provide incentive rewards for environmental behavior.

Faisal Al-Marzouqi's [3] in his study on "Challenges Facing the Application of Green Human Resources Management Practices in the Saudi Government Sector" dealt with: [4] aims to examine the challenges encounter practicing green human resource management application in government institutions in the Kingdom of Saudi Arabia. This study is important as it provided insight about the hindrances to effective implementation of these policies in Saudi government institutions. The findings indicate that the most important problems are the lack of environmental awareness among employees, poor supportive government policies, and lack of adequately intended training programs. The study recommended increasing environmental awareness among employees and developing supportive policies to promote a culture of sustainable work in the government sector.

Jackson's study [5] "The Current State and Future Trends of Green Human Resource Management: A Global Perspective" explored the global practices of green human resource management (GHRM) and their role in generating environmental sustainability and upgrading durable environmental behavior among employees in world-class organizations. The significance of this study lies on providing a comprehensive view of how these practices are operated in organizations and their effects on environmental performance and environmental behaviors of employees, as it was found that organizations that adopt GHRM practices witness a greater commitment by their employees towards sustainable environmental behaviors, which contributes to improving overall environmental performance. The study findings revealed that the integration of GHRM practices helps in to improve the regulatory environment and raise the effectiveness of sustainability efforts within organizations. The study recommended the importance of developing clear comprehensive strategies that incorporate GHRM practices at every stage of the organizational process, from recruitment and training to performance appraisal, in order to enhance the achievement of environmental goals for organizations on a global level [6].

3 Research Objective

1. Explore the impact of green HRM practices on the environmental behavior of employees in the Omani government
2. Analyze the challenges facing the implementation of green human resource management practices in the Omani government
3. Provide recommendations enhancing the integration of green human resource management practices in Omani government institutions

4 Research Objective

1. Reinforcing environmental awareness in the government sector.
2. Supporting Oman's Vision 2040 objectives of sustainability.
3. Ameliorating environmental policies in government institutions.
4. Activating green human resources management practices.
5. Felling the research gap in the application of GHRM practices in Oman.

5 Research Problem

Despite the growing interest in green human resource management (GHRM) practices globally, and the success of many countries in implementing these practices in the public and private sectors, there is a noticeable lapse in studies that emphasizes the application of these practices in the government sector in the Sultanate of Oman. However, government institutions in Oman assuming environmental policies under Oman Vision 2040, which seeks to achieve sustainable development, the application of green human resource management practices remains limited in some respects, which affects the level of environmental behavior of employees in these institutions. The research problem stems from the inability of many Omani government institutions to incorporate environmental considerations into their HR management policies. Training staff on eco-friendly behaviors encouraging them to come up with green innovations and cutting down on natural resource use are practical ways to promote sustainable environmental behavior in the workplace. However, the absence of specific well-organized programs to push these practices might hurt employees' dedication to sustainable environmental behavior.

On the other hand, locally related literature and previous studies lack in-depth research on the relationship between green human resource management practices and environmental employee behavior in the Omani government sector. There are not enough studies exploring how practices such as environmental recruitment, environmental training, environmental performance assessment, and green motivation can directly affect improving environmental awareness and behavior among employees within government organizations. The current study is not based on accurate information or data on how these practices are applied in the Omani context, which makes the need for this study even more urgent.

Hence the research problem emerges, it is how can green human resource management practices enhance the environmental behavior of employees in the government sector in the Sultanate of Oman?

6 Research Questions

1. What is the effect of green human resource management practices on enhancing the environmental behavior of employees in the government sector in the Sultanate of Oman?
2. What are the most influential green HR management practices in enhancing the environmental behavior of employees in the Omani government sector?
3. What are the challenges facing government institutions in the Sultanate of Oman in implementing green human resource management practices?

7 Theoretical Framework

This section investigates the key concepts related to green human resource management, environmental behavior, and their relationship to upgrading sustainability in official institutions, with an emphasis on the local context in the Sultanate of Oman.

7.1 Green HRM Practices

Green HRM practices refer to a set of policies and practices undertaken by organizations to integrate environmental values into all human resource management processes, including recruitment, training, performance appraisal, and motivation. These practices aim to reinforce environmental understanding among employees and encourage them to adopt environmentally friendly work behaviors, such as reducing the consumption of natural resources, improving energy efficiency, and reducing waste. Green HRM practices also include motivating employees to innovate in environmental solutions and offering rewards and incentives to contribute to achieving sustainability goals [2].

7.2 Environmental Behavior of Employees

The environmental behavior of employees refers to the set of actions and patterns carried out by individuals in the workplace that aim to guard the environment and minimize negative impacts on natural resources. This includes behaviors such as using energy efficiently, reducing water consumption, proper waste disposal, and participating in environmental activities inside and outside the work environment. The environmental behavior of employees can be the result of environmental awareness or due to incentives and programs offered by organizations to encourage employees to adopt environmentally friendly behaviors [1].

7.3 The Relationship of Green (HRM) Practices to the Environmental Behavior of Employees

The relationship between green HRM practices and employees' environmental behavior is embodied in the way HR policies influence employee behavior towards environmental issues. By providing environmental training, providing incentives to participate in environmental activities, and including environmental standards in performance appraisals, organizations can promote environmental awareness among employees and motivate them to adopt sustainable environmental practices. Studies show that employees who are exposed to environmental training programs and practice environmentally friendly behaviors within the work environment show a higher level of environmental awareness that extends to their daily lives [7].

7.4 Sustainable Development in the Sultanate of Oman

The Sultanate of Oman has undertaken sustainable development strategies in various fields, including in the government sector. Oman Vision 2040, a long-term strategic plan, focuses on balancing economic development and environmental conservation. In this context, the Omani government sector is one of the main pillars to achieve this vision, by undertaking durable environmental practices that contribute to enhancing environmental protection and achieving sustainable development goals. Despite these efforts, the implementation of green human resource management practices in Omani government institutions is still evolving, which calls for studying how these practices affect the environmental behavior of employees [8].

7.5 Challenges in the Application of Green Human Resources Management Practices in the Omani Government Sector

Despite an increasing attention to environmental sustainability issues being paid by Sultanate of Oman, there are challenges in implementing green human resource management practices in the government sector. The most prominent of these challenges is the lack of adequate environmental awareness among employees, and the lack of training programs that focus on environmental issues. Moreover, the absence of clear regulatory policies in some government institutions may lead to poor implementation of green HR practices effectively. The government sector also faces challenges related to training resources and funding to implement comprehensive environmental programs, which leads to difficulty in achieving the desired changes in the environmental behavior of employees [9].

7.6 Theory of Environmental Behavior

The study is based on environmental behavior theory that explains how environmental awareness and motivation affect individuals' behavior towards the environment. This theory proposes that environmental behavior is framed by many factors such as environmental education, organizational incentives, and leadership guidance in an organization. Through green HRM practices, government organizations can reinforce these factors and push employees to adopt environmentally friendly behaviors in the work environment [5].

8 Research Methodology

The descriptive and deductive approach was applied in this study to analyze green human resource management practices in the Omani government sector and study their impact on the environmental behavior of employees. The curriculum relied on deriving hypotheses from previous literature on the relationship between environmental policies in organizations and employees' environmental behavior, such as the hypothesis that "practices such as environmental training and motivation to adopt environmental behaviors contribute to improving employees' environmental behavior." The data was then collected through various research tools such as questionnaires and interviews with employees and government officials and analyzed using statistical methods to analyze quantitative data and test hypotheses related to the impact of environmental practices on employees' environmental behavior. In addition, interviews and field observations were analyzed to extract the factors affecting the application of green human resource management practices and identify the challenges facing Omani government institutions. Through this process, conclusions were reached that support the deductive hypotheses and help explain the relationship between green human resource management practices and the environmental behavior of employees, which contributed to providing practical recommendations to improve the application of these practices and enhance environmental sustainability in the Omani government sector.

9 Conceptual Framework /Literature Review

The conceptual framework of this study includes a set of key concepts that form the theoretical basis for understanding the relationship between green human resource management practices and the environmental behavior of employees in the government sector in the Sultanate of Oman. The conceptual framework consists of three main axes: Green Human Resource Management Practices (GHRM), Environmental Behavior of Employees, and Sustainable Development:

9.1 Green Human Resource Management Practices

Green HRM practices refer to a set of policies and procedures undertaken by organizations to encourage employees to adopt environmentally friendly behaviors. These practices include, among others, green hiring [10], where employees are selected based on their environmental awareness, environmental training, which enhances environmental awareness in employees and encourages them to improve their work practices, and motivates employees by offering incentives that encourage sustainable environmental behaviors such as reducing energy and resource consumption [11].

9.2 Employee Behavior of Employees

Environmental behavior of employees refers to the group of behavioral patterns undertaken by individuals in the workplace with the aim of reducing environmental impact, such as reducing the consumption of natural resources, using energy efficiently, and proper waste disposal. The environmental behavior of employees is influenced by several factors, including the organizational environment, training, motivation, as well as an organizational culture that encourages environmental sustainability [12].

9.3 Sustainable Development

Sustainable development is a concept that refers to a development that caters for the ambitions of present generations without bargaining the ability of future generations to achieve their needs. In the context of this study, sustainable development is about the application of environmentally friendly practices in the government sector to improve environmental performance and enhance sustainability. These practices help achieve the goals of Oman Vision 2040, which include enhancing the sustainability of natural resources and minimizing negative environmental impacts [13].

10 The Relationship Between Variables

Green Human Resource Management Practices (Independent Variable) includes a set of policies and practices applied by government institutions such as green recruitment, environmental training, and motivation for positive environmental behavior. These practices aim to motivate employees to adopt sustainable environmental behaviors in the work environment. The environmental behavior of employees is the "dependent variable", which is tied to the behavioral patterns adopted by employees at work to achieve environmental sustainability, such as reducing the consumption of natural resources, improving energy efficiency, and adopting environmentally friendly practices in daily activities. The relationship between these two variables is a direct impact relationship, as the study indicates that green human resource management practices positively affect the environmental behavior of employees; that is, the application of environmental policies such as environmental training and motivation for sustainable behaviors enhances employees' environmental awareness and motivates them to adopt sustainable behaviors within government institutions. In turn, by promoting these sustainable environmental behaviors, government institutions can achieve the sustainable development goals at the environmental and social levels, thus contributing to improving the overall environmental performance of enterprises and thus achieving the country's strategic goals such as Oman Vision 2040.

11 Results

The study concluded the following results:

1. Positive impact of green HRM practices: The study showed that practices such as environmental recruitment, environmental training, and green motivation contributed to enhancing environmental behavior among employees in the Omani government sector.

2. Increased environmental awareness: The results showed that employees who underwent environmental training programs showed higher levels of environmental awareness, which helped them adopt environmentally friendly behaviors.
3. Motivating employees for sustainable environmental behavior: Environmental incentives and recognition have been shown to motivate employees to reduce the consumption of natural resources and innovate in environmental solutions.
4. Challenges in the application of practices: Several challenges facing the implementation of green human resource management practices have been identified, such as lack of sufficient awareness among employees, and the absence of some integrated environmental policies.

12 Discussion

Within the framework of the study findings, the impacts gained by investigating green human resource management practices (GHRM) in the Omani government sector can be discussed. First, practices such as environmental recruitment, environmental training, and green motivation have proven effective in promoting environmental behavior in employees. These findings are consistent with previous literature that has shown that integrating environmental standards into HR policies helps improve environmental performance within organizations. It has been found that employees who undergo environmental training programs develop a higher level of environmental awareness, which reflects positively on their work behaviors and encourages them to adopt environmentally friendly practices. For example, they adopt behaviors such as reducing energy consumption, using resources more efficiently, as well as participating in environmental activities within the organization.

As regarding environmental incentives, the results showed that encouraging employees to adopt sustainable environmental behaviors through rewards and incentives leads to increased commitment to environmental practices, which enhances their interaction with the organization's green policies. These results confirm that employee motivation is a powerful tool to ensure the continuity of sustainable environmental behaviors, reinforcing their commitment to contributing to the achievement of corporate sustainability goals.

However, the challenges demonstrated by the study in implementing green human resource management practices cannot be overlooked. The findings indicate gaps in the implementation of environmental policies within Omani government institutions, such as a lack of environmental awareness among some employees and the lack of adequate training programs. In addition, the study showed that the absence of some integrated environmental policies in some institutions is a major obstacle to the effective application of these practices. Adopting green HRM practices requires not only the support of senior leadership, but also the development of clear and integrated environmental policies to ensure that all employees are committed to achieving sustainability goals.

An update to environmental policies periodically is obviously needed to keep pace with the progressive changes in the global and local environmental domains. The environmental challenges facing the world today, such as climate change and the increasing scarcity of natural resources, require government institutions in Oman to adopt sustainable strategies that are flexible and constantly evolving. Continuously developing and rethinking green HR policies is essential to ensure their effectiveness.

Regarding the above findings, it can be argued that green human resource management practices are now a necessary tool for many countries seeking to achieve the SDGs, especially in the government sector. However, despite the importance of these practices, the study highlights the need for the Omani government sector for more institutional support and guidance to accelerate the effective adoption of these environmental policies, thus achieving a greater impact on the environmental behavior of employees.

13 Recommendations

1. Developing comprehensive environmental training programs to enhance environmental awareness among employees in the government sector.
2. Including environmental criteria in recruitment policies to select environmentally conscious employees.
3. Motivating employees to adopt environmentally friendly behaviors through environmental incentives that encourage environmental innovation.
4. Set Establishing units or work teams dedicated to green human resources within government institutions to implement environmental policies.
5. Enhancing cooperation between the government sector and civil society to implement environmental sustainability practices within institutions.
6. Update environmental policies to keep up with new developments and meet local and global environmental needs.

References

1. Ali, F., Sami, A.: The impact of green human resource management practices on environmental performance in the Egyptian government sector. J. Adm. Environ. Stud. **24**(1), 73–89 (2020)
2. Al-Marzouqi, F.: Challenges facing the implementation of green human resource management practices in the Saudi government sector. J. Public Adm. **35**(2), 102–118 (2022)
3. Al-Mahrouqi, D., Al-Busaidi, N.: Challenges of implementing green human resource management practices in the public sector of Oman. Int. J. Public Adm. **43**(8), 689–700 (2020)
4. Boiral, O., Paillé, P.: Organizational citizenship behavior for the environment: measurement and context. J. Bus. Ethics **109**(4), 431–445 (2012)
5. Jackson, S.E., Renwick, D.W., Jabbour, C.J.C., Muller-Camen, M.: State-of-the-art and future directions for green human resource management: a global perspective. Int. J. Hum. Resour. Manag. **22**(4), 2263–2287 (2011)

6. Daily, B.F., Huang, S.: Achieving sustainability through environmental human resources. Int. J. Oper. Prod. Manag. **21**(7), 837–862 (2001)
7. Mohamed, G.-D., El-Attar, A., Rahab, S.: Green human resource management practices in government institutions in the United Arab Emirates. J. Environ. Manag. Sustain. Dev. **19**(3), 45–62 (2021)
8. Jabbour, C.J.C., Santos, F.C.A.: Environmental training in companies: the influence of human resource management. Int. J. Hum. Resour. Manag. **19**(10), 1848–1864 (2008)
9. Jackson, S.E., Seo, J.: The greening of strategic HRM scholarship. Organ. Manage. J. **7**(4), 278–290 (2010)
10. Renwick, D.W.: Green human resource management: a strategic approach. Int. J. Manag. Rev. **10**(1), 16–35 (2008)
11. Renwick, D.W., Redman, T., Maguire, S.: Green human resource management: a review and research agenda. Int. J. Manag. Rev. **15**(1), 1–14 (2013)
12. Schroeder, P., Richey, R.: Sustainability and corporate governance: a global perspective. J. Bus. Environ. Ethics **8**(3), 225–245 (2010)
13. Stern, P.C.: Toward a coherent theory of environmentally significant behavior. J. Soc. Issues **56**(3), 407 (2000)
14. Al-Rawahi, N., Al-Salti, Z.: Environmental sustainability and green practices in Oman's government sector. Oman J. Environ. Stud. Res. **22**(3), 118–130 (2017)

Towards Developing Sustainability-Driven Strategic Framework in ICT SMEs in Oman: Case Study

Aleksandra Minic, Ahmed Mohamed Al Naamany, Khubayb Ahmed Al Naamany, and Sanja Filipovic

Abstract ICT SMEs in emerging economies encounter a range of systemic challenges that affect growth. Some of these include inadequate leadership, insufficient market research and vulnerability to external factors. Despite their potential to be the drivers of economic growth in developing countries, small and medium ICT companies challenges in integrating sustainable development into their strategies. This paper proposes a strategic framework that integrates sustainability-driven practices: the Triple Bottom Line (TBL) and the United Nations Sustainable Development Goals (SDGs). By embedding sustainable leadership, stakeholder engagement, non-financial reporting and continuous innovation into core business strategies, ICT SMEs can simultaneously achieve operational efficiency, financial stability, and societal impact. Through a case study of an ICT SME in Oman, the paper demonstrates how adopting these sustainable practices enhances resilience and competitiveness in rapidly changing markets. The paper concludes by recommending future research to validate this framework across more SMEs, offering pathways for broader application.

Keywords ICT SMEs · Sustainable development · Competitive advantage · Triple bottom line

A. Minic (✉) · A. M. Al Naamany
Modern College of Business and Science, Muscat, Oman
e-mail: alexandra@mcbs.edu.om

K. A. Al Naamany
International Networking Company, Muscat, Oman

S. Filipovic
Singidunum University and Institute of Social Sciences, Belgrade, Serbia

B. Awwad (ed.), *AI and IoT: Driving Business Success and Sustainability in the Digital Age*, Studies in Systems, Decision and Control 590,
https://doi.org/10.1007/978-3-031-88874-8_52

1 Introduction

Small and Medium Enterprises (SMEs) in the Information and Communication Technology (ICT) sector have been an important pillar of technological innovation and thus the basis of the economic growth of developing countries. United Nations report [1] emphasizes the benefits that SMEs bring to the emerging economies in terms of innovation, employment opportunities, reducing poverty and overall economic growth of these countries. They also point out that one of the biggest challenges in growing and staying afloat is access to finance. The issue is no longer prevalent in developed countries. Although their potential is significant, these organizations encounter challenges that adversely affect their progress. Inadequate market research, the external factors and ineffective leadership have been identified as critical obstacles [2]. In response to these challenges, this short paper proposes sustainable development (SD) as a strategic tool not only to maneuver through these difficulties but also to help small IT companies establish their competitive advantage.

Transformative leadership is fundamental in a culture of sustainability in businesses [3]. This study examines the importance for equipping leaders with SD awareness and integrating sustainable practices to enhance competitiveness and long-term financial sustainability of small IT companies while benefiting society and the environment. Focusing on SD practices attracts environmentally conscious investors and sets ICT firms apart in a crowded market [4].

This paper also explores the impact of sustainability pillars—financial, social, and environmental—on the success of tech SMEs using the UN SDGs [5], circular economy principles [6] and the Triple Bottom Line (TBL) framework [7] as its backbone. By integrating these perspectives, digital SMEs may achieve sustainable growth that benefits economic viability, society and the environment. The study also proposes the importance of non-financial reporting for enhancing transparency and accountability [8] and emphasizes stakeholder proactivity as a key to long-term success.

Although these companies' potential to drive economic development in emerging markets is high, small and medium ICT Organisations face challenges in understanding sustainable development concepts and the importance of integrating them into their business strategies. They also lack the systematic ways of weaving sustainability frameworks into their strategic plans. To fill this gap, the authors propose an SDG-driven framework to improve resilience and competitiveness, tested through a case study of an SME ICT firm in Oman.

2 Literature Review

In addressing the role of SD in the competitiveness and success of ICT SMEs in developing countries, a thorough review of existing literature has been done. The idea that there is a need to explore ICT SMEs survival and success through various venues was brought to different stakeholders' attention among the private as well as public sector especially given that, according to the World Economic Forum, global economy heavily depends on them, nearly 70% of jobs and GDP globally have been created thanks to the SMEs [9]. A small country in Africa, Botswana has seen an SME failure rate of 80% [10], a trend that all major decision makers agree must not continue. The integration of SD practices into the ICT sector, the authors believe, is a strategic approach for dealing with challenges and securing a niche area that can distinguish them in an extremely competitive market and at the same time genuinely contribute to the better world for the generations to come. Literature has been systematized and organized according to the prevailing themes.

2.1 *Challenges and Potential Solutions*

Research presented by [2] discusses challenges facing ICT SMEs, including deficiencies in market analysis, external threats, and issues with leadership. The backdrop for this argument was set much earlier by [3] who emphasized the importance of ethical leadership in combination with spiritual, transformational and authentic leadership in order to develop an encouraging a culture within organizations. This type of leadership aligns with one P (people) in the TBL [7] shown in Fig. 1. This viewpoint is reinforced by the findings of [11] who selected a sample of 300 employees working in SMEs in Jordan. Applying Structural Equation Modelling method, they found significant impacts that well-planned business strategies have on the competitive advantage of SMEs, emphasizing the importance of leadership in guiding ICT SMEs towards success and longevity. Further, a study of SMEs in Bosnia and Hercegovina [12] states that in this small Southern European country, SMEs represent 99% of all business in a non-financial sector. Despite their high importance for the economies of developing countries, almost half of all SMEs globally close down in the first five years. Another research dealing with the question of why SMEs lack competitive advantage [13] delves into the matter by implying that the solution is managing knowledge, innovation and intellectual capital. Additionally, [14], identified several factors that affect SMEs' technological development—social capital, Transnational Corporations (TNCs), Multi-National Corporations (MNCs), innovation, networking, IT, and adoption of productivity-enhancing technology. The authors also state that it is necessary to have adequate human resources, the latest technology as well as policies in place for the success of SMEs.

2.2 *SD Practices Within SMEs*

The literature further emphasizes the need for ICT SMEs to adopt sustainable practices, focusing on energy management and efficiency. Hart and Milstein [4] explain

Fig. 1 Triple bottom line framework. *Source* Open-source license: CC BY 4.0

the strategic benefits of such practices, explaining how they may attract environmentally and cost-conscious investors and consumers. This strategic alignment with sustainability not only differentiates SMEs in crowded markets but also aligns with broader global initiatives, such as SDGs and principles of the circular economy, advocated by sources like the Ellen MacArthur Foundation. Further exploration into non-financial reporting, as discussed by Brown et al. [8] reveals its significance in fostering transparency and accountability within ICT SMEs. Such practices are integral to building trust among stakeholders and reinforcing the company's commitment to SD. Creating sustainable value in fact brings value to the firm's business, claim [4] who explain that the framework is simple, but the execution must be taken seriously and done systematically. Moreover, [15] presents an argument that ICT sector, apart from its major contributions to business, has adverse effects on the environment. The researcher therefore proposes, through a case study, investigating the impact, through measuring e-waste and other factors, and focusing on incorporating environmental sustainability in the ICT SMEs practices.

A concept called coopetition has been researched [16] which discusses an approach by combining competition with cooperation. Drawing on previous literature and observing practical approaches, the author discusses the long tendency of encouraged competition (driving down prices) and discouraged collaboration (monopoly). They conclude that, if both competition and cooperation are beneficial for the consumer (all stakeholders) this is then encouraged. SMEs may take this concept, which would promote SDG 17 to a higher level by collaborating in innovation technologies on a global level. Mottonen et al. emphasize the importance of taking internal stakeholders requirements into consideration when making decision in ICT sector. Benchmarking these requirements with a Japanese automotive company, the authors believe that ICT sector needs fresh ideas in dealing with their employees. It is important to note that there are few research papers that state digital innovation, Internet of Things (IoT) and other technological development in general as needed for the advancement and success of SMEs. Yousaf et al. [17] exploring these matters and interviewing CEOs and General Managers of SMEs in Pakistan observe that the lack of resources results in inability to collect relevant data and advise that management should invest in technologies that will allow them to collect big data successfully. For the purpose of this paper, however, the advantages of technological innovation are taken into account as a part of Continuous Quality Improvement (CQI) as it is assumed that ICT SMEs, by nature of their business and expertise, are capable of responding to most if not all of the challenges of this nature.

In summary, the literature review analyses ICT SMEs in developing countries and sheds light towards integrating sustainable development into their core operations. The insights from the reviewed literature propose a strategic direction—implementing SD actions, choosing sustainability literate leadership, and stakeholder engagement. These are seen as the success agents of ICT SMEs in the developing world.

3 Methodology

3.1 Review of Literature

Using databases like Scopus, Web of Science Google Scholar, over 20 peer-reviewed articles and relevant sources were selected. Inclusion criteria:

(a) Time Frame: Only the paper published after the year 2000 were selected with the priority given to those published after 2010, though seminal works on TBL and sustainability were used regardless of publication date.
(b) Publication Type: Focus on renowned books, industry reports, and peer-reviewed articles and sources like the Harvard Business Review and Ellen MacArthur Foundation.
(c) Language: For clarity, only English sources were used.
(d) Context: Global studies were used with the emphasis on those relevant to ICT SMEs in developing countries.
(e) Themes: SDGs in strategic planning; Sustainable strategies for marketing; CSR, and competitive advantage; TBL's link to environmental, social, and economic growth.

3.2 Data Collection and Analysis

The selected literature papers were reviewed to understand trends, challenges, and opportunities in using sustainability frameworks strategically. The focus was on strategies that align with SDGs and TBL, a case study that provide practical insights into partially successful SDG implementation and gaps and challenges found in existing frameworks for ICT SMEs.

The purpose was to gain a deeper understanding of how sustainable practices can enhance the resilience and growth of small IT firms.

3.3 The SDG and TBL Mapped Framework

The insights from the literature review allowed for an SDG-driven strategic framework which integrates principles from the Triple Bottom Line (TBL) to be developed. The framework's aim is to provide action-oriented strategies for ICT SMEs to use sustainability to be competitive in their market.

3.4 Case Study Analysis

The final stage involved an analysis of one SME in the ICT sector in Oman as a case study. This analysis gave insight into evidence of how integrating sustainability into a strategic framework can drive growth in the ICT sector of an emerging economy.

4 Framework Development and Discussion

4.1 SDG and TBL Framework

This paper has explored the integration of SD principles, specifically TBL which combines people, planet and profit (also known as three Ps model) as opposed to the traditional bottom line, profit only which most companies aim for. The SDGs are used together with TBL as this research aims to offer five pillars, Table 1 to be incorporated into the strategies of digital SMEs in developing countries. The TBL framework [7] provides a model that SMEs can use to measure their impact on social, and environmental and economic practices and can guide them towards more holistic approach to success. It is recommended that future research continue to investigate meaningful relationships between ICT innovations and SD practices thus enhancing further sustainable growth in the digital age. Having systematized relevant literature, this study finds that if the ICT SMEs include the five components below

Table 1 SMEs components for SD implementation, benchmarked with SDGs and TBL

Main Pillars	SDGs	TBL
Progressive leadership (Sustainability and innovation conscious)	SDG 4: Quality Professional Development SDG 5: Gender Equality SDG 17: Partnerships	People
All stakeholder engagement and collaboration	SDG 12: Responsible Consumption and Production SDG 17: Partnerships for the Goals	Profit and people
Ethical and sustainable development measures	SDG 13: Climate Action SDG 8: Decent Work and Economic Growth SDG 12: Responsible Consumption and Production	Planet and profit
Non-financial reporting	SDG 7: Affordable and Clean Energy SDG 11: Sustainable Cities and Communities SDG 16: Peace, Justice, and Strong Institutions	People and planet
Continuous quality improvement (CQI)	SDG 3: Good Health and Well-being SDG 8: Decent Work and Economic Growth	People, Planet, Profit

from the outset of their strategic design, they will have a comprehensive competitive edge which would lead to success of their business. The components are: progressive leadership, stakeholder engagement, SD practices, enhanced non-financial reporting, and continuous quality improvement (CQI).

4.2 Discussion

The progressive leadership element is believed to be essential in implementing and driving SD integration within SMEs. Leaders who have a substantial understanding of SD were seen to incorporate these practices into the organizational culture and strategy. That way they were able to provide environment for innovation and success to their employees [2]. This particularly aligns with people in TBL as the internal stakeholders are inspired and motivated to find innovative solutions as transparency and inclusiveness is practiced by their management team. They are also provided opportunities for constant growth through regular professional development sessions which is further motivating and inspires growth on a personal and professional level. This strategic differentiation is crucial in standing out in saturated markets, corroborating [4], assertion of sustainability as a competitive value driver. This pillar is aligned with People in TBL and SDGs 4, 5 and 17. The leaders' responsibility is to get familiar with and manage sustainable practices, then lead all the stakeholders to follow suit.

Lead by the management team all internal stakeholders, mid-management and all employees in ICT SMEs should buy into SD and start working on action plans with clear Objectives and Key Results (OKRs) and measured by Key Performance Indicators (KPIs). Once this becomes a part of the strategy, reports are written and presented to the external stakeholders such as investors. Managing intellectual capital well within ICT companies in Serbia [18] proves to have a highly positive effect on a company's performance and business sustainability. Further, although some environments are especially challenging for the competitive performance of SMEs, for instance Bosnia and Hercegovina [19], the authors suggest that the management of the SMEs attempts to encourage research and development, risk taking and continues effective monitoring of companies' performance.

Regarding sustainable and ethical measures, [20] discuss implementation options and challenges of another SD model, circular economy (CE) within the ICT and Consumer Electronics sectors. CE in its core has zero waste [6] which presents an obstacle for ICT companies—there must be a comprehensive strategy on how to reduce the materials and energy use; nonetheless, this is also an opportunity for the circular, sustainable practices which in the current market climate present an essential advantage. The adoption of green business strategies in energy management and efficiency specifically is seen to attract both environmentally conscious and cost-conscious customers and investors. Thus, adopting green practices would not only directly help the environment (in reducing CO2 emission, for instance) but distinguish small ICT organizations in saturated markets [21] and ideally, secure sustainable investments.

When it comes to Sustainability Reporting, SMEs which engaged in such non-financial reporting managed to track and communicate their sustainability efforts easier and more transparently leading to improved brand value and stakeholder trust [22]. They mostly adhered to Global Reporting Initiative (GRI), an internationally used mechanism by organizations to report on their environmental and social practices. The focus on non-financial reporting is shown to be an essential factor in enhancing liability and transparency within ICT SMEs [8]. This tool can show all stakeholders, employees, clients and potential investors the ethical nature of an organization. Used as fourth tool in our model above, this practice is in line with SDGs 7, 11 and 16, while in TBL it fits with both people and planet as it reports on how well a company actions non-profit KPIs. Arena and Azzone [22] argue, however, that this ethical and holistic practice can attract more clients and financing, resulting in in the final P in TBL—company's profitability.

4.3 Case Study

Based on systematized literature, the key issues were identified and solutions proposed presented as five pillars benchmarked with TBL and SDG. A case study of a small ICT company in the Gulf Cooperation Council (GCC) region shows that implementing some of the pillars stated by the authors brings more sustainable business and competitiveness to an ICT SME. Namely, having found various reasons for ICT SMEs' failure or inability to sustain business and financial success within the competitive market, [2] proves that Stakeholder Engagement and CQI, the second and fifth pillars in Table 1, indeed contribute to the success of ICT SMEs in a developing country, Oman. The study finds insufficient capital as one of the main challenges which hinders implementation of all parts of the projects simultaneously. Another reason mentioned for inadequate business functioning has to do with planning or a lack of appropriate strategic vision which are connected to the lack of strong leadership. As a solution, the author proposes a new platform that would help the ICT SMEs internal and external processes flow smoothly and ensures continuous monitoring and quality control adhering to SDG 8 and all three Ps in TBL. Moreover, the study addresses the stakeholders' involvement through training of employees (internally) and collaboration with external partners which directly aligned with Partnership for the Goals (SDG 17) as well as people and profit within TBL. In summary, the company implemented, unknowingly, the two pillars out of five thus proving that they directly contributed to successful outcomes.

5 Conclusion and Future Directions

5.1 Conclusion

As a result, a framework has been developed by the authors aiming to support ICT SMEs to navigate the challenges utilizing SD strategies through SDGs and TBL. The framework consists of five pillars—Progressive Leadership (Sustainability and Innovation Conscious), All Stakeholder Engagement and Collaboration, Ethical and Sustainable Development Measures, Non-financial Reporting, and Continuous Quality Improvement (CQI) and is tested through a case study of one ICT company in Oman. It is proposed that, should the ICT SMEs incorporate all or most of these pillars into their strategic approach, it will give them an edge especially in the markets of developing economies. This approach promises a path to sustainable competitive advantage emphasizing the crucial role of sustainability in today's unpredictable market landscapes. Consequently, the conclusion was made and demonstrated in this paper that incorporating sustainable development through its paradigms [23], plays a crucial part in advancing ICT SMEs in developing countries.

5.2 *Future Directions*

The major limitations encountered were the limited number of studies focusing specifically on small and medium-sized ICT businesses. Therefore, it is suggested that further research is conducted in different areas and with different goals in mind, such as:

(a) Conducting research using many case studies comparatively in Oman to understand the success of the application of the above framework.
(b) Selecting several ICT SMEs in emerging economies in various countries in MENA region and the South East Asia and analyzing their sustainability awareness level and application of the above pillars in their strategic approach. This would help with comprehensive understanding of sustainability awareness across regions.
(c) Selecting a narrow research focus to specific ICT sectors, for example, telecommunication or software development and explore how SD strategies vary in different ICT companies.
(d) Assessing the government directives' and policies' role, incentives and repercussions, in promoting sustainability among small ICT firms. In Oman, it might be beneficial to explore the impact of Oman Vision 2040 in the sector.
(e) Using a mixed method, interviews and surveys, a research can be conducted to have a deeper understanding of how sustainability is integrated and what more needs to be done to successfully use SD frameworks in enhancing ICT SMEs competitiveness.
(f) Analyzing how Green IT concept, designing sustainable computing solution such as energy efficient data centers, e-waste management, cloud computing, green procurement, can be implemented for the benefit of small IT companies and beyond.

References

1. United Nations: Improving the competitiveness of SMEs in developing countries the role of finance to enhance enterprise development. United Nation Conference on Trade and Development. United Nations, New York and Geneva (2001). Retrieved 12 Apr 2024
2. Al Naamani, K.: Digital Transformation as a Mean to Enable SME Sustainability and Growth. Master Thesis (n.d.)
3. Brown, M.E., Treviño, L.K.: Ethical leadership: a review and future directions. Leadersh. Q. **17**(6), 595–616 (2006)
4. Hart, S.L., Milstein, M.B.: (2003) Creating sustainable value. Acad. Manag. Perspect. **17**(2), 56–67 (2003)
5. The 17 Goals: Sustainable Development (n.d.). https://sdgs.un.org/goals
6. Ellen MacArthur Foundation: How to Build a Circular Economy (n.d.). https://ellenmacarthurfoundation.org/
7. Elkington, J.: Enter the Triple Bottom Line. NA, 23–38 (2013a)

8. Brown, H.S., de Jong, M., Levy, D.L.: Building Institutions Based on Information Disclosure: Lessons from GRI's Sustainability Reporting (2009)
9. Small Business, Big Problem: New Report Says 67% of SMEs Worldwide Are Fighting for Survival. World Economic Forum (n.d.)
10. Temtime, Z.T., Pansiri, J.: Small Business Critical Success/Failure Factors In Developing Economies: Some Evidences from Botswana (2004)
11. Alharafsheh, M., Ezmigna, A.A.R.: The Impact of Business Strategy on Competitive Advantage and Performance of Small & Medium Enterprises in Jordan (2023)
12. Petković, S., Sorak, S.: Effects of the Establishment of Entrepreneurial Orientation on the Performances of Small and Medium Enterprises in Transition Countries: Empirical Evidences from Bosnia and Herzegovina (2019)
13. Ngah, R., Abd Wahab, I., Salleh, Z.: The Sustainable Competitive Advantage of Small and Medium Enterprises (SMEs) with Intellectual Capital, Knowledge Management and Innovative Intelligence: Building a Conceptual Framework (2015)
14. Prasanna, R., Jayasundara, J., Naradda Gamage, S.K., Ekanayake, E., Rajapakshe, P., Abeyrathne, G.: Sustainability of SMEs in the Competition: A Systemic Review on Technological Challenges and SME Performance (2019)
15. Elliot, S.: Developing Organizational Capabilities in SMEs: Enabling Environmentally Sustainable ICT. AIS Electronic Library (AISeL) (n.d.)
16. Walley, K.: Coopetition: An Introduction to the Subject and an Agenda for Research (2007)
17. Yousaf, Z., Radulescu, M., Sinisi, C.I., Serbanescu, L., Păunescu, L.M.: Towards Sustainable Digital Innovation of SMEs from the Developing Countries in the Context of the Digital Economy and Frugal Environment (2021)
18. Lekić, N., Carić, M., Soleša, D., Vapa Tankosić, J., Rajaković-Mijailović, J., Bogetić, S., Vučičević, M.: Employees' Perceptions on the Relationship of Intellectual Capital and Business Performance of ICT Companies (2021)
19. Petković, S., Jäger, C., & Sašić, B.: Challenges of small and medium sized companies at early stage of development: insights from Bosnia and Herzegovina. J. Contemp. Manage. Issues **21**(2) (2016)
20. Baporikar, N., Baragde, D.: ICT Sector Perspective for Circular Economy in the Digital Era, vol. 12, no. 1, pp. 54–71 (2021)
21. Andreas, F.M., Cooperman, E.S., Gifford, B., Russell, G.: A Simple Path to Sustainability: Green Business Strategies for Small and Medium-sized Businesses (2011)
22. Arena, M., Azzone, G.: A Process-Based Operational Framework for Sustainability Reporting in SMEs (2012)
23. Minić, A., Filipović, S.: Underpinning Sustainability Paradigms: Circular Economy and Enhanced Triple Bottom Line (2023)

Transformational Leadership and Its Impact on Organizational Commitment at the Oman Chamber of Commerce and Industry

Abdulaziz Al-Ismaili

Abstract This study aims to examine the impact of transformational leadership on organizational commitment at the Oman Chamber of Commerce and Industry. It seeks to investigate the relationship between the different dimensions of transformational leadership, such as idealized influence, inspirational motivation, intellectual stimulation, and individualized consideration, and the level of organizational commitment among employees. The study employed a descriptive-analytical approach using a case study methodology, where data was collected through detailed surveys, in-depth personal interviews, and analysis of relevant official documents. The results revealed a statistically significant positive relationship between transformational leadership and organizational commitment, with transformational leadership contributing to enhancing emotional commitment and fostering positive interactions among employees. Several factors influencing the enhancement of commitment were identified, including effective leadership motivation and the presence of a participatory work environment. The study demonstrates that transformational leadership is a key driver for improving institutional performance and strengthening organizational loyalty, reflecting its active role in achieving institutional sustainability and growth. Based on these findings, the study recommends the development of continuous training programs for leaders to enhance transformational leadership practices within the Chamber of Commerce, which would contribute to improving organizational commitment and supporting the sustainable growth of the organization.

Keywords Transformational leadership · Organizational commitment · Oman chamber of commerce and industry · Leadership impact

A. Al-Ismaili (✉)
Modern College of Business and Science, Muscat, Sultanate of Oman
e-mail: 20247403@mcbs.edu.om

B. Awwad (ed.), *AI and IoT: Driving Business Success and Sustainability in the Digital Age*, Studies in Systems, Decision and Control 590,
https://doi.org/10.1007/978-3-031-88874-8_53

1 Introduction

1.1 Research Background

The world today is undergoing significant changes in various aspects, compelling both private and public organizations to adjust quickly and effectively. One of the most crucial approaches to adapting to these shifts is emphasizing the human aspect, which plays a pivotal role in achieving organizational success and meeting objectives. This requires cultivating a fair and healthy work environment that motivates employees to be innovative, engaged, and to perform at their highest potential. Despite advancements in technology and science, a major challenge for leaders across various sectors remains the retention of committed talent and the continuous development of their skills.

Transformational leadership is a fundamental concept in management, focusing on a leader's ability to inspire and motivate individuals to achieve goals that go beyond typical expectations. This leadership style is distinguished by its capacity to bring about significant changes in the values and behaviours of individuals, enhancing their commitment to the organization and contributing to improvements in their performance both personally and professionally. In the Sultanate of Oman, the Oman Chamber of Commerce and Industry plays a key role in strengthening the national economy and improving the competitiveness of the commercial sector in response to global challenges. Leadership within this chamber is crucial in guiding individuals and institutions toward achieving organizational goals while maintaining institutional values.

Recently, organizational commitment has gained increasing importance in business settings, as research has demonstrated a clear connection between commitment, individual performance, and overall institutional productivity. This highlights the importance of studying the impact of transformational leadership on organizational commitment, particularly within business institutions like the Oman Chamber of Commerce and Industry. Transformational leadership can play a key role in motivating individuals to become more dedicated and creative in the workplace, ultimately enhancing the organization's adaptability and growth. While many studies have explored the effects of transformational leadership in the broader commercial sector, focusing specifically on its influence within the Oman Chamber of Commerce and Industry offers a unique perspective on how leadership can shape individual behaviors, attitudes, and work performance.

The impact of transformational leadership on organizational commitment within the Oman Chamber of Commerce and Industry provides valuable insights into how leadership fosters strong relationships between individuals and the organization, increasing engagement and aligning personal goals with organizational objectives.

1.2 Problem Statement

Transformational leadership is considered one of the fundamental pillars that enhance an organization's ability to achieve success and excellence, particularly in the face of challenges arising from rapid changes in the business environment and increasing competition. This type of leadership plays a key role in enabling organizations to adapt to continuous transformations and innovate in response to accelerating challenges. With the economic and social developments occurring in the Sultanate of Oman, there is a growing need for an effective leadership model that can enhance organizational commitment among employees across various institutions. Given the role of the Oman Chamber of Commerce and Industry as a key institution supporting the national economy, it requires a leadership style that fosters innovation and strengthens employees' sense of belonging and loyalty.

However, there is a lack of studies that explore the relationship between transformational leadership and organizational commitment in the Omani context. This raises questions about how transformational leadership practices affect the improvement of commitment levels among employees in this institution. High organizational commitment can contribute to better performance and increased productivity, while its absence may lead to decreased morale and higher employee turnover.

Therefore, this research seeks to explore the impact of transformational leadership on organizational commitment at the Oman Chamber of Commerce and Industry, and to identify the factors that enhance the effectiveness of this leadership style in achieving employee commitment. Through this study, we hope to provide insights into how leadership practices at the chamber can be improved, contributing to a better work environment and overall institutional performance. The study aims to answer the following main question: How Does Transformational Leadership Affect Organizational Commitment and Employee Engagement at the Oman Chamber of Commerce and Industry?

1.3 Research Gap

1. The research gap in the topic of transformational leadership and its impact on organizational commitment within the Oman Chamber of Commerce and Industry lies in the lack of studies that specifically explore the relationship between these two concepts in the Omani context. While transformational leadership has been widely studied in Western environments, its application in Oman has not received the same level of attention. Oman has a unique cultural and social context that significantly influences leadership styles, with values such as mutual respect, collaboration, and paternal guidance prevailing, which may require adjustments to transformational leadership practices to better align with these values.

2. Furthermore, Omani institutions, such as the Chamber of Commerce and Industry, face increasing economic challenges due to global shifts and technological transformations, making it essential to explore how transformational leadership can contribute to enhancing organizational commitment in such a dynamic environment. Additionally, there is a need to understand how government policies and economic conditions impact the interaction between transformational leadership and organizational commitment, which may differ from Western contexts due to varying external and internal factors in Oman.
3. Therefore, a clear gap remains in the literature regarding the actual impact of transformational leadership on organizational commitment within specific Omani institutions. This gap highlights the need for further research into how modern leadership styles can improve performance and foster organizational loyalty amid the challenges faced in today's rapidly changing environment.

2 Theoretical Framework

2.1 Transformational Leadership

The concept of transformational leadership emerged in the late twentieth century, introduced by the American researcher James Burns, in response to organizations' need for a type of leadership that aligns with the spirit and demands of the times. Transformational leadership works to create a shared vision and mission for the organization, contributing to enhancing competitive advantage and commitment to ethical values. It aims to increase motivation among employees by developing their capabilities and empowering them to make decisions freely, which helps improve organizational relationships, boost productivity and creativity, and deepen loyalty to the organization. According to Al-Ma'ani, transformational leadership seeks to achieve the organization's goals by motivating individuals and encouraging them to exceed expectations.

The definitions of transformational leadership in academic literature vary, as follows:

1. Bass's Definition: Bass defines transformational leadership as a leadership style that aims to inspire and motivate individuals to achieve collective goals and promote positive change within the organization. The transformational leader is distinguished by their ability to positively influence members through motivation and ongoing support [1].
2. Avolio's Definition: Avolio considers transformational leadership as a tool to enhance the effectiveness of individuals and teams by promoting inspiration and critical thinking. It focuses on the leader's role in developing and empowering their followers to achieve success [2].

The researcher views transformational leadership as a leadership model characterized by the ability to inspire and motivate individuals toward achieving collective goals and fostering positive change within organizations. This type of leadership focuses on transforming organizational culture and driving individuals to improve their performance and contribute effectively to achieving the organization's objectives.

2.2 *Organizational Commitment*

Organizational Commitment: Organizational commitment is considered one of the key factors that significantly impact the success of organizations. It reflects the degree of an employee's attachment to the organization's goals and values, and their willingness to put forth their maximum effort to achieve these objectives. According to Dahmash, organizational commitment is defined as "the employee's integration with their organization and the extent of their attachment to its goals and values," while Erlan defines it as "a strong bond between the employee and the organization's goals, and a strong desire to stay and work within it." Yildirim considers organizational commitment as "a psychological state that forms a personal relationship between the employee and their organization, affecting the employee's desire to continue in their job" [3].

The Relationship Between Transformational Leadership and Organizational Commitment

It is clear from the table that there is a strong relationship between transformational leadership and organizational commitment, as transformational leadership positively impacts organizational commitment through several mechanisms. The results show that transformational leadership contributes to enhancing employee satisfaction and boosting morale, which increases their engagement and participation in achieving the organization's vision and goals. Furthermore, transformational leadership motivates individuals to improve collective performance and promotes a culture of teamwork, which helps strengthen loyalty and belonging to the organization. Transformational leadership also encourages employees to be more creative and innovative, which enhances problem-solving capabilities and fosters organizational excellence. It is important to note that strengthening organizational commitment through transformational leadership not only improves job satisfaction but also boosts employee morale in general, leading to improved performance and productivity within the organization. Therefore, the relationship between transformational leadership and organizational commitment plays a significant role in achieving organizational goals and enhancing its effectiveness in the workplace.

2.3 Research Objectives

1. Exploring the Relationship Between Transformational Leadership and Organizational Commitment.
2. Identifying Factors that Enhance Organizational Commitment.
3. Providing Strategic Recommendations.

2.4 Research Questions

1. How does transformational leadership affect organizational commitment and employee loyalty at the Oman Chamber of Commerce and Industry?
2. How do the dimensions of transformational leadership (such as inspirational motivation, idealized leadership, intellectual challenge, and individualized consideration) affect the effectiveness of organizational commitment among employees at the Oman Chamber of Commerce and Industry?
3. What is the role played by factors in supporting the ability of transformational leadership to improve organizational performance and enhance effective collaboration and fruitful interaction among employees at the Oman Chamber of Commerce and Industry?

3 Literature Review and Previous Studies Related to the Study

A survey was conducted to study the administrative literature related to the topic of the research. The previous studies were divided into three main areas, as follows.

First Axis: Studies on Transformational Leadership and Its Impact on Various Other Variables

A study by Shamsa Salem Al-Balooshi, titled "The Role of Transformational Leadership in Enhancing Organizational Performance: A Case Study of the Ministry of Education in the United Arab Emirates" [4]: This study aimed to explore the impact of transformational leadership on improving performance in public organizations, focusing on the Ministry of Education in the UAE. The study used a descriptive-analytical approach and a case study method to collect data through questionnaires directed at the ministry's employees. The study concluded that transformational leadership, through its four dimensions (idealized influence, inspirational motivation, intellectual stimulation, and individualized consideration), positively impacts employees' behavior and performance. Idealized influence primarily contributes to changing the values and behaviors of subordinates, enhancing their loyalty and satisfaction, and motivating them to prioritize the public interest. The study also emphasized the importance of transformational leadership in creating a work environment that fosters interaction and trust, which in turn strengthens employees' commitment to organizational goals.

A study by Maatin et al., titled "Comparing Transformational Leadership in Successful and Unsuccessful Companies" [5]: aimed to compare leadership styles in several successful and unsuccessful American companies to determine whether there is a relationship between company success and their leadership styles. The study found a significant relationship between successful companies and the presence of transformational leadership. Transformational leadership plays an important role in motivating employees, increasing their creativity, and helping them develop new values, goals, and needs.

A study by Asghari et al., titled "The Relationship Between Transformational Leadership Behavior, Organizational Justice, Leader-Member Exchange Relations, Perceived Organizational Support, Trust in Management, and Organizational Citizenship Behavior" [6]: aimed to examine the relationship between transformational leadership behavior and organizational justice behaviors and organizational citizenship behavior. The study was conducted on a sample of 162 employees in five ministries in Malaysia. The study found a direct and positive relationship between transformational leadership behaviors and organizational justice (distributive, procedural, and interactional) and organizational citizenship behaviors.

A study by Al-Azzam and Al-Jadayah, titled "The Impact of Transformational Leadership on Organizational Learning in the Commercial Banking Sector in Northern Jordan" [7]: concluded that the idealized influence aspect of transformational leadership had the greatest impact on organizational learning, followed by intellectual stimulation and role-playing.

Raja's study, titled "How Transformational Leadership Can Lead to Employee Engagement in Service Organizations in Pakistan" [8]: indicated that applying transformational leadership can contribute to increasing employee engagement and motivating them to better serve their organizations.

A study by Si et al., titled "The Relationship Between Transformational Leadership, Organizational Climate, and Creative Performance in Multinational Companies in China" [9]: concluded that transformational leadership has a positive impact on employees' creative performance in Chinese companies. The study also confirmed that organizational climate serves as a mediating factor between transformational leadership and creative performance.

The Second Axis: Studies Addressing the Impact of Organizational Commitment on Various Other Variables

Al-Afeef Study titled "The Impact of Human Resource Management Practices on Organizational Commitment: An Analytical Study at the Ministry of Culture in Jordan" [10]: This study revealed that human resource management practices have a significant impact on organizational commitment across its various dimensions (continuance, normative, and affective commitment).

Dina Al-Nashili Study titled "The Role of Organizational Agility in Achieving Organizational Commitment: A Field Study" [11]: The study aimed to examine the relationship between organizational agility variables and organizational commitment, using a sample of employees at the Paint and Chemical Industries Company (Baking). The study used a descriptive analytical approach and employed a two-part questionnaire: the first part to measure organizational agility and the second part to measure organizational commitment.

Beyah Study titled "The Impact of Human Resource Management Practices on Employee Productivity in Jordanian Banks: Organizational Commitment as an Intermediary Variable" [12]: This study revealed that human resource management practices statistically affect organizational commitment, which in turn contributes to increased employee productivity.

Al-Nuwaika Study titled "The Impact of Leadership Styles on Achieving Organizational Commitment among Employees of the Ministry of Interior in Jordan" [13]: The results showed that leadership styles have a positive impact on enhancing organizational commitment across its various dimensions (affective, continuance, and normative commitment).

Bani Issa and Abazid Study titled "The Role of Organizational Commitment in Improving Employee Performance in the Jordanian Banking Sector" [14]: The study concluded that organizational commitment positively influences employee performance in Jordanian banks, with affective commitment having the most significant impact.

The Third Axis: Studies Addressing the Impact of Transformational Leadership on Organizational Commitment

Dr. Samar Mohamed Al-Taher Yassin's Study titled "The Impact of Transformational Leadership on Organizational Commitment among Employees at King Khalid University" [15]: The study aimed to examine the impact of transformational leadership on organizational commitment at King Khalid University. Using a descriptive analytical approach and a questionnaire for data collection, the study surveyed a random sample of 50 employees (lecturers and administrators). The results showed that the university's practice of transformational leadership (including idealized influence, individual consideration, inspirational motivation, and intellectual stimulation) was perceived by employees as high, and that organizational commitment among employees was also high.

Gao Study titled "The Impact of Transformational Leadership Behaviors in Family Businesses in China on Organizational Commitment" [16]: This study found that transformational leadership behaviors, including idealized influence and intellectual stimulation, positively impacted organizational commitment in family-owned businesses.

Saruwas Study titled "The Relationship Between Transformational Leadership, Managerial Creativity, and Organizational Culture in the Australian Private Sector" [17]: The study revealed a positive relationship between transformational leadership and managerial creativity, with key leadership traits such as clarity of vision and individual support playing a significant role in fostering creativity.

What Distinguishes This Study from Previous Research?

The main distinction of this study lies in its focus on addressing the research gap related to "how transformational leadership contributes to achieving organizational commitment." While prior studies have explored the impact of transformational leadership on variables such as organizational citizenship behavior, organizational conflict, organizational change, and managerial creativity, none have specifically examined its direct effect on organizational commitment. Although Gao's (2013) study is the only one that explored the relationship between transformational leadership and organizational commitment, it was limited to family businesses in China.

In contrast, this study focuses on a different research context, involving all employees in leadership and supervisory roles at the Oman Chamber of Commerce and Industry. This provides a unique setting, distinct from previous studies conducted in various organizational environments, and offers a broader perspective on the impact of transformational leadership in a key institution that plays a pivotal role in the national economy.

3.1 Conceptual Framework

The Conceptual Framework for the Topic of Transformational Leadership and Its Impact on Organizational Commitment in the Oman Chamber of Commerce and Industry aims to build a theoretical framework that explains the relationship between the concepts of transformational leadership and organizational commitment within the context of the Oman Chamber of Commerce and Industry, while considering the local cultural and economic specifics.

1. **Transformational Leadership**

Transformational leadership is a leadership style that focuses on inspiring and motivating individuals to achieve higher levels of performance and innovation by building relationships based on trust and support. This type of leadership differs from traditional leadership, as the transformational leader seeks to change the values and beliefs of individuals, encouraging them to excel and innovate. Transformational leadership consists of four main components:

- **Inspirational Motivation**: Motivating individuals and directing them toward shared goals.
- **Individualized Consideration**: Providing individual support and guidance to employees (Avolio and Bass 2004).

2. **Organizational Commitment**

Organizational commitment refers to the degree to which individuals are connected to the organization's goals and values and their willingness to work towards achieving these objectives. It reflects the level of loyalty and attachment individuals feel towards the organization and can significantly influence job performance, discipline, and productivity. Organizational commitment can be divided into several types:

- **Affective Commitment**: When individuals feel a strong attachment to the organization and believe in its values.
- **Continuance Commitment**: When individuals remain committed to staying in the organization due to the benefits and resources available.

3. **The Relationship Between Transformational Leadership and Organizational Commitment**

Studies have shown that transformational leadership has a positive impact on organizational commitment, as transformational leaders help foster a desire to stay with and belong to the organization by creating an environment that encourages personal growth and continuous improvement. Leaders who adopt this style are able to motivate their employees not only to achieve the organization's goals but also to accomplish their personal objectives, thus strengthening their emotional and institutional commitment. Research also indicates that transformational leadership can be a key factor in increasing organizational commitment in dynamic, ever-changing environments [18].

4. **The Omani Context**

In the Omani context, where traditional values such as mutual respect and close collaboration dominate the work environment, transformational leadership can play a pivotal role in enhancing organizational commitment. These deeply rooted cultural values are an integral part of the Omani identity, making it essential to understand how these values interact with modern leadership practices that focus on motivation, inspiration, and sustainable development. In a work environment that fosters participation and understanding, transformational leadership can help drive Omani institutions toward achieving higher levels of organizational commitment, which in turn positively impacts overall performance and productivity.

However, there remains a pressing need to explore the relationship between these traditional values and modern leadership styles, and how this interaction can shape the development and adoption of effective leadership strategies within Omani institutions. A prime example of this is the Oman Chamber of Commerce and Industry, which serves as a key pillar of the national economy, where transformational leadership can contribute to improving commitment levels within such institutions, enhancing their capacity to face current economic challenges [19].

4 Research Methodology

The researcher relied on the descriptive-analytical method.

Descriptive-Analytical Method: The study adopted the descriptive-analytical method to examine the relationship between transformational leadership and organizational commitment. The descriptive method was used to collect data on the dimensions of transformational leadership (idealized influence, inspirational motivation, intellectual stimulation, and individualized consideration) and the levels of organizational commitment (affective, continuance, and normative) among employees through surveys and interviews.

5 Results

1. The Influence of Transformational Leadership on Organizational Commitment.
2. Key Factors Shaping the Strengthening of Organizational Commitment.
3. Psychological Impacts of Transformational Leadership.

6 Recommendations

1. Investing in the development of transformational leadership.
2. Conducting comprehensive and in-depth future studies.
3. Adopting a flexible and dynamic work environment that fosters adaptability and innovation.

References

1. Bass, B.M.: Bass & Stogdill's Handbook of Leadership: Theory, Research, and Managerial Applications, 3rd edn. Free Press (1990)
2. Avolio, B.J.: Full Leadership Development: Building the Vital Forces in Organizations. Sage Publications (1999)
3. Barbuto, J.E., Wheeler, D.W.: Scale development and construct validation of a scale to measure servant leadership. Int. J. Leadersh. Stud. **1**(1), 61–75 (2006)
4. Al-Balooshi, S.S.A.: The Role of Transformational Leadership in Enhancing the Performance of Public Organizations: An Applied Study on the Ministry of Education in the United Arab Emirates. Ph.D. Dissertation, Faculty of Economics and Political Science, Cairo University (2009)
5. Zarei Matin, H., et al.: Comparing transformational leadership in successful and unsuccessful companies. Proc. World Acad. Sci. Eng. Technol. **31**, 464 (2008)
6. Asgari et al.: The Relationship Between Transformational Leadership (1) Behaviors, Organizational Justice, Leader-Member Exchange, Perceived Organizational Support, Trust in Management and Organizational Citizenship Behaviors
7. Al-Azzam, A., Al-Judaya, M.: The effect of transformational leadership on organizational citizenship behavior among employees at the University of Jordan. Zarqa J. Res. Stud. Hum. **15**(2), 30–51 (2015)

8. Raja, M.: Does transformational leadership leads to higher employee work engagement. Int. J. Acad. Res. Bus. Soc. Sci. (2012)
9. Si, S., Wei, F.: Transformational and transactional leadership, empowerment climate and innovation performance: a multilevel analysis in the Chinese content. Bus. J. (2012)
10. Al-Afeef, A.: The impact of human resource management practices on organizational commitment: an analytical study at the Ministry of Culture in Jordan. Arab J. Adm. **38**(3), 123–152 (2018)
11. Al-Nishili, D.: The role of organizational agility in achieving organizational commitment: a field study. Arab J. Adm. **40**(3), 163–182 (2020)
12. Bey'a, A.: The Impact of Human Resource Management Practices on Employee Productivity in Jordanian Commercial Banks: Organizational Commitment as a Mediating Variable. Unpublished Master's Thesis. Al-Bayt University, Mafraq, Jordan (2016)
13. Al-Nuwaika, A.: The impact of leadership styles on achieving organizational commitment among employees of the Ministry of Interior, Jordan. Dirasat Adm. Sci. **42**(1), 45–63 (2015)
14. Bani Issa, A., Abazeid, R.: The role of organizational commitment in improving employee performance in the Jordanian banking sector. Dirasat Adm. Sci. **41**(2), 362–372 (2014)
15. Yassin, S.M.A.T.: The Impact of Transformational Leadership on Organizational Commitment Among Employees at King Khalid University, vol. 32, no. 5, pp. 163–182 (2023)
16. Gao, Y.: The impact of transformational leadership on organizational commitment: evidence from family businesses in China. J. Bus. Res. **66**(1), 136–144 (2013)
17. Sarros, J.C., Cooper, B.K.: Building the organizational culture of a transformational leadership style: a practical approach to leadership. J. Leadersh. Organ. Stud. **12**(2), 88–100 (2006)
18. Judge, T.A., Piccolo, R.F.: Transformational and transactional leadership: a meta-analytic test of their relative validity. J. Appl. Psychol. **89**(5), 755–768 (2004)
19. Al-Alawi, A.I.: Leadership styles in the Omani public sector: a study of the relationship between leadership styles and organizational commitment. Int. J. Public Adm. **41**(4), 324–336 (2018)

An Investigation on the Utilization of Natural Gas Infrastructure for Hydrogen Transportation in Oman Using Machine Learning: A Conceptual Framework

Amjed Sid Ahmed, Eman Al-Naamani, and Wadhah Al Mandhari

Abstract Advancing technology for successful utilization of hydrogen as energy carrier is the subject of investigation for different countries in the world. The potential of replacing natural gas and fossil fuels by hydrogen is proven effective for many applications. Designing this transformation is, however, unique to each case. Necessitating understanding the available infrastructure and investigating optimum ways of re-directing hydrogen energy. This document represents a research proposal to study the feasibility of using natural gas infrastructure in Oman to transport hydrogen. One of the major concerns related to using natural gas infrastructure to transport hydrogen is the incompatibility of this infrastructure with hydrogen gas in terms of material thermodynamic properties. Therefore, synergy of hydrogen within natural gas seems to be a promising solution if challenges can be identified and risks are minimized. In the current research proposal, we plan to use simulation and machine learning algorithms to predict the range of hydrogen to natural gas concentration ratio allowable for safe injection into the natural gas system in Oman. The investigation begins with data collection of infrastructures specification, identification of critical parameters, prediction of concentration ratio, and finally to perform optimization for best approaches that can be considered for future implementation of hydrogen energy in Oman.

Keywords Renewable energy · Sustainability · Green hydrogen · Blue hydrogen

A. S. Ahmed (✉) · W. Al Mandhari
Modern College of Business and Science, Bawshar, Muscat, Oman
e-mail: amjed.sidahmed@mcbs.edu.om

E. Al-Naamani
Global College of Engineering and Technology, Bawshar, Muscat, Oman

B. Awwad (ed.), *AI and IoT: Driving Business Success and Sustainability in the Digital Age*, Studies in Systems, Decision and Control 590,
https://doi.org/10.1007/978-3-031-88874-8_54

1 Introduction and Problem Domain

Hydrogen energy is emerging rapidly worldwide as a potential and sustainable alternative to natural gas and petroleum fuel. Hydrogen is energy carrier which can be utilized as fuel in automobiles, producing electricity and in oil refineries. It can be produced using natural gas or renewable resources and therefore can be referred to as grey or blue hydrogen (if CO_2 emission is controlled and minimized) or green hydrogen, respectively. Depending on the end use of hydrogen the transportation method can differ in terms of local use or exporting purposes. While transportation for local use can be easily performed using trucks, it is more complicated and need sophisticated transporting system and storage facilities for exporting overseas.

The infrastructure development to utilize and transmit hydrogen fuel is yet at early stages. This project aims for computational prediction to analyze and optimize cost-effective transport and storage facilities for hydrogen. The study will investigate the compatibility of transforming the available natural gas infrastructure in Oman for future utilization in hydrogen energy. The proposal is concerned to align hydrogen energy storage and transportation requirements into natural gas infrastructure in terms of material and process specifications, safety, efficiency, and cost. In addition, the study will examine the uncertainties around the future demand for hydrogen energy in local and international markets.

2 Literature Review and Analysis of Related Work

Hydrogen is an energy carrier which releases amount of energy equivalent to 122 kJ/g when reacted with oxygen (ignited). This amount of energy is higher than hydrocarbons by about 2.75 times [1, 2]. Nonetheless, hydrogen storage in gaseous or compressed liquid forms is more challenging due to its low density. A number of case studies were conducted to determine the most suitable approach for hydrogen transportation depending on hydrogen production, demand and end-users. Simulating the transport process of hydrogen are conducted to obtain optimization between demand and production cost. Such studies concern more on modeling the hydrogen supply chain to be utilized locally as fuel for transportation systems [3–5]. These studies suggested transporting hydrogen in liquid form via trucks which is more viable for such cases.

Nevertheless, adopting natural gas infrastructure is one of the promising approaches investigated in many countries like Belgium, Pakistan, Canada, and US. The existing natural gas infrastructure can be used for transporting hydrogen depending on each case. The compatibility of using the currently available infrastructure for hydrogen is subject to its durability and safety. The National Renewable Energy Laboratory [6] estimated the blending concentration as low as 5–15% by volume for hydrogen to natural gas to maintain durability of pipelines. While the appropriate concentration for the European system was found to be not exceeding

12% [6]. Successful synergies require case-by-case investigation as it significantly varies depending on natural gas composition and pipeline specifications [7].

The critical key factor needs to be identified to allow hydrogen into the natural gas infrastructure is the ratio of hydrogen to natural gas. For the case study conducted in Belgium, it was theoretically verified that injecting hydrogen up to 17 vol% should be possible [8]. The study demonstrated challenges associated with increasing hydrogen content which can cause significant implications related to pressure drop in the pipeline. Another factor highlighted was hydrogen leakage due to permeability caused by the material of the pipeline. In a study conducted by Mukherjee et al. based on Ontario natural gas infrastructure found that adding a storage facility leads to possibility for increasing hydrogen concentration in the pipelines [9]. Although the paper considered low hydrogen concentration of up to 5 vol% it was noticeable that price of power would inversely relate to the amount of hydrogen produced. This observation raises a question of what would be the economic implications of the predicted concentration fraction in Oman.

The end user market, on the other hand, in case to be used locally as hydrogen enriched natural gas (rather than exporting as hydrogen energy) is important to determine the quality of gas resulted from the blending. Gondal [10] showed that blending 2 vol% of hydrogen in the natural gas improves the calorific value. Whereas, injecting 10 vol% hydrogen causes a drop in the calorific value. The calorific value of the blend is another key factor for the gas if to be used in local markets. On the contrary if the aim is to export hydrogen energy then the process phases would include H2 production, blending facility, transportation, and separating facility, as illustrated in the diagram below. Based on literature the added cost of the extra facilities can be significantly high which again depends on the concentration ratio [6]. To accurately perform estimation to optimize the trade-off between hydrogen content, pipeline durability and added cost, it is required to understand the governmental directions toward the extent of implementing hydrogen energy.

The scenario of blending hydrogen into natural gas streams has the potential to be feasible to carry the hydrogen in the mid- to down-streams. The main question this proposal aim to address is to estimate the hydrogen to natural gas ratio compatible with the available infrastructure. Assuming renewable pathways for green hydrogen production, the current study uses computer simulation for the compatible hydrogen to natural gas ratio prediction. Implications on the natural gas pipelines resulting from thermodynamic properties alteration [11] due to blending hydrogen into the natural gas will also be investigated. We use simulation to answer the following questions focusing on mid to down streams:

- How far could we elongate natural gas life span if hydrogen enriched natural gas is to be utilized?
- What fractions of hydrogen to natural gas can be adopted by current NG pipelines and storage facilities?
- What will be the economic implications of the estimated fraction?
- What is the feasibility of using NG as carrier of hydrogen in the NG infrastructure?

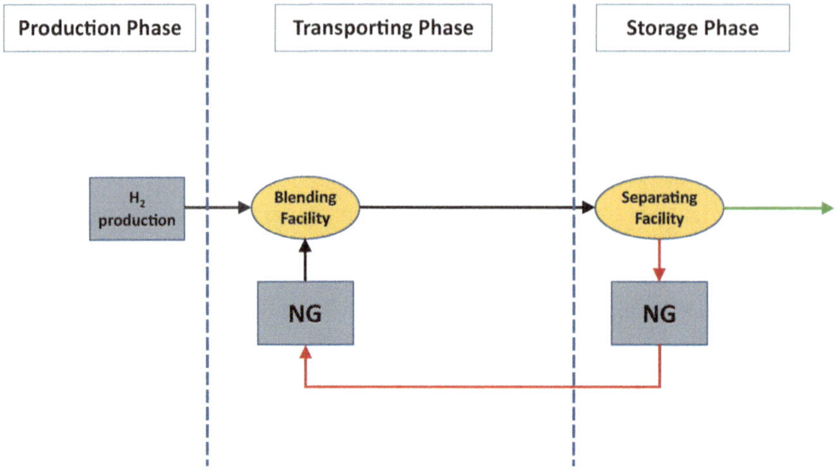

Fig. 1 Hydrogen lifecycle

- If natural gas is not to be used as a medium of transporting hydrogen what are other options for the NG infrastructure to be successfully implemented?

  Information required (see Fig. 1).

- What are the NG transportation system used in Oman LNG pipelines or CNG (compressed)?
- What are the specifications of NG pipelines used in Oman?
- What is the NG composition in mid-stream pipeline in Oman?

3 Research Conceptual Framework

3.1 Research Objectives

- Objectives of this project is to identify current and future challenges of hydrogen energy utilization within Oman.
- Findings will provide an insight on the future improvement requirements to adopt hydrogen energy.
- Through computer aid simulation this project will predict the suitable hydrogen concentration that can be injected into the natural gas network.
- This investigation aims to identify the most feasible hydrogen carrier to ensure safety in storage and transporting facilities.
- The project also aims to systematically address the question of which is more feasible and cost-effective for Oman to proceed with blue hydrogen or green hydrogen depending on the transporting scenarios.

- The project will also highlight the modification needs to transform liquefied natural gas (LNG) infrastructure to liquefy Hydrogen for export purposes.

3.2 Research Methodology

In the following the proposed methodology that will be followed to complete the project, which consists of two main phases.

Phase 1: Investigation of Using Existing NG Infrastructure

This phase will study and investigate the possibility of using the existing infrastructure of natural gas in Oman to transport the hydrogen.

SLR and Comparative Analysis, (Other Countries): A systematic literature review (SLR) will be conducted in this phase. The need to conduct research on this area is particularly important, several countries started working in this area and provided results that shown the feasibility/applicability of such project based on their infrastructures. A comparative study based on the existing literature will add to the body of knowledge and help researchers to predict future research directions either with positive or negative outcomes.

Infrastructure Specifications and Details Collections: In the second part of this phase a comprehensive data collection of Oman oil and gas infrastructure will be carried out. OQ Company will be the data provider and its sites' infrastructure specification will be the subject of study. Mid- to down-stream will be recorded for further transformation into computer-based simulation.

Phase 2: Computerization and Simulation

In this phase collected specifications and parameters will be converted to computerized based simulation. A model that emulates real infrastructure using computer simulation and machine learning algorithms will buildout. Different testing scenarios will be implemented based on real circumstances of the infrastructure. Machine Learning algorithms will be used to predict the outputs based on factors for each scenario. A range of hydrogen to natural gas composition ratios will be loaded into the algorithm to predict the maximum allowed hydrogen injection based on the specifications of the available infrastructure. The scenarios to be investigated are as follows:

First Scenario: Temperature

Assumption: Set the temperature range based on the case of Oman.

Process: Emulation of pipeline under different temperature levels will be carried out for the range of hydrogen concentration to measure the thermodynamic properties and their influence on the safety aspects.

Output: The maximum allowable hydrogen concentration to be injected with the natural gas at the studied set of temperatures. For the predicted maximum hydrogen concentration, the added cost for blending and separating facilities will be discussed and the optimum ratio will be suggested.

Second Scenario: Pressure

Assumption: Set the pressure range based on the case of Oman.

Process: Same as the process above with the set of pressure for a range of hydrogen concentration.

Output: The maximum allowable hydrogen concentration to be injected with the natural gas at the studied set of pressure. Based on which the same argument as mentioned in the temperature scenario will be discussed.

Third Scenario: Flow Rate

Assumption: Set the flow rate range based on the case of Oman.

Process: Same as the process above with using the set of flow rate for a range of hydrogen concentration.

Output: The maximum allowable hydrogen concentration to be injected with the natural gas at the studied set of flow rate. For the predicted maximum hydrogen concentration, the number and added cost for pump stations and pumping rates will be discussed and the optimum ratio will be suggested.

Fourth Scenario: Chemical Reactions and Influence in Corrosion

Assumption: Set the expected chemical reactions that might occur due to presence of hydrogen with the natural gas.

Process: Emulation of pipeline under the different expected chemical reactions will be carried out to measure the effects on corrosion.

Output: The maximum allowable hydrogen concentration to be injected with the natural gas for the studied set of chemical reactions. For the predicted maximum hydrogen concentration, the added cost for maintenance will be discussed and the optimum hydrogen ratio will be suggested.

Fifth Scenario: Chemical Reactions and Influence in Flammability

Assumption: Set the expected chemical reactions that might occur due to presence of hydrogen with the natural gas.

Process: Emulation of pipeline under the different expected chemical reactions will be carried out to measure the consequences on flammability.

Output: The maximum allowable hydrogen concentration to be injected with the natural gas for the studied set of chemical reactions. Safety concerns are to be investigated.

Sixth Scenario: Combine all the Above Scenarios

Assumption: Set the ranges of the parameters above based on the acceptable hydrogen ratio found from the individual scenario.

Process: Emulation of pipeline under set range of parameters will be carried out for the range of hydrogen concentration to confirm results of each scenario and to identify their influence on the applicability and the expected safety risks.

Output: The consequences of adding hydrogen to the natural gas infrastructure in terms of quality of the resulted combination, environmental aspects, and feasibility.

3.3 Research Timeline and Milestones

In Table 1 the timeline of different phases of the project along with its sub-tasks is shown.

4 Academic and Scientific Significance

Based on the literature, many case studies on hydrogen energy feasibility of large-scale infrastructure were conducted in countries such as Germany, Republic of South Korea, Netherlands, UK and US. There is a very wide area of investigation for Oman

Table 1 Research timeline and milestones

Phase	Tasks	Period
Phase One	**Task 1: Perform SLR and Comparative analysis**	October-December 2024
	Task 2: Data collection on natural gas infrastructure in Oman	January-March 2025
Phase Two	**Task 3: Identify critical parameters required for hydrogen energy infrastructure**	April-June 2025
	Task 4: Perform computing simulation to predict feasibility of using natural gas infrastructure	June-October 2025
	Task 5: Optimize findings for best adoption of natural infrastructure in Oman for the future hydrogen energy	October-December 2025
	Task 6: Write-up and publication	December-February 2026
	Task 7: Use the same algorithm to predict the implications of blending hydrogen with other carriers to fit into the available infrastructure in Oman	March-July 2026
	Task 8: Write-up and publication	August-October 2026

to support national vision of 2040 towards sustainability in energy. The knowledge to be gained from this study will benefit the hydrogen energy researcher community as well as will contribute to the local industrial decision makers. Building a successful computer algorithm to predict hydrogen concentration for real life parameters will provide an opportunity of using machine learning approaches to minimize inaccuracies and support more practical solutions. As an outcome, the study in the second phase will investigate the possibility of using carriers other than natural gas that would be compatible with the available infrastructure through machine learning and depending on the thermodynamic properties of blending hydrogen with other gases. We hope the research outcome to be considered as a potential future plan and a contribution to deeper scientific analysis that will benefit similar projects in the region.

4.1 Benefits to Oman

It is a national and a worldwide interest to find cleaner and sustainable energy resources alternative to the diminishing fossil fuels. This project will contribute to the Oman 2040 vision [12] in moving toward sustainability in energy sector. Producing hydrogen from electrolysis of sea water can open the door to water-food-energy nexus, stream 5 in the Oman energy master plan of 2040 [13]. The geographic location of Oman in the long coastal strip of the Indian Ocean and sea of Oman can play a major role in securing and sustaining energy and opening new horizon for clean water. It will provide a scientific and an academic overview to support the economical and engineering aspects of utilizing natural gas infrastructure for hydrogen energy from a perspective of machine learning. Furthermore, this work will present a case study investigation on the technical and economic feasibility of hydrogen specific to the mid to down streams infrastructure in Oman. In the context of Oman, the proposed research plan aims to achieve the following:

- Hydrogen to natural gas concentration ratio suitable for Oman.
- Possibility of using hydrogen enriched natural gas in local market and its implication on the extension of natural gas lifetime in Oman.
- The economic aspects of the predicted concentration ratio.
- Investigate other alternative carriers for hydrogen in the Omani gas infrastructure specifications.
- The study outcome will help in identifying the appropriate locations of the hydrogen blending stations with reference to the natural gas infrastructure.

4.2 Other Benefits

1. This project will contribute to the Body of Knowledge (BoK) on the area of renewable energy and green hydrogen. The research outcome can potentially

help other similar studies for counties of same topographies aiming to produce and transport hydrogen.

2. The study is in alignment with the UN Sustainable Development Goal number seven of Affordable and Clean Energy as well as goal number thirteen of Climate Action.

Acknowledgements The research leading to these results has received funding from the Ministry of Higher Education, Research and Innovation (MoHERI) of the Sultanate of Oman under the Block Funding Program. MoHERI agreement number [MoHERI/BFP/MCBS/2024].

References

1. Kapdan, I.K., Kargi, F.: Bio-hydrogen production from waste materials. Enzyme Microb. Technol. **38**, 569–582 (2006). https://doi.org/10.1016/J.ENZMICTEC.2005.09.015
2. Balat, M.: Potential importance of hydrogen as a future solution to environmental and transportation problems. Int. J. Hydrogen Energy **33**, 4013–4029 (2008). https://doi.org/10.1016/J.IJHYDENE.2008.05.047
3. Almansoori, A., Betancourt-Torcat, A.: Design of optimization model for a hydrogen supply chain under emission constraints—a case study of Germany. Energy **111**, 414–429 (2016). https://doi.org/10.1016/J.ENERGY.2016.05.123
4. Han, J., Ryu, J.-H., Lee, I.-B.: Multi-objective optimization design of hydrogen infrastructures simultaneously considering economic cost, safety and CO_2 emission. Chem. Eng. Res. Des. **91**, 1427–1439 (2013). https://doi.org/10.1016/J.CHERD.2013.04.026
5. Kamarudin, S.K., Wan Daud, W.R., Yaakub, Z., et al.: Synthesis and optimization of future hydrogen energy infrastructure planning in Peninsular Malaysia. Int. J. Hydrogen Energy **34**, 2077–2088 (2009). https://doi.org/10.1016/J.IJHYDENE.2008.12.086
6. Melaina, M.W., Antonia, O., Penev, M.: Blending Hydrogen into Natural Gas Pipeline Networks: A Review of Key Issues. National Renewable Energy Laboratory, Golden, Colorado (2013)
7. Ogden, J.M., Jaffe, A.M., Scheitrum, D., et al.: Natural gas as a bridge to hydrogen transportation fuel: insights from the literature. Energy Policy **115**, 317–329 (2018). https://doi.org/10.1016/J.ENPOL.2017.12.049
8. Haeseldonckx, D., D'haeseleer, W.: The use of the natural-gas pipeline infrastructure for hydrogen transport in a changing market structure. Int. J. Hydrogen Energy **32**, 1381–1386 (2007). https://doi.org/10.1016/J.IJHYDENE.2006.10.018
9. Mukherjee, U., Elsholkami, M., Walker, S.B., et al.: Optimal sizing of an electrolytic hydrogen production system using an existing natural gas infrastructure. Int. J. Hydrogen Energy **40**, 9760–9772 (2015). https://doi.org/10.1016/J.IJHYDENE.2015.05.102
10. Gondal, I.A.: Hydrogen integration in power-to-gas networks. Int. J. Hydrogen Energy **44**, 1803–1815 (2019). https://doi.org/10.1016/J.IJHYDENE.2018.11.164
11. Hassanpouryouzband, A., Joonaki, E., Edlmann, K., et al.: Thermodynamic and transport properties of hydrogen containing streams. Sci. Data **7**, 222 (2020). https://doi.org/10.1038/S41597-020-0568-6
12. Oman Vision 2040. Ministry of Commerce
13. Oman Energy Master Plan: Progress Report 2015–2017 (2017)

The Impact of Power and Influence on Organizational Performance and Job Satisfaction in the Construction Sector in the Sultanate of Oman

Nasser Al Badi and Fadi Abdelfattah

Abstract This research paper dealt with the concept of power and influence in leadership, and focused on how they affect organizational performance and job satisfaction in organizations operating in the construction sector in the Sultanate of Oman and aimed to explore the different types of power and their sources such as official power, news power, charismatic power, reward power, coercive power and how to use it in the work environment. Want to understand these elements enables leaders to improve their performance and enhance their influence within organizations. The qualitative approach to data collection was used by conducting interviews for 200 leaders and employees in a variety of construction industries to measure the impact of power and influence on organizational performance and job satisfaction, and through data analysis, the results showed that leaders with personal influence could motivate their followers. Further, resulting in improved performance. The data also revealed that official power, while important, does not promote job satisfaction as much as personal influence. It follows that effective leadership requires a balance between formal power and personal influence, which is advisable, as well as further research to explore the impact of these concepts in diverse contexts.

Keywords Strategic leadership · Power and influence · Organizational performance · Job satisfaction · Construction sector · Charisma · Formal power · Motivation · Leadership balance · Leadership influence

---

N. Al Badi (✉) · F. Abdelfattah
Modern College of Business and Science, Muscat, Sultanate of Oman
e-mail: 20247135@mcbs.edu.om; Nasser.albadi@albadigroup.com

F. Abdelfattah
e-mail: Fadi.Abdelfattah@mcbs.edu.om

N. Al Badi
Albadi Group, Saham, Sultanate of Oman

B. Awwad (ed.), *AI and IoT: Driving Business Success and Sustainability in the Digital Age*, Studies in Systems, Decision and Control 590,
https://doi.org/10.1007/978-3-031-88874-8_55

1 Introduction

The concept of power in leadership refers to the ability to influence others and bring about change in behavior and thoughts, while influence is defined as the ability to positively influence others by building strong relationships.

Leaders who possess a combination of power and influence are more effective in guiding and motivating their teams and the importance of this research is highlighted in understanding how power and influence can be used strategically to improve organizational performance and its impact on job satisfaction, where power and influence play a pivotal role in this context through leaders dealing with their teams in multiple ways and different styles of power and influence. The importance is also growing the desire to study power and influence in leadership given the continuous and complex changes in the work environments in the construction sector, where complex dynamics require effective communication and the ability to influence, and the research aims to understand how leaders can use these elements strategically to achieve positive results within their organizations. This paper was based on three hypotheses:

1.1. **Hypothesis 1**: Formal power (such as authority and positions) enhances the effectiveness of leadership, but to a limited extent.
1.2. **Hypothesis 2**: Personal influence (such as charisma and the ability to influence) contributes more to employee motivation and increased job satisfaction.
1.3. **Hypothesis 3**: Balancing official power and personal influence is the key to leadership success.

This study raises some of the research questions:

1.4. How do different types of force affect performance within teams?
1.5. What is the impact of personal influence on employee satisfaction and motivation?
1.6. How can leaders balance the use of formal power with personal influence?

The bottom line is to seek to provide a comprehensive framework for understanding the importance of power and influence in leadership, defining goals and hypotheses and highlighting the research questions that will guide this study, paving the way for a more detailed discussion of future findings and recommendations.

2 Literature Review

Previous studies show that there are different types of power, such as formal power (authority) and personal power (charisma and persuasion). According to a study [1], personal power has a greater impact on positive performance than formal power. A study [2] also found that leaders who use multiple influence styles tend to achieve better results.

The scientific literature on power and influence in leadership addresses a wide range of studies that explore how these elements affect performance and satisfaction within organizations. The literature review is divided into several main axes:

1. Sources of power.
2. Tactics of leadership influence.
3. Navigate regulatory policies.
4. The relationship of the interlocutor to each other.

2.1 Sources of Strength in Leadership

Power is an essential element that enables leaders to influence employee behavior. Studies show that force can be classified into several types, according to the French and Raven [1] model, which identifies five main types of force:

2.1.1. **Formal Power**: Depends on the position and formal authority granted to the commander. It includes the ability to make decisions and distribute tasks. This force is effective in guiding teams towards goals.

2.1.2. **News Power**: Reflects the leader's strength in relationships, abundance of knowledge, and skill in exploiting those with experience to influence others.

2.1.3. **Charismatic Strength**: Highlighting the leader's personality and his ability to have a charismatic influence, inspire others, and motivate them to change visions and values to serve the organization.

2.1.4. **Coercive Force**: The use of authority to punish individuals, and it may sometimes be an effective tool, although it may create an uncomfortable work environment.

2.1.5. **Rewarding Power**: Enhancing commitment and quality of performance through motivation and in-kind and material rewards.

2.2 Tactics of Leadership Influence

These are the methods used by leaders to influence groups and individuals, including the following:

2.2.1. **Effective Communication**: Creating an interactive work environment through transparency, relationship building, and direct communication with individuals.

2.2.2. **Inspiration and Motivation**: This enhances internal motivation by spreading a spirit of encouragement, appreciation, and enthusiasm in the hearts of employees.

2.2.3. **Appreciation and Recognition**: It is an appreciation of the achievement that celebrates the individual in front of everyone by enhancing job satisfaction.

2.2.4. **Active Listening**: Making the employee's voice important and valued by listening to opinions and respecting viewpoints.

2.2.5. **Negotiation and Persuasion**: Possessing the skill of understanding and negotiating for useful solutions.

2.3 *Navigate Regulatory Policies*

This is called job rotation, continuous distribution of roles, and the ability of leaders to build policies and organizational structures. It includes the following:

2.3.1. **Understand the Organization Structure**: Achieving strategic goals through a clear understanding of tasks and roles, how to use resources, and optimal direction of efforts.
2.3.2. **Understanding Political Dynamics**: Influencing decisions through the possibility of movement of individuals within work teams and understanding their common interests.
2.3.3. **Adapting to Changes**: This is responding and adapting to administrative changes and updating policies to ensure the speed and ease of leaders' influence on work teams.

2.4 *The Relationship of the Interlocutor to Each Other*

2.4.1 The Effect of Strength on Performance

Previous studies show the power of formal authority in making decisions, managing work teams, and organizing tasks. However, this authority may not be effective in increasing production, as indicated by Yukl's study [2], where she stated that the combination of formal authority with personal influence has a greater impact on performance and increasing production. It was found that Kirkpatrick and Locke [3] found that personal influence or charismatic leadership has a greater role in motivation and productivity.

2.4.2 Influence and Its Relationship to Motivation

Previous literature has addressed the extent to which job satisfaction is affected by the strength of the leader's personal influence, and the study of Podsakoff et al. [4] that the power of inspiration and encouragement used by the leader stimulates the individual's management and improves the relationship between employees. Therefore, it is important for the leader to have positive influence skills within the organization.

2.4.3 Challenges Associated with Power and Influence

A positive outlook on the use of power and influence may not always benefit organizations, as there are many challenges, as highlighted by Dvir and Shamir in his study, which showed that the excessive use of power undermines trust within teams and reduces innovation, so leaders must avoid dominating Authority and enhancing a collaborative work environment.

2.4.4 Current Contribution to Research

This paper contributes to expanding knowledge about leadership by providing new insights into how power and influence affect modern business contexts. Although there have been many previous studies, focusing on the impact of these elements in changing and diverse work environments is an important contribution to the current literature.

Sources of power and influence tactics are therefore an essential part of effective leadership. Understanding these elements enables leaders to improve their performance and enhance their impact within organizations. Navigating organizational policies is a vital skill for leaders to ensure goals are achieved and facilitate collaboration between individuals and teams. The relationship between the mentioned axes shows the importance of reviewing the literature, studying experiences in different work environments, and forming a broad understanding that serves this study.

The conclusion of the literature review confirms that power and influence are critical elements in leadership and contribute to improving performance and enhancing employee satisfaction and the literature also shows that effective leaders are those who combine formal power and personal influence, which enhances the effectiveness of leadership in organizations.

3 Methodology

This study relied on the qualitative methodology through deep interviews for a sample of 200 leaders and employees in different institutions working in the construction sector. Some questions and spaces have been designed to take ideas and opinions at specific points in order to reach results that support this research and help answer research questions and solve the research problem. The data was analyzed using special methods.

4 Results

The results showed that leaders with formal power and personal influence had higher levels of motivation and satisfaction among individuals. The data also showed that personal influence is strongly correlated with organizational performance. Official power, while important, had less impact on performance than personal influence.

The results also showed the relationship between power and influence in leadership and how it affects performance and job satisfaction. The results are further illustrated below.

4.1 The Relationship Between Formal Power and Performance

The results indicated that formal power, represented by the authority and privileges given to leaders, plays a role in achieving organizational goals. However, their impact was limited when measuring the actual performance of employees. Analysis showed that teams with leaders who only benefit from formal power did not achieve the same levels of performance as teams with leaders with personal strength and influence.

4.2 The Impact of Personal Influence on Job Satisfaction

The data showed that personal influence, such as charisma and the ability to influence, is positively associated with job satisfaction. The results showed that employees who work under charismatic leadership feel a higher level of commitment and belong to the organization. For example, 78% of respondents indicated that their leaders have the ability to motivate and inspire them.

4.3 Motivation and Collective Performance

The results show that teams led by leaders with personal influence achieve higher levels of motivation. Study participants who reported that their leaders use effective influence styles (such as positive motivation and appreciation) scored higher on collective performance. This shows that personal influence not only enhances job satisfaction, but also competence and collective performance.

4.4 Diversity in the Effects of Power and Influence

The results of the study indicated that there is a difference in the strength of being influenced by the use of power and influence according to the different organizational contexts of the organizations. It was found that formal authority has a greater role in the orientation of individuals in traditional environments, while the influence decreases in environments that require flexibility and innovation.

4.5 Challenges Associated with Power and Influence

The study revealed the presence of concerns among participants about the excessive use of official authority, indicating that this may sometimes lead to reducing innovation and creativity, and it is always advised to strike a balance between the use of authority and charismatic power.

The results also indicate the need for leaders to develop their personal influence skills, as power and influence are essential elements in achieving the organization's goals, as well as interest in job satisfaction and motivating employees.

5 Discussion

There is agreement between the literature and the results of the study, as it showed that charisma and personal influence are an important element in enhancing organizational performance. Therefore, it is very necessary to work on developing the skills of leaders in the field of personal influence, charisma, and fruitful communication.

The discussion points to the importance of clarity of strategies to strike a balance between the use of official power and the power of charisma or personal influence to achieve the best results. Here it is clear that this clause is interpreted as the results of the study and puts them in the context of literature and previous studies with a deep discussion of practical implications and future goals.

5.1 Interpretation of Results

The balance between the use of formal authority and personal influence has a significant role in effective leadership and shows much better results than those that use only formal authority or personal influence alone and showed that each element has an advantage that the other element misses, for example, leaders who have charisma have a strong influence on inspiring employees and reaching job satisfaction while the holder of authority lacks it.

5.1.1. **Formal Power**: It remains necessary to manage and direct teams, but total reliance on it reduces the positive interaction of the employee and thus reflects on creativity and innovation. This is what was mentioned in the study by Kirkpatrick and Locke [3] about the role of effective communication and good relations and its importance among the work team.

5.1.2. **Personal Influence**: The results showed that enhancing collective performance lies in the use of multiple methods of influence by leaders, such as positive motivation, inspiration, and appreciation of individuals' efforts. This understanding reflects the essential role of motivational activities in promoting commitment and job satisfaction.

5.2 Comparison with Previous Literature

These findings are consistent with Yukl's [2] stating that charismatic leadership plays a crucial role in enhancing organizational performance. In addition, the results support the study of Podsakoff et al. [4], which stated that positive influences strengthen relationships between leaders and employees, resulting in performance motivation.

5.3 Practical Effects

The study concludes that organizations should seek to enhance their leadership abilities. Leadership training programs should encompass the development of communication skills, relationship building, and the ability to inspire.

5.3.1. **Leadership Training**: It is recommended to design training programs that focus on improving personal influence skills, assisting leaders in achieving greater impact on their teams.

5.3.2. **Promote a Positive Work Environment**: Organizations must foster a work culture that promotes collaboration and innovation, making it easier for leaders to utilize their personal influence effectively.

5.4 Future Challenges

Despite the positive outcomes, there are challenges to consider. For instance, there should be strategies to ensure that employees are not overly reliant on charismatic leadership. Concerns related to the use of formal force must also be addressed, to ensure that there is no negative impact on creativity and innovation.

5.5 Future Recommendations

The paper recommends further research to explore the relationship between power and influence in diverse business contexts, including:

5.5.1. **Various Industrial Fields**: Study how power and influence affect various sectors such as education, health, and technology.
5.5.2. **Different Generations**: Learn how the new generation of leaders and employees interact with patterns of power and influence.

6 Conclusions

The study shows that power and influence are essential elements of leadership. Based on the results of the study and discussion, the research recommended the need to train leaders to develop their personal influence skills and conduct more research in various fields in order to explore the influence of power, influence and personal influence enjoyed by leaders and how to multiply skills to achieve better performance and commitment. Accordingly, we address the most important conclusions as follows:

6.1 Power and Influence Are Crucial Elements of Leadership

The results showed that power and influence are essential elements of effective leadership and that leaders who have personal influence skills and possess authority have a more comprehensive impact on performance and job satisfaction.

6.2 The Importance of Personal Influence

The study showed that formal authority and reliance on it may not enhance job satisfaction among employees, so it is important to have an aspect of personal influence. This confirms the necessity of developing influence, charisma, and communication skills.

6.3 Balance Between Formal Authority and Influence

From the results, it is clear that personal influence and total dependence on authority cannot be neglected, so a balance must be achieved between them in order to raise employee morale through motivation, cooperation, and listening to them.

6.4 Challenges Related to Power and Influence

One of the most important challenges is the inability of leaders to balance power, personal influence, and excessive power. Therefore, the results indicate the necessity of developing strategies that guarantee the sustainability of performance and develop creativity and innovation.

6.5 Recommendations for Future Research

The study recommended the necessity of studying the impact of organizational culture on the use of power, as well as conducting more research in various fields.

6.6 Conduct Future Research

6.6.1. Study the impact of power in different fields and environments such as education and industry.
6.6.2. Study the culture of generations and know the methods of interaction with different leadership strategies.
6.6.3. Develop training programs to develop the skills of leaders and establish systems for continuous evaluation and measurement of the level of success of these programs.

Acknowledgements This research was funded by the Modern College of Business and Science and conducted under the supervision of Dr. Fadi.

References[1]

References in English

1. French, J.R.P., Raven, B.: The bases of social power. In: Studies in Social Power (1959)
2. Yukl, G.: Leadership in Organizations, 6th edn. Prentice Hall, Upper Saddle River, NJ (2006)
3. Kirkpatrick, S.A., Locke, E.A.: Direct and indirect effects of three core traits on performance and effectiveness. J. Appl. Psychol. (1996)
4. Podsakoff, P.M., MacKenzie, S.B., Paine, J.B., Bachrach, D.G.: Organizational citizenship behaviors: a critical review of the theoretical and empirical literature and suggestions for future research. J. Manag. (2000)

[1] Below are some studies and references on which the study was based.

The Importance of Feasibility Study in Achieving Sustainable Economic Development in the Sultanate of Oman: A Conceptual Paper

Zaid Al Nabhani and Mohamed Elgeddawy

Abstract This conceptual paper aims to fill the lack of applied studies related to the Omani economy and provide a comprehensive analysis of the importance of the feasibility study in achieving sustainable economic development in Oman, by focusing on the period 2020–2024, and studying the follow-up of implemented projects, in addition to exploring the role of modern technologies and artificial intelligence in improving the quality of feasibility studies. The paper also aims to analyze how feasibility studies affect non-oil sectors and examine the possibility of linking feasibility studies with sustainable development plans in the framework of Oman Vision 2040. This conceptual paper raises several questions about the role of the feasibility study in achieving the success of sustainable economic projects in Oman, in addition to questions related to the possibility of using artificial intelligence to improve the efficiency and quality of feasibility studies. The main hypothesis is that the feasibility study can contribute to achieving sustainable economic returns and ensuring the success of development projects. The sub-hypotheses were since the feasibility study can ensure the implementation of sustainable economic projects, that modern technologies such as artificial intelligence can enhance the efficiency and accuracy of the feasibility study, and that linking feasibility studies with development plans can contribute to achieving sustainable goals in the Sultanate of Oman. In this conceptual paper, the researcher relied on a descriptive and deductive approach by describing and analyzing economic feasibility studies, extracting the relationship between them, and achieving sustainable economic development. Previous studies have shown the existence of research gaps in this field, such as the lack of clear investment policies and the impact of external factors on the success of projects. There is also a need to develop more comprehensive and accurate evaluation tools for feasibility studies.

Z. Al Nabhani (✉) · M. Elgeddawy
Modern College of Business and Science, Muscat, Oman
e-mail: 20247115@mcbs.edu.om

M. Elgeddawy
e-mail: Mohamed.Elgeddawy@mcbs.edu.om

B. Awwad (ed.), *AI and IoT: Driving Business Success and Sustainability in the Digital Age*, Studies in Systems, Decision and Control 590,
https://doi.org/10.1007/978-3-031-88874-8_56

Keywords Feasibility study · Economic development · Sustainability · Project follow-up · Modern technologies · Artificial intelligence

1 Introduction

Studies related to investment and development projects are very important, which are also known by different names such as (national or social feasibility study); this is because they study the feasibility of the project from the perspectives of benefits and returns on society or the national economy [1].

The feasibility study is one of the vital tools that contribute to achieving sustainable economic development by enhancing economic competitiveness. It is also used to differentiate and diagnose investment opportunities through the alternatives available to implement projects, minimize waste in economic resources, and avoid the failure of projects that may have modest returns [2].

Oman Vision 2040 is a gateway to overcoming challenges, keeping pace with local, regional, and international changes, as well as generating and seizing opportunities to enhance economic competitiveness, raise social welfare, and stimulate growth by building confidence in most social, economic, and developmental relationships across Oman [3].

Due to the lack of national cadres that specialize in conducting feasibility studies, the Supreme Planning Council, in cooperation with the Institute of Public Administration, launched on 20-11-2017 the National Project to Prepare National Cadres in the Field of Economic Feasibility Studies and Development Projects, which is one of the initiatives of the National Program to Promote Economic Diversification (Tanfeedh), to promote local economic diversification [4]. From this point of view, it is important to prepare research that supports the importance of feasibility study in achieving sustainable economic development.

2 Research Gap

Research gaps are one of the essential elements in the formation of scientific research, through which it is possible to reveal knowledge aspects that have not been adequately addressed or studied, which calls for more studies and developmental research. Oman Vision 2040, represented in the economy and development axis, aims to achieve the development goals and priorities stipulated in the vision, especially with regard to the economic side, represented in the industrial, commercial, tourism, urban, and other sectors that support the Omani economy [5]. By highlighting the title of the study (The importance of feasibility study in achieving sustainable economic development in the Sultanate of Oman), the researcher seeks to explore a number of themes that he expects have not been extensively addressed in the literature and previous studies, especially with regard to local studies, including:

The lack of applied studies on the Omani economy, as there are many studies that deal with the feasibility study in foreign investment environments and may be in a generalized form, which leaves a research gap regarding the feasibility study on Omani economic projects.

Limited research on follow-up on projects implemented in Oman, there is a paucity of research that examines the performance of implemented projects based on feasibility studies, through which the extent to which projects achieve long-term economic development goals can be measured.

The impact of modern technologies and artificial intelligence on feasibility studies, with the rapid technical and technological development, many tools such as artificial intelligence have emerged, which has created a gap in research on how these technologies can improve the efficiency of feasibility studies and their potential to achieve the required sustainability in the Sultanate of Oman's projects.

Lack of focus on linking feasibility studies with development plans, despite the existence of several studies that highlighted the importance of economic feasibility studies, but (to the best of the researcher's knowledge) they focused on financial aspects related to costs and profits. A research gap has emerged that examines the importance of linking feasibility studies and development plans to achieve sustainable economic development.

3 Research Objectives

First, the research aims to fill the lack of applied studies related to the Omani economy, which limits a deep understanding of the challenges facing the Sultanate and capitalizing on its unique opportunities. Secondly: To study the methods of accurate follow-up of implemented projects, to provide data that can be utilized to develop and improve future projects. Third: Exploring the impact of modern technologies such as artificial intelligence on feasibility studies and utilizing the great potential of these technologies to improve the efficiency and accuracy of studies. Finally, examining the possibility of linking feasibility studies with sustainable development, leading to the complementarity of achieving the goals related to some elements of Oman Vision 2024.

4 Importance of Research

From this perspective, the researcher seeks to prepare a comprehensive study on the importance of the feasibility study in achieving sustainable development in the Sultanate of Oman, by focusing on the dimensions of the Omani economy for the period (2020–2024), following up the implemented projects, as well as exploring the role of modern technologies and artificial intelligence in improving the feasibility study, in addition to its impact on non-oil sectors, and linking it to the sustainable development of the Sultanate in line with the pillars and objectives of Oman Vision 2040.

5 Research Questions

Therefore, the main question formulated from this issue is: How important is the feasibility study in achieving success and avoiding the failure or closure of sustainable economic projects in the Sultanate of Oman?

From which the following questions were derived:

*RQ*1: How can the feasibility study follow up on the implementation of sustainable economic projects in Oman?

*RQ*2: Can modern technologies and artificial intelligence contribute to improving the efficiency and quality of the feasibility study, and enhance the possibility of achieving sustainability for economic projects in Oman?

*RQ*3: What is the importance of linking the feasibility study with development plans in Oman?

6 Research Hypotheses

The researcher formulated the main hypothesis that (the feasibility study can achieve sustainable economic returns and ensure the success of development projects).

The following sub-hypotheses emerged from it:

The first sub-hypothesis: The feasibility study can ensure the implementation of sustainable economic projects in the Sultanate of Oman.

Second sub-hypothesis: Modern AI technologies can significantly contribute to improving the efficiency and quality of the feasibility study, thus achieving sustainability for economic projects in the Sultanate of Oman.

Third sub-hypothesis: By linking the feasibility study with development plans, it can ensure the realization of sustainable economic development goals in the Sultanate of Oman.

7 Research Methodology

Due to the nature of the study and to achieve its objectives, the researcher used the descriptive-deductive method, by describing what a feasibility study is based on sources and studies, analyzing them, deducing the most important findings, and finding the relationship between the feasibility study and its importance in achieving sustainable economic development.

8 Literature Review

The study [6] aimed to highlight the role of feasibility studies in evaluating projects and making investment decisions, and emphasizing the importance of addressing marketing, engineering, financial, and technical considerations, as the results, feasibility studies are essential tools in reducing uncertainty in making investment decisions, after making a comprehensive assessment of the project's potential by relying on the feasibility study. They also help decision makers to allocate resources effectively and ensure that investments are compatible with economic rationality.

The study [7] aimed to know the reality of the feasibility study in various departments of the State of Iraq and its impact on the success or failure of investment projects, in addition to studying, identifying and analyzing the factors affecting the feasibility study in investment projects through the study of many field terms, interviewing engineers and experts, in addition to analyzing previous research in the Arab world. The findings revealed that five categories of factors (political, legal, economic, financial, consultant, owner, and contractor) that affect the feasibility study are the absence and lack of clarity of the investment policy, the presence of many obstacles, slow and complex administrative procedures, lack of field knowledge, and finally failure to meet the scientific standards accurately.

Raharjo [8] also showed that feasibility study is not the only thing to be considered when deciding on the establishment of any project, but it is considered one of the auxiliary tools in decision-making. It depends on assumptions and data that may change depending on the specific purpose of the project. The study showed that although the project may not be financially feasible, there is a need for further research to see if it has other returns from economic growth. This shows that the feasibility study is only part of the decision-making process.

The study [9] also highlighted the importance of economic feasibility studies as a critical mechanism for the success of investment projects, and emphasized the importance of conducting a feasibility study for the project well to provide a realistic picture before starting, which may increase the likelihood of achieving it. The results of the study showed that the success of investment projects depends greatly on the success of preparing a realistic feasibility study, through the availability of comprehensive and accurate data and information and emphasizing the need to prepare the study by an experienced and qualified person in project management and preparation of economic studies.

The researcher concluded from these previous studies that there are research gaps that include several aspects, including the lack of clear investment policies that affect the accuracy of the feasibility study [7], and the impact of external factors (political, legal, and economic) on the success of projects. Raharjo [8] also showed the need to evaluate non-financial returns in addition to financial returns in feasibility studies. Studies also indicate the need to develop more comprehensive and accurate financial evaluation tools, such as improving the use of discounted cash flow. Bin Sha'a et al. [9] pointed out the importance of developing better mechanisms for collecting and analyzing data to ensure more reliable and accurate results in the feasibility study.

9 Conceptual Framework

9.1 The Concept of Feasibility Study

An economic feasibility study is a study that includes all aspects related to the project (technical, financial, legal, and marketing), which is done by collecting data and information, and analyzing them to make a decision that suits the idea of implementing the project or not [10]. The economic feasibility study is prepared to determine the feasibility of the project idea, by studying the legal aspects, technical aspects, and economic aspects [11].

The feasibility study evaluates projects and opportunities from all financial, economic, technical, commercial, commercial, environmental, social, and other aspects to ensure the feasibility of the project or opportunity. The feasibility analysis process comes once the proposed investment opportunity is recognized and before starting the project work, so the feasibility of the investment is studied according to the criteria and parameters of the feasibility study, which is one of the most important steps in decision-making [12].

Through what was mentioned earlier, it can be said that the economic feasibility study is a tool that enables stakeholders to make investment decisions that ensure the achievement of the desired returns, after studying and analyzing the legal, technical, financial, market and other aspects accurately in the interest of the investment project.

9.2 The Importance of Feasibility Studies

The importance of the feasibility study lies in helping investors in achieving many of the planned goals, as well as helping them evaluate many technical, legal, economic, financial and marketing aspects, in addition to determining the size and trends of the market, identifying competitors and studying their strengths and weaknesses, developing appropriate practical plans to make projects successful, convincing funding institutions such as banks to finance the project, as well as convincing investors of the profits the project will earn worth investing in Al-Muhrrammi [11]. If we talk about the importance of the feasibility study for projects, it is considered as the investment pillar of any country that can be relied upon in achieving development goals, and it also contributes to identifying the appropriate source of funding for project needs and thus achieving returns on the investment of that funding.

9.3 Oman's Economy

In 2022, the Omani economy witnessed growth in various economic indicators, as evidenced by the reliable performance of public finances, rising oil and non-oil revenues and GDP, improved global trade and local markets, in addition to controlling inflation at a safe level of 2.8%. The performance results came because of fiscal and economic policies that focused on recovering from the effects of the COVID-19 pandemic, improving the management of oil and gas revenues, supporting the private sector, and increasing investments and development projects in line with Oman's Vision 2040. Oman Vision 2040 is a roadmap for achieving sustainable development and diversifying the economy by promoting knowledge and innovation [13]. According to global and regional economic forecasts, in 2024, the growth rate of the global economy is expected to slow down to 2.9%, because of the effects of monetary policies set in the face of inflation, according to International Monetary Fund forecasts. As for Oman, the IMF expects the Omani economy to achieve a growth rate of about 3.6% in 2024 [14].

9.4 Project Tracking

The Oman Investment Authority works to build a diversified investment portfolio that includes different investment instruments and assets that are geographically distributed, to minimize risks and ensure sustainable and long-term returns. It also relies on a precise and flexible strategy, based on the best international practices, through a periodic review of asset allocation and development in line with market changes. In 2022, the agency developed an action plan that extends until 2026, aiming to enhance the performance of its subsidiaries by increasing their investments, supporting economic and social development, and increasing their contribution to the GDP of the state's general budget. Since its establishment, the agency has achieved an annual return of 10.3% [13].

9.5 Modern Technologies and Artificial Intelligence

Modern technologies, especially artificial intelligence (AI), contribute to improving decision-making processes by providing accurate and timely information, which enhances the quality of economic feasibility studies. These technologies allow decision makers to analyze massive amounts of data, enabling them to make informed decisions regarding resource allocation, cost management, and risk assessment. AI applications in improving supply chain management and financial fraud detection enable organizations to accurately identify potential challenges and opportunities, supporting more reliable and effective feasibility studies [15].

Modern technologies, particularly artificial intelligence (AI), play a prominent role in improving productivity and efficiency in various industries, enhancing the effectiveness of feasibility studies. AI enables companies to analyze vast amounts of data to identify potential challenges and opportunities and make informed decisions on a project schedule. The integration of AI improves the measurement and management of supply chain performance, which is an essential component of a complete business case [16].

Through previous studies, the researcher concluded that modern technologies, especially artificial intelligence (AI), play a pivotal role in enhancing economic feasibility studies, and if I may say so by improving decision-making through accurate and efficient data analysis, supporting resource allocation, cost management, and risk assessment. In addition, it contributes to improving operational efficiency through fault prediction and cost reporting, thus enhancing supply chain management through transparency, demand forecasting, and inventory management. This contributes to supporting sustainable development by increasing resource efficiency and ensuring comprehensive feasibility studies that support the success of projects in a competitive environment.

9.6 Economic Development

The study of Levine and Tantardini [17] highlighted the definition and measurement of economic development through a literature review, due to the complexities and lack of a universally accepted definition of economic development, as economic development is a multifaceted concept. This research study confirms that there is no specific definition of economic development, nor is there an agreed upon comprehensive measure, as there is still a wide range of different methods and definitions used to conceptualize and measure economic development. The study concluded that the common orientation of economic development toward improving material conditions in human societies can be used to distinguish between policies and ideas that work and those that do not work at all.

The study of Nastu et al. [18] aimed to differentiate between the concepts of economic growth and economic development, highlighting that economic growth is the quantitative component of economic development, which includes qualitative elements related to quality of life and living conditions. The study concluded that economic growth is a quantitative measure usually represented by GDP, while economic development encompasses both quantitative and qualitative aspects including quality of life and living conditions.

The study of Soaduan Pohan [19] aimed to explore the dynamics of economic development with a focus on achieving high economic growth to improve the quality of life and well-being of the community. The study emphasized that the primary goal of national development is to achieve high economic growth, and that such growth is critical to enhancing the quality of life and well-being of the entire society, by increasing the production of goods and services, which in turn leads to higher national income, usually measured by Gross Domestic Product (GDP).

Through previous studies, the researcher concludes that the concept of economic development is multidimensional, encompassing both quantitative and qualitative aspects to improve the quality of life and well-being of societies, despite the continuing challenges related to defining economic development and measuring its dimensions in a unified manner. While economic growth is the quantitative component of development, usually measured by GDP, development encompasses the improvement of living conditions and quality of life. Economic development aims to increase production and enhance national income, to ensure the well-being of society, while emphasizing the importance of justice and equity in reducing economic gaps.

9.7 Sustainability

Sustainability is defined as a state of dynamic equilibrium, where the timing and amount of resource replenishment is sufficient to continue performing a given task without destabilizing the current and future availability or replenishment of other resources. The concept of sustainability emphasizes the importance of balancing the use of resources to ensure the well-being of current and future generations without depleting or damaging the environment [20].

Sustainability refers to the ability to meet the needs of the present without affecting the ability of future generations to meet their own needs, considering the following dimensions (environmental, social, and economic). Sustainability aims to achieve long-term growth that balances the dimensions to improve the quality of life [21].

Sustainability means enabling harmonious coexistence between the natural environment and human civilization, by emphasizing that resources should be used sparingly [22].

From previous studies, the researcher concludes that sustainability is a dynamic process of balancing the use of current resources and ensuring their continuity to serve future generations. To achieve sustainability, it is necessary to use resources sparingly to meet the needs of the present without affecting the needs of future generations.

10 Concept Paper Results

Feasibility Study Concept: It is a comprehensive analysis of the technical, financial, legal, and marketing aspects of the project to determine whether it is feasible. The feasibility study contributes to making the appropriate investment decision based on a comprehensive analysis that includes the legal and technical conditions to achieve the success of the project.

Importance of the Economic Feasibility Study: The feasibility study helps investors evaluate the various aspects of the project, such as (financial feasibility, technical feasibility, and economic feasibility), in addition to analyzing the market and competition, identifying sources of funding, and convincing beneficiaries and funding institutions of the feasibility of the project.

Omani Economy: The Omani economy witnessed growth in 2022, with improved revenues and GDP, thanks to the financial and economic policies followed. The Omani economy is expected to achieve a growth of 3.6% in 2024.

Follow-Up on Implemented Projects: The Oman Investment Authority is working to diversify its investment portfolio and reduce risks with a focus on non-traded investments in various sectors. It aims to increase its contribution to the national economy and achieve sustainable returns.

Modern Technologies and Artificial Intelligence: Modern technologies and artificial intelligence contribute to improving feasibility studies by analyzing data very quickly and accurately, which enhances informed decision-making that will lead to reducing costs and increasing operational efficiency.

Economic Development: Economic development includes quantitative and qualitative aspects, such as improving the quality of life and living conditions. While economic growth is measured by GDP, development focuses on improving society in general.

Sustainability: It means meeting the needs of the present without affecting the ability of future generations to meet their needs. It focuses on balancing the optimal use of resources and achieving long-term growth while preserving the environment.

11 Recommendations

The researcher reached the following recommendations through this conceptual paper: The importance of conducting comprehensive feasibility studies: The researcher recommends preparing comprehensive economic feasibility studies that take into account all technical, financial, legal, and marketing aspects, through which projects can be successfully implemented, and provide an accurate database and information for investors and stakeholders to make their decisions.

Application of artificial intelligence techniques in the feasibility study: The researcher recommends the need to use modern technologies such as artificial intelligence (AI) to improve analysis and decision-making processes, as these technologies help in analyzing data more accurately and effectively, which contributes to improving and allocating resources, in addition to reducing costs and achieving higher efficiency.

Supporting economic growth and diversifying sources of income: The researcher recommends the importance of developing strategies that support Omani economic growth, with a focus on economic diversification and strengthening the private sector, to reduce dependence on oil revenues, and thus enhance investments in non-oil sectors.

Focus on sustainability in projects: The researcher recommends the need to adopt sustainability strategies in all investment projects, in order to ensure achieving a balance between the use of current resources and the needs of future generations, with a focus on the following dimensions (environmental, social, and economic) to ensure sustainable development.

References

1. Al-Muharrammi, S.M.: Feasibility Study of Projects Between Commercial Profitability and National Returns, 1st edn. Academic Publication and Outreach Department, Sultan Qaboos University (2022)
2. Shabib, H.S., Kazem, M.J.: The concept of feasibility study and its economic dimensions. J. Kufa Stud. Center **1**(60A), 281–308 (2021)
3. Oman Vision 2024 (2019) [Online]. Available: https://andp.unescwa.org/ar/plans/1200. Accessed 13 Oct 2024
4. Hamza, M.R.: Specialization in preparing feasibility studies, 25 Nov 2017 (Online). Available: https://www.omandaily.om. Accessed 18 Nov 2024
5. Implementation Follow-up Unit Oman Vision 2040: Implementation Follow-up Unit, Oman Vision 2040 Report (2024)
6. Noon, M.R.: The Role of Financial Feasibility Study in Investment Decision-Making: A Case Study of an Investment Project at the Level of the Bank of Agriculture and Rural Development. Qalmah (2024)
7. Alhamami, B.R., Nassar, Y.S., Qendeel, L.M.: Feasibility studies and their effects on the success or failure of investment projects. Najaf Governorate as a Model, De Gruyter (2023)
8. Raharjo, D.S.: How feasibility study guides an execution of project development. Int. J. Bus. Appl. Soc. Sci. **8**(1), 43–55 (2022)
9. Bin Sha'a, W., Almawi, A., Bin Odhina, B.: Feasibility studies as a mechanism for the success of investment projects. Al-Muntada J. Econ. Stud. Res. **3**(2), 132–143 (2020)
10. Small and Medium Enterprises Development Authority: Feasibility Studies Basics Guide for Small and Medium Enterprises, 1st edn (2022)
11. Al-Muhrrammi, S.M.: Preparing and Analysing the Economic Feasibility Study. Academic Publication and Outreach Department, Sultan Qaboos University (2020)
12. Yousef, A., Marhej, M.: Project Evaluation and Feasibility Study. Syrian Virtual University (2018)
13. Information Portal: Omani Economy (Online). Available: https://www.omaninfo.om/ar/pages/229/show/930. Accessed 18 Nov 2024

14. National Center for Statistics and Information: Features of the Omani Economy 2023 (Online). Available: https://ncsi.gov.om (2024)
15. Xiao, Y.: The multifaceted relationship between AI and economics: impacts, challenges, and insights. J. Econ. Manag. Sci. **6**(3), 1 (2023)
16. Tripathi, V.M.: Role of modern technology in industry 4.0: an empirical study in the context of artificial intelligence. Turk. Online J. Qual. Inq. **11**(2), 854–861 (2020)
17. Levine, E.J., Tantardini, M.: Defining and measuring economic development: a literature review and outlook. J. Public Nonprofit Aff. **9**(3), 405–436 (2023)
18. Nastu, A., Dumitrache, A., Melian, D.: Economic development and economic growth: case study. J. Econ. Financ. Manag. Stud. **5**(6), 1604–1612 (2022)
19. Soaduan Pohan, M.E.: Economic development and its measurement. Al-Mutharahah: Jurnal Penelitian Dan Kajian Sosial Keagamaan **18**(1), 80–90 (2021)
20. Ramirez, G.A.: Sustainability (Online). Available: https://www.qeios.com/read/6P7VQT (2024)
21. Adula Harif V, Natasha P (2024) A review on green accounting practices and sustainable financial performance. J. Bus. Manag. Inf. Syst. **11**, 82–88 (2024)
22. Yang, J., Zeng, Y.: Sustainability: design, making, and technologies. J. Integr. Des. Process Sci. Arch. **24**(2), 1–3 (2022)

Women Empowerment and Gender Diversity on Boards—A Review

Anitha Narayanan Nair, Dimple Deoli, and Ayyappan Palanissamy

Abstract Keeping up with the rapidly changing corporate environment and modernizing the composition of the board of directors has been one of the latest hot themes in boardroom discussions. The efficacy of the boardroom may be strengthened and contributed to by increasing gender diversity. Even though empirical evidence does not conclusively show that adding more women to boards of directors would directly improve companies' financial performance, many nations and business organizations have this issue at the top of their agendas because it can show a strong commitment to good corporate governance practices. Gender diversity can be understood as an equitable or nondiscriminatory representation of people from different genders. The ratio of men to women on corporate boards, which facilitates better decision-making and increases company performance, is the main aspect of corporate gender diversity. The importance of gender diversity in corporate leadership roles is being emphasized in a growing body of literature. Many research studies have shown the benefits of gender heterogeneity on firms' financial performance, factors influencing the gender composition of boards as well as firm's policies, a decrease in corporate risk-taking behavior, improved governance excellence, better health corporate environmental practices, and decreased corporate environmental violations, and women on boards and their association with high CSR ratings. By emphasizing the crucial fact and strength of women's tech experience, which is regarded as necessary to combat the pace of present and future technology disruption, this study reflects the future role of women in the boardroom.

Keywords Gender diversity · Sustainability · Boardroom · Corporate governance

A. N. Nair (✉) · D. Deoli · A. Palanissamy
Modern College of Business and Science, Muscat, Oman
e-mail: anitha@mcbs.edu.om

B. Awwad (ed.), *AI and IoT: Driving Business Success and Sustainability in the Digital Age*, Studies in Systems, Decision and Control 590,
https://doi.org/10.1007/978-3-031-88874-8_57

1 Introduction

Organizations around the world are grappling with the imperative of inclusivity. The presence of women in leadership roles has emerged as a critical determinant of organizational success and societal advancement. Gender diversity on Boards is not merely a matter of equity; it is intrinsically linked to better decision-making, innovation, and financial performance.

Historically, male perspectives have dominated corporate governance, often leading to a homogeneity that stifles creativity and limits the breadth of insights necessary for effective leadership. However, the landscape is shifting. The increased presence of women in boardrooms reflects a growing recognition that diverse perspectives foster efficient discussions and lead to more robust strategies [1]. The gender of a leader serves as a significant intervening variable, influencing how different personality traits manifest and subsequently affect cultural and performance outcomes for both men and women leaders [2, 3]. In today's era, which is increasingly centered on sustainability, organizations are expected to actively involve and discuss with diverse stakeholders [4]. As such, the characteristics of CEOs, specifically those that are related to gender, play a significant role in addressing the needs of these stakeholders, especially when it comes to reporting corporate activities like carbon emissions that contribute to climate change.

Here are some of the theories based on which the literature has been conducted.

Feminist Theory is one of the theories that have a perspective that underscores the systematic inequalities faced by women and the need for structural changes to achieve gender equality. Feminist theory is a diverse and multifaceted field of study that explores the social, political, and economic inequalities between men and women [5]. It encompasses a wide range of perspectives and approaches, from liberal feminism to radical feminism, each with its unique focus and goals [6]. Moving on to liberal feminism, this is one of the earliest forms of feminist thought. It promotes equal rights and opportunities for women within the existing social and political frameworks [7]. Liberal feminists argue that women ought to possess the same legal and political rights as men and should be empowered to engage fully in every facet of society [8].

Another theory is the social identity theory which highlights the influence of social categories, such as gender, on an individual's self-perception and behavior. According to this theory, people get their self-confidence and "identity" from their social groups [9]. Stets and Burke [10] point out that people often tend to compare their groups with other groups, and this can have an impact on their self-esteem as well as their behavior. If a person feels that their social group is better than the other, then it will instill positive feelings in the individual. But, if the person feels that their social group is inferior to other groups, then it can lead to negative feelings [11]. Another popular theory that has gained a lot of importance in the past few years is the Resource Dependence Theory. This theory argues that firms depend on the resources that they get from the environment, and this includes human capital too.

These resources vary from capital and labor to raw materials and technology [12] and by adopting gender diversity, the firms can have more diverse resources.

The concept of gender diversity has become so popular nowadays, that having women in leadership roles not only ensures equality but can even better the firm's overall performance. This paper attempts to study the impact of gender diversity on corporate Boards and the challenges that women possibly face while treading on this path. Just like men, all women are not the same; their experiences and challenges differ based on their culture and socioeconomic backgrounds. Moreover, a thorough understanding of these dynamics is needed for developing strategies that promote inclusivity and empower women from diverse backgrounds.

2 Literature Review

As firms have increasingly started to realize the importance of having gender diversity, the inclusion of women at the highest levels of decision-making is not merely a matter of equity alone; it is also a strategic move that can enhance the business's overall performance. Women bring unique viewpoints and experiences that can challenge routine and mundane activities, and increase problem-solving abilities [13].

A study by Singhania et al. [14] reported that as the number of women directors and the number of independent female board members increase, the quality of sustainability reporting is expected to increase.

Bukalska et al. [15] illustrated a positive correlation between operating efficiency and the percentage of females on the boards and overall board gender diversity. However, they did not observe a statistically significant link between the representation of women on the management board and market performance.

Griffin et al. [16] thought that Boards which supported gender diversity are linked to higher levels of innovation across countries. Moreover, the authors felt that the Boards were more inclined to incorporate women in various nations where gender disparities are minimal, along with increased female labor ratio participation and less patriarchal culture. Firms with gender-diverse Boards tend to secure more patents, including innovative patents, and show greater efficiency in their innovative endeavors. Additionally, such gender-diverse boards were linked with more resilience and long-term incentives for CEOs, along with an innovative corporate culture with various investors. All these attributes pave the way for an enhanced overall innovative performance.

Another study conducted by Chen et al [17] found that firms that gave importance to innovation enjoyed greater success and had improved performance when their boards were gender diverse. The authors also opined that the benefit of gender diversity on innovation is more evident in situations where there is a lower rate of market competition. It is also common among managers who are more firmly established in their roles. This may be because, increased oversight provided by female directors, can motivate managers to pursue innovative initiatives.

A study conducted by Kim and Starks [18] found that gender diversity on corporate boards can increase a company's value by bringing in unique functional expertise that improves decision-making and strategic direction. Moreover, having female directors brings in specialized skills that are often lacking in corporate boards. This increases board diversity and boosts the overall value of the firm.

Najaf et al. [19] conducted a study that showed that the market performance of firms that had adopted CSR initiatives had increased by "3.1%" while having gender diversity on boards increased market performance by "4.8%". Another interesting finding in their study was that women board members helped reinforce the relationship between CSR and market performance by "1%". The authors also pointed out that environmental and social issues like climate change and gender equality could be tackled if the firms had gender diversity on their boards.

Luh et al. [20], in their research, noted that banks in Ghana that had more women in leadership roles had more transparency when it came to disclosure of environmental, social, and governance activities. This indicates the reason why gender diversity on boards should be encouraged in banks so that ESG activities can be handled more efficiently.

Another study by Wiersema and Mors [21] showed that women usually arrived at board meetings fully prepared. It was also observed that women were more open to acceptance when they were not very familiar with certain topics. They also never hesitated to ask questions and engage in open discussions which helped in having more communication. They are more likely to admit when they are unfamiliar with a certain topic and encourage open discussions. This helped in having a free flow of conversations within the boardrooms. The authors observed that the male counterparts had a sort of hesitation to recognize their female peers as equal to them. This was seen as a limiting factor in women's effectiveness in decision-making on corporate boards.

Nae [22] in his research studied gender diversity on corporate boards across countries like North America, Europe, and Asia. The author found that though women were exposed to higher education and active participation in the workforce, their path to leadership was often blocked by cultural, corporate, and legal barriers. These issues limit women in moving forward in their careers.

Srivastava et al. [23] in their study analyzed that gender diversity on boards accounted for a little more than "5.5% of the change in a firm's cost of equity". At the same time, having gender diversity on boards had about "45% on the firm's return on assets". The authors observed that having more women directors helped in bringing down the "cost of equity".

3 Role of Women on Boards in Promoting CSR

Having women on boards has been increasingly viewed as a crucial component of promoting corporate social responsibilities. Women carry with them a unique blend of behavioral attitudes like empathy and sensitivity, which are needed for ensuring ethical and socially responsible practices. Having women on boards helps in increasing CSR activities that can lead to more efficient overall corporate performance.

Muhammad et al. [24] noted that having gender diversity on boards tends to give more importance to socially responsible activities. This, in turn, would lead to having a better impact on the CSR performance of a firm. The authors also noted that women leaders were more inclined to uphold ethical values as they were more interested in increasing CSR activities and reducing unethical behaviors. The authors were of the view that if firms included gender diversity on boards along with CSR initiatives into their core strategies, thereby mitigating the risks related to the management of earnings.

Jizi et al. [25] observed that in the GCC, having more women on boards was considered to be an important feature since it helped in executing the firm's social agendas and ensured more efficient sustainability reporting. The study also pointed out that having gender diversity on boards led to a quicker and more efficient implementation of policies related to the environmental, ethical, and health aspects.

Alawi [26] in his research pointed out that firms registered an increase in their firm performance when CSR initiatives increased as investors preferred organizations that had a responsible connection with the society. The author also recommended that having more women members on the corporate boards would help in improving the firm's overall performance.

Jin et al. [27], in their study, noted that firms with more women leaders tend to prioritize internal CSR activities as compared to external CSR initiatives. The study also pointed out that the impact of women leaders on internal and external CSR activities was more when the major proportion of the boards' "independent directors" had political relations.

4 Women on Boards and Sustainability

In recent times, sustainability has become the new "buzz" word. Adopting gender diversity on corporate boards has been viewed as a driving factor for achieving sustainability [28]. Having women in leadership roles helps in achieving better corporate governance, in addition to having different perspectives on a particular agenda. Women tend to be more considerate towards ethical and moral aspects and at the same time, align these aspects with the stakeholders' interests. This helps boost the firm's goodwill and reputation among investors too. There have been numerous studies that depict the positive impact on the environment and sustainability when organizations have gender diversity on their boards. This will not only benefit the environment but also improve the company's reputation and overall look more appealing to investors.

Enciso-Alfaro and García-Sánchez [29] in their research, noted that firms that had more women as leaders on their boards were able to shift to circular business models easily. Circular models happen to be a core element of sustainability. The authors also observed that women directors had creative perspectives when it came to decision-making, especially in areas related to the environment, personnel development, and economic well-being. Having gender-diverse boards in organizations was found to have a better impact on social responsibility, including better labor practices and community engagement activities.

Trireksani et al. [30] conducted a study which was based on a sample of "5087 firms" across "50 countries". The authors, in their analysis, noted that having women on their boards helped in improving their overall sustainability performance. However, the authors also reported that the intensity of this relationship depends on the country in which the firm is set up and it can vary from country to country. Countries that favored equality among genders and "individualism", had more gender diversity in their boards and better sustainability performance. Moreover, having a strong legal framework within the country and the country's ability to solve economic issues helped in empowering women to a greater extent.

Research done by Claudia and Lindrianasari [31] observed that a firm's value does not significantly depend on green accounting practices and carbon emission disclosures adopted by a firm. On the contrary, a firm's value increases if the firm has a larger proportion of women on its boards. Moreover, having women on corporate boards serves as moderators in maintaining and improving the relationship between carbon emission disclosures, green accounting practices, and a firm's value.

Ramos et al. [32] in their study stated that having gender diversity on boards helped in improving the "environmental value" of the firm. It led to better transparency in environmental reporting especially in "family-owned" businesses. The authors concluded that there existed a strong positive relationship between gender diversity on corporate boards and the degree of environmental sustainability.

5 Challenges Faced by Women

The path for women to achieve their dreams has always been ridden with obstacles and challenges in the form of social expectations, discrimination within the firms, and the existing organizational cultures.

One of the main biases faced by women is cultural bias. A study conducted by Carrasco et al. [33] opined that the number of women hired on corporate boards has always been low and so is the rate of new female appointments to these positions. The authors also found that countries that tolerated power inequalities and those that favored males in traditional roles had a lower proportion of women in leadership roles.

Another study conducted by Moraka [34] on the challenges faced by South African Women, showed that gender stereotyping is a common issue faced by women. Despite having higher educational qualifications than their male counterparts, very few women were appointed only to satisfy the minimum requirement as stipulated by legislation. Gender Stereotypes and Bias are the traditional gender roles and stereotypes can hinder women's advancement to board positions. Implicit biases can affect perceptions of women's leadership abilities [35].

Another problem is the lack of mentorship and sponsorship. Women often lack access to influential mentors and sponsors who can guide their career progression and advocate for their board appointments [28]. Women also face a major challenge while working in demanding board jobs as they are often balancing work and domestic chores [36]. Moreover, institutional barriers in corporate governance structures and practices may unintentionally exclude women from board positions [37].

Several countries have implemented gender quotas to increase female representation on boards [38]. While effective in some cases, quotas can also be controversial and may not address underlying systemic issues [39]. Providing diversity training and education can be one of the programs that promote diversity and inclusion help to challenge stereotypes and create a more exclusive corporate culture [40]. Mentorship and Sponsorship Programs also take initiatives that connect women with experienced mentors and sponsors that can accelerate their career advancements. In addition to this, the process for nominating to the Boards should be transparent, and enabling a merit-based board nomination will ensure getting more competitive women leaders on Boards.

6 Women Directors and IT Expertise

Women directors are known for their creativity and innovative ideas in the boardrooms. They combine their leadership skills with problem-solving skills, risk management, and strategic perspectives, all of which are needed to solve the complex intricacies of technology. In today's era, more women have started entering the IT field and most of them are adept in handling IT systems and supervising digital strategies. In addition, women directors with a background in IT are more versatile in adopting innovative methods thereby enabling firms to survive in today's competitive technical world.

Other than technological innovation, women directors with IT expertise ensure maintaining a strong cybersecurity without compromising on ethics or moral values. Women are very well versed in matters related to data protection and are therefore very efficient in leading firms through complex regulations like GDPR (General Data Protection Laws) and Privacy laws. They often pay attention to details and along with it, women are also very skilled at risk management, thus ensuring that any investments in technology sync with the goals of the corporate firms. Women directors ensure that firms function responsibly by ensuring the ethical use of technology, be it in the field of AI or data analytics. Gender diversity, when combined with expertise in the IT field, helps in achieving improved corporate governance, flexibility, and sustainability for firms.

7 Conclusions and Recommendations

Many studies have been conducted globally to portray the impact of gender diversity on the composition of corporate boards. Not only does the problem of gender diversity help to narrow the gender equality gap, but it also upholds the principles of good corporate governance. Throughout, traditional boardrooms have always been dominated by men. As the problem of gender diversity is gaining more attention, the need for board renewal has begun to play an important role. Therefore, research on female directors in boardrooms is also gaining attention and is being highly investigated. The benefits of having women and submitting to gender equality in boardrooms are stressed in this study. According to the scholarly studies mentioned throughout the research, there are many advantages to having women on boards. For instance, as listed above, with a better focus on making strategic decisions for the firm, women get better and careful oversight fostering positive connections among the stakeholders of the organization. Women also typically provide additional skill sets that are needed in the boardroom. In addition, the last section highlights the growing responsibilities of women directors, particularly those who have an IT background.

In terms of AI connection women have shown higher soft skills and proficiency. Moreover, as opposed to male directors, females tend to have great professional IT experience, which is a crucial skill for organizations to evolve within today's technological growth. As high-tech innovation continues to disrupt businesses, they must adopt an agile approach to deal with this ever-evolving disruption. Boards of directors must keep up with this transformation and be able to represent a realistic dimension of a company's development plan and potential hazards due to the global competitive environment and the rapid change in digital technologies. Governments ought to concentrate on measures that foster a welcoming workplace for female executives. Countries with legislation and mandates have improved gender balance on corporate boards. The private sector would gain by offering leadership development programs to its talented female employees and by doing away with unfair workplace rivalry. Diversity has a favorable effect on business value for the following reasons: To keep up with the present trend of more diverse marketplaces, business diversity must also diversify.

Finally, Diversity encourages innovation and creativity, and companies with greater diversity can have a wide range of perspectives and make better decisions regarding different matters in addition to the ability to rapidly grow and succeed within their industry.

References

1. Kong, L., Usman, M., Yue, W., Yasmin, F., Sokolova, M.: Leadher: role of women leadership in shaping corporate innovation. Hum. Soc. Sci. Commun. (2024)

2. Awwad, B.S., Binsaddig, R., Kanan, M., Al Shirawi, T.: Women on boards: an empirical study on the effects on financial performance and corporate social responsibility. Compet. Rev. Int. Bus. J. **33**(1), 147–160 (2023)
3. Macrae, Z., Baur, J.E.: Female CEO leadership at the Fortune 500 level: personality, culture, and profit. Res. Pers. Hum. Resour. Manag. 115–144 (2024)
4. Elkington, J.: Partnerships from cannibals with forks: the triple bottom line of 21st-century business. Environ. Qual. Manag. 37–51 (1998)
5. Zeigler, S.L.: Feminist Theory. Free Speech Centre (2023)
6. Crossman, A.: Feminist Theory in Sociology: An Overview of Key Ideas and Issues. ThoughtCo (2024)
7. MasterClass: Liberal Feminism: A History of Liberal Feminism. MasterClass (2022)
8. Baehr, A.R.: Liberal Feminism. Stanford Encyclopedia of Philosophy (2007)
9. Madeson, M.: Social identity theory: I, You, Us & We. Posit. Psychol. (2024)
10. Stets, J.E., Burke, P.J.: Social comparison in identity theory. Communal Funct. Soc. Comp. **1**, 39–59 (2014)
11. Wang, D., Zhu, L., Maguire, P., Liu, Y., Pang, K., Li, Z., Hu, Y.: The influence of social comparison and peer group size on risky decision making. Frontiers (2016)
12. Heatley, S.: The Low-Down on Resource Dependence Theory. Perkbox (2018)
13. Giovnannoni, C.: Top 10 Reasons to Have Women on Company Boards-Girls Day School Trust (2018)
14. Singhania, S., Singh, J., Aggarwal, D., Rana, S.: Board gender diversity and sustainability reporting quality: a generalized ordered logit approach. Kybernetes **53**(8), 2679–2699 (2024)
15. Bukalska, E., Sosnowski, T., Wawryszuk-Misztal, A.: Women on management board and firm performance: evidence from the Visegrad Group Companies. Entrep. Bus. Econ. Rev. **12**(2), 43–59 (2024)
16. Griffin, D., Li, K., Xu, T.: Board gender diversity and corporate innovation: international evidence. J. Financ. Quant. Anal. 123–154 (2019)
17. Chen, J., Leung, W.S., Evans, K.P.: Board gender diversity, innovation, and firm performance. J. Empir. Financ. (2018)
18. Kim, D., Starks, L.: Gender diversity on corporate boards: do women contribute unique skills? Am. Econ. Rev. 267–271 (2016)
19. Najaf, R., Chin, A., Chin, A., Najaf, K., Thuraisingham, J.: Women on board and business performance. Foresight **26**(5), 844–866 (2024)
20. Luh, P.K., Arthur, M., Vera Fiador, B.A.: Gender of firm leadership and environmental, social and governance (ESG) reporting: evidence from banks listed on Ghana Stock Exchange. Gend. Manag. **39**(6), 778–795 (2024)
21. Wiersema, M.F., Mors, M.L.: Women directors and board dynamics: qualitative insights from the boardroom. J. Manag. **50**(7), 2413–2451 (2024)
22. Nae, N.: Gender diversity on corporate boards—still a distant dream? Cogito **14**(1), 53–69 (2022)
23. Srivastava, V., Das, N., Pattanayak, J.K.: Women on boards in India: a need or tokenism? Manag. Decis. **56**(8), 1769–1786 (2018). https://doi.org/10.1108/MD-07-2017-0690
24. Muhammad, H., Migliori, S., Berardino, D.D.: CSR as a catalyst: examining the effects of board gender diversity on earnings management. Corp. Soc. Responsib. Environ. Manag. (2024)
25. Jizi, M., Nehme, R., Melhem, C.: Board gender diversity and firms' social engagement in the Gulf Cooperation Council (GCC) countries. Equal. Divers. Incl. **41**(2), 186–206 (2022)
26. Alawi, S.M.: The role of corporate social responsibility, ownership structure, and gender diversity in firm performance. Int. J. Econ. Financ. Issues. Issues **14**(2), 97–110 (2024)
27. Jin, R., Jiang, X., Hu, H.W.: Internal and external CSR in China: how do women independent directors matter? Asia Pac. J. Manag. **40**(1), 169–204 (2023)
28. Adams, R.B., Ferriera, D.: Women in the boardroom and firm performance. J. Financ. Econ. (2009)
29. Enciso-Alfaro, S.-Y., García-Sánchez, I.-M.: Are innovative firms more circular when women hold upper positions? J. Innov. Knowl. **9**(4) (2024)

30. Trireksani, T., Djajadikerta, H.G., Kamran, M., Butt, P.: The impact of country characteristics on board gender diversity and sustainability performance: a global perspective. Sustainability (Switzerland) **16**(7) (2024)
31. Claudia, A., Lindrianasari: The role of women on board and firm value in the association of Southeast Asian nations polluting firms. Corp. Board Role Duties Compos. **20**(3), 20–29 (2024)
32. Ramos, A., Jayantilal, S., Sardo, F.: Exploring gender dynamics and environmental sustainability in family firms. Proc. Int. Conf. Gend. Res. **7**, 331–338 (2024). https://doi.org/10.34190/icgr.7.1.2045
33. Carrasco, A., Francoeur, C., Labelle, R., Laffarga, J., Ruiz-Barbadillo, E.: Appointing women to boards: is there a cultural bias? J. Bus. Ethics **129**(2), 429–444 (2015). https://doi.org/10.1007/s10551-014-2166-z
34. Moraka, N.: Is mining corporate boardrooms still a male club? Experiences and challenges faced by South African women. J. South. Afr. Inst. Min. Metall. **124**(1), 25–32 (2024). https://doi.org/10.17159/2411-9717/2789/2024
35. Kalkstein, Y., Kopack, K.: How Implicit Biases Complicate Female Mentorship. The Decision Lab (2020)
36. Lee, J.: How Can Women Leaders Balance Family & Career? CEIBS (2022)
37. Branson, D.M.: No Seat at the Table: How Corporate Governance and Law Keep Women Out of the Boardroom. Research Gate (2006)
38. Terjeen, S., Aguilera, R.V., Lorenz, R.: Legislating a Women's Seat on the Board: Institutional Factors Driving Gender Quotas for Board of Directors. Research Gate (2014)
39. Nayar, J.: Equal representation? The debate over gender quotas. Harward Int. Rev. (2022)
40. Plaza, C.H., Robotham, D., Windon, S.: Diversity Training in the Workplace. PennState Extension (2024)

The Role of Emotional Intelligence in Improving Employee Performance (Conceptual Paper)

Bader Al-Majarafi and Mohamed Elgeddawy

Abstract The research aims to emphasize the importance and role of emotional intelligence in its five dimensions according to the Goleman scale (1995) which are (self-awareness, self-control, motivation, emotional awareness and social skills). Its impact on improving the performance of employees at work in the organization under study, and based on the importance of the relationship between the research variables between emotional intelligence and employee performance. And contributes to raising the level of performance of employees in the government sector. Accordingly, this research concludes by confirming the result of the positive association between emotional intelligence and its role in improving job performance at work by studying the five dimensions of emotional intelligence and their impact on improving the level of employee performance in the institution. The research methodology will focus on the qualitative study by using the descriptive deductive approach, and the results will confirm the existence of a correlation between emotional intelligence and its impact on improving the level of employee performance.

Keywords Emotional intelligence · Improving job performance · Public sector · Oman

B. Al-Majarafi · M. Elgeddawy (✉)
Modern College of Business and Science, Muscat, Oman
e-mail: Mohamed.Elgeddawy@mcbs.edu.om

B. Al-Majarafi
e-mail: Bader@mol.gov.om; 20247408@mcbs.edu.om

B. Awwad (ed.), *AI and IoT: Driving Business Success and Sustainability in the Digital Age*, Studies in Systems, Decision and Control 590,
https://doi.org/10.1007/978-3-031-88874-8_58

1 Introduction

1.1 Research Background

Many studies have proven the importance of emotional intelligence. For example the study of Afsh (2011) in [1, 2], which said that emotional intelligence is one's ability to focus, perceive and good awareness of feelings and emotions. Also, the ability to control his feelings and emotions that are not positive and reflect them into positive feelings, develop them and direct them towards desired goals. As well as the study of Gibbon et al. (2020) in Al-Shammari [2] lead to major challenges, as negative emotions dominate the relationships in the organization or between workers, or in their relations with the employer.

As the Sultanate of Oman is part of this world, organizations suffer from lack of understanding of the elements of emotional intelligence. Especially in government institutions, specifically service institutions that affect a large segment of the community. No family that does not need the services provided by the institution in question. Therefore, regarding the institution in question, it did not achieve an advanced level in the Institutional Excellence award emanating from the Institutional Excellence System. It honors the best performing government units within the framework of the government's orientation towards enhancing work mechanisms, raising efficiency and adopting new methods to achieve the objectives of Oman Vision 2040. In addition, the researcher asked the stakeholders regarding the challenges they face. One of them said that employees in the research organization do not deal tactfully with reviewers and do not understand their feelings in an appropriate way. Another added that some employees do not have high adaptability to work conditions and do not cooperate with others to accomplish work. While, another said that his colleagues do not control personal feelings and have a weakness in dealing with work pressures. In same point, another mentioned that some employees do not use emotional intelligence as a motivation to achieve achievement at work by practicing social relations with colleagues and officials at work.

1.2 Problem Statement

Emotional intelligence is related to the performance of workers in organizations as performance depends on successful interaction with others. Also, the person's ability to prevent adverse behavior such as violence in the workplace that arises from the individual's lack of emotional discipline or emotional control, and stress. And, it is an appropriate predictor of motivation, procedural skills, motivation, ability to focus, and maintaining a level of motivation in workers (Melhem, Ghosh, Harahsha, Sen, and Kanaan) in Al-Shammari [2]. The main motivation for researchers' interest in emotional intelligence is that it includes several different skills that contribute to achieving success. Such as knowing and managing one's emotions, the ability to empathize and motivate motivation. And the most important characteristic of emotionally intelligent personalities is the ability to control moods and control emotions. In addition to the ability to empathize with others and feel for group members. As well as the ability to empathize with others and feel for group members. (Khan and Al-Bashiti) in Al-Shammari [2]. Here, many questions are raised about the extent of the use of emotional intelligence in the organization under research. Accordingly, this study came to shed light on the issue of emotional intelligence. And its relationship with the level of performance among the employees of the Ministry under research in the Sultanate of Oman.

1.3 Research Gap

In the researcher's mission during the search for sources, references and scientific literature related to the subject of the study under research on emotional intelligence. And its impact on improving the performance of employees, he noticed the lack of studies on emotional intelligence in government institutions in the Arab world. Especially in the Sultanate of Oman, and here the research gap stands out in the scarcity of scientific papers that dealt with emotional intelligence locally. In addition, the methodology used was not a qualitative study and the performance was not studied effectively about the existence of a study on emotional intelligence in government institutions. And based on the above mentioned, we will state the issue and based on which this study was prepared.

2 Theoretical Framework

In this part of the study, the researcher will shed light on emotional intelligence and its impact on improving the performance of employees working in the governmental organization in the Sultanate of Oman. Based on the studies, published literature and gray literature that addressed this topic, as follows.

2.1 Emotional Intelligence

The government of the Sultanate of Oman started implementing the Ijada system to measure organizational performance. Which is one of the main enablers of Oman Vision 2040 to raise the efficiency of the state's administrative apparatus. The system supervised by the Ministry of Labor seeks to find the necessary standards to build a culture of excellence in job performance. And improve performance evaluation systems. This new system benefits more than (193) thousand employees in the public sector and improve the quality of government services provided to individuals and institutions. Based on the above mentioned the government institution under research is part of the government sector and in this research the researcher will focus on the role of emotional intelligence in improving employee performance in the institution under research [1].

2.2 Emotional Intelligence

The actual interest in emotional intelligence began with the studies of Thorndike (Thorndike, 1920) who searched for the best understanding of human intelligence, and these studies resulted in the theory of social intelligence. As this theory assumed that people have four different types of intelligence. Abstract intelligence, which is the ability to understand and deal with verbal and mathematical symbols, sensory intelligence, which is the ability to understand and deal with sensory or physical objects. Second type, social intelligence, which is the ability to understand and belong to people. Third type, emotional intelligence, which focuses on the emotional and psychological aspect with its feelings, emotions and emotions in the behavior of people and their interaction with each other, perhaps more than mental intelligence (Sukkar, 2009) in Al-Nashash [3]. The opinions of researchers varied in the names given to the concept of emotional intelligence, some of them called it emotional intelligence. It refers to unsatisfactory matters in emotions such as fear, sadness, and anger, and emotional intelligence focuses on pleasant emotions. Such as pleasure, love and admiration, and emotional intelligence contains pleasant emotional things and their opposite. Emotional intelligence contains the sensory aspect of the emotional experience, such as feeling warm or comfortable. Emotional intelligence can be defined as the ability to create positive outcomes in the individual's relationship with himself and others, by knowing the emotions of the individual and the emotions of others. It including positive outcomes such as joy, optimism, and success in school, work and life (Hussein and Hussein 2006) in [3, 4].

2.3 Ijada Organizational Performance Measurement System

The Ijada Institutional Performance Measurement System is a tool for the governance of government performance by measuring the most important criteria. It indicators related to institutional performance and creating a distinctive shift in the performance of government units in the Sultanate of Oman. Also, it fosters a spirit of positive competition between government units to achieve better performance rates. The application of this system comes based on some justifications, which is one of the possibilities of Oman Vision 2040. As well as to raise the competitiveness of the government and raise the Sultanate's rating in some international indicators. In addition to creating an effective environment and better quality of government services provided to customers. Moreover, to creating a national system for governance and institutional performance, it is worth noting that the system is expected to achieve several results. The most important point to achieve excellence in providing government services to achieve increased customer satisfaction [2, 5].

The system is expected to achieve many results. Such as achieving excellence in the provision of government services to increase the satisfaction of beneficiaries. Also, raise the Sultanate's competitive indicators in accordance with Oman Vision (2040). Moreover, encourage the spirit of competition and innovation, reward glorious units according to the approved standards. In addition to the existence of a national system of governance and institutional performance. Also, to strengthening the role of institutionalism and teamwork in the performance of government units and consolidating the principle of control and accountability and benefiting from the expected results from the application of the system in developing strategic plans for government units. Noting that the number of employees working in units of the administrative apparatus [6].

2.4 Improving Employee Performance

Performance is an important factor in achieving the growth and development of the organization. It focuses on a set of elements that are considered a requirement of the job, including competencies related to the profession. Including skills, knowledge, values, abilities and the duties and tasks it requires. In addition to the quality of work and includes the employee's performance of his work with a high degree of quality that minimizes his errors. Which are the mental images that the employee has about his job. In addition to the amount of work done is considered an element of task performance, because it focuses on speed of achievement with few errors. While perseverance and confidence are vital elements, because the employee bears responsibility for doing the jobs entrusted to him with excellence. Employee performance can be defined as the result of the total expected value that he contributes to the organization through his work. Which includes the performance of job tasks, as well as the behaviors he performs according to the job description that each employee performs [7].

Performance is a multifaceted construct that describes the behavioral component of an individual, including the actions that a person takes to complete a particular task. While the outputs describe the results obtained from this behavior (Pardhan and Jena 2017) in Sobh [1]. Task performance and contextual performance, according to the theory developed by Motowidlo and Borman in 1993. Where the task performance component refers to the ability of employees to carry out tasks that achieve organizational goals (Bang and Reio 2017) in Sobh [1]. As for contextual performance, it is concerned with behaviors that are unrelated to the job description assigned to the employee. But useful to the organization and includes actions that serve the social network and improve the psychological safety environment. Such as helping employees to follow regulations, making extra effort to work and meet the goals of the organization. And, volunteering to perform tasks in addition to the job description [1].

Emotional intelligence is integrated with the performance of employees in different organizations. It focuses on successful interaction with others, and the individual's ability to prevent adverse behavior. Such as violence in the workplace that results from a person's lack of emotional discipline or emotional control. And the results associated with stress, and it is an appropriate predictor of motivation, procedural skills, motivation, ability to focus, and maintaining a level of motivation in workers [2]. The main motivation for researchers' interest in emotional intelligence is centered on the fact that it contains several different skills that contribute to achieving success. And it is considered the personality's knowledge of his feelings and good management of them. The ability to empathize and motivate motivation. And the most important characteristic of emotionally intelligent personalities is the ability to control mood and complete discipline in emotions. In addition to the ability to empathize with others and feel for group members and build distinct social relationships full of trust and mutual respect, along with flexibility, adaptation and issue resolution (Khan and AljBashiti, 2020) in [2]. Based on the above mentioned, we conclude how important emotional intelligence is to improve the performance of the organization [8].

3 Research Objectives

1. Determining the degree of availability of emotional intelligence among the employees of the local governmental entity in Oman.
2. Identify the level of performance of the employees of the local government agency in the Sultanate of Oman.
3. Determining the nature of the relationship between emotional intelligence and the level of performance of government employees [5].

4 Research Questions

1. What is the degree of availability of emotional intelligence among the employees of the local government agency?
2. What is the level of performance of the employees of the local government agency?
3. What is the nature of the relationship between emotional intelligence and the level of performance of government employees?

5 Literature Review and Previous Studies Related to the Study

The importance of reviewing previous literature is due to the importance of understanding and grounding emotional intelligence. The components of organizational performance measurement to develop the performance of employees working in the organization under research by knowing the objectives and finding appropriate solutions to the research issue and reaching the results and recommendations that were built on to find appropriate solutions to that issue. From this perspective, the researcher found that there is grounding this issue not only through gray literature and stakeholder opinions, but through the literature and published studies mentioned above and related to the research under study. Therefore, the research examined some of the previous literature related to emotional intelligence, such as (Birole et al. 2009) in [2]. Which focused on analyzing the ratio of emotional intelligence and performance management perceptions of teachers in secondary schools, and the study tool was represented by Daniel Goleman's emotional intelligence scale, as well as Wolmarans' FCP scale, where the results confirmed that there is a significant difference between teachers' perceptions of performance management for the variable of gender and years of experience, and it is noted that there is no significant difference between teachers' perceptions of performance management and emotional intelligence for the variable of age. While there is no significant difference between teachers' perceptions of performance management and emotional intelligence for the variable of age. Significant difference between teachers' perceptions on performance management and emotional intelligence for the variable of age. While the significant difference between teachers' perceptions on performance management and emotional intelligence for the variable of school type. Moreover, there was a correlation between performance management in general and two components of emotional intelligence—self-motivation and empathy—and between performance management in general and two components of emotional intelligence—self-motivation and empathy. While the study of Al-Resini (2020) in Al-Shammari [2] sought to identify the level of emotional intelligence among the leaders of private schools in the city of Buraydah, and the descriptive survey method was used. The result of the study was that the leaders of private schools in the city of Buraydah, enjoy a high degree of emotional intelligence. While the study of Al-Shaibi and Al-Dhafri (2020) in Al-Shammari [2] was concerned with identifying the relationship of emotional intelligence with the collective competence beliefs of mathematics teachers in the Sultanate of Oman. Also, the study used the descriptive method, where the result was the presence of a high level of emotional intelligence. In addition to the superiority of females over males in emotional intelligence.

Razali et al. (2022) in Sobh (2024) [6] studied the relationship between emotional intelligence and work performance and concluded that the level of emotional intelligence of workers is average and a significant relationship was found between emotional intelligence and workers' performance. While Sabie et al. in [1] studied the impact of emotional intelligence on workers' performance and found a significant correlation between emotional intelligence and performance, while Vrontis et al. (2021) in [1] studied the impact of managers' motivation, empathy, self-awareness, self-organization and social skills while examining the relationship between managers' emotional intelligence and workers' performance and the results showed that leaders' emotional intelligence had a significant impact on workers' performance (2021) studied the impact of managers' motivation, empathy, self-awareness, self-organization and social skills while examining the relationship between managers' emotional intelligence and workers' performance and the results showed that the emotional intelligence of leaders had a clear impact on the performance of workers and the performance of leaders is positively affected by the level of social skills, empathy, emotional control in addition to self-motivation. While Alonazi (2020) in [1] looked at how emotional intelligence affects work performance during the Covid-19 crisis and the results showed that emotional intelligence had a statistically significant impact on employee performance. While Afifa (2019) in [1] study was concerned with studying.

While the research also shown in same study of Abu Al-Khair and Abu Shaira (2018) in [1] focused on knowing the level of emotional intelligence of principals and its relationship in improving the level of performance from the point of view of supervisors in lower basic schools. And the result of the research was a statistically significant correlation between the level of emotional intelligence of principals and improving their level of performance. Salahuddin (2016) in [1] searched for the level of emotional intelligence of leaders in organizations and institutions in the private sector in Egypt and its relationship with the level of job performance and used the questionnaire as a data collection tool and concluded that the level of emotional intelligence in the research sample was average and concluded that job performance was average.

This study came to confirm this relationship with a focus on the qualitative dimension in the data collection process to know the level of performance when studying the presence and absence of emotional intelligence and measuring the impact between them on the ground. And, to identify emotional intelligence accurately with its various dimensions. Goleman's theory is considered the most famous theory of emotional intelligence and one of the first theories that were able to analyze and measure the emotional intelligence of people according to the researcher's knowledge and accordingly the research questions were formulated based on it.

6 Conceptual Framework

It is recognized that the conceptual framework consists of three basic elements that affect each other. In this study, the independent variable is emotional intelligence with its five dimensions according to Goleman's scale, which are (self-awareness, self-control, motivation, emotional awareness and social skills). The independent variable affects the dependent variable, which is improving employee performance [7].

Whereas the basic concepts of emotional intelligence are the theory of emotional intelligence. According to the Goleman scale (1995), which explains how emotional intelligence affects job performance through five dimensions. It is including self-awareness, which is the knowledge and understanding of one's own feelings in addition to self-management, which is the ability to control one's feelings and response to them. Also, social awareness, which is the ability to understand the feelings and needs of others and interact effectively with them, including empathy and the ability to recognize the dynamics of groups and social relationships, motivation, which is the internal motivation to achieve goals, excellence, achievement and optimism [8].

Also, job performance theories emphasize that different elements such as skills, knowledge, motivations and emotions play a role in improving employee performance, as employee performance, which is the dependent variable, includes a set of elements such as productivity, which symbolizes productive work. In addition to the quality of work and the extent of mastering the tasks assigned to the employee. Also, to punctuality and time management and others. So, the conceptual framework includes emotional intelligence as an independent variable and performance as a dependent variable. As it is expected that there will be a positive relationship between emotional intelligence and employee performance. As the emotional intelligence scale will be used to measure the emotional intelligence of employees (Al-Jawhar et al. 2023).

The conceptual framework is characterized by its governance through frameworks and mechanisms defined by the Ministry of Labor. These mechanisms were generalized to all institutions of the Government of the Sultanate of Oman applying the organizational performance measurement system, and therefore it is affected by those regulations, standards and indicators imposed by the supervising body of the project on the relevant parties, including the institution under research, and since the impact caused by the organizational environment and the culture of the institution under research was omitted. It is known that every organization has an environment and culture that directly affects and influences employees based on the above-mentioned Fig. 1 shows the conceptual framework for the study as follows.

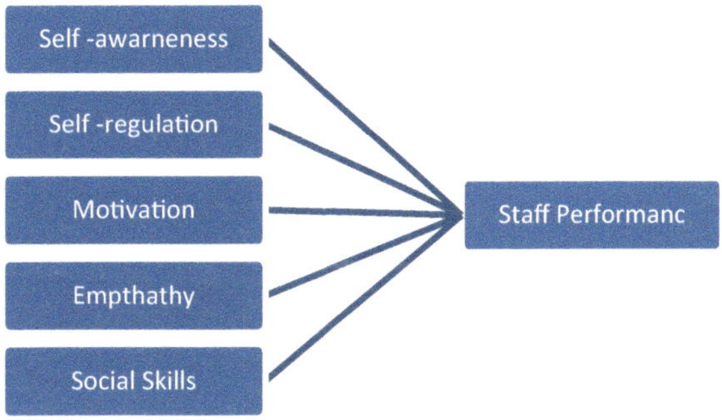

Fig. 1 Conceptual framework

It can be concluded from the above, that the conceptual framework is considered dynamic and that it interacts and evolves with the surrounding environment and culture. So, when setting goals and associated outcomes, it is necessary to consider the organizational environment of that institution. The researchers and those interested in this topic are urged to study research in depth into the extent of the impact of emotional intelligence as an independent variable and its impact on the performance of employees in the institution under research.

References

1. Sobh, A.S.A.H.: Work relatedness as a mediating variable in the relationship between emotional intelligence and the performance of employees in information technology companies in Egypt. Scientific Journal of the Faculty of Commerce, Assiut University (2024)
2. Al-Mukhtari Al-Shammari, N.R.: Emotional Intelligence and Its Relationship with the Level of Performance Among Early Childhood Teachers in Hafar Al-Batin, Saudi Arabia. Faculty of Education, Hafar Al-Batin University, Saudi Arabia (2023)
3. Al-Nashash, F.M.: Emotional intelligence and its relationship with leadership practices among the principals of private schools in the University Brigade from their point of view. Jordan. Educ. J. (2024)
4. Al-Shurafat, S.S.B., Al-Shura, M.S.K.: The Impact of Transformational Leadership on Job Integration: The Modified States of Emotional Intelligence in Jordanian Islamic Banks. Unpublished Doctoral Dissertation, University of Islamic Sciences, Amman, Jordan (2024)
5. Atheer Newspaper: Eight Best Performers and Implementers of Organizational Excellence. Sultanate of Oman (2024)
6. Ministry of Labor: National System for Organizational Excellence (2024)
7. Abdul-Samad, K.Y.: The impact of emotional intelligence on the impact of emotional intelligence on work pressure among the nursing staff at Al-Azhar university hospitals. Sci. J. Econ. Commerce (2023)
8. Khalifa, S., Salim, S., Nimr, G.: Emotional Intelligence of Public Relations Managers and Its Reflection on Reducing Work Pressure in Governmental Organizations from the Perspective of Employees. Unpublished Master's Thesis, Arab American University, Jenin (2023)

The Oil Prices Impact on GCC Stock Markets: The Dynamic Connectedness Approach Based on QVAR

Muhammad Naeem, Sulaf Alsabti, Reem Al Mahruqi, Nassor Al Jahwari, and Josephmani Puthenparampily

Abstract This study examines the dynamic connectedness among WTI crude oil, Tadawul, QSE, and MSM.30 indices using a quantile-based framework to differentiate between average (50th percentile) and extreme (95th percentile) market conditions. The results highlight significant differences in the transmission and reception of shocks across these financial markets depending on the market state. Under average conditions, markets exhibit high self-dependence, with over 93% of variance in each market explained by its own shocks. Cross-market spillovers remain low, with total connectedness indices (TCI) reflecting limited systemic interdependence (7.20%). MSM.30 emerges as the most influential transmitter of shocks, while WTI acts as a net receiver. In contrast, extreme conditions reveal a sharp increase in cross-market spillovers, as evidenced by a substantially higher TCI (76.07%). Markets exhibit reduced self-dependence, with a larger proportion of variance attributed to external shocks. Notably, Tadawul and QSE transition into net transmitters, while WTI becomes a significant net receiver, underscoring its vulnerability during crises. The elevated cross-market connectedness during extremes emphasizes systemic risks and heightened interdependence, with implications for portfolio diversification and risk management. The findings provide valuable insights for policymakers and investors in mitigating contagion risks and formulating strategies to enhance financial stability during turbulent times.

Keywords Dynamic connectedness · Oil price volatility · GCC stock markets · QVAR

M. Naeem (✉) · S. Alsabti · R. Al Mahruqi · N. Al Jahwari · J. Puthenparampily
Modern College of Business and Science, Muscat, Oman
e-mail: Muhammad.naeem@mcbs.edu.om

B. Awwad (ed.), *AI and IoT: Driving Business Success and Sustainability in the Digital Age*, Studies in Systems, Decision and Control 590,
https://doi.org/10.1007/978-3-031-88874-8_59

1 Introduction

Some scholars were debating that oil markets and stock markets are connected while others denied this argument, and recent sharp movements in global oil prices raised the reincarnation of this debate again. Fluctuations in supply and demand affect both oil prices and stock market. There is a relationship between supply chain and demand supply in determining price of oil as well as stock market. But these markets respond to news (of all sorts) and there is hardly any dearth of news which leads to daily oscillations in the oil and stock prices. Consequently, energy portfolio managers are left to cope with either more or less liquid asset with which to hedge against high dimensional risks in the energy portfolio: governments of oil-rich countries must regulate public funding for projects during the height of the oil prices to insulate themselves from 'low price' periods. The gulf cooperation council (GCC) countries have more than 40% of the global proved reserves of crude oil which supports it's the relative significance of the GCC countries in international oil market. Recently, Saudi Arabia's production reached 12,088,000 barrels per day, Qatar's reached 2,068,000 barrels per day, and Oman's production was at 980,000 barrels per day. As a result, the role of the oil in the structure of those states has a significant strategic importance. Affectation by this pressure fluctuation is experienced by all individuals such as you the reader, small to medium sized as well as enormous corporations and governments, inside these economies because of the globalization of the economic entwinement with the oil producing and exporting nations.

The relationship between oil prices and financial markets is a well-established topic in economic research, particularly in regions where oil is a significant contributor to economic performance. The Gulf Cooperation Council (GCC) countries, consisting of Bahrain, Kuwait, Oman, Qatar, Saudi Arabia, and the United Arab Emirates, are among the largest oil exporters globally. For these economies, oil revenues constitute a substantial portion of government income and play a critical role in driving economic growth and financial stability. Consequently, fluctuations in oil prices have profound implications for the stock markets in these countries.

Previous studies have consistently shown that the impact of oil price changes ultimately affects stock market returns through 3 channels, a reduction of corporate earnings, government spending and investor sentiment [8, 12, 13]. These dynamics are particularly sensitive to GCC markets as they have a high reliance on oil revenues. Arouri et al. [3] exhibit that GCC economies experience considerable fluctuations of market return of stocks due to oil price movements, reflected in the asymmetric behaviour during the booms and busts periods. In fact, a plethora of studies have investigated the linear associations between oil prices and stock performance However, other research suggest that financial markets behave in a nonlinear way especially in extreme market conditions.

Such interactions have inherent non-linear behaviour and require sophisticated methods to describe. The model used in this study, the Quantile Vector Autoregression (QVAR) model, is capable of analyzing spillover effects at different quantiles than existing models, which both reveals market dynamics under 'normal' and extreme conditions. Dynamic connectedness which is a relatively new framework first developed by Diebold and Yilmaz [6] has become a standard way to assess spillover between markets. More recent research, however, has expanded this framework to include both time-varying [1] and quantile approaches [4], whereby the nature of co movements between financial and commodity markets can be better understood. Using this methodology, this research aims at investigating the time-varying and asymmetric interconnectedness between oil prices and GCC stock markets in order to understand the shock transmission mechanisms in these dynamic systems. These findings are important for policymakers and investors in the GCC region. This could help policymakers in crafting oil price strategies for addressing systemic risk.

2 Literature Review

There has been considerable scholarly research on the interrelationships between oil prices and financial markets, especially in those economies that rely heavily upon oil. Hamilton [8] was really the first to identify oil price shocks as a key driver of economic cycles. Later research extended this relation to financial markets stressing the position of oil as a system risk variable in an equity pricing context [12]. Numerous studies have found oil prices a big driver of stock market performance in the context of GCC economies. For instance, Arouri et al. [3] provided support for a high degree of dependence between oil prices and stock markets in the GCC countries, but the level of dependence is market specific. In a similar manner, Jouini [11] found that GCC markets also respond asymmetrically to the oil price changes, i.e., negative shocks have larger impact than positive ones.

Recently, dynamic connectedness approaches have become popular tools for analyzing spillovers and interdependencies. In 2012, Diebold and Yilmaz developed a framework for measuring spillover effects, which has been widely adopted in energy and financial markets. In this context, Antonakakis et al. [1] extended the method to allow for time-varying spillovers and increase the granularity of market state depending on perturbation. This QVAR model was used in the present study to identify nonlinear relationships at different quantiles. Some similar quantile-based approaches have analyzed oil-stock dynamics [4, 15] split the 4 states, bull market and bear market on stock and oil price shock high return and low return. However, there is scarce research on extreme market conditions and their effects on GCC economies. The novel contribution of this study is that, while analysing the connectedness due to oil price and GCC stock markets, So far, no investigated average connectedness along with extreme (tail) connectedness.

3 Methodology

The Quantile Vector Autoregression (QVAR) approach generalizes the conventional vector autoregression (VAR) model by bringing in quantile regression methods. In particular, one of the strengths of QVAR is its capability in evaluating asymmetric and non-linear connection between variable over different quantiles of various distributions. It is used frequently in finance and economics to examine dynamics under different market conditions (e.g., extreme volatility or for stress scenarios). There are other methodologies used to capture the extreme dependence such as copula as illustrated in the research of Naeem et al. [14]. In general, the methodologies are given in more detail from research papers of Diebold and Yilmaz [6] and Ji et al. [10].

4 Descriptive Properties of the Data

In Table 1 summary statistics provide insights into the distributional properties of the returns of West Texas Intermediate (WTI) crude oil prices and GCC stock indices. Returns are close to zero, consistent with typical financial return distributions, while variances indicate higher volatility in GCC stock indices (e.g., Tadawul) compared to WTI. All series exhibit significant deviations from normality.

WTI and QSE have pronounced negative skewness, while MSM.30 shows positive skewness, indicating asymmetric return distributions. Excess kurtosis values highlight the presence of fat tails, particularly in Tadawul.

The large JB statistics confirm the non-normality of returns across all markets. All series reject the null hypothesis of a unit root, confirming stationarity. Significant Q1 and Q2 statistics for WTI imply autocorrelation in returns and volatility, while other series show weaker dependencies.

Table 1 Summary statistics

	WTI	Tadawul	QSE	MSM.30
Mean	0.002	0.134	0.18	−0.16
Variance	6.451	21,481.016	11,674.832	614.79
Skewness	−3.172***	−18.845***	−3.965***	1.988***
Ex.Kurtosis	201.359***	583.076***	71.017***	44.792***
JB	2,803,792.269***	23,584,890.332***	352,758.603***	139,696.279***
ERS	−11.138***	−13.864***	−3.728***	−14.855***
Q(10)	75.362***	11.502**	26.322***	64.266***
Q2(10)	353.276***	0.004	0.577	1.711

Note Q(10) and Q2(10) refers to the Ljung-Box test for the autocorrelation of returns series. The symbols ** and *** indicate, respectively, significance at the 5% and 1% levels

5 Results and Discussion

Table 2 provides insights into the average connectedness of the financial markets and crude oil (WTI) and GCC stock markets (Tadawul, QSE, MSM.30) under normal conditions (50th percentile). The values along the diagonal represent the percentage of variance in each variable explained by its own shocks. For instance, WTI explains 95.10% of its own variance, Tadawul explains 93.91%, and so on. These values show that each market is mostly influenced by its internal factors, with high self-dependence. The off-diagonal elements reflect how much one market is influenced by shocks from others. For example, WTI transmits 1.75% of its variance to Tadawul and QSE receives 1.66% of its shocks from WTI. These numbers are relatively small, indicating low interconnectedness under average market conditions. FROM columns measures the total percentage of shocks a market receives from others for instance WTI receives 4.90% of its shocks from other markets. TO columns measures the total percentage of shocks a market transmits to others for example WTI transmits 4.02% of its shocks to others.

The NET column calculates the difference for instance a positive value (e.g., MSM.30, 0.96) indicates a net transmitter of shocks and negative value (e.g., WTI, −0.88) indicates a net receiver of shocks. This reflects the total degree of connectedness across the system. The 0.072/0.054 ratio suggests a relatively low level of interconnectedness under average conditions, with the system predominantly driven by own-market dynamics. NPT (Net Pairwise Transmission) represents the rank of net transmission for example MSM.30 ranks highest (3.00), indicating it is the most active transmitter in this context. The results from Table 2 (50th percentile) reveal that financial markets and crude oil are predominantly self-driven during average conditions, with high levels of own variance explained for each variable (e.g., WTI: 95.10%, MSM.30: 95.30%). Cross-market spillovers remain minimal, as reflected in the low Total Connectedness Index (TCI) of 7.20%. This low interconnectedness aligns with findings in the literature that under normal market conditions, financial markets typically exhibit regional or sectoral independence [6]. The net transmission metrics indicate that MSM.30 is the most significant transmitter of shocks, with

Table 2 Average dynamic connectedness (50th Percentile)

	WTI	Tadawul	QSE	MSM.30	FROM
WTI	95.10	1.75	1.66	1.48	4.90
Tadawul	1.51	93.91	2.68	1.90	6.09
QSE	1.28	2.35	94.09	2.28	5.91
MSM.30	1.23	1.78	1.69	95.30	4.70
TO	4.02	5.88	6.03	5.66	21.60
Inc.Own	99.12	99.79	100.12	100.96	cTCI/TCI
NET	−0.88	−0.21	0.12	0.96	7.20/5.40
NPT	0.00	1.00	2.00	3.00	

a net value of 0.96, whereas WTI acts as a net receiver (−0.88). These dynamics suggest that MSM.30, potentially due to its economic drivers, plays a more prominent role in influencing regional markets under normal conditions. Similar patterns have been observed in studies analysing equity indices in stable economic environments, emphasizing the localized impact of shocks [2].

Table 3 captures the relationships during extreme conditions (e.g., market crises, high volatility). These values drop significantly compared to the average condition. For instance, WTI's own connectedness reduces to 45%, suggesting that during extreme conditions, external shocks play a much larger role. The influence of markets on each other increases substantially. For instance, WTI transmits 18.20% to Tadawul (up from 1.75%) and QSE transmits 19.32% to WTI (up from 1.66%). This indicates heightened interdependence during market turmoil. FROM columns indicates the proportion of shocks a market receives from others increases significantly under extreme conditions for instance WTI now receives 55% of its shocks from external sources. TO row shows the proportion of shocks transmitted by each market also increases dramatically for example WTI now transmits 53.09% of its shocks to others. This shift highlights the vulnerability and interconnectedness during crises. NET columns represent that Tadawul and QSE become net transmitters of shocks under extreme conditions (0.20 and 1.57, respectively). While WTI turns into a net receiver (−1.91), indicating its increased sensitivity to external shocks during crises. Further, the connectedness increases significantly to 0.76/0.57, indicating that external influences dominate the system in turbulent times. Moreover, QSE ranks highest (3.00), reflecting its role as the most influential transmitter during extreme conditions as shown by NPT.

As shown in Table 3 (95th percentile), there has been a clear change in the dynamics of connectedness during tail events. The TCI, on the other hand, rises up to 76.07%, indicating that systemic risk and interdependence across markets is intensifying. Such increase in spillovers corroborate with the literature which find that financial networks get more connected during stress periods such as financial crises [9] or oil price shocks (Baur et al. 2022). Extremes reduce self-dependence

Table 3 Extreme Dynamic connectedness (95th Percentile)

	WTI	Tadawul	QSE	MSM.30	FROM
WTI	45.00	18.20	19.32	17.48	55.00
Tadawul	17.47	41.92	21.78	18.84	58.08
QSE	18.54	21.11	40.70	19.65	59.30
MSM.30	17.08	18.97	19.77	44.18	55.82
TO	53.09	58.28	60.87	55.97	228.20
Inc.Own	98.09	100.20	101.57	100.15	cTCI/TCI
NET	−1.91	0.20	1.57	0.15	76.07/57.05
NPT	0.00	2.00	3.00	1.00	

drastically, with WTI's own variance contribution sinking to just 45.00%. Cross-market spillovers increase dramatically i.e., WTI transmits 18.20% of its variance to Tadawul while receives 19.32% from QSE. Such results reveal how in times of crisis the dynamics between crude oil prices and regional markets are particularly sensitive, aligning with prior findings on oil price spillovers [16]. Importantly, QSE is the largest net transmitter in extreme cases, with an overall transmission value of 1.57. Such a change is indicative of the increasing importance of QSE in transmitting shocks across markets which may be due to regional characteristics, such as economic ties or investor sentiment. The higher interdependence at extremes is in agreement with the contagion reason [7], which argues that stress over one market increases fragilities of other markets.

The absence of average conditions among extremes carries important policy relevance. During times of crisis, and in particular after a period of booming markets, regulators need to center their efforts on countering systemic risks expressed—avoiding globalization of problems by using better regulatory capacity to fortify financial buffers and provide best fit market oversight. The results highlight the need for adaptive portfolio strategies that reflect amplified connectedness at times of extreme events from an investors perspective. This research illustrates that during periods of crisis assets tend to be more correlated, which makes diversification strategies less useful. This study uses a quantile-based measure for the dynamic connectedness, which adds to the existing studies using static measures. This study not only informs the literature on the asymmetric and nonlinear nature of market interactions [5], but it also provides a much finer understanding by distinguishing between normal and extreme market conditions. This result emphasises the importance of quantile-based approaches in risk measurement, as they account for extreme risks that may go undetected using a standard linear estimation framework.

6 Conclusion and Policy Implications

Under average conditions, markets exhibit high self-dependence and cross-market spillovers remain low while, extreme conditions reveal a sharp increase in cross-market spillovers, as evidenced by a substantially higher TCI (76.07%). Markets exhibit reduced self-dependence, with a larger proportion of variance attributed to external shocks. Policymakers should promote risk management approaches to contain spillover during stressed financial market conditions. Potential measures may include strengthening market diversification, establishing stricter regulations or upgrading crisis management policies in order to protect GCC economies from the risk of fluctuating oil prices. GCC markets should be freeing themselves from oil dependence as much as possible by encouraging other industries and having a wider base. Better monitoring of market conditions, in particular during oil price shocks may help to avoid risks associated with extreme interconnectedness. Relying even more on GCC financial and economic relations will help dilute risks, contain losses, and favour collective resilience. Stabilizing markets in times of volatility spurts requires implementing solid financial policies that emphasize calmness in the short run to achieve desired objectives in the long run.

References

1. Antonakakis, N., Chatziantoniou, I., Filis, G.: Dynamic spillovers of oil price shocks and economic policy uncertainty. Energy Econ. **87**, 104730 (2020)
2. Antonakakis, N., Gabauer, D.: Refined measures of dynamic connectedness based on TVP-VAR. J. Risk Financ. Manage. **10**(2), 1–20 (2017). https://doi.org/10.3390/jrfm10020012
3. Arouri, M.E.H., Jouini, J., Nguyen, D.K.: On the impacts of oil price fluctuations on European equity markets: volatility spillover and hedging effectiveness. Energy Econ. **34**(2), 611–617 (2012)
4. Balcilar, M., Gupta, R., Wohar, M.E.: Oil price shocks and stock returns predictability: evidence from nonparametric causality-in-quantiles tests. Energy Econ. **81**, 693–707 (2019)
5. Barunik, J., Krehlik, T.: Measuring the frequency dynamics of financial connectedness and systemic risk. J. Financ. Economet. **16**(2), 271–296 (2018). https://doi.org/10.1093/jjfinec/nbx028
6. Diebold, F.X., Yilmaz, K.: Better to give than to receive: predictive directional measurement of volatility spillovers. Int. J. Forecast. **28**(1), 57–66 (2012)
7. Forbes, K.J., Rigobon, R.: No contagion, only interdependence: measuring stock market comovements. J. Financ. **57**(5), 2223–2261 (2002). https://doi.org/10.1111/0022-1082.00494
8. Hamilton, J.D.: Oil and the macroeconomy since World War II. J. Polit. Econ. **91**(2), 228–248 (1983)
9. Huang, W., Zhuang, X., Wu, L.: Dynamic spillover effects between oil prices and stock markets across different quantiles. Energy Econ. **85**, 104596 (2020). https://doi.org/10.1016/j.eneco.2019.104596
10. Ji, H., Naeem, M., Zhang, J., Tiwari, A.K.: Dynamic dependence and spillover among the energy related ETFs: from the hedging effectiveness perspective. Energy Econ. **136**, 107681 (2024)
11. Jouini, J.: Return and volatility interaction between oil prices and stock markets in GCC countries. Econ. Model. **35**, 1–9 (2013)
12. Kilian, L., Park, C.: The impact of oil price shocks on the US stock market. Int. Econ. Rev. **50**(4), 1267–1287 (2009)
13. Naeem, M., Ahmed, S.: Hedging effectiveness of currency ETFs against WTI crude oil price fluctuations. In: Intelligent Systems and Applications in Business and Finance, pp. 189–216. Springer International Publishing, Cham (2022)
14. Naeem, M., Umar, Z., Ahmed, S., Ferrouhi, E.M.: Dynamic dependence between ETFs and crude oil prices by using EGARCH-Copula approach. Phys. A **557**, 124885 (2020)
15. Nusair, S. A., Al-Khasawneh, J. A.: Oil price shocks and stock market returns of the GCC countries: empirical evidence from quantile regression analysis. Econ. Chang. Restruct. **51**(4), 339–372 (2018).
16. Zhang, D., Broadstock, D.C., Ruan, B.: Oil price shocks and equity market linkages: evidence from top oil consumers. Energy Econ. **59**, 253–259 (2016). https://doi.org/10.1016/j.eneco.2016.07.023

Promoting the Achievement of the United Nations Sustainable Development Goals Through the Efforts of Young Social Entrepreneurs

Mohamed Elgeddawy

Abstract Recently, it is argued that social entrepreneurship as a process for solving community-based problems has great potential in supporting individuals, communities, and governments in achieving the United Nations' Sustainable Development Goals (SDGs). More specifically, there is a considerable amount of literature that varies across contexts and conceptually establishes a desired link between social entrepreneurship and the achievement of SDGs. However, understanding how social entrepreneurship can play a significant role in achieving these goals at the practical level remains challenging. This paper aims to explore the role social entrepreneurship plays in advancing the UN SDGs at a practical level. The paper critically analyzes pedagogical models that relate social entrepreneurship to the UN SDGs. This is followed by the development of a practical model that can contribute to the achievement of SDGs. Then the paper explores the development and presentation of a conceptual model for assessing the impact of social entrepreneurship on the development of SDGs. It ends with a list of tips for designing and developing effective assessment framework to assist in implementing the proposed practical model that links social entrepreneurship to the achievement of SDGs.

Keywords Social entrepreneurship · United Nations · Sustainable development goals (SDGs)

1 Introduction

Over the last fifteen years, the story of mobilizing college students to engage in societal development and preparing skilled workforces capable of assessing and addressing national, regional and international problems and crises has been narrated many times by many stakeholders such as businesses, academicians, scholars, and officials. Each stakeholder highlights certain facets while veiling others. It is observed

M. Elgeddawy (✉)
Modern College of Business and Science, Muscat, Oman
e-mail: Mohamed.Elgeddawy@mcbs.edu.om

B. Awwad (ed.), *AI and IoT: Driving Business Success and Sustainability in the Digital Age*, Studies in Systems, Decision and Control 590,
https://doi.org/10.1007/978-3-031-88874-8_60

from prior research that countries across the world consider mobilizing students to engage in societal development as strategic and a key factor for sustainable development. In this paper, the concept 'sustainable development' is used to indicate that any growth that humans achieve in any field (environmental, social, economic, financial, and ecological) should be done in a way that sustains the world with all its species [1]. This concept is inclusive in nature; it suggests that all individuals have a responsibility to sustain and preserve the present and future of the world. The sustainable growth of the planet needs all humans to participate in exploring and solving all types of problems that impact the present and future existence of our planet. For this to happen, human beings need to be beings of care. In its endeavor to protect the planet and create a world culture of care, the United Nations proposed the adoption of 17 goals reflecting certain domains of action [2].

2 First Dimension of the Sustainable Development Goals

The first domain of action represents four of the 17 goals and calls for actions to end global poverty (goal 1), end global hunger (goal 2), achieve gender equality (goal 5), and reduce inequality in general (goal 10). These four goals concern global problems related to poverty and inequality (See Fig. 1).

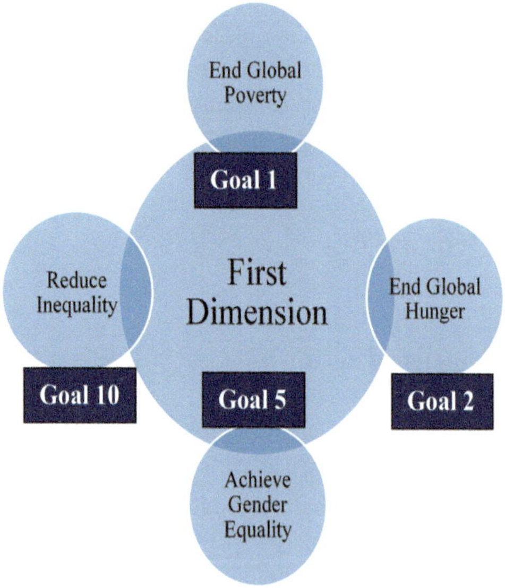

Fig. 1 First dimension of the United Nations sustainable development goals

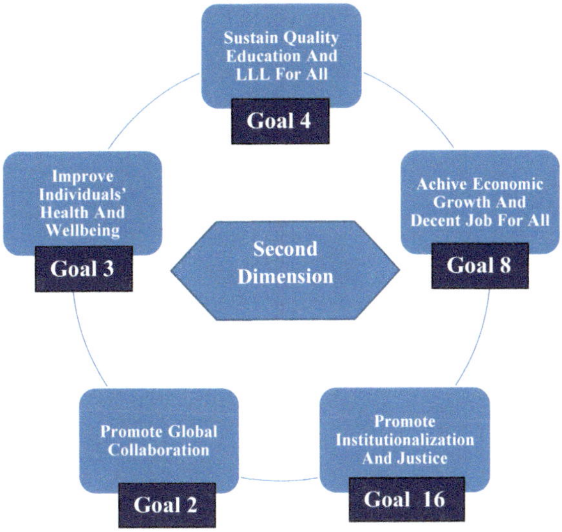

Fig. 2 Second dimension of the United Nations sustainable development goals (SDGs)

3 Second Dimension of the Sustainable Development Goals

The second domain concerns improving the quality of life and work. This domain includes goal three which calls for improving individuals' health and wellbeing, goal four calls for improving education and humans' lifelong learning skills, goal eight calls for creating decent work that advances economic development, goal sixteen calls for the existence of peace and justice, and goal seventeen calls for promoting global collaboration and partnership for sustainable development [3] (See Fig. 2).

4 Third Dimension of the Sustainable Development Goals

The third domain includes goal six which calls for ensuring the availability of clean water and sanitation, goal seven calls for assuring the availability and affordability of clean energy, goal thirteen calls for combating climate change and global warming, goal fourteen calls for protecting water species, and goal fifteen calls for protecting life on land (See Fig. 3).

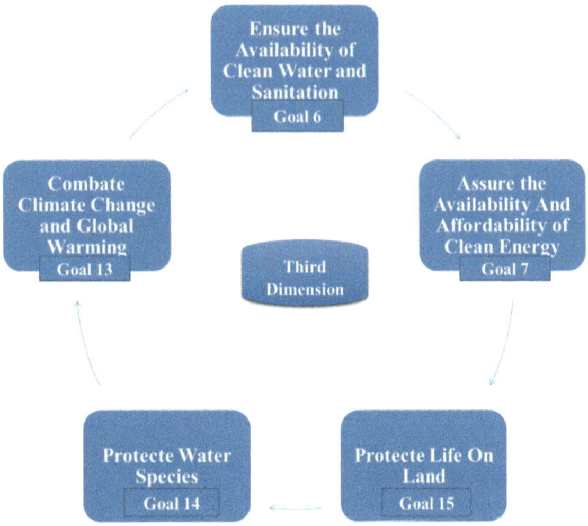

Fig. 3 Third dimension of the United Nations sustainable development goals (SDGs)

5 Fourth Dimension of the Sustainable Development Goals

The fourth dimension of the SDGs calls for managing a sustainable economy. This dimension includes three goals: goal nine which calls for assuring innovative and sustainable industrialization, goal eleven calls for ensuring the environmental development of sustainable green and safe cities and communities, and goal twelve calls for assuring the growth and sustainability of production (See Fig. 4). Table 1 depicts the four dimensions of the UN's SDGs.

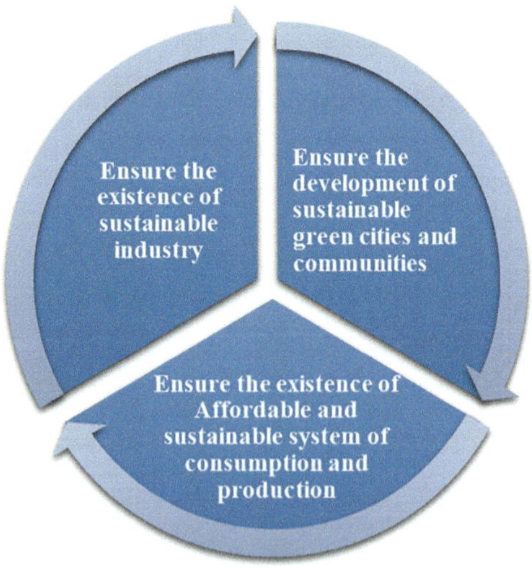

Fig. 4 Fourth dimension of the United Nations sustainable development goals (SDGs)

Table 1 UN's four dimensions

Dimension 1	Dimension 2	Dimension 3	Dimension 4
End of poverty and inequality	Improve quality of work and life	Sustain the environment	Sustain the economy
G1: No poverty **G2**: No hunger **G5**: Gender equality **G10**: Reduce inequality	**G3**: Good health and wellbeing for all **G4**: Sustain quality education and LLL for all **G8**: Economic growth and decent job for all **G16**: Institutionalization and justice **G17**: Partnerships	**G6**: The availability of clean water **G7**: The affordability of clean energy **G13**: Take action to combat climate problems and global warming **G14**: Protection and sustainability of water species **G 15**: Ensure a sustainable use of ecosystem	**G9**:Ensure the existence of sustainable industry **G11**:The development of sustainable green cities and communities **G12**: Affordable and sustainable system of consumption and production

6 Developing a Global Mindset

Considerably, engaging students in societal development is to enable them to play an active role in assessing and addressing local problems in relation to global issues. Research indicates that students' development of global entrepreneurial competencies and skills enables them to be receptive to local, national and international disasters and problems [4, 5]. The literature in this domain of research indicates that engaging students in societal development requires an entrepreneurial mindset that allows them to explore, analyze and address national, regional and international problems with local insights. They also need to develop a global mindset (the ability to think globally and act locally) that is flexible, adaptable and responsive to change.

7 Rhetorical Questions Need Answers

Yet, the questions that arise at this point are: Why should educators and university students care about, for instance, world hunger or environmental issues? Does the Higher Education system empower graduates with the required knowledge, skills and attitudes that enable them to live sustainably? Do we teach college students about environmental sustainability and the promotion of preserving life on land or below water? Does higher education system enable university graduates to promote equity and eliminate poverty, squalor, disease, corruption and hunger at the local, regional and international levels?

Does the higher education system generate beings of care with a sustainable passion and willingness to acquire global survival skills that enable them to solve local, regional and international problems? If the answer is yes for the abovementioned questions, we wouldn't have poverty, squalor, hunger, wars, inequality, corruption, oppression, narrow-mindedness, carelessness and aggression to be the reality of this world. The truth of the matter is that millions of people around the world are starving to death and lining up waiting for a box of food. Many African and Middle East countries are currently facing hunger, diseases, and squalor due to lack of food supply resulted as an inevitable outcome of the existing wars. The United Nations is indicating that a global food crisis is imminent. More recently, UN Secretary–General Antonio Guterres indicated that the current wars are increasing global food crises. The inevitable questions that arise in this domain of influence are: why should we care about the sustainability of our planet? Why should stakeholders such as policy makers, educators, students and others care about, for instance, environmental issues? Why do we have to increase our awareness of issues related to global problems?

8 Why Should We Care?

As beings of care, who should deal with all these problems? Who would have the luxury to develop a culture of care that sustains all species of our planet while millions of people across the globe are facing existential threats? Would people living in poverty, squalor, disease and corruption be able to be part of an ecological sustainable movement to better the planet and develop solutions to world's crises? Are world countries really committed and resilient to achieving the UN's SDGs and bring about a balance between humanity system and planetary system? Can we sustain our planet and create a world culture of care, justice, equality, and human rights, while nuclear threats prevail the media as a result of the existing war? Will we be able to create a sustainable globe with no poverty, no hunger, clean water and energy, quality education, health care systems and economic growth?

The truth of the matter is that war creates disastrous breakdown of the economic, social, and environmental structures of nations and hinders all efforts for sustainable development. It is noticeable that realistically, by and large, governments, organizations and civic agencies are theoretically committed to achieving the NU's SDGs [6, 7]. Yet in practice, there is now resilient commitment to create and put into practice one unified vision to sustain the planet at all levels (social, economic, and environmental). As being of care, university students and graduates, educators, curriculum designers, policy makers and other stakeholders' responsibility is to contribute to sustaining our planet and empowering young people with the required knowledge, skills and attitudes necessary for human existence.

As educators, our primary goal should be promoting students to develop a sustainable global mindset capable of exploring possible solutions to world's problems and crises [8, 9]. What is needed is an in-depth practical understanding of the magnitude of the problems our planet is currently facing. We need to instill in young people the knowledge, skills and attitudes (KSAs) in living in harmony with themselves,

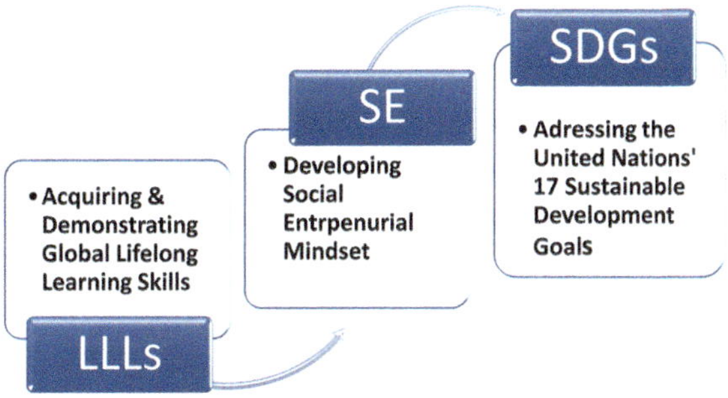

Fig. 5 Proposed pedagogical model

with their natural environment and with the different intellectual other. We need to develop a culture of care. Our global awareness becomes an inevitable necessity.

9 A Proposed Conceptual Model

That said, the other important questions would be: How do we instill in our students the required KSAs that allow them to develop a culture of care and responsiveness to world's crises? How can we embed current and future ecological and global problems into higher education curriculum? How can we engage university students in entrepreneurial skills and competencies for sustainable development? What follows is a practical way of the type of pedagogical model that has the potential to achieve this goal. The pedagogical model that the author of this paper promotes has the potential to promote the UN's SDGs and empower college graduates with global competencies. The proposed model relates the concept of Social Entrepreneurship (**SE**) to the **UN's SDGs** and lifelong learning skills. The model combines Social Entrepreneurship (SE) knowledge and the UN's SDGs education to advance sustainably global competencies (see Fig. 5).

In conclusion, the proposed conceptual model is premised on the social capital theory and supports the argument that SE is a practical and operational process that startup companies and entrepreneurs use to assess, address, and solve socio-cultural and environmental problems to bring positive change to society. Also, social entrepreneurs are those who have the will, passion, perseverance, and the required global competencies that allow them to take calculated risk to analyze, design, develop, implement and evaluate businesses that have socio-economic and environmental value. The aim of competent social entrepreneurs goes beyond achieving profit out of their business to have a positive social and ecological change while using innovative solutions to social problems. Acquiring and practicing Global Lifelong Learning Skills (LLLs) enable students to develop a Social Entrepreneurial (SE) mindset which motivates them to address the United Nation's Sustainable Development Goals (SDGs).

References

1. Eichler, G.M., Schwarz, E.J.: What sustainable development goals do social innovations address? A systematic review and content analysis of social innovation literature. Sustainability **11**(2), 522 (2019)
2. Turyakira, P., Sendawula, K., Nanyanzi, M., Nantale, H., Tamale, J.N.: Social entrepreneurship: empirical evidence on its contribution to the realization of the sustainable development goals in Uganda. J. Work-Appl. Manage. (2024)
3. Apostolopoulos, N., Al-Dajani, H., Holt, D., Jones, P., Newbery, R.: Entrepreneurship and the Sustainable Development Goals, pp. 1–7. Emerald Publishing Limited, Leeds (2018)
4. Tomasella, B., Wylie, A., Gill, D.: The role of higher education institutions (HEIs) in educating future leaders with social impact contributing to the sustainable development goals. Soc. Enterp. J.Enterp. J. **19**(4), 329–346 (2023)
5. Bărbulescu, O., Tecău, A.S., Munteanu, D., Constantin, P.: Innovation of startups, the key to unlocking post-crisis sustainable growth in Romanian entrepreneurial ecosystem. Sustainability **13**(2), 671 (2021)
6. Crecente, F., Sarabia, M., del Val, T.: Sustainable entrepreneurship in the 2030 horizon. Sustainability **13**(2), 909 (2021)
7. Malhotra, S., Kiran, R.: Integrating cognitive competency, social competency and risk propensity with the theory of planned behaviour to attain sustainable-development-goal-8-driven sustainable entrepreneurial intentions. Sustainability **16**(10) (2024)
8. Akkuş, Y., Çalıyurt, K.: The role of sustainable entrepreneurship in UN sustainable development goals: the case of TED talks. Sustainability **14**(13) (2022)
9. Tunçalp, D., Yıldırım, N.: Sustainable entrepreneurship: mapping the business landscape for the last 20 years. Sustainability **14**(7) (2022)

A Conceptual Model for Promoting Sustainable Entrepreneurship Education

Mohamed Elgeddawy

Abstract This paper provides a clarifying and hermeneutically ontological reading of what it means to be an entrepreneur. It stimulates new thinking on Entrepreneurship Education (EE), while solidifying the intellectual discourse that underscores its assumptions. The paper voices the ontological and pedagogical implications of perceiving the entrepreneurial process as a public, interpretive and situated act. The first section critically reviews the divergent strands of conceptualizing the entrepreneurial mindset in relation to the fields of behavioral and cognitive psychology. In the second part, hermeneutically ontological concepts are applied to the entrepreneurial process to clarify our understanding of the nature of this process and point out the implications of this understanding for EE. The paper ends with a pedagogical vision for the development of students' entrepreneurial mindsets, pointing out the importance of viewing the entrepreneurial process as a human value, and a mode of existence and inquiry.

Keywords Entrepreneurship education · Hermeneutics · Ontology

1 Introduction

Given the challenges of globalization, digitalization and the socioeconomic problems that cut across world countries, it is more compelling than ever before for Higher Education Institutions (HEIs) and societies to promote the development of entrepreneurial skills and attitudes among young people. Also, given the employment crisis that followed the emergence of digitalization, the creation of new jobs, new skills, and new business opportunities has become a necessity. The uncertainty of work environments caused by the consequences of digitalization suggests that university graduates need to be more entrepreneurial and able to generate innovative ideas that enable them to be employable and start their business if needed. In

M. Elgeddawy (✉)
Modern College of Business and Science, Muscut, Oman
e-mail: Mohamed.Elgeddawy@mcbs.edu.sa

B. Awwad (ed.), *AI and IoT: Driving Business Success and Sustainability in the Digital Age*, Studies in Systems, Decision and Control 590,
https://doi.org/10.1007/978-3-031-88874-8_61

this regard, recent studies indicate that Entrepreneurship Education (EE) advances entrepreneurs' skill development and employability [1].

1.1 Needs for Entrepreneurship Education

The paper responds to the growing interest in Entrepreneurship Education (EE), which is assumed in the literature to enable students to demonstrate their acquisition and application of the knowledge, skills, and attitudes (KSAs) of successful enterprise tasks and activities. It addresses the importance of designing, developing, and implementing a competency-based EE that enables students to develop the KSAs necessary for empowering their self-employability potential as a solution to combat youth unemployment which spans the globe. Digitalization and its evident impact on the world of work has accelerated youth unemployment and the underemployment crisis. Considerably, young people in both developed and developing countries have been facing challenging and unstable labor market conditions over the last few years, particularly with the beginning of digitalization [2]. In this regard, EE helps in creating intervention strategies that generate practical solutions for this crisis as it has the potential to assist various stakeholders (government agencies, employers, academicians, and community members) in enlightening the process of developing and implementing self-employment strategies that can sustain the economic growth of nations.

1.2 A Critical Review of Available EE Models

"Before there can be entrepreneurship there must be the potential for entrepreneurship" [3]. The question that emerges in response to this statement is: what needs to be done to help college students develop their entrepreneurial potential. Kouakou, Li, Akolgo, and Tchamekwen tried to answer this question by conceptualizing the entrepreneurial mindset process within the framework of behavioral and cognitive psychology [4]. The behavioral approach focuses on molding and shaping the behavior of individuals in such a way that reflects the expected needed behavior of an entrepreneur. Behaviorists concentrate on equipping individuals with a set of standardized skills and attitudes that allow them to act and develop their entrepreneurial mindset models which are constant and never change over time. On the other hand, cognitivists put a premium on the motivational and cognitive process that individuals need to go through to be effective entrepreneurs. In line with this approach, personality traits would determine entrepreneurs' business performance. According to the behavioral and cognitive approaches, individuals would demonstrate either a fixed mindset or a growth mindset. Individuals with a fixed mindset would deal with business opportunities inflexibly in accordance with a set of drilled patterns that they already memorized. The limitation of this approach is that these exercise patterns refrain students from developing their abilities to cope with emerging situations that do not fit within the acquired business patterns.

Also, advocates of the cognitive approach believe that individuals would be able to learn how to develop their business goals when they could be engaged in a learning process for intelligence enhancement. In fact, both the behavioral and cognitive approach ignore the socio-cultural situated factors and constructs that impact the process of developing an entrepreneurial mindset. In addition, the literature embraces entrepreneurship programs that fall within three categories of approaches reflecting (1) reproduction practices where the educator is the main center of the teaching–learning process, (2) construction practices which consider guided research as an instrumental tool for students in developing entrepreneurial projects, and (3) co-construction practices which perceive learning as a constructive shared process between the educator and students.

Another entrepreneurship model considers offering students awareness courses for creating startups and overcoming the challenges that face the process of creating a new business. This model endorses the integration of specialization courses in EE for students who want to be qualified advisors and consultants in the field of entrepreneurship.

On the other hand, by and large, the EE literature [5] centralizes the importance of instilling into students the motivation to develop entrepreneurial skills such as creativity, innovation, risk taking, proactivity, flexibility, opportunity seeking, commitment, self-efficacy, design thinking, systematic planning, negotiation, networking, persuasion, critical thinking, information seeking, problem solving, self-learning, goal setting, technological knowledge, leadership, quality performance and professionalism.

Noticeably, EE gained a great deal of currency in business studies; the literature in this domain of influence reflects a myriad of approaches. Design thinking methodology, for example, is perceived as a skill that enables students to convert problems into initiatives and business opportunities that benefit their lives. Another approach perceives EE as an add on and extracurricular activities that help students develop entrepreneurial mindsets. The limitation of this approach is that students trivialize entrepreneurial activities as they do not carry any credit weight. The literature also reveals another pedagogical model that is action-oriented which engages students in real life social activities while working with others collaboratively to achieve specific goals. Though this model is project-based, it does not take into consideration the different learning styles that cut across the members of the group. The model is not truly individualized; it considers all the learners of the group to learn the same way at the same time. It ignores the fact that different learners have different learning styles [6].

Unfortunately, these models are epistemologically oriented and prescribe a set of courses and competencies that students need to acquire to start their business. They conjoin around a rigid codified pedagogy. An epistemological perspective would mean that all students, regardless of their culture, education background, setting and political orientation can process and develop entrepreneurial mindsets the same way and at the same time. This process decontextualizes the dynamic act of developing entrepreneurial mindsets and has its pedagogical ramifications.

1.3 Limitations of Codifying Entrepreneurial Knowledge

Viewing EE as a set of pedagogical regulations reflects a monolithic and prescriptive mode of learning associated with epistemological thoughts. This epistemological view is built on the psychology of behaviorism which intends to reshape entrepreneurs' mindsets through internalizing the required knowledge, storing it and retrieving it in a systematic way. The limitation of this pedagogical approach is that it emphasizes the adoption of entrepreneurial models that may not be applicable to various audiences. The epistemological model puts a premium on the mechanical learning of a set of procedures to shape entrepreneurs' patterns of behavior. This traditional model leaves no place for personal interpretations of the social domain under inquiry. Also, the assessment process of the epistemological model yields well-tested learning outcomes that decontextualize the discourse of entrepreneurship. Students in this teacher-centered model are perceived as empty vessels waiting to be filled with the right entrepreneurial model that allows them to develop the required homogenous entrepreneurial mindset. This conventional pedagogical style dehumanizes the entrepreneurial process and does not meet the expectations of today's students who are born in a digital era that values diversity of learning styles. In fact, current college students, by and large, look forward to experiencing life-contextualized and meaningful entrepreneurial experiences that enable them to be self-employed and relate what they already know to what they want to know. It is important to note here that the entrepreneurial process is not static; rather, it is dynamic, changeable and context specific. Therefore, the meaning of EE remains vague to many stakeholders such as educators, employers, policy makers and curriculum designers. Hence a pressing need for a comprehensive vision exists to guide the instructional strategies that have the potential to enhance students' development of entrepreneurial mindsets.

2 A Hermeneutically Ontological EE Model Development

The author of this paper proposes a hermeneutically oriented EE conceptual model premised on hermeneutics [7, 8]. The model stimulates new thinking on entrepreneurship education while highlighting the ontological discourse that underscores its assumptions, towards the goal of understanding what it means to be an entrepreneur and the pedagogical ramifications of this understanding. The proposed model shifts attention from a focus on the question of how we teach entrepreneurship and what EE is to another hermeneutically ontological question of what it means to be an entrepreneur. In this respect, the paper will be of a substantial use to many practitioners, researchers, and educators as the field of EE continues to grow and reconstruct a pedagogy that is responsive to the challenges of today's labor market. Ontologically, it is important for students to understand and practice what it means to be entrepreneurs. Understanding entrepreneurship from an ontological diverse perspective allows both students and their coaches (educators) to consider entrepreneurship as a ***public, interpretive and situated*** process (**PIS** will be used as an acronym for the entrepreneurial process as **P**ublic, **I**nterpretive and **S**ituated act).

2.1 Entrepreneurship as a Public Act

Entrepreneurship as a Public Act presupposes the existence of other stakeholders who are dynamically engaged in the entrepreneurial process. Students as entrepreneurs are always already with others in the world and the meaning they develop of the entrepreneurship process is socio-culturally constructed and reconstructed because of their dialogic interactions with other stakeholders who are capable of understanding and interpreting their actions and reactions. The public aspect of the entrepreneurial process, which the author promulgates, has its ontological roots in Bakhtin's dialogism [9], where the focus is on advancing dialogic interactions among entrepreneurs from various cultures, backgrounds, and social conventions. Viewing EE as a public enterprise provides students with ample opportunities to anticipate the reactions of other stakeholders. Ontologically speaking, the entrepreneurial process is an engaging, dynamic, public, and dialogic act with other stakeholders. It is through this dialogic process that new meaning and new business opportunities emerge. Perceiving the entrepreneurial process as a public act considers meaning making as an interpretive process of being in the world with others. In this sense, students need to be continuously engaged in a meaning making process where they can construct and develop their discursive entrepreneurial mindsets, while anticipating their stakeholders' interpretive actions and reactions. Likewise, the anticipated stakeholders (community members, policy makers, and other individuals who have interests and stakes in the entrepreneurial process) are assumed to react responsively and interpretively while publicly addressing and assessing what concerns the entrepreneurial processes of their community. Educationally speaking, the entrepreneurial process, as such, requires interpretive and public interactions with others. In other words, the entrepreneurial process is always already public and includes an assumed interdependent relationship among the entrepreneurs and other stakeholders within a specific context. As such, the entrepreneurial process needs effective education that takes into consideration three interrelated entities, mainly, the context, the entrepreneur, and the stakeholders. The triangulated public process of entrepreneurship requires entrepreneurs to develop three interrelated and interdependent types of knowledge, mainly knowledge of the context, knowledge of the self and knowledge of other stakeholders. It is important to note here that the entrepreneurial process of meaning making is a result of the dynamic interactions that take place among the three types of knowledge. That is, entrepreneurs' collaborative interactions with other stakeholders within a specific context generate meaning and an understanding of what it means to be an entrepreneur. So, the concept of entrepreneurship as a public act takes place within this triangulation.

Entrepreneurship as triangulation, in this respect, is a dialogic enterprise of communicative interactions that necessitate entrepreneurs to match their needs, aspirations and goals with other stakeholders' expectations within a specific context. This process (entrepreneurship as triangulation) is not an internal or cognitive process; it is an external, public, and dynamic act of communicative triangulated interactions that stimulate entrepreneurs to make sense of their motives, goals and needs in relation to others in the world within a specific context. As such, EE requires entrepreneurs to explore the market needs in relation to their goals and the motives of other stakeholders in their community. The key factor in this ontological and public entrepreneurial process is the relationship that needs to take place between the three types of knowledge. Essentially, effective EE should assist students in exploring the expectations of the three interrelated worlds.

2.2 Entrepreneurship as an Interpretive Act

The second main principle of the PIS model is that the entrepreneurial process is interpretive in nature. This process requires both the entrepreneurs and the stakeholders to be engaged in interpreting their contextualized events, business opportunities and problems. The key factor in this respect is for entrepreneurs to develop a desired interest in understanding and interpreting the motives and needs of other stakeholders and vice versa.

2.3 Entrepreneurship as a Situated Act

The third concept of the model considers the situatedness of the entrepreneurial process. From a hermeneutically ontological perspective, the indication here is that we are who we are as entrepreneurs because of being in the world with others within a specific context. That is, prospective entrepreneurs construct and reconstruct their goals and attitudes from a pre-disposition perspective. The stance taken here is that the entrepreneurial process as a communicative dialogic act is possible only because as entrepreneurs, we know our beliefs, values, traditions, and worldviews in relation and contrast to other stakeholders' beliefs and values. It follows that students as prospective entrepreneurs are always already situated in relation to others, and they draw on their socio-cultural beliefs and worldviews to understand and interpret other people's belief systems. Being situated also indicates that the entrepreneurial process is always changing and changeable as the context changes. Therefore, EE process cannot be dogmatically systematic, but rather interpretive and anticipatory in accordance with the setting in which it takes place. As such, the entrepreneurial process, which is a process of making meaning of the available business opportunities, per se, is an outcome of a triadic association among the three interrelated concepts.

3 Pedagogical Ramifications

Perceiving the entrepreneurial process as public, interpretive and situated act has pedagogical ramifications. One of these ramifications is that a codified body of knowledge or a set of entrepreneurial courses would not be enough to generate graduates with entrepreneurial mindsets. Also, the acquisition of entrepreneurial skills can help but would not guarantee that university graduates will be successful entrepreneurs. My argument is that the design, development, implementation, and assessment of EE should take into consideration an understanding of the entrepreneurial process as a public, interpretive and situated act. It is important to note here that the discourse of the PIS model does not negate the importance of the discourse of mastery with its conventional set of courses and skills. Yet, it moves beyond it to promote a practical understanding of the entrepreneurship process. The PIS discourse allows practitioners of the EE to have global and glocal outlooks while developing their entrepreneurial mindsets. Promoting glocalization concepts and embedding their attributes into EE has the potential to advance local insight and equip students with the required entrepreneurial skills necessary for developing globalized as well as glocalized mindsets.

4 The Dynamicity of the Entrepreneurial Process

The entrepreneurial process in this respect would be dealt with as a dynamic, socio-cultural, and ideological enterprise within emancipatory contexts that remove the boundaries between the personal, the social and the international. Perceiving the entrepreneurial process as a public, interpretive and situated act raises entrepreneurs' awareness of being socio-culturally, politically, and ideologically constructed. This perception opens for practitioners of EE the likelihood of seeking new business opportunities and requires open-ended pedagogical approaches that give them the freedom to represent their constructed worlds.

4.1 Embedding Open-Ended and Hermeneutical Entrepreneurship Pedagogy

The PIS conceptual model promotes an open-ended entrepreneurship pedagogy that may draw on management education, digital literacies, design thinking, culturalism, multiculturalism, dialogism, or any other suitable pedagogical content that allows students to construct their entrepreneurial identities. The PIS process aims to assist entrepreneurship practitioners in extending the available codified pedagogies. In other words, the PIS model is not a radical rejection of current entrepreneurship pedagogies. It is an extension of the departmental concepts of EE. The PIS concepts promote an interpretive stance that engages students in activities that allow them to solve local problems of significance. The focus of an open-ended pedagogical model is to enable students to critically perceive and understand their motives in relation to the needs of the different intellectual Other. The model allows its practitioners to explore and solve in socially responsible ways the complex problems that impact their public lives. The model embraces digital communicative interactions where entrepreneurs engage each other as well as others from around the world in goal-oriented dialogue about current social issues and problems that concern the local and global domains. This type of pedagogical understanding is integral to the discourse of EE. It helps students identify and recognize their personal and social domain in relation and contrast to other people's domains. The model is ontologically hermeneutical and promotes open-ended activities and projects of interpreting the word and the world. It stimulates students to interpret reflectively the available business opportunities and take calculated risks in starting their business. It is important to note that the interpretively of the entrepreneurial act entails that entrepreneurs have the will and freedom to decide on the business opportunities that are meaningful to them.

Being ontologically hermeneutical, the PIS vision and its pedagogical ramifications provide insights for EE practitioners in understanding that the entrepreneurial process, as an act of situated interpretation, is a matter of give-and-take dialogic interactions between entrepreneurs and other stakeholders in relation to the problem under inquiry. According to this line of logic, since the entrepreneurial process is a human enterprise that is socio-culturally situated, it follows that there is no grand pedagogical narrative that enables all entrepreneurs, regardless of their setting, to master the entrepreneurial process through a set of pedagogical procedures. That is, entrepreneurship is context-specific and depends on entrepreneurs' understanding of their past, present, and future business opportunities. Considerably, the PIS model allows students to dialogically enter the constructed world of different entrepreneurs and explore how similar or different their entrepreneurial mindset is. In the PIS model, students are scaffolded to analyze various local, regional, and international entrepreneurship case studies and projects. These constructivist-oriented activities allow them to enter into the otherness of the different, intellectual others which makes it possible for them to construct and reconstruct their entrepreneurial mindsets. Engaging students in experiencing the act of entrepreneurship through real life projects that involve other stakeholders' views allows them to develop their own beliefs and theories of what it means to be entrepreneurs within a specific context.

5 An Experiential EE

An experiential EE that is grounded on the conceptual framework of the PIS expands students' horizons of expectations, each according to his/her constructed world-view. These individually constructed realities make students decide for themselves what is meaningful and what is meaningless, each according to his/her interpretation of the addressed entrepreneurial experience. In a learner-centered environment, entrepreneurs will be able to self-regulate their own entrepreneurship knowledge, skills, and attitudes, each according to their own interpretation of the situated entrepreneurial experience he/she is exposed to. The PIS model endorses independent thinking, collaborative design thinking, collaborative problem solving, inquiry-based learning, and learning by-doing. Within this conducive and scaffolded learning environment, both students and educators function as co-learners and co-explorers while developing their entrepreneurial mindsets. In other words, students learn to develop their entrepreneurial mindsets by practicing entrepreneurial thinking within their communities and social fields. It is important to note here that though the PIS thought does not specify a particular pedagogy to be followed, it stimulates its practitioners to go beyond the traditional constraints of the academic discourse conventions. The revolutionary discourse of the PIS is enacted by hermeneutical ontologists such as Hiedeggar [8], Gadamer [7] and Bakhtain dialogical approach [9]. This discourse does not mean the removal of the traditional way that characterizes EE with its focus on embedding a set of entrepreneurship courses into higher education curriculum. The underlying view here is that the PIS model enhances rather than diminishes what is already in place. Perceived as such, the entrepreneurial process, being a public, interpretive and situated act, elevates entrepreneurs' awareness of being socially constructed in a community with others. The discourse of the PIS model looks for innovative pedagogical initiatives to address and assess changes taking place nowadays in the labor market due to digitalization. In accordance with the triadic concepts of the PIS, EE is a process that enables students to interpret global events and business opportunities through intentional goal-oriented exercises and projects. Entrepreneurs, as such, are always in a state of thinking dialogically about what they experience and observe in the world in which they live. During this triadic entrepreneurial process, students think while perceiving and perceive while thinking of taking risk in starting their business. Entrepreneurs' perceptions work referentially and differentially. That is, they get engaged in mental exercises through which they compare similar businesses and draw on their pre-understanding to make sense of the available business opportunities. The PIS practitioners acknowledge the importance of the social domain in constructing entrepreneurs' choices. The model is grounded on open-ended pedagogical techniques to provide students with business opportunities through which they can broaden their understanding beyond their communal private world. The model also empowers its practitioners with the techniques of developing their own entrepreneurial theories of practice which promote knowledge of the self, knowledge of the other and knowledge of the world. This pedagogy of triangulation creates a classroom ecology that values dialogic interactions among entrepreneurs. The pedagogy of triangulation: **creates** a learning environment responsive to local and global issues raised by various stakeholders worldwide, **engages** students in dialogic interactions with different stakeholders, **stimulates** entrepreneurs to go beyond the limits of personal community, **embraces** interest in making sense of the world with its concerns and problems, **puts** a premium on the development of entrepreneurial competencies, and **endorses** inquiry-based learning and assessment where educators play the role of facilitators. The pedagogical vision that the PIS endorses is consistent with the changes that have taken place in the ways entrepreneurs currently communicate with each other and across the world. The evolution of digitalization and globalization provides entrepreneurs with digital tools that facilitate the practice of entrepreneurship as a public, interpretive and situated human activity that promotes an understanding of multiple perspectives regarding local and global businesses. The PIS vision also stimulates its practitioners to co-develop thought-provoking and enlightening entrepreneurial activities that connect students as entrepreneurs with others worldwide so that they can reflectively and dialogically collaborate and lean from one another.

Essentially, the model enables educators to be committed to preparing entrepreneurs who can function receptively and responsibly in a global society. This pedagogical approach lends itself to a great deal of scholarly inquiry. In this regard, the embedding of global studies into EE has the potential to enable students to create and design their business and think globally and act locally. Also, all social studies that connect the local to the global would provide EE with a plethora of content. After all, in my endeavor to articulate and voice an alternative vocabulary of talking about the nature of EE, this paper has focused on pointing out the significance of understanding the entrepreneurial act as a public, interpretive and situated human activity and a mode of inquiry. My reading of EE has yielded an understanding that the entrepreneurial act is publicly oriented and values life contextualized and dialogic interactions. Viewing EE through the lenses of hermeneutical ontology empowers its practitioners with a sense of what it means to be entrepreneurs whose inevitable existence with other stakeholders embraces the possibility of understanding and identifying the available business opportunities that benefit them as well as their communities.

6 Assessing EE

The proposed open-ended pedagogy suggests a performance-based assessment approach which is grounded on the development of e-Portfolios. Digital portfolios function as measuring instruments that assess entrepreneurs' degree of developing entrepreneurial competencies and understanding of the entrepreneurial process as a public, interpretive and situated act. Graded digital portfolios also are considered part of direct assessment which is essential for achieving the targeted performance through a myriad of learner-centered activities and innovative entrepreneurial projects. Students' e-Portfolios establish a link between the perceived entrepreneurial knowledge and the competencies and learning outcomes that students are supposed to demonstrate. The assessment strategy is project and problem based. Students' e-Portfolios aim to **document** entrepreneurs' reflective, academic, professional and community service achievements in the field of entrepreneurship, **point out** their practical and applied understanding of the entrepreneurial process as a public, interpretive and situated act, **identify** the market needs in relation to their abilities in generating entrepreneurial ideas that are innovative and benefit their community, and **demonstrate** their abilities in taking timely action and calculated risks in identifying beneficial business opportunities.

Conclusively, engaging students in thought-provoking situated entrepreneurial projects and activities enable them to actively and interactively start their businesses. In the digital era of the twenty-first century, practicing the entrepreneurial process as a public, interpretive and situated act empowers individuals with the required practical knowledge for assessing and addressing the challenges of this process. The Web, with its gigantic capabilities of providing telecommunication tools, gives potential entrepreneurs the prospect to network with a broader variety of audiences. Digital communications, in this sphere of influence, bring the entrepreneurial act alive, giving entrepreneurs ample opportunities to be knowledgeable and skilled in starting their business. In fact, the perception of the entrepreneurial process as a public, interpretive and situated act inspires EE practitioners to provide their students with digital tools that enable them to contribute intelligently and thoughtfully with their innovative entrepreneurial ideas in the world they always already share with others. Certainly, effective pedagogical integration of the entrepreneurial process as a public, interpretive and situated act has the likely of opening new prospects for entrepreneurs to experience each other and the world.

References

1. Kalyoncuoğlu, S., Aydintan, B., Göksel, A.: The effect of entrepreneurship education on entrepreneurial intention: an experimental study on undergraduate business students. J. Manag. Res. **9**, 72–91 (2017)
2. International Labour Organization. Global Employment Trends for Youth 2020: Technology and the Future of Jobs. https://www.ilo.org/sites/default/files/wcmsp5/groups/public/@dgreports/@dcomm/@publ/documents/publication/wcms_737648.pdf. Accessed 11 Nov 2024
3. Krueger, N., Brazeal, V.: Entrepreneurial potential and potential entrepreneurs. Entrep. Theory Pract. **18**, 91–104 (1994)
4. Kouakou, C., Akolgo, G., Tchamekwen, M.: Evolution view of entrepreneurial mindset theory. Int. J. Bus. Soc. Sci. **10**(6), 116–129 (2019)
5. Portuguez Castro, M., Gómez Zermeño, M.: Identifying entrepreneurial interest and skills among university students. Sustainability (Switzerland) **13**(13), 1–19 (2021)
6. Knowles, M.: Andragogy in Action. Jossey-Bass, San Francisco (1984)
7. Gadamer, G.: Philosophical Hermeneutics. University of California Press, Berkeley, CA (1976)
8. Heidegger, M.: Being and Time. Harper & Row, New York, NY (1995)
9. Bakhtin, M.: Art and Answerability. University of Texas, Austin, TX (1990)

A Conceptual Framework for Blending STEAM Education Principles with Entrepreneurial Thinking for Lifelong Learning: Towards Preparing Omani College Graduates for the Labor Market

Mohamed Elgeddawy

Abstract Recently, there has been a clear focus in the literature on the accelerated changes and challenges that automation and digitalization are bringing to the labor market of Sultanate Oman. In all business and education sectors, new solutions for promoting skills related to these changes have been sought out persistently. Within this domain of influence, it is increasingly acknowledged that higher education curriculums need to respond to the challenging demands of digitalization. There is a need for curriculum designers, policymakers and educators to look ahead, be futuristic and empower college graduates with the required knowledge, skills and attitudes that make them employable, and ready to deal with jobs that necessitate knowledge of handling big data and complex problems that require skills in Science, Technology, Engineering, Arts and Mathematics, acronymed as STEAM skills. It is also evident in the literature that preparing college graduates for a dynamic innovative future requires the development of entrepreneurial thinking. Despite their interrelatedness and interdependence, the STEAM and Entrepreneurship Education are dealt with in Higher Education compartmentally, leading to several limitations among which is graduates' inability of competence development. This paper proposes a contextualized conceptual framework that blends the two entities (STEAM and Entrepreneurial Education) together as interrelated variables moderated by lifelong learning. The proposed conceptual framework has been guided by the following question: How does blending STEAM and Entrepreneurship Education contribute to graduates' preparedness for the labor market within the context of Sultanate Oman?

Keywords STEAM education · Lifelong learning competencies · Entrepreneurial education

M. Elgeddawy (✉)
Modern College of Business and Science, Muscut, Oman
e-mail: Mohamed.Elgeddawy@mcbs.edu.om

B. Awwad (ed.), *AI and IoT: Driving Business Success and Sustainability in the Digital Age*, Studies in Systems, Decision and Control 590,
https://doi.org/10.1007/978-3-031-88874-8_62

1 Introduction

Based on the assumption that most of the future jobs require STEAM skills and Entrepreneurial thinking and to avoid falling short in preparing college graduates to face the business challenges when integrating new technologies in their jobs, higher education institutions (HEIs) initiate programs and courses that instill in students a multidisciplinary mindset. In this domain of influence, research indicates that success in the workplace, by and large, depends on graduates' preparedness in Science, Technology, Engineering, Arts and Mathematics competencies [1].

1.1 Integrating STEAM Skills into the Curriculum

The literature [2] highlights a myriad of approaches that guide the integration of STEAM concepts into HE curriculum. The integrating process of STEAM skills takes place through (1) a standalone approach where a set of elective core courses in science, technology, engineering, social sciences and mathematics are implemented across the schooling years of the student, (2) an integrative approach where STEAM skills are integrated across the assignments and activities of disciplinary based courses, and (3) a blended method which makes use of both the standalone and the integrative approach in designing, developing, implementing STEAM skills.

The limitation of the stand-alone model is that the subjects that constitute the model are taught and practiced in isolation. The integrative approach is thematically based and focuses on solving contextualized problems. The blended approach and integrative model are commended in the literature for removing the artificial boundaries that cut across disciplines. Despite the diligent endeavors on the part of HEIs in addressing the needs of rapidly changing labor market needs, research points out that there is a mismatch between the supply of and demand for employability skills. Recent studies [3] show that employers report that the supply of STEAM skills does not meet the required skill demands of the labor market.

It is important to mention that instilling in college students the motivation to acquire and demonstrate STEAM skills is important for graduate employability, but not enough to function successfully in a technology–advanced industry as graduates need other transferable skills that are in alignment with market realities. Graduates need to develop the twenty-first century Lifelong Learning Entrepreneurial Competencies (LLLEC) such as communication, teamwork, leadership, problem solving, critical thinking, entrepreneurial thinking, technology, professionalism, innovation, and creativity) in addition to STEAM skills.

This paper presents a conceptual framework that blends LLLEC and STEAM skills to represent the right mix of skills which need to be embedded into higher education curriculum using stand-alone and integrated pedagogical approaches. The underlying assumption of the proposed conceptual framework is that blending STEAM and

Fig. 1 Blending STEAM with EE through lifelong learning

Entrepreneurship education through the moderation of lifelong learning competencies contributes to graduates' preparedness for the labor market. Figure one represents a visualization of the proposed conceptual model which is premised on constructivist thinking (Fig. 1).

STEAM education promotes the development of lifelong learning competencies such as critical thinking, problem solving, creativity, negotiations, and creativity, which are all important skills for the success of college graduates as entrepreneurs. That is, what relates to the two interdependent entities (STEAM and EE) is a set of lifelong learning skills. The two entities such are interrelated so much that you cannot separate one from the other.

Considerably, the proposed conceptual framework blends and integrates the two entities through two pedagogical approaches. The first one is the stand-alone approach, which refers to the development of a mix of lifelong learning elective core curriculum, skill-oriented and entrepreneurial courses such as system design, project management, oral communication, critical thinking, problem solving, design thinking, emotional intelligence, and STEAM courses such as Science, Technology, Engineering, Arts and Mathematics. These skill-based courses need to be part of student degree plan and all university students, regardless of their majors, have to choose the right mix that is in alignment with their preferences and degree plan. The STEAM elective courses intend to equip students with thinking skills related to science, technology, engineering, social sciences and mathematics. The integrative approach refers to the embedment of the Lifelong Learning Entrepreneurial Competences (LLLEC) and STEAM skills across disciplines. The approach refers to the integration of these skills through course assignments, projects and activities.

1.2 Beyond the Pedagogical Divide

Traditionally, STEAM and LLLEC are perceived as a curriculum that specifies a set of courses to be offered and taught sparely. The proposed conceptual model aims to promote a worldview that goes beyond perceiving STEAM and LLLEC as a set of discipline specific courses that are taught separately. Building on previous works [1–3] the proposed conceptual model promotes the fusion of STEAM and LLLEC mind-sets into HE curriculum. According to the proposed model, STEAM and LLLEC are interdisciplinary learning approaches and ways of thinking for creatively and contextually assessing and addressing real life problems. Figure 2 visually demonstrates the two main variables of the proposed conceptual model (STEAM and Entrepreneurial Education) moderated by a Lifelong learning Variable.

The outcome of blending STEAM education and EE allows college students to acquire and demonstrate lifelong learning entrepreneurial competencies (such as critical thinking, problem solving, leadership, technological thinking, digital literacy, numerical literacy, service orientation literacy, socio-cultural intelligence, emotional intelligence, project management, and resilience) and STEAM competencies (such as design thinking, computational thinking, analytical thinking, ethical thinking, scientific thinking, mathematical thinking, innovation thinking, and social Sciences thinking).

Developing a STEAM way of thinking goes far beyond the implementation of a set of science and or mathematics courses in HE curriculum. The content of these inter-related disciplines (Science, Technology, Engineering, Social Sciences and Mathematics) needs to be thematically integrated across the curriculum of any academic program which requires students to think, act and react at multiple levels of abstraction. In other words, the framework that STEAM represents is perceived in this context as a learning approach for integrating knowledge, skills, and attitudes from various disciplines such as, science, engineering, mathematics, computer science, technology, arts and social sciences. Similarly, LLLEC is a curriculum approach

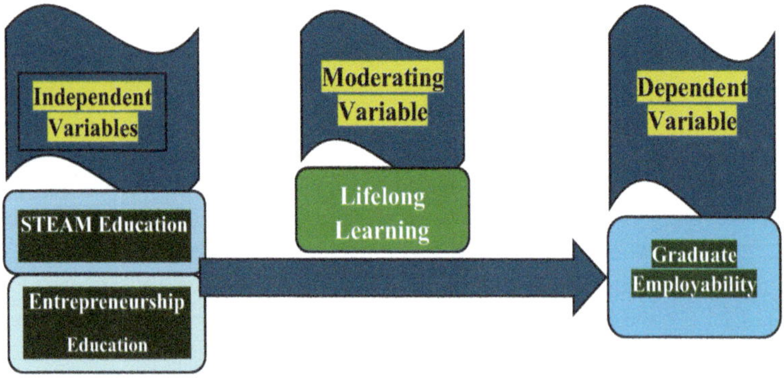

Fig. 2 Visualization of the conceptual framework variables

that focuses on mastering, applying and practicing basic learning techniques of lifelong learning and entrepreneurial skills which college graduates need to acquire and demonstrate to be job ready.

2 STEAM and LLLEC as Interdisciplinary Learning Approaches

The assumption taken here is that what characterizes today's world is the accelerated change in all facets of life. In a world of increasingly sophisticated digitalization, integrating STEAM and LLLEC across the curricula has the potential to prepare future workforces capable of using digital devices effectively. It is also important to note that as automation, digitalization, globalization and leading innovative technologies adoption transform the workplace at the national, regional and international levels, so must higher education curriculum. It has become more compelling than ever before for colleges and universities to integrate innovative learning approaches that meet the learning styles of a generation growing up with digital technologies.

Research [4, 5] suggests that new types of college programs need to be established to meet the demands of the digital era. The transitions that automation and digitalization bring to the workplace challenge the traditional education models that value memorization and transmission of knowledge from one individual to another. Towards this goal, higher education curricular should imbed activities that reflect the usage of digitalization, big data, deductive and inductive reasoning, and a mix of lifelong learning skills. In this domain of influence, employing constructivist oriented pedagogical strategies to competence development would have a great value in increasing graduate employability and ability to solve community problems. Effective implementation of the proposed conceptual framework requires constructivist pedagogical strategies that value experiential learning. Blending STEAM and EE has the potential to promote the integration of cross-disciplinary teaching materials and practical exercises that are technology-oriented and could advance students' understanding of artificial intelligence, coding, complex problem solving and complex data analysis. As a transformative framework, effective implementation of the proposed model within a scaffolded and constructivist learning environment has the potential to empower Omani graduates and workforces with digital business skills and data analytics which enable to get a job, sustain it, and move to another if needed.

3 Perceived Value of Blending STEAM with EE

Blending STEAM and LLLEC aims to generate citizens with the right mix of competencies that allow them to be socially and culturally responsive to the problems of their society and the world. The right mix of competencies also helps students to articulate their understanding of social problems while seeking well-thought-out sustainable solutions. For example, analytical thinking needs knowledge and understanding of mathematical, numerical and design thinking. Also, the LLLEC such as critical thinking and communication skills are needed to articulate mathematical and engineering data. These types of skills are deeply interrelated so much so that you cannot separate one from the other. The two interrelated entities (LLLEC and STEAM skills) are required to effectively assess and address authentic and contextualized problems. It would be impossible to address problems related to big data and computational concepts, for example, without making good use of the right mix of skills. Scientists, technologists, engineers, social scientists, historians, businessmen, mathematicians, and physicists use critical thinking, communication, decision-making strategies, and technology to solve problems. Developing a mathematician, engineering or historian mindset allows students to think, act, and react creatively and scientifically.

The "A" in the Arts of the STEAM approach allows student to enhance their creativity while visually and artistically articulating and communicating their interpretation of the data and the solution of problems. STEAM, as an interdisciplinary pedagogical approach, allows students to use arts, humanities and social sciences knowledge creatively while dealing with STEM subjects. The idea is how to make a link between arts and STEM subjects. Art is how to make learning more exciting and motivational. The function of art is to make STEM subjects and knowledge more engaging. Understanding and adopting the mindset of a scientist provides practitioners with ample opportunities to observe phenomena, construct and reconstruct, and test hypotheses that bring about viable solutions to community problems.

The "**E**" refers to the engineering skills that students need to acquire, practice and demonstrate. For students to develop the engineering habits of mind, they need to practice system and design thinking. The STEAM approach allows students to adapt an engineering mindset while assessing and addressing life-contextualized problems. The engineering mindset provides individuals with abilities to identify and define the background of the problem and apply the knowledge, skills and computational thinking they learned collaboratively to develop a blueprint for designing, implementing the best and well-thought-out solution. Computational thinking and analytical skills allow students to maximize the probability of choosing the best solutions. It helps them think, act and react to the proposed solution 'recursively' [2, 3]. Also, computational thinking enables students to reasonably consider multiple scenarios in solving a problem. It is this skill that allows students to think like a computer scientist. Computational thinking provides its users with the ability to make good use of relevant big data to back track their steps.

The question that always needs to be addressed is how to simplify computer science terminologies in such a way that makes them part of our daily life and vocabulary. In other words, computational thinking, which traditionally belongs to the computer science field, allows us to understand and explore the patterns of behaviors that we adopt and adapt while solving daily complex problems. Computational thinking, which the STEAM approach promotes, is changing the way historians, chemists, biologists, physicists, businessmen, and economists think. This type of thinking allows students to consider a multiplicity of world views and levels of abstraction while solving problems. It is the arts and social sciences knowledge that enables students to imaginatively and innovatively tackle and solve problems. Computational thinking also enables students to make good use of mathematical and engineering thinking. The STEAM approach allows students to identify patterns and variables that generate solutions for contextualized problems. It enables them to collect logically relevant data, analyze and identify patterns and codes to understand problems.

The truth of the matter is that computational thinking allows students to practice a logical process and a set of procedural instructions in addressing a problem. Computational thinking is more inclusive in solving problems as it includes all the components of the STEAM approach. This ability allows students to learn how to synthesize information, connect patterns, examine context and its background, and find solutions. Through the acquisition of computational thinking, students can learn how to visualize the design of practical solutions, move from abstract to concrete, develop sketches, drills, models, prototypes and various scenarios, generate interdisciplinary ideas and solutions, and test assumptions reflectively.

Acquiring and implementing successfully this type of design thinking allows students to be more resilient, resourceful, reflective, open-minded, ethical, curious, pragmatic, analytical, evaluative, realistic and effective team players. Adopting a scientific attitude while addressing problems is essential for bringing about viable solutions. Using the available digital tools to communicate solutions of authentic problems enables practitioners of the right mix of skills in articulating abstract ideas. Effective integration of the right mix of skills enables students to acquire and demonstrate the ability to design, innovate, create, communicate, lead, analyze, scientifically investigate, formulate problems, represent solutions, collaborate, design prototypes and models, and solve problems. Essentially, the acquisition of this mix of skills helps college students to develop entrepreneurial mindsets that generate innovative solutions conducive to local, national and international communities. Developing the right mix of skills allows students to adopt the mentality of researchers who can analytically and systematically identify patterns that help in interpreting data.

4 Pedagogical Perspective Towards STEAM and LLLEC

The STEAM approach to learning identifies the required knowledge, skills, and attitudes that HE students, regardless of their majors, use to identify, understand, and solve cross disciplinary community problems. STEAM, as a contextual pedagogical approach, perceives various disciplines as interrelated. The pedagogical stance that the proposed conceptual framework promotes resonates with works by other scholars in this area (1), who have argued for a similar construction. I argue that a thematic interdisciplinary approach towards the implementation of STEAM and LLLEC stimulates student to explore contextualized problems that concern their communities and goes far beyond the traditional approach that teaches these subjects separately. Through the proposed conceptual framework, I argue that the proposed unconventional pedagogical approach allows students to collectively get together and reflectively connect and share concepts.

This stance articulates a vision of what an integrative and transdisciplinary approach to skills development can do to promote students interdisciplinary understanding of the various aspects of skills learning. An integrated approach to STEAM and LLLEC development enriches the environment where students' local and global insights can develop while authentically practicing technological, mathematical, and computational thinking within a community of practice. Pedagogically speaking, an interdisciplinary approach to skills development plays a positive role in providing STEAM and LLLEC practitioners with challenging thinking tools that support and enhance connections across disciplines, meaning making and knowledge creation which are all vital for problematizing community issues in such a way that enables students to develop deductive and inductive reasoning necessary for creatively solving real life problems.

Constructivist—oriented pedagogical approaches such as learning-by-doing and inquiry-based learning guide STEAM and LLLEC practitioners during the learning process and allow them to perceive the world through the lenses of the intersubjectively and interdisciplinary constructed worlds they grow up in. The core of the integrative stance is to engage college students in interdisciplinary interactions with what is unfamiliar to them. The insight I promote here is that an interdisciplinary thematic approach stimulates students to have the likelihood of thoughtfully making learning more contextually meaningful and impactful while developing the acquired thinking and design skills that prepare them for their future work and life. The integration of STEAM and LLLEC across the curriculum through thematic assignments and projects represents the main components of a competency- based approach (CBA).

4.1 Principles of the CBA

The proposed conceptual framework reflects a competency-based approach (CBA) to teaching and learning. The main principles of the CBA emphasizes that:

- The vision, mission and institutional learning outcomes of HEIs need to reflect an embedment of the CBA concepts across curricula and extra curricula of all disciplines.
- Program specifications and learning outcomes should include a statement specifying that all students regardless of their majors are required to practice STEAM and LLLEC through course assignments, projects, and other activities.

- The identified STEAM and LLLEC need to be aligned with the program learning outcomes to facilitate the assessment process of the targeted competencies.
- Course assignments and activities should reflect this alignment process. For example, students' collaborative projects could be aligned with skills related to the development STEAM and LLLEC.
- Assessment of students' performance in these skills need to be reflected at the course, program and institutional levels.

5 Assessing STEAM and LLLEC

One way to assess students' acquisition of STEAM and LLLEC is through peers and self- assessment. Through these two types of assessment, students independently can collect, analyze, synthesize, revise, evaluate, reflect on and improve their own accomplished artifacts that represent their understanding of what it means to develop and demonstrate STEAM and LLLEC knowledge, skills and attitudes (KSAs). Through the self-assessment processes, students will have the opportunity to identify their strengths, weaknesses and the type of intervention strategies that they need to develop to improve their performance. Self-assessment gives students the ability to identify the resources that help them achieve their targeted goals; meanwhile, peer assessment helps them in developing a mechanism to monitor the results of the intervention improvement plan to further identify the KSAs that they need to master.

Self and peer assessment of learning engage learners of STEAM and LLLEC in a process of self-regulating and self-personalizing their learning progress. During this process, students measure their performance against a set of key performance indicators (KPIs) which reflect specific quality criteria. Also, students' achievement of the stated institutional learning outcomes (STEAM and LLLEC) could be formatively assessed through assignments, projects, exams and e.Portfolio development.

Electronic Portfoilos hold a significant potential for digitally assessing students' competence development over their schooling years. As digital learning tools, students' e.Portfolios make the acquisition of STEAM and LLLEC more visible. They give students the opportunity to reflect on their academic, professional, social and entrepreneurial achievements. Well-developed and competence based e.Portfolios help students create and demonstrate an understanding of who they are and what they want to be. E.Portfolios represent evidence-based assessment products that allow students to artistically document and archive their inquiry personalized modes of learning through which they address contextualized problems. They also showcase the type of societal transformation that students resiliently look to achieve. Guided by the constructivist philosophy, e.Portfolios function as real evidences of the extent to which learners have become STEAM and LLLEC professionals. This evidence is important for students' future career and employability. In addition, students' e.Portfolios mark the institution's commitment to using STEAM and LLLEC development as enablers to develop highly responsive and responsible citizens who can participate in solving community problems.

The stance that this paper promotes is that when STEAM and LLLEC educators lead their students to integrate interdisciplinary concepts while solving contextualized problems, they would have the likelihood of entering a desired zone of understanding and interpretation where conflicting disciplines converge and interact, differences are recognized, and the prospective for connection can flourish. In other words, interdisciplinary interactions, which STEAM and LLLEC students practice in an engaging and constructivist learning environment, opens a desired captivating horizon for them to establish inspiring concepts that enhance the process of solving community problems for bringing about positive societal transformation. As such, the evolving interdisciplinary and communicative discourse that results in integrating STEAM and LLLEC across the curriculum opens new possibilities for exploring, analyzing and solving community problems. When students are guided to document this shared horizon of aligning interdisciplinary skills, values and attitudes, they have the aptitude to develop an enlightening mode of responsiveness towards contextualized and world crises.

After all, assessing students' academic and professional performance in applying STEAM and LLLEC in life through e.Portfolio development is one form of assessment. Other formative forms of assessment would have the potential to enhance students' development and demonstration of these skills. Engaging students in developing inquiry projects and activities allows them to practice these skills while measuring their competence level in practice.

Conclusively, the discourse of embedding STEAM and LLLEC across higher education curriculum speaks to the concept of knowledge that transcends the artificial boundaries that cut across disciplines. In this respect, this paper provides wealth of insights into the significance of understanding that the knowledge, skills and attitudes we intersubjectively and interdisciplinary construct and reconstruct are deeply rooted in the human experiences that we encounter in our daily life. It is vital to acknowledge that the perceived skill gap will widen soon due to automation, digitalization and globalization. Also, due to the advances in digital technology, the job market is changing fast, and employers are seeking new entry level employees with STEAM and LLLEC. So, promoting lifelong employability through a combination of lifelong learning skills and STEAM skills would be inevitable.

References

1. UKCES Bosworth.: The Supply of and Demand for High-level STEM Skills. Ukces 1–24 (2011)
2. Ng, B.: Exploring STEM competences for the 21st century. In-Progress Reflection **30**, 1–53 (2019)
3. Asunda, P.: A conceptual framework for STEM integration into curriculum through career and technical education. J. STEM Teach. Educ. **49**(1) (2014)
4. Seikkula-Leino, J., Salomaa, M.: Entrepreneurial competencies and organisational change-assessing entrepreneurial staff competencies within higher education institutions. Sustain. (Switz.) **12**(18) (2020).

5. Kakouris, A., Molina, V., Liargovas, P.: Assessing the impact of entrepreneurship education on entrepreneurial beliefs and conceptualizations. Proceedings of the European Conference on Innovation and Entrepreneurship, ECIE, 2020-Sept, pp. 320–329 (2020)

Awareness of Sustainable Development Goals Among Students from an Omani Business College

Bashir Ahmad Fida, Umar Ahmed, AlFadhal Al-Maskari, and Sauda Al Marhoobi

Abstract This paper attempts to assess the awareness level, formal and informal means of information, and learning level through classrooms regarding the Social Development Goals (SDGs) of students of a business college. A structured questionnaire was constructed and distributed to students of various study levels. The results reveal that only 42% of respondents received information about the SDGs through training, formal education, and traditional and social media. Similarly, a considerable number of respondents showed indifference or limited access to education or training towards the SDGs at their educational place. This implied an active role of HEIs in promoting awareness about SDGs among students and providing them with the necessary education and training to achieve Vision 2040 and SDGs.

Keywords Awareness · Sustainable development goals · Students · Omani business college

1 Introduction

The Sustainable Development Goals (SDGs) are 17 global goals established by the United Nations, New York, in 2015. They are designed to reduce poverty, protect the planet, and ensure peace and prosperity for all. These goals, often referred to as the "5Ps," Have replaced the Millennium Development Goals (MDGs) to address previously unattainable targets and cover all areas. The SDGs are universal, indivisible, and applicable to both developing and developed countries. With 169 specific targets, they cover a wide range of areas, including peace, security, gender equality, environment, and education [1, 2].

The SDG goal 4 is related to education. The opportunity to provide education to all and emphasize the quality of education are essential components of the goal. The role of higher education in achieving the SDGs is critical, and many universities and

B. A. Fida (✉) · U. Ahmed · A. Al-Maskari · S. Al Marhoobi
Business and Economics Department, Modern College of Business and Science, Muscat, Oman
e-mail: bashir@mcbs.edu.om

B. Awwad (ed.), *AI and IoT: Driving Business Success and Sustainability in the Digital Age*, Studies in Systems, Decision and Control 590,
https://doi.org/10.1007/978-3-031-88874-8_63

colleges worldwide are actively integrating the SDGs into their curricula, research, and community engagement activities, fostering a sense of global connection. These goals can only be achieved with the involvement of relevant stakeholders.

Research studies have continuously identified the vital role of universities, business colleges, and other institutions in promoting awareness about the SDGs among students, faculty, and the community. This can be done through offering relevant study programs, incorporating relevant learning materials, and equipping students with innovative and sustainable ideas and skills needed to achieve these goals [3, 4].

Countries have shown a slower pace for adoption of SDGs than expected. Oman, being an exception, has embedded SDGs into the Vision 2040 framework and has identified achievable indicators with follow-up units in place. In achieving all 17 SDGs, assessing awareness among all stakeholders of HEIs, particularly among students, is considered a significant step. In this regard, the level of awareness is assessed through the extent to which relevant information has been conveyed through curricula, formal and informal education and training, and academic and non-academic activities. Such efforts are particularly significant for SDG 4, which focuses on the importance of inclusive and equitable quality education and lifelong learning opportunities for all [4, 5].

This study aims to assess the self-reported understanding of students, the sources through which they obtain information, learning experience, and incorporation of material on SDGS in a business college in Oman. As a quantitative inquiry, the study is expected to contribute to Oman's *Vision 2040*, which aligns with the SDGs and serves as a roadmap for a sustainable future.

2 Literature Review

Sustainable development, grounded in the principle of intergenerational equity, is defined as development that meets the needs of the present without compromising the ability of future generations to meet their own needs [6]. As a global priority, it has led to the formulation of the Sustainable Development Goals (SDGs), a comprehensive framework of targets aimed at eradicating poverty, protecting the environment, and promoting peace and security. The successful realization of these goals necessitates coordinated efforts from governments, businesses, and civil society organizations [7].

Sustainability is founded on three interconnected dimensions: environmental, social, and economic. These pillars collectively serve as the basis for addressing global challenges, as outlined in the 17 Sustainable Development Goals (SDGs), which provide a comprehensive framework to achieve sustainable growth and economic development [8]. Social Development Goal one is about the elimination of poverty in all forms; SDG Number two is about ending hunger, provision of food security and nutrition and enhancing sustainability in the agriculture sector. SDG No. 3 talks about the health sector, while SDG No. 4 focuses on education and promoting equal chances for education. Likewise, other SDG goals cover entire human and non-human lives, spanning sustainable water creation, gender equality, women's empowerment, clean energy, fostering innovation, production processes, climate change, managing forests, and many other areas [6].

The Sustainable Development Goals (SDGs) provide a clear and accessible framework for addressing complex sustainability challenges. Despite this, global awareness of the SDGs remains limited, with fewer than half of respondents recognizing them. Priorities among the SDGs vary, with climate action, quality education, and good health and well-being consistently ranked as top concerns, though regional differences are evident. Younger respondents tend to prioritize climate action, while older individuals place greater emphasis on broader environmental issues. Notably, gender disparities are apparent in perceptions of gender equality, with women exhibiting higher levels of awareness and concern compared to men [9].

Yuan et al. [4] surveyed senior high school students in China to assess their cognition, motivation, and readiness regarding the Sustainable Development Goals (SDGs). The study found that students had limited knowledge of the SDGs and few information sources. No significant gender differences were observed in terms of knowledge, learning environments, or the impact on personal life and career planning. While students showed basic awareness of the SDGs, their primary information sources were formal education and traditional media, highlighting the need for improved SDG engagement and awareness initiatives.

3 Research Method

This study employs a quantitative research approach, utilizing a survey to achieve its objectives. A structured questionnaire, adopted from Zamora-Polo [7], serves as the research instrument. The target population comprises business students from a higher education institution in Oman. A convenience sampling technique was employed, with questionnaires distributed through students' contacts. A total of 200 questionnaires were distributed to business students across various disciplines, yielding a response rate of 80%. The questionnaire is divided into three sections: Section One Captures the demographic profile of the respondents. Section Two: Examines higher education students' perceptions regarding the United Nations Sustainable Development Goals (SDGs). Section Three: Provides space for respondents to offer suggestions. The questionnaire was designed using Microsoft Forms and distributed via email and social media platforms, including WhatsApp. Descriptive statistics, such as frequencies, means, and standard deviations, were used to analyze the collected data.

4 Findings

4.1 Demographic Background of the Respondents

The data presented in Table 1 reveals several demographic characteristics of the respondents. A majority (75%) of the participants were female, while only 25% were male. In terms of age distribution, more than half (57%) fell within the 21–23 age bracket, followed by 31% who were below the age of 20, and 11% who were above 24 years old. The respondents were evenly distributed across various business disciplines: 26% specialized in accounting, economics, and finance, while another 26% represented aviation, logistics, transport, health, and safety majors. Additionally, 25% were from management and law disciplines, and 23% were from computer science and mathematics. Regarding academic standing, almost half (50%) of the respondents were at the sophomore level, while 27% were freshmen, 18% were juniors, and only 6% were seniors. Furthermore, the majority (74%) were enrolled in bachelor's degree programs, with 22% pursuing associate degrees and a small proportion (4%) engaged in postgraduate or foundation programs.

Table 1 Demographic profile of the respondents

Demographic Information		Frequency n = 162	Percentage
Gender	Male	40	25
	Female	122	75
Age	20 & below	50	31
	21–23	93	57
	24–26	12	7
	Above 26	7	4
Majors	Accounting, economics and finance	43	26
	Management and law	41	25
	Computer science and mathematics	35	23
	Aviation, logistic, health and safety	43	26
Student level	Freshmen	44	27
	Junior	29	18
	Sophomore	80	49
	Senior	9	6
Education level	Bachelor's degree	119	74
	Associate degree	36	22
	Postgraduate and Foundation	6	4

4.2 Descriptive Analysis of Students Awareness on Sustainable Development Goals

Table 2 highlights a mixed level of awareness regarding the Sustainable Development Goals (SDGs). While 46% of respondents reported some familiarity with the SDGs, 14% indicated no knowledge, and 40% expressed indifference. Awareness of the SDGs' 2015–2030 implementation timeframe is similarly limited, with only 40% demonstrating awareness, 23% unaware, and 37% indifferent. Additionally, although 40% recognize that all UN member states are obligated to achieve the SDGs, 36% remain indifferent to this aspect. In terms of specific knowledge of the SDGs and their targets, 47% are familiar with all goals and targets, while 19% are unaware, and 34% are indifferent. These findings underscore the need for enhanced public education and engagement strategies to improve awareness and understanding of the SDGs.

Table 2 illustrates a diverse pattern in students' primary sources of information regarding the Sustainable Development Goals (SDGs). The bold numbers in the table 2 shows the evarge of all components of the criterion. Notably, 38% of respondents reported gaining knowledge through informal channels, such as workshops and sustainability-related activities. However, a significant proportion (62%) indicated no exposure to such informal avenues.

When considering formal education, 37% of respondents acquired information about SDGs through structured academic instruction. In contrast, 23% reported no exposure to SDGs within formal educational settings, and 40% expressed indifference towards the topic. Traditional media, including television, radio, and print, served as a source of SDG awareness for 44% of the respondents. Similarly, social media platforms, such as Facebook and Instagram, contributed to SDG information for 48% of participants.

The results presented in Table 2 highlight the limited integration of Sustainable Development Goals (SDGs) into higher education curricula, despite their significance as a global agenda with a defined timeframe. The findings reveal significant gaps in embedding SDG-related content into academic programs.

- **SDG 1 (Poverty Eradication)**: Only 31% of respondents indicated that their courses addressed poverty eradication, while the majority (69%) were either indifferent or unaware of such content in their curriculum.
- **SDG 2 (Zero Hunger)**: Twenty-five percent of respondents reported that their courses provided knowledge and skills to combat hunger. In contrast, 75% were either indifferent or unaware of related content.
- **SDG 3 (Good Health and Well-being)**: A relatively higher proportion (60%) noted that their courses taught health and wellness concepts, while 40% were indifferent or unaware.
- **SDG 4 (Quality Education)**: Fifty-five percent of respondents acknowledged being taught about inclusive education and student-centered learning, whereas 45% remained indifferent or uninformed.

Table 2 Descriptive analysis of students awareness on sustainable development goals

Criterion	Statement on students awareness on SDGs	Disagree to somewhat strongly disagree	Neutral	Agree to somewhat strongly agree
Knowledge 1	I know the number of SDGs and could indicate one of their goals	14	40	46
Knowledge 2	I am aware of the time framework for achieving the sustainable development goals	23	37	40
Knowledge 3	I am aware of the SDGs at all UN member states	25	36	39
Knowledge 4	I am familiar with all the sustainable development goals	19	34	47
Knowledge	**Average score of knowledge criteria**	**20**	**37**	**43**
Source Info. 1	I have gained knowledge about SDGs through institutional workshops and activities	30	32	38
Source Info. 2	I have received information about SDGS in formal education	23	40	37
Source Info.3	I have learned about SDGs through traditional media channels, including the press, radio, and television	24	32	44
Source Info.4	I have received information about SDGs through social media platforms, such as Instagram, Facebook, and others	19	33	48
Source of Info	**Average of source of information criteria**	**24**	**34**	**42**
Learning SDG1	College courses provided me with knowledge on poverty eradication	42	27	31
Learning SDG2	College courses equipped me with the knowledge and skills to combat hunger	44	31	25
Learning SDG3	College courses provided me with information on healthcare and wellness	30	10	60
Learning SDG4	Through college courses, I learned about the importance of inclusive education and student-centered learning."	18	27	55
Learning SDG5	College courses equipped me with the knowledge and skills to advocate for gender equality	11	33	56
Learning SDG6	College courses trained me in implementing sustainable water management practices	34	37	29
Learning SDG7	College courses provided me with information on clean energy	29	36	35

(continued)

Table 2 (continued)

Criterion	Statement on students awareness on SDGs	Disagree to somewhat strongly disagree	Neutral	Agree to somewhat strongly agree
Learning SDG8	Through college courses, I learned about labor rights, fair wages, and safe working conditions	20	34	46
Learning SDG9	Through college courses, I learned about the importance of sustainable infrastructure development	21	37	42
Learning SDG10	College courses equipped me with the knowledge and skills to promote equity and reduce inequalities	28	37	35
SDGS	**Average score of SDGs Criteria**	**28**	**31**	**41**

- **SDG 5 (Gender Equality)**: More than half (56%) reported that their courses equipped them to advocate for gender equality, while 44% were either indifferent or uninformed.
- **SDG 6 (Clean Water and Sanitation)**: Only 29% of respondents stated that their courses addressed sanitation and sustainable water management practices, while a vast majority (81%) were indifferent or unaware.
- **SDG 7 (Affordable and Clean Energy)**: Thirty-five percent of respondents acknowledged learning about energy affordability, accessibility, and sustainability, whereas 65% expressed indifference or unawareness.
- **SDG 8 (Decent Work and Economic Growth)**: Forty-six percent of respondents reported being taught about labor rights, fair wages, and safe working conditions, while 54% were indifferent or unaware.
- **SDG 9 (Industry, Innovation, and Infrastructure)**: Forty-two percent stated their courses covered sustainable infrastructure development, whereas 58% were indifferent or uninformed.
- **SDG 10 (Reduced Inequalities)**: Thirty-five percent of respondents noted that their courses provided knowledge and skills to promote equity and reduce inequalities, while 65% were either indifferent or unaware.

5 Conclusion

The researcher shows an overall analysis that there's lack of investigations to SDG-related expertise among students, and only 42% accessing information through a blend of informal training, formal education, traditional media, and social media. A substantial proportion of students confirmed either indifference or a lack of access to such information, emphasizing the urgent need for more effective alertness and engagement initiatives.

Furthermore, the results emphasize the limited progress made by academia in incorporating SDGs into higher education curricula. The insufficient exposure to SDG principles within college courses points to a missed opportunity to encourage students with the knowledge and skills essential for addressing global sustainability challenges.

To recognise the ambitious goals of the 2030 Agenda, significant efforts are required to embed SDG principles into educational frameworks. As a foundation of societal transformation, education must take centre stage in fostering awareness, instilling values, and preparing students to contribute meaningfully to sustainable development.

References

1. Da Cal Seixas, S.R., de Moraes Hoefel, J.L.: Environmental Sustainability: Sustainable Development Goals and Human Rights. CRC Press, Boca Raton, FL, USA, (2021); Jeffrey, D.S.
2. Aziz, F., Tahir, F., Qureshi, N.A.: Millennium development goals (MDGs-2000–2015) to sustainable development goals (SDGs-2030): a chronological landscape of public sector health care segment of Pakistan. J. Pak. Med. Assoc. **71**, 596–601 (2021)
3. Mehmood, U.: Contribution of renewable energy towards environmental quality: the role of education to achieve sustainable development goals in G11 countries. Renew. Energy **178**, 600–607 (2021)
4. Yuan, X., Yu, L., Wu, H.: Awareness of sustainable development goals among students from a Chinese senior high school. Educ. Sci. **11**, 458 (2021). https://doi.org/10.3390/educsci11090458
5. UN.: Quality Education: Why It Matters. Available online: https://www.un.org/sustainabledevelopment/wp-content/uploads/2017/02/4_Why-It-Matters-2020.pdf
6. Halkos, G., Gkampoura, E.-C.: Where do we stand on the 17 sustainable development goals? an overview on progress. Econ. Anal. Policy **70**, 94–122 (2021). https://doi.org/10.1016/j.eap.2021.02.001
7. Zamora-Polo, F., Sánchez-Martín, J., Corrales-Serrano, M., Espejo-Antúnez, L.: What do university students know about sustainable development goals? A realistic approach to the reception of this UN program amongst the youth population. Sustainability **11**(13), 3533 (2019). https://doi.org/10.3390/su11133533
8. Desa, U.N.: Transforming Our World: The 2030 Agenda for Sustainable Development (2016). Available online: https://stg-wedocs.unep.org/bitstream/handle/20.500.11822/11125/unepswiosm1inf7sdg.pdf?sequence=1
9. Global Survey.: Report of Results Global Survey on Sustainability and the SDGs Awareness, Priorities, Need for Action. https://www.globalsurvey-sdgs.com/wp-content/uploads/2020/01/20200205_SC_Global_Survey_Result-Report_english_final.pdf

Harnessing the Power of Swarm Intelligence in IoT: Optimization Techniques and Future Directions

Hothefa Shaker Jassim, Rami J. Oweis, Khalid Shaker, and Arwa Alqudsi

Abstract The Internet of Things (IoT) has been a major significant trend in recent years, and it is the explosive growth of devices connected and controlled by the Internet. It has been widely used and is an integral part of people's lives where it has been applied in almost all fields including home, building automation, customer market, healthcare, education, transportation, agriculture, and others. However, various serious challenges and limitations in the growth of IoT need to be solved. Thus, Artificial Intelligence (AI) has been applied to reduce IoT issues and optimize its challenges. One of the main AI solutions used in IoT is Swarm Intelligence (SI) where natural biological algorithms are used the inspire natural beings like bird flocking, ant colonies, hawks hunting, bacterial growth, animal herding, fish schooling, and microbial intelligence. This paper discusses an overview of the Internet of Things and reports a brief aspects of SI algorithms and and how they are adapted to enhance some IoT challenges. This review will guide future studies, giving a clearer way to help determine the suitable SI-based algorithm for IoT-based systems.

Keywords Internet of Things · Swarm intelligence · Optimization techniques · Bio-inspired algorithms

---

H. S. Jassim (✉) · R. J. Oweis
Modern College of Business and Science, Muscat, Oman
e-mail: hothefa.shaker@mcbs.edu.om

R. J. Oweis
Jordan University of Science and Technology, Irbid, Jordan

K. Shaker · A. Alqudsi
Department of Information Systems, College of Computer Sciences and Information Technology, University of Anbar, Ramadi, Iraq

B. Awwad (ed.), *AI and IoT: Driving Business Success and Sustainability in the Digital Age*, Studies in Systems, Decision and Control 590,
https://doi.org/10.1007/978-3-031-88874-8_64

1 Introduction

The Internet of Things (IoT) integrates a worldwide network of interconnected physical objects such as sensors, actuators, machines, and smart devices that communicate based on standard protocols. This smart device impacts human daily life through automation, intelligence, and smart transportation. Recently, IoT applications, such as healthcare, smart grids, smart homes, online item tracking, and environmental monitoring, are rapidly evolving. These innovations generate massive amounts of data, requiring extensive computing power, storage capacity, and communication bandwidth [1]. Cisco, in 2020 estimated numerous wearable devices, smart meters, wireless sensors, connected vehicles, and other smart devices will be 50 billion connected to the internet and it will increase to 500 billion by 2025. Therefore, a large IoT platform requires intelligent computational capabilities to process a bunch patch of data generated by both fixed and mobile smart devices that cover a wide geographic area and that are related to heterogeneous domains and several use cases. Consequently, new data-driven (IoT) services are raising problems related to supporting a high degree of mobility and real-time data analysis with decision-making abilities as well as low latency [2, 3]. Hence, Artificial intelligence (AI) helps to make sense of this data with its capabilities of analyzing, optimizing, machine learning, and making a decision. AI is a key assistant to the growth of the IoT revolution and takes it to a new level. AI and IoT are combining to provide greater visibility, reliability, real-time monitoring, and control of the broad array of devices and sensors connected to the internet. It uses several techniques: Metaheuristic, Machine Learning, Fuzzy Model, and Probabilistic Model. However, Swarm Intelligence (SI) is one of the Population-based methods under the Metaheuristic category which is currently widely used to optimize the IoT challenges and meet its requirements [4]. The rest of this paper is divided into 3 main sections. Section 2 gives an overview of IoT, SI algorithms. Section 3 will present SI models. The fourth section merges the IoT and SI in a way that explains how SI algorithms optimize some IoT challenges.

2 Related Work

2.1 Internet of Things IoT

The Internet of Things (IoT) is a new revolution of the Internet. This concept was invented in 1999 by a member of the Radio Frequency Identification (RFID) development community. Recently, It has become increasingly relevant to the practical world with the rise of mobile devices, cloud computing, data analytics, and universal communication [5]. IoT is a concept and a paradigm consisting of various things/objects connected through wireless and wired connections. All these objects are prepared computing devices, digital and mechanical devices, objects, or persons that are provided with unique identifiers and can move information over the network without requiring communication.

2.2 *Swarm Intelligence*

Swarm intelligence is a subset of artificial intelligence that is considered a bio-inspired computation that was presented for the first time in 1989 by Jing Wang and Gerardo Beni in the context of developing the systems of cellular robotics [6]. The development of swarm intelligence algorithms is driven by the collective actions of natural swarms as they are vastly utilized in addressing problems related to optimization. SI is a self-organized and non-centralized system. SI-based algorithms have represented one of the nature-inspired groups that can produce strong, cost-effective, and less time solutions to many intricate optimization problems [7].

Swarm intelligence (SI) is acquiring significance progressively, as more as high intricate problems require solutions that might not be the best optimal solution, but they can be achieved within a sensible period. The capability of adaptability and independent learning to extrinsic divergences are the main features offered by the algorithms which have enticed massive interest and identified many fields of application. Recently, the popularity of SI has noticeably increased with the increase of the complex problem where getting the global optimal becomes almost unattainable in a real-time scenario [8]. The growing popularity of such SI-based algorithms was because of several reasons, but the most vital reason is the ability to adapt and be flexible in SI-based algorithms. SI is an important optimization technique category that is inspired by simple behaviors of interacted agents such as ant colonies foraging, animal herding, bacterial growth, honeybees, fish schooling, bird flocking, and so on. The illustration of the chart of bio-inspired algorithms classification is presented in Fig. 1.

The specific behaviors of these Bio-inspired SI-based algorithms in finding a solution have a variety of support for getting the best solutions, therefore the performance of each SI algorithm is varied for different problems. The delineation of the overarching conceptual framework of Swarm Intelligence is clarified in Fig. 2 through a comprehensive flowchart.

3 Swarm Intelligence Based Internet of Things (IoT)

Many researchers have introduced several SI algorithm-based IoT applications. However, Table 1 summarizes the most recent reviewed works. It is composed of five columns used to show: the algorithm on which the specific work is SI algorithm-based, the name of the proposed algorithm (Mechanism), the aim of the research (Aim), the type of experimentations (Experimentation), and the related works taken from (Reference).

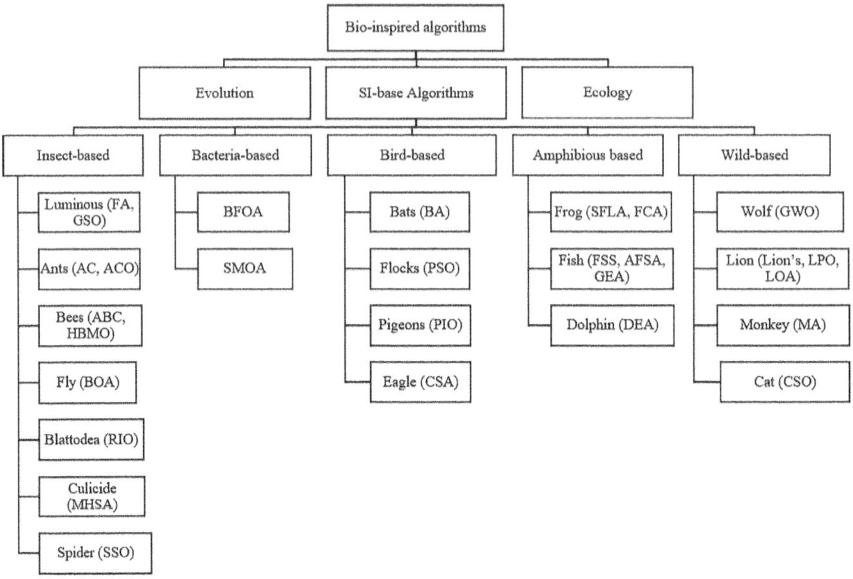

Fig. 1 Categorization of bio-inspired algorithms

Fig. 2 Swarm intelligence flowchart

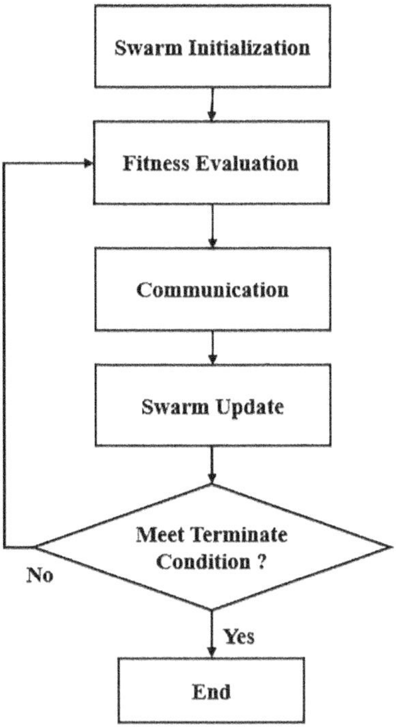

Table 1 SI-based IoT application

SI algorithm	Mechanism	Aim	Experimental	References
ABC	ABC-DC	Develop an energy-efficient clustering mechanism in data transmission	MATLAB	[9]
	RABC	Solving energy consumption for the travel path and uplink route	–	[10]
	ABC-GSA	Enhance the lifetime of IoT nodes	–	[11]
	RST-IoT	Providing an efficient and robust scheme for data gathering	MATLAB	[12]
	HABCA-EST	Enhancing the lifetime of a wireless smart devices network for target coverage application	MATLAB	[13]
	H-MOABC	Reduce the RNP optimization complexity	Cd100 & Rd500	[14]
	ABC System architecture	Address the challenges faced by a massive volume of data	MapReduce & Hadoop	[15]
BFO	WBFA	Maintain the utilize of power energy	MATLAB	[16]
	BFO-DV-HOP	Optimize the indoor positioning accuracy	MATLAB	[17]
	BFOA-R	Route optimization for resource constraint	NS-3	[18]
	OSEAP + IBFO	Saving and securing energy among nodes	MATLAB	[19]
	HBG	Balancing a load demand-side management in smart buildings	MATLAB	[20]
	RF-BFO-SVM	Identify the optimal parameters for the diagnosis of Parkinson's disease	MATLAB	[21]
BA	WHCBA	Enlarging the global search space	CEC2013	[22]
	BLDC	Provide flexible and long-distance connectivity	MATLAB	[23]

(continued)

Table 1 (continued)

SI algorithm	Mechanism	Aim	Experimental	References
FA	FFA-based SS	Substantially enhance the sensing time and utilization of IoT communication channels	BPSK-modulated LU signals	[24]
	FiCA	Providing a clustering approach for IoT applications	Java-based	[25]
GWO	GAB-GWO	Applying two new crossover and mutation operators to try to find the most relevant traffic features and eliminate worthless ones, to improve the LIDS performance	AWID	[26]
	FG-GWO	Multipath data transmission and higher performance of nodes energy and alive	MATLAB	[27]
	SVM-GWO	Characterize the exactness of infection utilizing the gained information	Heart disease dataset15	[28]
	MOGWO	Achieve efficiency and reduce the number of anchor nodes	Extensive simulations are carried out	[29]
	MOGWOLA	A node localization in smart parking with the consideration of distance and topological constraints	Extensive simulations are carried out	[30]
ACO	ARAH-IOT (Ant colony based routing algorithm)	Obtain the reliable routing path and reduce the network delay time, network overhead, packet loss, and consideration to other such primary factors	C#	[31]
	ACO-SS	Get an optimal solution quickly and efficiently	MATLAB	[32]
	Optimized Ant Colony Algorithm	Determine the best shortest path for remote hospitals data sharing	C++	[33]
	STDACO	Optimize the path and city traffic efficiently while avoiding congestion in all circumstances using up-to-date city traffic data	Otcl	[34]

(continued)

Table 1 (continued)

SI algorithm	Mechanism	Aim	Experimental	References
	Proposed algorithm based on the ACO	Address the IoT task allocation issues by distributing the capabilities of the task allocation to the best corresponding nodes	MATLAB	[35]
	Proposed algorithm based on the ACO	Improve cultural media communication efficiency	GNACA	[36]
PSO	SPOCC-HSA-PSO	Ensuring optimal routing with reduced energy consumption	MATLAB	[37]
	PIPSO	Speed up the optimization process	MATLAB	[38]
	SPSO	Cluster and secure data	NS-3	[39]
	FODPSO-FLC	Reliability, gives high speed, and accuracy to extract utmost power point hence, the power and speed fluctuations are minimized in MPPT	MATLAB	[40]
	ICPSO	Find the position optimal VM in the cloud environment with the minimum response time	MATLAB	[41]
	MST-PSO	Solve the problem of energy consumption and path cost of Zigbee cluster tree network in a smart home environment	MATLAB	[42]

This paper categorizes its findings of optimization works in IoT based on several SI algorithms into five general categories, which are resources, data, network, security and management. Figure 3 presents these findings of each category in general terms.

4 Analysis of SI Algorithms on IoT Applications

This subsection presents an analysis exploration of swarm intelligence-based IoT applications. According to a detailed study in Sect. 3 of SI algorithms and comprehensive literate review work in subsections 4.1, a comparison of SI techniques-based IoT applications that have recently been used to enhance, optimize, and improve IoT applications is displayed in Table 2. It is composed of ten columns and used nine columns to represent different IoT applications and algorithms used in these IoT

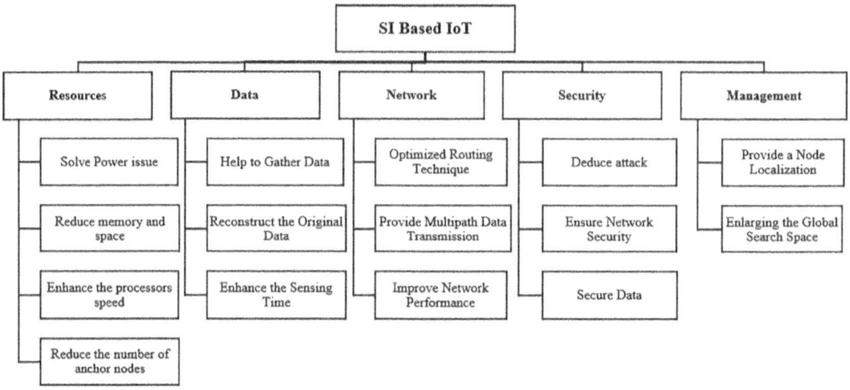

Fig. 3 The optimizing categories of SI-based IoT

Fields (Algorithm). It illustrates that ACO, PSO, and GWO are widely applied in IoT-based applications. Furthermore, FA and CSO need to take more consideration to be exploited. While FSS is not yet applied and much more work needs to be done on it. Thus, it must be avoided in IoT-based systems until it is clearer.

The presentation of Fig. 4 is the best range of SI practices for IoT-based applications. Wireless Sensor Network takes the lion's share in which all SI algorithms are applied to enhance it and optimize its challenges followed by the smart grid. Moreover, Smart City and Healthcare applications are implemented 5 different types of SI models. On the other hand, the connected car application is the least.

Table 2 SI algorithms based IoT applications

Application / Algorithm	Smart city	WSN	Smart Grid	Smart Building	Healthcare	Industrial	Connected Car	Smart retail	Smart Farming
ABC	✓	✓	✓	✓	✓				
BFO	✓	✓	✓	✓	✓				
BA		✓	✓			✓			
CSO		✓							
FA		✓	✓						
FSS									
BOA		✓	✓						
GWO	✓	✓		✓	✓	✓		✓	
ACO	✓	✓	✓		✓	✓	✓	✓	✓
PSO	✓	✓	✓	✓	✓				✓

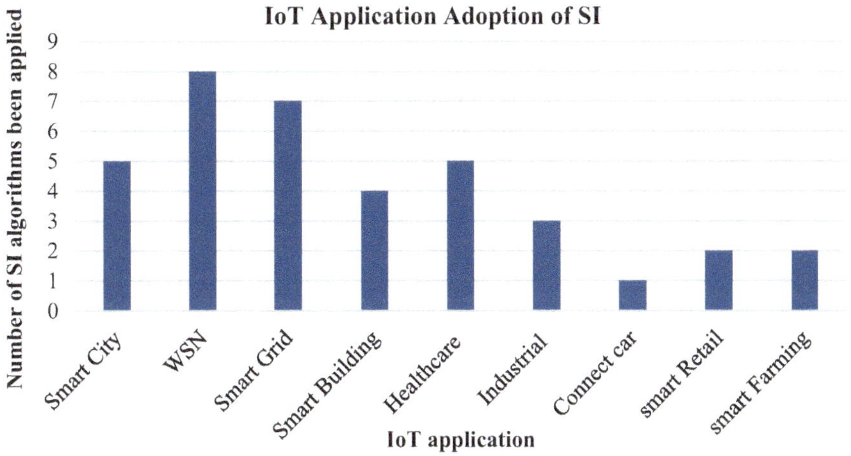

Fig. 4 IoT application adoption of SI

5 Future Directions

Through decentralized optimization and adaptive control, Swarm Intelligence (SI) in conjunction with the Internet of Things (IoT), also promising solutions in biomedical engineering and nanotechnology. The prospect of this notion for the future can be viewed in various directions. In biomedical engineering, SI enhances diagnostic precision, including MRI imaging, and optimizes IoT-enabled medical equipment for patient monitoring [43, 44]. SI also helps with procedures involving micro/nanorobots [45]. It is inevitable to mention that future approaches for SI integration with IoT systems also include developing real-time adaptive treatments, simulating biological systems, and enabling collaborating nanodevices for activities like illness detection and tissue healing leading to the improvement of scalability, efficiency, and creativity in healthcare [43, 44]. To accomplish its aims, SI utilizes its organization and flexibility. Similarly, SI makes it possible for nanotechnology applications, including energy harvesting, self-assembly of materials, and targeted medication delivery to be implemented through nanorobotics [45]. SI aids in the self-assembly of nanomaterials with features and directs nanorobotic swarms for targeted drug delivery in nanotechnology. Future advancements will concentrate on developing responsive materials for tissue engineering and drug delivery, as well as combining SI with IoT for personalized medicine [44].

6 Conclusion

The convergence of the Internet of Things (IoT) with Swarm Intelligence (SI) has proven to be a transformative approach to overcoming critical challenges in IoT systems, such as scalability, energy efficiency, and system optimization. This paper provided an overview of IoT, its limitations, and the role of SI-based algorithms, including Ant Colony Optimization, Particle Swarm Optimization, and Grey Wolf Optimization, in addressing these issues.

By leveraging bio-inspired algorithms, SI offers a robust, flexible, and scalable solution to many of the architectural and operational requirements of IoT. These techniques have demonstrated their ability to optimize resource management, enhance security protocols, and improve network performance across various applications, such as smart cities, healthcare, and industrial systems. Despite the advancements, several gaps and challenges remain, particularly in areas such as real-time adaptability, integration with emerging technologies, and security enhancements. Addressing these areas will be crucial for the continued evolution of SI-driven IoT solutions. SI-based methods can unlock the full potential of IoT systems, paving the way for a more connected, efficient, and intelligent future.

References

1. Nižetić, L.P., et al.: Internet of Things (IoT): opportunities , issues and challenges towards a smart and sustainable future. J. Clean. Prod. **274** (2020)
2. Noor, M., Hassan, W.H.: Current research on Internet of Things (IoT) security: a survey. Comput. Networks 283–294 (2019)
3. Mei, G., et al.: A survey of internet of things (IoT) for geo-hazards prevention: applications, technologies, and challenges. IEEE Internet Things J. **7**(5), 4371–4386 (2020)
4. Alrubaiei, M., et al.: Internet of things in cyber security scope. In: Blockchain Technology and Computational Excellence for Society 5.0, pp. 146–187. IGI Global (2022)
5. Al Hilali, R., et al. (2021). Blockchain technology's status of implementation in Oman: empirical study. Int. J. Comput. Digital Syst.
6. Zhou, B.Y., et al.: UAV swarm intelligence: recent advances and future trends. IEEE Access **8**(IEEE), 183856–183878 (2020)
7. Jumani, T.A., et al.: Swarm intelligence-based optimization techniques for dynamic response and power quality enhancement of AC microgrids: a comprehensive review. IEEE Access **8**, 75986–76001 (2020)
8. Chen, X., et al.: A survey of swarm intelligence techniques in VLSI routing problems. IEEE Access **8**, 26266–26292 (2020)
9. Zannou, A., et al.: A task allocation in IoT using ant colony optimization. Proceeding—2019 International Conference Intelligence System Advanced Computer Science ISACS (2019)
10. Yousefi, S., et al.: An energy-efficient artificial bee colony-based clustering in the internet of things R. Comput. Electr. Eng. **86** (2020)
11. Zhang, X.I.U., et al.: An energy efficient internet of things network using restart artificial bee colony and wireless power transfer. IEEE Access **7**, 12686–12695 (2020)
12. Kumar, P.R., et al.: Energy aware cluster head selection for maximizing lifetime improvement in internet of things. Int. J. Grid High Perform. Comput. **9**(4) (2017)

13. Najjar-ghabel, S., et al.: Reliable data gathering in the Internet of Things using artificial bee colony. Turkish J. Electr. Eng. Comput. Sci. **26**(4), 1710–1723 (2018)
14. Chouhan, D., et al.: Selection of optimal path for the communication of multimedia data in internet of things. In: 2020 6th International Conference on Advanced Computing and Communication Systems (ICACCS), pp. 477–481 (2020)
15. Murali, S., et al.: A lightweight intrusion detection for sybil attack under mobile RPL in the internet of things. IEEE Internet Things J. **7**(1), 379–388 (2019)
16. Muhammad, Z., et al.: Hybrid artificial bee colony algorithm for an energy efficient internet of things based on wireless sensor network. IETE Tech. Rev. **34**, 39–51 (2017)
17. Ma, L.X., et al.: Two-level master-slave RFID networks planning via hybrid multi-objective artificial bee colony optimizer. IEEE Trans. Syst. Man, Cybern. Syst. **99**, 1–20 (2017)
18. Hafeez, G., et al.: Efficient energy management of IoT-enabled smart homes under price-based demand response program in smart grid. Sensors **20** (2020)
19. Huang, H., et al.: An improved DV-HOP algorithm for indoor positioning based on bacterial foraging optimization. 2016 8th International Conference Wireless Communication Signal Processing, pp 1–5 (2016)
20. Rajesh, G., et al.: A multi-objective routing optimization using swarm intelligence in IoT networks. Intelligence Computer Innovation Data Science, pp 603–613 (2020)
21. Reddy, P.K.: An evolutionary secure energy efficient routing protocol in Internet of Things. Int. J. Intell. Eng. Syst. **10**(3), 337–346 (2017)
22. Khalid, A., et al.: Towards dynamic coordination among home appliances using multi-objective energy optimization for demand side management in smart buildings. IEEE Access **6**, 19509–19529 (2018)
23. Balamurugan, K., et al.: Mathematical Analysis of Bldc Motor Parameter Estimation Using Internet of Things (IoT), vol. 14, pp. 2429–2436 (2018)
24. Reda, H.T.: Firefly-inspired stochastic resonance for spectrum sensing in CR-based IoT communications. Neural Computing Application, no. 1–13 (2019)
25. Jabeur, N., et al.: Toward a bio-inspired adaptive spatial clustering approach for IoT applications. Futur. Gener. Comput. Syst.. Gener. Comput. Syst. **107**, 736–744 (2020)
26. Davahli, A., et al.: A lightweight anomaly detection model using SVM for WSNs in IoT through a hybrid feature selection algorithm based on GA **7**(1), 63–79 (2020)
27. Dhumane, A.V.: Fractional gravitational grey wolf optimization to multi-path data transmission in IoT. Wirel. Pers. Commun. (2018)
28. Gupta, S.: Employability of support vector machines with gray wolf optimisation (Svm-Gwo) algorithm to optimise data classification from IoT enabled devices to develop an. Int. J. Invent. Eng. Sci. Technol. (2), 186–194 (2016)
29. Ghorpade, S.N., et al.: GWO Model for Optimal Localization of IoT-Enabled Sensor Nodes in Smart Parking Systems, pp. 1–8 (2020)
30. Agnihotri, S., et al.: An enhanced routing algorithm for internet of things that resolves heterogeneity issues. Proceeding 2019 6th International Conference Computer Sustainable Global Development INDIACom 2019, pp. 542–547 (2019)
31. Zhang, C., et al.: A Multi-Objective Service Selection Method Based on Ant Colony Optimization for QoE Restrictions in the Internet of Things, vol. 284, no. 15. Springer International Publishing (2019)
32. Priyanka, P.: Ant colony optimization based routing in IoT for healthcare services. In: In 2018 Second International Conference on Intelligent Computing and Control Systems (ICICCS), no. Iciccs, pp. 155–1159 (2018)
33. Rehman, A., Rathore, M.M., Paul, A., Saeed, F., Ahmad, R.W.: Vehicular Traffic Optimisation and Even Distribution Using Ant Colony in Smart City Environment, pp. 1–8 (2018)
34. Zhang, Q.: Improvement of culture media efficiency in Internet of Things based on global numerical ant colony algorithm. Pers. Ubiquit. Comput.Ubiquit. Comput. **24**(3), 347–361 (2020)
35. Hemalatha, R., et al.: Analysis on energy consumption in smart grid WSN using path operator calculus centrality based HSA-PSO algorithm. Soft. Comput.Comput. **24**(14), 10771–10783 (2020)

36. Abdelaziz Mohamed, M., et al.: Modeling and simulation of smart grid integrated with hybrid renewable energy systems. Stud. Syst. Decis. Control **121**, 1–10 (2018)
37. Chandnani, N., et al.: A novel secure data aggregation in IoT using particle swarm optimization algorithm. Int. Conf. Adv. Comput. Telecommun. ICACAT **2018**, 1–6 (2018)
38. Jasmine Gnana Malar, A., et al.: IoT based sustainable wind green energy for smart cites using fuzzy logic based fractional order Darwinian particle swarm optimization. Meas. J. Int. Meas. Confed. **166**, 108208 (2020)
39. Shukr, R.M.: Efficient integer categorical particle swarm optimization algorithm and cloud IoT for health service **XIII**(Vi), 17–31 (2020)
40. Lana Fernando, S.: (2017) IoT : Smart Homeusing Zigbee Clustering Minimum Spanning Tree and Particle Swarm Optimization (MST-PSO) IoT: Smart Homeusing Zigbee Clustering Minimum Spanning Tree and Particle Swarm Optimization (MST-PSO), vol. 3, no. October, pp. 50–55
41. Liu, L., et al.: Swarm intelligence algorithms for medical and biological applications: a review. Expert Syst. Appl. **126**, 174–194 (2019)
42. Soni, M., Soni, A.: Swarm intelligence and its applications in biomedical systems. J. Med. Syst. **44**(6), 112 (2020)
43. Prasad, R., et al.: Nanotechnology in biomedical applications: a review. Biotechnol. Adv.. Adv. **37**(2), 189–204 (2019)
44. Biniwale, R., Chavan, R.: Nanotechnology for drug delivery and biomedical applications. BioNanoScience **8**(1), 11–23 (2018)
45. Yang, X., Yen, G.G.: Swarm intelligence for bioinformatics and nanomedicine: recent advances and future directions. Comput. Biol. Chem.. Biol. Chem. **63**, 8–16 (2016)

Enhancing Business Sustainability Through Digital Legal Compliance and Regulatory Frameworks

Ibrahim Yaro Gambo, Benson Ruzive, Siraj Kariyilaparambu Kunjumuhammed, and Hisham Kamil Madi

Abstract The era of rapid digital transformation has presented both opportunities and challenges for businesses. This development demands businesses to be legally complaint with the requirements for sustainability. Digitally legal compliance requires adherence to laws and regulations governing digital activities. Effective compliance ensures that businesses mitigate risks related to data breaches, cyber threats and illegal activities which are increasingly prevalent in the digital age. Long-term viability of businesses ventures is enhanced by a view that sustainability extends beyond environmental considerations to include economic and social dimensions. AI is required in the integration of digital legal compliance within the legal frameworks of businesses. This paper focuses on sustainable business practices as well as the intersection of digital legal compliance and the regular regulatory frameworks. Regulatory frameworks such as the General Data Protection Regulatory (GDPR) and the Digital Markets Act (DMA) are essential in promoting transparency, accountability and ethical behaviours among businesses. The study underscores the importance of adaptive compliance strategies and the integration of regulatory considerations into corporate governance models. The role of technological advancement such as blockchain and AI in enhancing compliance mechanism and fostering sustainable business is investigated. Findings of this research will enhance the proactivity of businesses in as far as digital legal compliance and adherence to stringent regulatory standards is concerned. The paper contributes to understanding how digital compliance and regulatory frameworks can serve as enablers of business sustainability, providing insights for policymakers, business leaders, and scholars on fostering resilient and ethically responsible digital economy.

Keywords Digital transformation · Digital challenges and opportunities · Digital legal compliance · Digital laws and regulations

I. Y. Gambo · B. Ruzive (✉) · S. K. Kunjumuhammed · H. K. Madi
Department of Business and Economics, Modern College of Business and Science, Muscat, Oman
e-mail: benson.ruzive@mcbs.edu.om

B. Awwad (ed.), *AI and IoT: Driving Business Success and Sustainability in the Digital Age*, Studies in Systems, Decision and Control 590,
https://doi.org/10.1007/978-3-031-88874-8_65

1 Introduction

Adelakun et al. [1] define digital economy as a system in which economic activities are aided by digital technologies and the use of internet. This development has led to researchers and policymakers re-visit the existing regulatory frameworks in digital economies. Artificial intelligence has brought about rapid changes in the digital age, but these changes have not been met with the same speed in reforming regulations that are aimed at controlling business operations. Pouri and Hilty [2] claim that one of the characteristics of digital economies is the extensive use of digital technologies in economic activities thereby revolutionizing traditional business economies leading to the emergence of new paradigms such as the circular economy and sharing economy. There is need for countries to collaborate in tax enforcement laws in the digital age to avoid and reduce tax evasion [1]. Bornman and Ramutumbu [3] point out that complying with tax laws is of greater importance due to several problems caused by digital economy to the traditional international tax system. There is greater need for those operating in digital economies to possess knowledge on tax requirements that ranges from general, procedural to legal tax knowledge [4]. Focus has been put on areas such as taxation, environmental protection and energy distribution to examine the impact of regulatory framework on digital economies [5]. Transformation of the digital economy posed great impact on both environmental and economic systems resulting in reformations of traditional approaches to the financial sector such as taxation systems [3]. There have been fundamental shifts in business strategies, consumer behavior, and regulatory frameworks thereby leading to re-examination of taxation approaches [6].

The components of regulatory framework include but are not limited to digital compliance management, regulatory frameworks, risk management, stakeholder engagement and training and development. The rise of digital transactions has called for regulations that are aimed at addressing compliance with tax laws and regulations including their enforcement and associated penalties [3]. Some of the challenges posed by digital transactions are that they span across state boundaries making it difficult to identify the jurisdictions under which they fall thereby creating loopholes that businesses exploit and evade tax. It is imperative that businesses comply with statutory regulations that will be prevailing at any moment [7]. A clear understanding of applicable regulations, codes of conduct and risk assessments is a must for corporate organizations if penalties and image tarnishing are to be avoided. Corporate organizations operate in the digital economy reduce risks of being found on the wrong side of the law by effectively implementing compliance management policies. Seeking legal advise if in doubt is the best practice as it ensure proper compliance with regulations that are applicable to a particular interrogation [7]. The process of management is synonymous with the traditional four pillars of management which are planning, organizing, leading and controlling [8]. Renz et al. [9] maintain that planning involves problem identification, decision-making, and coming up with explicit and measurable plan of action. The organization function of management involves dividing tasks and responsibilities, efficiently allocating resources, and team building

[7]. Leading refers to motivating team members and individuals to achieve organizational goals, while controlling part refers to measuring performance, monitoring results, evaluation of set goals and reviewing and giving feedback to stakeholders.

World Bank [10] defines law as a set of rules and principles that society recognizes and is applied by authority when regulating behavior and how individuals, organizations and countries relate. Law is aimed at maintaining social order, protecting individuals' rights and giving those individuals freedom of association while at the same time promoting justice and providing a clear method of resolving disputes. Hailbronner [11] opines that the application and understanding of law differs from one country to the other based on the prevailing legal systems. Smith [12] points out that every individual is bound to be subservient to a particular law and this includes corporate organizations because they are referred to as legal *persona*. Purba [13] argues that the increasingly complex and stringent legal challenges faced by organizations make it crucial for compliance with these laws.

This study aims at finding out the role of digital legal compliance in digital economies to be sustainable. To answer this question, the study seeks to (a) examine the impact of regulatory frameworks on digital transformation and business sustainability, (b) identify the challenges businesses face in integrating digital legal compliance into sustainable business models, (c) analyze the role of technology in streamlining digital legal compliance for sustainable business practices, and (d) propose digital compliance strategies that foster both legal adherence and sustainable growth in businesses. In addition, literature review will seek to identify benefits of digital legal compliance for sustainability.

2 Literature Review

2.1 Challenges and Considerations of Digital Legal Compliance

The study reviewed literature on digital laws and regulations challenges that companies face in their operations. The increased use of technologies and internet has exposed companies to several risks that always coming up. Proactive strategies are required to fight and thwart these risks as well as complying with regulations. Clarity on policies and procedures, employee training as well as regular risk assessment and audits are some of the strategies that help companies survive. Rakha [14] points out that rapid growth in digital technology exposes companies to new challenges. Navigating through the legal landscape to ensure compliance, risk mitigation and effective maintenance of corporate governance practices is essential [1]. Just like corporate social responsibility (CSR), corporate digital responsibility (CDR) demands explicit addressing regarding concerns that are raised by society [15, 16]. More than four decades ago, Kano et al. [17], pointed out that digital compliance is a must.

2.1.1 Data Privacy and Protection

Challenges of digital legal compliance and considerations have been highlighted as data privacy and security, complexity of regulations, resource constraints, and resistance to change [1]. Digital data is governed by stringent laws like the GDPR in the European continent and CCPA in California. Businesses are required to safeguard personal data, and non-compliance can lead to hefty fines. Prevention of unauthorized access to personal data can be achieved by heavy investment in robust data management systems and putting in place cybersecurity measures. Businesses must keep abreast of the digital regulations that are constantly evolving because of technological advancements as well as societal changes. Staying informed and embracing an agile position is a better strategy to be resilient and respond to these changes.

2.1.2 Cybersecurity

Businesses that handle sensitive data like finance, healthcare, government contracts and other forms of crucial data face threats of cyber security. Organizations must adhere to standards such as NIST or ISO 27 001 for mitigating cyber threats. Incidents must be reported, staff must be trained, data security audits must be conducted on a regular basis and data encryption practices must be put in place. These actions do not only meet compliance requirements, but they promote resilience. Maintaining compliance with data protection laws is crucial and this may require putting in place robust cybersecurity measures, regularly training employees and conducting regular audits. It is vital for companies to be proactive in navigating the digital technology landscape and put in place comprehensive compliance programs to deal with these challenges in including comprehensive employee training [3].

2.1.3 Ethical Considerations

Using artificial intelligence (AI) in business operations raises questions on ethics, discrimination and transparency therefore businesses must be able to explain the technologies they use and apply fair principles as they conduct their operations. It can be a challenge to balance ethics and corporate social responsibility together with sustainability goals. It is therefore imperative that organizations integrate legal compliance with corporate social responsibility initiatives as this enhances brand reputation and customer trust.

2.1.4 Intellectual Property (IP) Protection

Business operations that are involved in the development of digital assets (software, branding, patents) require intellectual property (IP) rights. Organizations face risks of intellectual property theft because of counterfeiting and increased data sharing, which can adversely impact on competitive advantage and innovation. Digital right management, trademarks and copyrights protection are some of the actions that businesses can take to mitigate IP theft.

2.2 *Benefits of Digital Legal Compliance for Sustainability*

Adelakun et al. [1] suggest benefits of digital legal compliance for sustainability such as enhanced efficiency, improved risk management, increased transparency, and competitive advantage. Adebayo et al. [18] point out that in procurement ISO 20400 gives guidelines for integrating sustainability into procurement processes, focusing on impacts of purchasing decisions on environmental, social and economic dimensions. ISO 20400 is there to encourage organization to consider the entire lifecycle of a product from way it is produced to the way it will be disposed including engaging with suppliers to address sustainable issues [19, 20]. Weinberger and Schrodler [21] complying with sustainable procurement practices assist organizations in mitigating risks, improving organizational reputation and creating long-term value. Compliance with sustainable procurement practices is vital in contemporary business because organizations recognize their roles in addressing global challenges like resource depletion, social inequality and climate change [22]. Research indicates that the achievement of sustainable development goals (SDGs) especially those that are related to production, climate action, economic growth and responsible consumption is made possible by sustainable procurement [23].

2.3 *Digital Legal Compliance Framework*

Digital transformation is increasingly impacting every aspect of our lives with the continuous development and advancement of technologies such as cloud computing, robotics, artificial intelligence (AI) and information sharing [24]. Shaik and Poojasree [25] prefer to define e-commerce as the ability of consumers to do their shopping requirements from any place at any time online, but ironically the development has paved way for vendors to put misleading advertisements on those online platforms. This deceptive behaviour calls for authorities to come up with legal compliance frameworks that are targeted at harnessing these untoward behaviours because they go against ethics and moral boundaries in commerce. The establishment of best practices in the digital economy has been brought about by the need for a security-oriented culture [26]. Even though standards do not guarantee legal compliance, at least the presence of key elements within their frameworks and an endorsement by governments as legal custodians of consumer protection is necessary. For example, the European Union encourages the operation of a single market while at the same time encouraging each country to have its own legal framework whether they are regulations or directives. It is necessary to reassess the level of legal compliance each time new laws are reshaped. The increase in the use of software information systems has been brought about by the shift towards information a knowledge-driven economies [27]. It is important for companies to ensure that the data processing and sharing complies with certain legislation such as the General Data Protection Regulation (GDPR), Open Digital Rights Language (ODRL) and the Health Insurance Portability and Accountability Act (HIPAA). Ahmed et al. [28] maintain that efficiency and accuracy in attainment of sustainable city management can be achieved through integrating blockchain and AI. In addition, Nguyen et al. [29] posit that blockchain and AI in tax enforcement can contribute to combating tax evasion by enhancing transparency and traceability in financial transactions.

3 Methodology

The study will adopt a mixed method approach to avoid relying on one methodology only. Sekaran and Bougie [30] prefer to call this method triangulation, and it gives confidence in the results. The qualitative approach will use in-depth interviews to solicit information from participants such as legal practitioners, sustainability managers and compliance officers who work in different industries. This is believed to aid in gathering detailed insights into the practical aspects of implementing legal compliance in the digital age. Data collected will then be analyzed in different themes to identify recurring patterns, key insights and themes. A structured questionnaire will be used to gather quantitative data on digital legal compliance, its perceived effectiveness in harnessing and improving business sustainability and business outcomes such as cost savings and risk reductions. Descriptive statistics will be used to analyze the variables used in the study. The study will ensure that all participants and respondents are informed about the purpose of the study and that they will remain anonymous and the information they provide will be treated confidentially.

4 Discussion

The speed at which digital technology has transformed business operations has not been equally met with the speed of reforming and compliance with legal regulatory frameworks. The financial sector has faced inefficiencies due to lack of adherence to digital legal regulations. Bornman [3] highlight that problems caused by digital transformations require compliance with tax laws. The components of regulatory framework include but are not limited to digital compliance management, regulatory frameworks, risk management, stakeholder engagement and training and development. Challenges of digital legal compliance and considerations have been highlighted as data privacy and security, complexity of regulations, resource constraints, and resistance to change [1].

5 Managerial Implications and Future Research

This study will assist managers to recognize the value of investing in compliance-oriented digital technologies like data management, cybersecurity and legal compliance softwares. This approach reduces the risk of non-compliance which may result in attracting non-compliance penalties. Regular training of employees should be fostered by managers with organizations because this strengthens employees' commitment and involvement as well as aligning individual and departmental objectives to organizational values. A proactive stance helps managers prepare for the ever-evolving regulatory digital framework which requires resources to be allocated, monitoring on a regular basis, auditing and reviewing compliance. This enables organizations to remain adaptable to changes and reduce shocks caused by unexpected events that demand new regulations. Compliance is a sign of a company's commitment to ethical data practices, sustainable digital operations and responsible use of artificial intelligence. It is vital for different departments within the same organization to collaborate towards the achievement of digital compliance and this usually calls for cross-functional teams that strive for compliance. These teams will require management support as they meet on regular basis sharing compliance goals and clarifying responsibilities. Digital compliance and sustainability should be integrated into business models as early as possible in the planning stages of any operations. This proactive stance reduces compliance costs and creates value by aligning the business goals with global sustainable goals.

Future research may explore the impact of digital legal compliance on financial and operational performance. This may involve investigating whether companies that highly invest in compliance measures are highly profitable. It is important to understand the potential of technology legal compliance in achieving business sustainability. Research may explore consumer attitudes towards companies that prioritize digital legal compliance. This assessment may bring insights into the reasons why customers remain loyal, trust the products or services of the organization, and consumer purchasing behaviour. It is important for future research to investigate how companies navigate digital legal compliance across the globe which include data localization, privacy regulations, and cross-border data flaws. Research can further investigate ethical implications of AI-driven compliance solutions, digging deep into how to balance regulatory efficiency with ethical considerations. It is vital for research to explore how sustained commitment to digital legal compliance impacts organizational reputation. The analysis of longitudinal data helps identify trends in stakeholder perceptions, market valuation and brand trust for those companies that exceed compliance standards. Identifying sector specific best practices, tailored compliance frameworks and the role played by regulatory bodies in facilitating digital regulatory compliance across different industrial sectors is also important. This is because different industries face unique challenges when implementing and maintaining compliance measures.

References

1. Adelakum, B.O., Nembe, J.K., Oguejiofor, B.B., Akpuokwe, C.U., Bakare, S.S.: Legal frameworks and tax compliance in the digital economy: a finance perspective. Eng. Sci. Technol. J. **53**(3), 844–853 (2024)
2. Pouri, M., Hilty, L.: Conceptualizing the digital sharing economy in the context of sustainability. Sustainability **10**(12), 4453 (2018). https://doi.org/10.3390/su10124453
3. Bornman, M., Ramutumbu, P.: A conceptual framework of tax knowledge. Meditari Accountancy Res. **27**(6), 823–839 (2019). https://doi.org/10.1108/medar-09-2018-0379
4. Bornman, M., Wassermann, M.: Tax knowledge for the digital economy. J. Econ. Financ. Sci. **13**(1) (2020). https://doi.org/10.4102/jef.v13i1.461
5. Cheng, C., Huang, H.: Evaluation and analysis of regional economic growth factors in digital economy based on the deep neural network. Math. Probl. Eng. **2022**, 1–10 (2022). https://doi.org/10.1155/2022/1121886
6. Hanisch M., Goldsby C.M., Fabian N.E., Oehmichen J.: Digital governance: a conceptual framework and research agenda. J. Bus. Res. 162 (2023)
7. Widjaja, G.: Managing legal and corporate compliance to induce and enhance business. Law Humanit. Q. Rev. **3**(2), 8–17 (2024)
8. Hugos, M.H.: Essentials of Supply Chain Management. Wiley (2024)
9. Renz, D.O., Brown, W.A., Andersson, F.O. (Eds.).: The Jossey-Bass Handbook of Nonprofit Leadership and Management. Wiley (2024)
10. World Bank.: Women, Business and the Law 2024. The World Bank (2024)
11. Hailbronner, K.: Immigration and Asylum Law and Policy of the European Union. BRILL (2024)
12. Smith, M.L.: History as Precedent: Common Law Reasoning in Historical Investigation, p. 27. University of Pennsylvania Journal of Constitutional Law (2024)
13. Purba, A., Mangkan, D., Pulungan, A.: Does tax amnesty program reduce cross-border tax evasion? A case study in Indonesia. J. Bus. Appl. Manage. **15**(1), 001 (2022). https://doi.org/10.30813/jbam.v15i1.3263
14. Rakha, N.A.: Exploring the role of blockchain technology in strengthening international legal guarantees for investment activity. Int. J. Law Policy **1**(3), 1–12 (2023)
15. Lobschat, L., Mueller, B., Eggers, F., Brandimarte, L., Diefenbach, S., Kroschke, M., Wirtz, J.: Corporate digital responsibility. J. Bus. Res. **122**, 875–888 (2021). https://doi.org/10.1016/j.jbusres.2019.10.006
16. Mihale-Wilson, C., Hinz, O., van der Aalst, W., Weinhardt, C.: Corporate digital responsibility: relevance and opportunities for business and information systems engineering. Bus. Inf. Syst. Eng. **64**(2), 127–132 (2022). https://doi.org/10.1007/s12599-022-00746-y
17. Kano, N., Seraku, N., Takahashi, F. and Tsuji, S.: Attractive quality and must-be quality. J. Jpn. Soc. Qual. Control **41**, 39–48 (1984)
18. Adebayo, V.I., Paul, P.O., Eyo-Udo, N.L.: Sustainable procurement practices: balancing compliance, ethics and cost-effectiveness. GSC Adv. Res. Rev. **20**(1), 98–107 (2024)
19. Animashaun, E.S., Familoni, B.T., Onyebuchi, N.C.: Advanced machine learning techniques for personalising technology education. Comput. Sci. IT Res. J. **5**(6), 1300–1313 (2024)
20. Scott, O.A., Amajuoyi, P., Adeusi, K.B.: Advanced risk management for supply finance. Finance Account. Res. J. **6**(6), 868-876 (2024)
21. Weinberger, B.E., Schrodter, A.: The institutionalization of digital compliance. Manag. Decis. (2024). https://www.emerald.com/insight/0025-1747.htm
22. de Oliveira, D.F., de Souza, R.G.: Life cycle sustainability impact categories for sustainable procurement. J. Clean. Prod. **383**, 135448 (2023)
23. Islam, M.M., Turki, A., Murad, M.W., Karim, A.: Do sustainable procurement practices improve organizational performance? Sustainability **9**(12), 2281 (2017)
24. Weinberger, B.E., Schrodter, A.: The institutionalization of digital compliance. Management Decision (2024). https://www.emerald.com/insight/0025-1747.htm

25. Shaik, D., Poojasree, V.: Consumer protection in e-commerce: a legal and compliance framework in the digital market. Adv. Soc. Sci., Educ., Humanit. Res. **549**, 18–23 (2024)
26. Bartonili, C., Giurgiu, A., Lenzini, G., Robaldo, L.: A framework to reason about the legal compliance of security standards (2020)
27. Agarwal, S., Steyskal, S., Antunovic, F., Kirrane S.: Legislative Compliance Assessment: Framework, Model and GDPR Instantiation (2018)
28. Ahmed, I., Zhang, Y., Jeon, G., Lin, W., Khosravi, M., Qi, L.: A blockchain- and artificial intelligence-enabled smart IoT framework for sustainable city. Int. J. Intell. Syst. **37**(9), 6493–6507 (2022). https://doi.org/10.1002/int.22852
29. Nguyen, D., Ding, M., Pathirana, P., Seneviratne, A.: Blockchain and AI-based solutions to combat coronavirus (covid-19)-like epidemics: a survey (2020). https://doi.org/10.36227/techrxiv.12121962
30. Sekaran, U., Bougie, R.: Research Methods for Business, 6th edn. Wiley (2013)

The Role of Exceptional Leadership in Regulatory Performance in Construction Sector Organizations in Oman: A Case Study

Shymaa Al Sinani and Fadi Abdelfattah

Abstract This research aims to examine the impact of exceptional leadership on improving the performance of a private construction company in the Sultanate of Oman. With the increasing challenges facing construction companies, such as technological and economic changes, the importance of effective leadership as a key factor in achieving success and sustainability in this vital sector is emerging by analyzing exceptional leadership practices and their impact on company performance. Study the relationship between leadership styles and operational efficiency. Assess the impact of exceptional leadership on enhancing employee satisfaction and motivation. Finally, make recommendations to improve institutional performance through the development of leadership skills. The case study approach was used to analyze the impact of exceptional leadership in a construction company, by collecting qualitative data through interviews with leaders and managers and analyzing various performance indicators of the company. The questionnaires were also used to survey employee opinions about the quality of leadership and its impact on work performance. The results showed that exceptional leadership in the considered company played a pivotal role in improving performance. Leaders who adopt integrated and inclusive leadership styles, such as transformational leadership, were able to motivate employees and increase their productivity. Exceptional leadership has also contributed to improving interaction between teams, fostering innovation, and reducing turnover rates. The study recommended enhancing the skills of leaders through training programs that focus on transformational and innovative leadership, encouraging leaders to adopt effective communication strategies to strengthen and motivate work teams, and enabling leaders to encourage a culture of innovation among employees to improve performance and productivity. Conduct a periodic evaluation of leadership practices to ensure the achievement of the company's goals and increase employee satisfaction. This study confirms that exceptional leadership is

S. Al Sinani (✉)
Modern College of Business and Science, Muscat, Oman
e-mail: 20247137@mcbs.edu.om; Shaima.alsinani@albadigroup.com

F. Abdelfattah
Albadi Investment Group LLC, Saham, Oman
e-mail: Fadi.abdelfattah@mcbs.edu.om

B. Awwad (ed.), *AI and IoT: Driving Business Success and Sustainability in the Digital Age*, Studies in Systems, Decision and Control 590,
https://doi.org/10.1007/978-3-031-88874-8_66

a key element to improve performance in construction companies, as it contributes to creating a stimulating work environment that pushes employees to achieve the highest levels of productivity and innovation. Adopting these leadership practices enhances the success of companies in achieving their strategic objectives in a competitive sector such as construction.

Keywords Exceptional leadership · Job performance

1 Introduction

To this day, the Department is still the focus of attention and attraction for many researchers and specialists of different intellectual and theoretical orientations. Considering the rapid transformations and diverse changes that the world is witnessing, management is an essential element for the success of organizations; it distinguishes developed societies from less developed societies. The higher the efficiency of management within an organization, the more effective the organization, and vice versa. Sound management is the one that cares about human resources and works to achieve its stability and development as an essential element in the production process and a vital tool. To ensure the continuity and success of the organization.

Hence, leadership has emerged as one of the main administrative functions affecting the effectiveness of organizations, as it represents a basic base in various disciplines and expresses the process of influencing the activities of individuals and the nature of their organizational relationships within the organization. The study of leadership has become a focus where many fields of knowledge intersect, which led to the increasing interest of researchers in studying and analyzing it through different fields such as sociology, psychology, and economics, where each field dealt with the subject of leadership with its own perspective and methodological tools for the purpose of analysis and interpretation.

Frederick Taylor has contributed through his ideas and studies to addressing the various shortcomings in management, where he developed a set of principles that had a great impact on defining and developing administrative leadership, through his perceptions of the manager's tasks and his reliance on scientific standards in evaluating the performance of workers, and working to develop their capabilities and closer cooperation with them to accomplish the work with high efficiency in terms of time, effort and cost.

In this context, leadership is an essential tool through which organizations can achieve their goals, as this function cannot be activated without a leader who is able to influence others and direct their efforts. This leader adopts a set of leadership styles aimed at raising the level of employee performance, whether the method used is participatory, liberal or even directive according to the needs of the organization [1].

2 Problem of the Study

Many companies in the construction sector suffer from low performance because of poor leadership or poor human resource management. This study addresses a key question: How can exceptional leadership contribute to improving the overall performance of a company? Can embracing cutting-edge leadership styles such as transformational leadership help achieve higher levels of performance and innovation?

3 Research Objectives

3.1 Identify the most effective leadership styles in motivating employees and increasing productivity.
3.2 Study the relationship between exceptional leadership and turnover rates in a company.
3.3 Analyze the impact of transformational leadership on the interaction of teams within the company.
3.4 Provide recommendations to foster exceptional leadership in Oman's construction sector.

4 Key Terms and Concepts

Exceptional Leadership

Exceptional leadership is concerned with superior leadership performance in the face of difficult external or internal circumstances facing the organization. An exceptional leader is the key to the success of any organization under unusual circumstances, as it requires him to provide leadership performance that exceeds expectations and deviates from the traditional pattern [2].

5 Previous Studies

5.1 Al-Madima and Al-Qutaibi [3]

This study aimed to assess the impact of strategic leadership in its various dimensions, namely: strategic direction, core capabilities, human capital development, effective organizational culture, emphasis on ethical practices, and balanced organizational control, on improving institutional performance in its four dimensions: financial performance, customer satisfaction, efficiency of operations, and learning and growth, in Yemeni banks in the capital, Sana'a. The study included a comprehensive survey of all employees from banking and administrative leaders in Yemeni banks, with a total of (758) leaders, and (678) of them. The study used the descriptive analytical approach and adopted the questionnaire as the main tool for data collection, with the analysis being performed using the Statistical Package for the Social Sciences (SPSS).

The results showed that strategic leadership has a positive impact on improving institutional performance in Yemeni banks (study sample), where strategic leadership showed remarkable interest in all its dimensions, and it appeared that the strategic direction had the greatest impact, while after balanced regulatory control came in the least influential place. The results also indicated the interest of banking leaders in improving institutional performance, where the financial dimension was the most achieved, followed by the operations dimension, while the learning and growth dimension was the least achieved. The study concluded with several recommendations, The most important of which are: the need for Yemeni banks (the study community) to continue to practice strategic leadership, retain talents and competencies, prepare qualified administrative leaders from their cadres, promote ethical practices by providing incentives and rewards to distinguished people, and provide banking services at the appropriate times.

5.2 Study Al-Somali et al. [4]

This study aimed to identify the role of creative leadership in its various dimensions (such as intellectual energy, mental flexibility, sensitivity to problems, risk acceptance, and the ability to analyze and link) in improving the performance of workers in private hospitals in Jeddah Governorate in the Kingdom of Saudi Arabia. To achieve the objectives of the study, the researchers relied on the descriptive analytical approach, where a questionnaire was distributed to a simple random sample of 350 members of the study population.

The results showed a strong positive relationship between creative leadership and employee performance, as it was found that each dimension of creative leadership contributes to improving performance. Based on these findings, the study recommended the development of a strategy based on precise criteria to discover creative and talented leaders and train them to qualify them as management leaders capable of investing their creative potential in improving job performance. The study also stressed the importance of simplifying systems and procedures, avoiding centralization and strictness in dealing with formal issues, and giving employees the freedom to develop procedures. Especially to accomplish their tasks, as this provides a space of freedom for employees, especially leaders, which contributes to showing their creativity.

5.3 Study Farraj [5]

Leadership is a vital and continuous effective process, which expresses the relationship between the boss and his subordinates, as this relationship allows the president to directly influence the behavior of individuals working under his management and provide them with the necessary information to support his decisions. Leadership is dynamic that stems from the constant changes in project conditions, which requires continuous adjustments in work plans and policies, as well as the continuous change in the behavior of individuals over time, which necessitates the continuous development of relationships between superiors and subordinates.

Leadership forms a pivotal link between the human resources in the organization and its future, through which various management concepts and strategies interact, enhancing positive forces and minimizing negative aspects as much as possible. It also enables leadership to control and resolve professional problems, in addition to resolving disputes, developing and training individuals and caring for them as the most important resource for the organization. Individuals take the leader as their role model, making leadership a key strategic factor in the development of organizations.

This study gains its importance from addressing a strategic topic concerned with the development of successful administrative leadership, as it focused on analyzing the role of leadership in improving performance in private sector institutions. Organizational development in business organizations is one of the basic processes that leaders and administrative supervisors should understand and adopt, to ensure effective and successful performance. The importance of organizational development stems from the changing nature of our time, as we are witnessing rapid and continuous developments that affect the external general environment and the environment of organizations both in various fields.

5.4 Study Al-Saadi [6]

This study sought to explore the nature of the relationship between transformational leadership and crisis management from the perspective of employees in the general administration of the electricity company in Benghazi. To achieve this goal, the study adopted the descriptive analytical approach, and the study population consisted of all employees in the company under study, which numbered (260) employees. To determine the sample size, a table (Krejcie and Morgan 1970) was relied upon. And the selection of a stratified random sample with a size of (155) individuals. After distributing the questionnaire, (100) valid forms for statistical analysis were retrieved, which were processed using the SPSS program.

The general results of the study showed that the level of practice of transformational leadership in its five dimensions was weak, and the dimensions were distributed in descending order as follows: exemplary influence, inspirational motivation, intellectual arousal, empowerment, and individual consideration. The level of practice of crisis management phases was also weak, and their dimensions were arranged in descending order as follows: reactivation, damage containment, preparedness and prevention, and then learning and detecting early warning signals. The study showed a strong and statistically significant positive relationship between transformational leadership and crisis management at the level of Significance (0.01).

5.5 Emere et al. [7]

Effective organizational success is linked to having adequate leadership, as leadership is an essential element to effectively manage project teams in construction to achieve the set goals. Some leadership practices contribute to enhancing the performance of stakeholders and workers on any project, which in turn supports the completion of the project within the specified time frame. This paper aims to examine the critical leadership styles needed to enhance the performance of employees in the construction sector in South Africa. Data when collected through a structured questionnaire that included primary and secondary data, The survey was conducted in Gauteng County, South Africa. Respondents included construction professionals selected using non-probability and available sampling techniques. The data were analyzed using SPSS version 25.0, where averages and standard deviations were calculated, and results were ranked.

The results showed that providing clear instructions and defining procedures contributes to completing tasks on time. The participation of team members in decision-making also enhances the level of trust and supports building positive working relationships. The study also showed that recognizing employees' contributions through individual performance recognition is essential to optimal performance. Other significant results include the provision of necessary resources with the opportunity for employees to perform their tasks, and the ability to make quick decisions. This study adds to the literature by identifying some leadership styles that will support effective leadership in the implementation of construction projects in South Africa.

5.6 Kalambayi et al. [8]

Currently, there is little research exploring the impact of project managers' leadership styles on the performance of workers in Cape Town construction companies. While this topic has been addressed in other studies at the national or international level, comprehensive data on construction companies in Cape Town is still not available, hindering local companies from understanding the role of leadership styles in achieving employee performance results.

This article aims to illustrate how leadership styles affect the performance of employees in construction companies, and to develop a conceptual framework for improving leadership and performance. Using quantitative methods, 456 questionnaires were distributed at various construction sites in Cape Town. The responses were then analyzed using the analysis of variance test (including descriptive and regressive methods), post-event testing (using the Bonferroni method) and correlation analysis.

The results revealed a statistically significant positive relationship between the leadership styles of project managers and the performance of subordinate employees. These findings were later used to develop a conceptual framework for leadership quality improvement tools, including training, empowerment, participation, and delegation. This research contributes to enriching knowledge related to the development of leadership skills of local project managers, leading to improved employee performance and increased chances of success for projects across the city.

6 Theoretical Frameworks

The Concept of Exceptional Leadership

The concept of exceptional leadership can be understood by the unique traits that an exceptional leader possesses, which enable him to convince others, belonging to certain groups, of the importance of teamwork to achieve the desired goals. The leader also plays the role of role model in providing productive and positive work, motivating others to follow his or her approach.

The term exceptional leadership reflects the ability of the leader to impose his views and apply the laws of the educational institution through persuasion rather than force, as he aims to achieve the desired interests with high efficiency without resorting to sanctions or formal dealings with teachers. Therefore, an exceptional leader must identify appropriate approaches to dealing with teachers, which contributes to building positive relationships inside and outside the organization [9].

Exceptional leadership refers to outstanding leadership performance in the face of difficult internal or external circumstances affecting an organization. An exceptional leader is the real force behind the success of any organization in exceptional times, as it requires him to deliver leadership performance that exceeds expectations and goes beyond the usual standards [10].

An exceptional leader can be defined as someone who uses his influence, authority, and skills to influence the behavior, performance, and attitudes of the people around him to achieve the set goals. It represents the true force for the success of any organization or company in exceptional circumstances that require leadership beyond traditional boundaries [2].

7 Qualities of an Exceptional Leader

An exceptional leader has several qualities that distinguish him from other leaders, including [11]:

7.1. Organization and accuracy: It is characterized by strength in organization and accuracy in all work, whether paper, administrative or supervisory.
7.2. Decision Making: Has the ability to make critical decisions and make events rather than wait for them, which makes him proactive in the face of challenges.
7.3. Positive impact: Has effective communication skills that contribute to guiding others towards teamwork and achieving goals.
7.4. Insight: Has a keen vision, where he accepts criticism and absorbs innovative ideas from his subordinates.
7.5. Motivation and inspiration: Contributes to enhancing the spirit of enthusiasm in his team and increasing their motivation to work hard.
7.6. Self-confidence: He has high confidence based on his knowledge and experience in leadership.
7.7. Planning skills: Possesses accurate and logical planning skills that leave no room for chance.
7.8. Decentralized management: Adopts a management style that focuses on delegating authority and investing the energies and knowledge of his team, with indirect follow-up of decisions.
7.9. Culture and Knowledge: Has a high level of culture and follows developments in the field of leadership and management.
7.10 Ethical Commitment: It is characterized by solid values, intelligence and the ability to analyze situations and solve problems to achieve the goals of the educational institution.

8 Duties of an Exceptional Leader in Creating Organizational Happiness

To promote organizational happiness, an exceptional leader should perform several key duties [12]:

8.1. Turning goals into achievements: Transforming the organization's goals into tangible results in collaboration with teachers and students.
8.2. Motivate teachers: Enhance teachers' motivation to achieve school goals and their professional and scientific development.
8.3. Adaptation: The ability to deal with variables affecting the school.
8.4. Future Foresight: Planning for the future of the school and its goals.
8.5. Supporting management functions: Strengthening the five elements of management: planning, organizing, recruiting, directing, and oversight.
8.6. Preparing Future Leaders: Developing the capabilities of the new generation of leaders.

8.7. Adopting new ideas: daring to adopt methods and changes that benefit the school, and not adhering to routines that hinder creativity.

9 Job Performance Criteria

Job performance standards are considered as an input or approach that reflects the efficiency of performance in terms of the effective use of available resources. Performance standards are defined as: "The measure that determines the level of performance required to be achieved by workers in the work entrusted to them, based on these criteria the performance of individuals is judged."

These criteria are generated through the process of job analysis, and job performance criteria are divided into the following types:

9.1 Humanitarian Standards

These criteria focus on the nature of the relationships that exist between the worker and his colleagues in the work environment, as well as his relationships with his superiors or individuals under his supervision. These criteria include the extent to which the worker can influence his subordinates by encouraging and motivating them to exert more efforts to achieve better performance, and promote teamwork and cooperation between them, which contributes to meeting their individual needs and achieving the goals of the organization. It can be said that an individual's relationships with those he deals with in his work environment are an important measure and indicator of the extent to which he succeeds or fails in performing his job duties.

9.2 Personal Criteria

These criteria relate to the amount of effort an employee puts into work, which is influenced by a set of factors that contribute to the formation of his personality. These criteria are relied upon according to the criteria of the individual's personal qualities, and among these qualities we find openness, sincerity, kindness, balance, emotion, and intention. These criteria are the ability of an individual to give and realize himself and highlight his abilities and skills while accomplishing his work or in his dealings with others. These skills and abilities depend on the factors that contributed to the formation of his personality.

9.3 Professional Standards

These criteria include the extent to which an individual can fully perform his/her tasks, duties and responsibilities without any defect. In addition, it is the ability of an individual to solve problems and deal with difficulties that may hinder his career, with the aim of reaching the desired goals, both for him and for the organization in which he works. It also includes other capabilities such as the ability to delegate authority to lower-ranking employees, ensuring that work is done on time and with proficiency.

9.4 Quantitative Criteria

These standards measure performance in terms of quantity and include various aspects that determine performance according to quantity and time. These criteria are evidence of the relationship between product quantity and performance-related time.

9.5 Qualitative Criteria

These standards indicate that an individual has achieved a certain level of quality and mastery in his production.

9.6 Standards of Conduct

These criteria are based on indirect metrics, as they are not directly related to an individual's performance, but rather to their behavior such as cooperation, perseverance and care. Based on the above, it can be said that an individual's performance is not limited to human or personal standards only, but extends to other aspects such as quality, quantity and behaviour [13].

10 Job Performance Elements

Job Performance consists of a set of elements, including:

A. **Knowledge of job requirements**: includes professional skills, technical knowledge, and general background about the job and related fields.

B. **Quality of work**: related to accuracy, organization and the ability to carry out work without errors.
C. **Quantity of work**: related to the volume of work performed under normal conditions and the speed of its completion.
D. **Perseverance and reliability**: includes dedication to work, seriousness, and the ability to take responsibility, in addition to the ability to complete work on time and the need for supervision and guidance.
E. **Activities and tasks**: The task is the intellectual or muscular activity carried out by the worker in certain conditions, and the implementation of formal operations to achieve a certain goal.

According to Hayter, performance elements are defined in three areas:

- Employee: in terms of knowledge of his skills, interests, values, attitudes, and motivations.
- Job: In terms of the challenges and job opportunities it provides, including the elements of administrative systems and organizational structure.
- Position: In terms of the characteristics of the organizational environment in which the job is performed, which includes the working environment and supervision methods [14].

11 Study Methodology

The qualitative approach was used to study the case of a private company in the construction sector in the Sultanate of Oman. The data was collected through personal interviews with the company's leaders and executives.

12 Data Collection

The data was collected by interviewing three leaders in the company.

12.1 Interview with the Senior Leader: Chief Executive Officer (CEO)

12.1.1 How do you view the role of exceptional leadership in achieving the company's strategic objectives?

Exceptional leadership, in my opinion, is essential to achieving strategic goals because it ensures that every member of the company understands and contributes to our vision effectively. I believe that a leader should be motivated, innovative, and set clear goals, so that they are aligned with team development and build a results-focused work environment.

12.1.2 What are the main challenges the company faces, and how does exceptional leadership help overcome them?

One of the most important challenges we face is maintaining efficiency and quality considering the pressures of time and the multiplicity of projects. Exceptional leadership helps us overcome these challenges by fostering a culture of teamwork, training employees to think creatively to solve problems quickly, and providing an environment that encourages continuous improvement of processes.

12.1.3 How do you balance the requirements of growth in the construction sector with quality assurance and high performance?

We believe that sustainable growth must be balanced with quality, and we adopt a leadership style that empowers mid-level leaders to achieve this balance. We focus on selecting projects that align with our capabilities and allocate teams with experience in executing projects with the highest possible quality.

12.1.4 What do you consider to be the most important indicators of organizational performance at the company level as a whole?

Our most important indicators of success are customer satisfaction, quality of execution, and commitment to budget and time. These indicators allow us to measure performance comprehensively, helping us assess how effective exceptional leadership is in driving the company towards success.

12.2 Interview with Second Commander: Operations Manager (COO)

12.2.1 How is exceptional leadership reflected in managing day-to-day operations and guiding teams?

Exceptional leadership provides the foundation for a flexible work environment in which everyone can have a say and contribute to innovative solutions to everyday problems. By stimulating independent thinking and providing confidence to employees, we perform better and increase productivity.

12.2.2 What methods do you use to motivate the team and raise their productivity?

I always try to be at the front of the team, listen to their feedback and provide the support they need to overcome challenges. One of the most important methods I use is to provide rewards when achieving achievements, setting realistic challenges to develop their skills, and providing positive feedback to support their performance.

12.2.3 How are performance measured, and operational objectives achieved within the company?

We have a follow-up system based on a set of indicators such as time spent, quality, and costs. We review these indicators periodically and evaluate performance against them to ensure that the goals are achieved efficiently.

12.2.4 What is the role of flexibility in making operational decisions, and how does it contribute to improving employee performance?

Flexibility is essential to managing operations in a changing sector such as construction, as there are always unforeseen circumstances. We rely on exceptional leadership to make quick decisions that fit the situation, which contributes to improved performance and gives employees confidence in the company's ability to meet project requirements effectively.

12.3 *Interview with the Third Leader: HR Director*

12.3.1 How does exceptional leadership help attract and develop high-caliber employees?

Exceptional leadership allows us to build a strong reputation as a company that values innovation and supports the personal growth of employees. We focus on attracting people with an ambitious career vision and providing them with an environment that allows them to innovate and develop their skills, contributing to building a strong and supportive team.

12.3.2 What motivation strategies do you use to ensure employee engagement and outstanding performance?

We have several strategies, including various reward programs, professional development, and continuous communication. We focus on recognizing outstanding efforts, providing opportunities for training and professional growth, as well as holding events to promote team spirit among employees.

12.3.3 How are potential leaders identified within the company, and what programs do you adopt to qualify them?

We periodically evaluate employees based on their performance and ability to take responsibility and provide them with opportunities to lead small teams within projects as a first step. We have training programs and continuous development courses that focus on leadership development, helping us develop visionary leaders who value teamwork.

12.3.4 How can employees express their opinions and share their ideas to ensure they positively interact with the company's goals?

We encourage employees to share their ideas and suggestions through monthly meetings and HR sessions. Through these sessions, we give them freedom of expression and make sure that management listens to all suggestions, which enhances their interaction and makes them partners in decision-making and achieving goals.

13 Interview Analysis

These interviews demonstrated the essential role that exceptional leadership plays in improving performance at a construction company. The CEO focuses on strategic vision and ensuring a balance of growth and quality, while the COO is concerned with enhancing operational efficiency by supporting the team and making flexible decisions. The HR manager motivates employees and develops future leaders, building an environment that supports performance and creativity.

14 Results and Discussion

The results of the interviews showed that exceptional leadership played a key role in improving the institutional performance of the company studied in the construction sector in the Sultanate of Oman. By adopting integrated leadership styles such as transformational leadership, leaders were able to make a positive difference in the work environment, directly impacting employee performance and overall productivity of the company. These findings can be further discussed as follows.

14.1 Motivate Employees and Increase Productivity

Company leaders who adopted transformational leadership styles were able to create a stimulating work environment that encouraged employees to work more effectively. Transformational leadership is based on four main elements: inspirational motivation, exemplary impact, individual consideration, and intellectual motivation. Taken together, these elements contributed to raising the level of enthusiasm and dedication among employees. The study showed that employees felt valued for their individual contributions and were an integral part of the company's success, resulting in increased productivity.

This finding makes sense given that inspirational motivation, which is part of transformational leadership, contributes to employees feeling that they are working towards higher collective goals, which strengthens their connection to the company. The exemplary influence of exceptional leaders makes employees see leaders as role models, further motivating them to perform better.

14.2 Improve Interaction Between Teams

Another important finding was that exceptional leadership helped improve interaction between teams. Transformational leadership styles encourage teamwork and appreciation of joint efforts, which promotes effective communication between team members and encourages fruitful collaboration. The adoption of these methods by leaders made work teams more aligned and better communicated, which contributed to achieving project goals faster and more efficiently.

Enhanced collaboration between teams leads to less internal conflicts and increased harmony in the work environment, which contributes significantly to improving the overall performance of the company. Teams that interact well can solve problems faster and innovate new solutions, and this pushes projects forward in a smoother way.

14.3 Fostering Innovation

One element that exceptional leaders in the company have focused on is intellectual stimulation, which is a key element of transformational leadership. Through this approach, employees were encouraged to think critically and innovate, developing innovative solutions to everyday business problems and increasing the efficiency of operations.

Innovation in the construction sector is a critical factor, as the sector faces increasing challenges such as changes in technology and environmental needs. Exceptional leadership has created a fertile environment for innovation by motivating employees to think outside the box, enhancing the company's ability to adopt new technologies and methods that increase the quality and efficiency of work.

14.4 Reduce Turnover Rates

The results also showed that exceptional leadership contributed to reducing turnover rates within the company. Transformational leadership makes employees feel cared for and appreciated by leaders, and this leads to increased job satisfaction and organizational loyalty. As a result, resignation rates have decreased significantly, contributing to the stability of the work environment and raising the level of performance in the long term.

Stability in the workforce enhances the productivity of the company, as retaining trained and experienced employees in the company reduces the time and cost required to train new employees, and it also maintains the level of knowledge and skills required to succeed in complex projects.

14.5 Discussion of the Results

The results of this study are in line with academic literature that transformative leadership is one of the most effective leadership methods in improving organizational performance. The ability to motivate employees, improve communication between teams, and encourage innovation are key elements to success in any organization, especially in a sector such as construction that requires high accuracy and great operational efficiency.

On the other hand, the study showed the importance of the psychological and social factors that come with transformational leadership. Paying attention to the development of individuals and appreciating their individual efforts contributes significantly to improving the organizational climate, which positively affects performance.

15 Recommendations Based on the Results

15.1 **Enhance leadership training**: The company should invest more resources in training leaders and managers in transformational leadership techniques and developing their skills in motivating teams and managing operations.

15.2 **Foster a culture of innovation**: Encourage a culture of innovation within the company by providing an environment that allows employees to express their ideas and try new methods to improve performance.

15.3. **Developing systems to motivate employees**: Develop motivational programs that enhance employees' engagement with the company and encourage them to give their best.

15.4. **Continue to improve interaction between teams**: Communication channels and collaboration between different teams must continue to be improved to ensure peak performance.

Acknowledgements This Paper Was Supported by Funding from the Modern college of Business and science and was Supervised by Dr. Fadi Abd Al-Fattah.

References

1. Biscuit, N., Bosfat, F.: The Role of Leadership in Improving the Performance of Employees—a Field Study in the Institution of the Catamia for Insulation Materials and Cork—Jijel. Mohammed Siddiq Bin Yahya University–Jijel–College of Humanities and Social Sciences (2022)
2. AbuYazid, H.: The relationship between transformational leadership behaviors and happiness and job burnout of workers considering the mediation of the role of managerial behaviors of managers, a field study. J. Fac. Commer., Benha Univ. **11**(2), 343–383 (2020)

3. Al-Madima, J., Al-Qutaibi, M.: The role of strategic leadership in improving institutional performance: a field study in Yemeni banks in the capital Sana'a. Sana'a Univ. J. Humanit. **1**(2), 587–616 (2024)
4. Somali, S., Bajunaid, A., Zaki, K.: The role of creative leadership in improving employee performance: an applied study on private sector hospitals in Jeddah, Saudi Arabia. IUG J. Econ. Bus. **28**(3) (2020)
5. Farraj, M.: The impact of leadership on the development of organizations in the private sector: an applied study on SADAFCO Food and dairy company in the Kingdom of Saudi Arabia. Sci. J. Econ. Commer. **49**(1), 267–316 (2019)
6. AlSaadi, N.: Transformational leadership and its relationship to crisis management a case study on the general management of the electricity company in Benghazi. University of Benghazi, Faculty of Economics (2017)
7. Emere, Aigbavboa, C., Thwala, W.: Critical Leadership Style Actions for Improving Construction Personnel Performance in South Africa. Periodica Polytechnica Architecture (2021). https://doi.org/10.3311/ppar.17505
8. Kalambayi, Onojaefe, D., Kasse, S., Tengeh, R.: The Influence of Leadership Styles on Employee Performance in Construction Firms. Social and Humanities, EUREKA (2021). https://doi.org/10.21303/2504-5571.2021.002037
9. Zenati, M.R.: Organizational Behavior, (Behavior of Individuals and Groups in Organizations), 1st Floor. Al-Mutanabbi Library, Saudi Arabia (2013)
10. Taleb, Student.: A Journey into the Affairs of Life. 1st Floor. Atlas for Publishing and Media Production, Cairo (2017)
11. Al Awar, M.: What Do You Know About Corporate Happiness, an End or a Means? 1st Floor. Al-Arabi for Publishing and Distribution, Cairo (2019)
12. Zain, A.A., Walid, K.: Institutional Happiness, 1st edn. Dar Al-Fikr for Publishing and Distribution, Saudi Arabia (2020)
13. Tafer, Baumgerk, Boulbina.: Administrative Leadership and its Role in Improving Job Performance (Doctoral dissertation) (2019)
14. AlNumyan, A.: Administrative Control and its Relationship to Job Performance, Master Thesis, Faculty of Graduate Studies, Naif Arab University for Security Sciences (2003)

The Benefits of Smart, Fully Automated Greenhouses Over Traditional Manual Greenhouses

Hamed Rashid Al Khanjari, P. Vijaya, and P. Praba

Abstract With the pressing need for sustainable, better quality food production, smart greenhouses offer a vital solution to the challenges faced in traditional agriculture. This research addresses the gap in understanding the full benefits of automated greenhouses, especially given rising global resource shortage and the increasing labor demands in farming. This study compares traditional manual greenhouses with fully automated smart greenhouses, emphasizing safety, resource efficiency, cost-effectiveness, and crop yield improvements. Key findings show that automated greenhouses increase health and safety conditions from 56 to 90%, reduce water waste from 30–40% to 5–10%, and significantly lower operational costs. This paper highlights why automated greenhouses are essential now, examining real data and case studies to understand the change or impact of automation on modern greenhouses. By identifying these advantages, this study contributes to the growing body of research recommending smart, resource-efficient farming solutions. Finally, we'll look at some of the challenges farmers face when making the switch to smart systems and consider the broader implications these technological advancements may hold for the future of agriculture.

Keywords Greenhouse · IoT · Sustainability

1 Introduction

Greenhouses have been one of the most important parts of modern agriculture, providing farmers with the ability to farm a diverse and big number of crops throughout the year, in controlled environments. While effective, traditional greenhouses demand substantial manual labor and careful oversight. Farmers are often

H. R. Al Khanjari · P. Vijaya (✉)
Modern College of Business and Science, Muscat, Sultanate of Oman
e-mail: 20202162@mcbs.edu.om

P. Praba
University of Technology and Applied Sciences, Muscat, Sultanate of Oman

B. Awwad (ed.), *AI and IoT: Driving Business Success and Sustainability in the Digital Age*, Studies in Systems, Decision and Control 590,
https://doi.org/10.1007/978-3-031-88874-8_67

required to manually and repetitively monitor factors like temperature, humidity, water levels, and ventilation by hand, tasks that are not only labor intensive but also sometimes lead to human error and high costs. As a result, 7 out of 10 farmers face difficulties with greenhouse farming, dealing with problems such as inefficient water usage, from either overwatering or underwatering, inconsistent environmental conditions, and labor shortages.

As the global demand for high-quality food continues to rise and natural resources become and the production of high-quality fruits or vegetables increasingly hard to find, the need for more consistent, sustainable and efficient farming methods is now required more than ever. Smart greenhouses offer a very good solution to these issues, involving technology into everyday farming operations. Equipped with sensors, automated systems, and Internet of Things (IoT) devices, these greenhouses can monitor and control every aspect of the growing environment in real time. This paper aims to showcase the many advantages of smart greenhouses, particularly how they enhance safety, conserve resources, increase yields quality, and lower operational expenses.

2 Literature Review

Existing research proves and shows some big challenges with traditional greenhouses—especially when it comes to water use, labor demands, and energy consumption. A 2019 study [1] found that manual greenhouses can waste up to 40% of their water because of inefficient watering systems, meaning they often either overwater or underwater these plants. This is a major issue, particularly in dry areas like California and Spain [2]. On top of that, running traditional greenhouses requires a lot of labor, which drives up costs and puts workers at risk. These risks include exposure to chemicals, handling equipment, and working in harsh conditions where human error can also play a role or also having to do daily repetitive work that will eventually lead to human error [3].

That's where smart greenhouses come in, using IoT technologies to tackle these issues immediately. Research by [4] showed that automated irrigation systems could cut water usage by as much as 60% compared to manual methods and still maintain or even improve crop yields and quality. Automation also helps stabilize growth conditions by controlling temperature, humidity, and light, which reduces stress on plants and boosts productivity by being more consistent [5]. However, while these benefits are promising, current studies fall short of providing detailed, side-by-side comparisons of manual versus smart greenhouses. Key aspects like cost-effectiveness, worker safety, and scalability remain underexplored. This paper aims to fill that gap by offering a thorough, data-driven comparison of these two types of greenhouse systems.

3 Methodology

This research used a mixed-methods approach, combining data from real case studies, IoT-based simulations, and industry reports. Data were gathered on water usage, crop yield, and labor costs from a sample of traditional greenhouses and automated greenhouses in California and Israel, regions with contrasting climates and resource availability. IoT devices, including water level sensors, temperature and humidity controllers, and automated watering and monitoring systems, were installed in smart greenhouses to collect data on resource usage and consistency.

A comparative analysis was done based on these real-time sensor readings and historical data from traditional greenhouses. Key metrics included water usage per plants requirement or need, yield per crop type, and labor hours per hectare. Data visualization techniques, such as bar graphs and line charts, were used to show all the differences in water usage and safety between traditional and automated systems. These comparisons were supported by statistical analysis to validate the significance of observed differences in resource usage, cost, and productivity.

Different monitoring and sensor and watering and climate control and IoT devices need to be used as shown in Fig. 1.

4 Results and Comparative Analysis

4.1 Health and Safety: A Priority in Modern Agriculture

One often Worker safety is often an overlooked challenge in traditional farming, especially in greenhouses. Manual greenhouses require workers to handle a variety of environmental factors, often putting them in unsafe conditions. For example, adjusting heating systems by hand can expose workers to extreme temperatures, while handling pesticides or fertilizers brings serious health risks. A 2022 study published in the American Journal of Agricultural Safety reported that manual greenhouse environments carry a 56% risk of injuries or long-term health issues. These risks range from heat-related illnesses and respiratory problems caused by chemical exposure to physical injuries from equipment handling or even mistakes coming from repetitive tasks.

Basically, smart greenhouses minimize these dangers by automating risky processes. Automated climate control systems eliminate the need for manual adjustments, while automated irrigation systems reduce direct contact with harmful chemicals. As a result, worker safety improves significantly up to 90% according to recent studies because human interaction with dangerous tasks is dramatically reduced. This highlights how smart greenhouses can create a much safer working environment for agricultural laborers [3].

Beyond reducing physical risks, smart greenhouses also contribute to better mental health for farmers. Traditional greenhouses demand constant vigilance, leading to

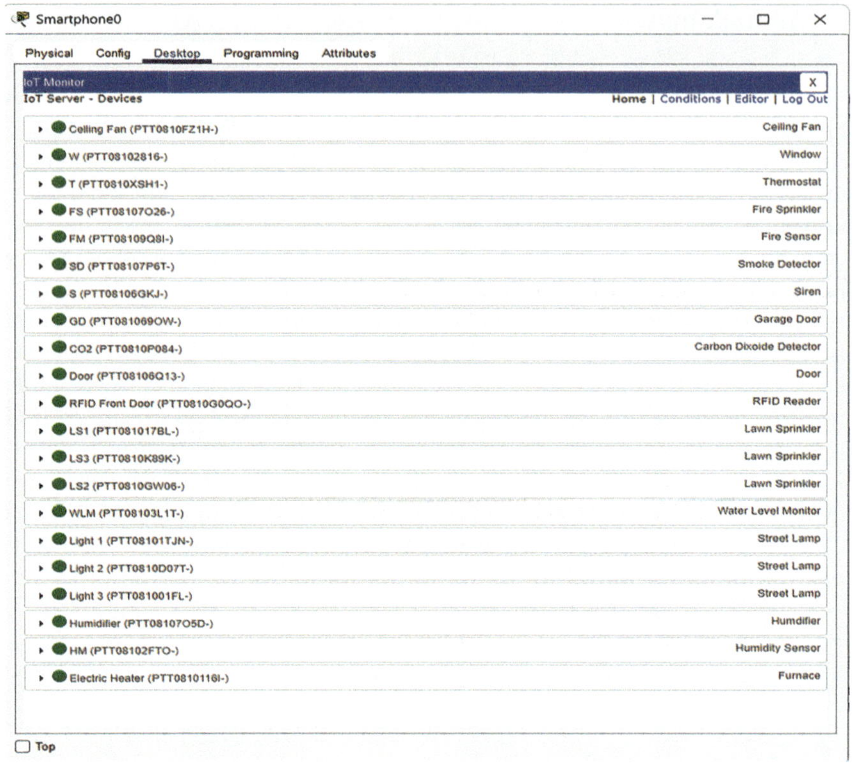

Fig. 1 Green house IoT devices

stress, fatigue, and the possibility of errors that can impact both worker safety and crop productivity, sometimes even because of the repetitive work they must go through. By automating many of these daily tasks, farmers can shift their focus to higher-level responsibilities like crop management and planning, rather than being consumed by boring day-to-day operations as shown in Fig. 2.

4.2 *Water Management: Efficiency at Its Best*

Water is one of the most valuable resources in agriculture, yet it's also one of the most poorly managed. In traditional greenhouse farming, water is often applied manually or through basic timed irrigation systems that fail to consider real-time soil moisture or weather conditions. This leads to significant water waste through overwatering, evaporation, or runoff. On average, manual greenhouses waste between 30–40% of their water, a particularly troubling figure in areas where water is scarce [2].

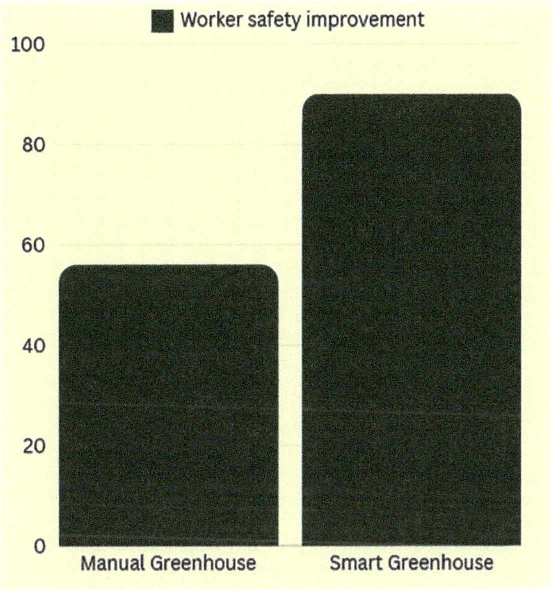

Fig. 2 Smart green house versus manual green house safety

Smart greenhouses tackle this problem by using precision irrigation systems equipped with water level sensors. These sensors continuously monitor the water levels, ensuring that water is only delivered when necessary. Thanks to this precise control, water waste in smart greenhouses drops drastically, down to as low as 5–10%. This not only conserves water but also reduces the energy required for pumping and distribution, contributing to further cost savings, from the amount of water that also gets saved.

Case studies from greenhouse farms in Spain and Israel, two of the world's most water shortage regions, demonstrate the transformative impact of smart irrigation systems. Farmers in these areas have reported up to a 50% reduction in water use without compromising crop yield. In fact, many have observed healthier plants, thanks to the consistent and efficient water delivery that avoids the common cycles of over and under watering or over watering typical of manual systems. The result is evident from Fig. 3.

4.3 Cost Savings and Economic Impact

The financial strain of operating a manual greenhouse is significant. Labor costs make up a large portion of the budget, as workers are needed to constantly monitor and adjust factors like temperature, humidity, and lighting. Additionally, the inefficiencies of manual systems, whether from overwatering, excessive energy use, or crop losses due to not good enough growing conditions further burden the farm's finances.

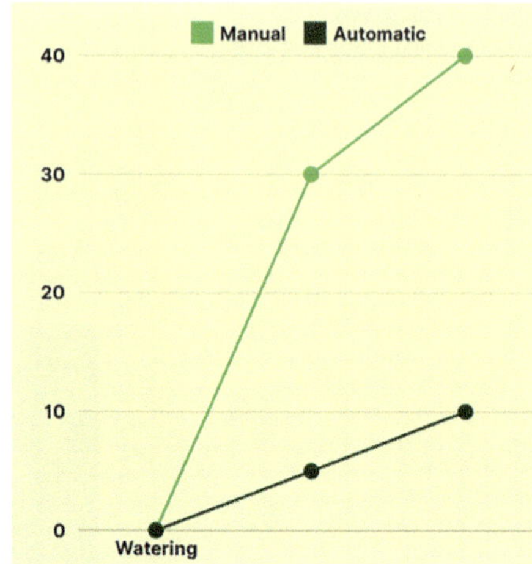

Fig. 3 Smart green house versus manual green house water consumption

Smart greenhouses, on the other hand, drastically lower labor costs. Automated systems take care of nearly every aspect of daily operations, from watering and temperature regulation to providing healthier plants. This allows farmers to manage multiple greenhouses with fewer staff, freeing up labor for other tasks or reducing workforce needs altogether. One California farmer remarked, "Before automation, I needed five people to run a small greenhouse. Now, with smart systems, two people can manage the same greenhouse, and we've even increased production and quality production."

In addition to labor savings, automated lighting systems operate only when needed, by understanding how much light each plant needs daily and providing it. Heating and ventilation systems respond in real time to weather changes, cutting down on energy consumption during extreme conditions. Farmers who adopt smart greenhouse systems often report a 20–30% reduction in overall operating costs, making the switch not only environmentally sustainable but also financially advantageous in the long term (see Fig. 4).

4.4 Enhanced Crop Quality and Yield

The precise control over environmental factors that smart greenhouses provide results in healthier plants and higher yields. Fruits and vegetables thrive when their growing conditions, temperature, humidity, water are kept within optimal ranges. However, in manual greenhouses, maintaining these conditions consistently is challenging, leading to inconsistent growth and, in some cases, crop failure.

In smart greenhouses, automated temperature and humidity controls adjust conditions with precision based on real-time data from sensors. For example, if the humidity drops below optimal levels, an automated system can start adjusting the

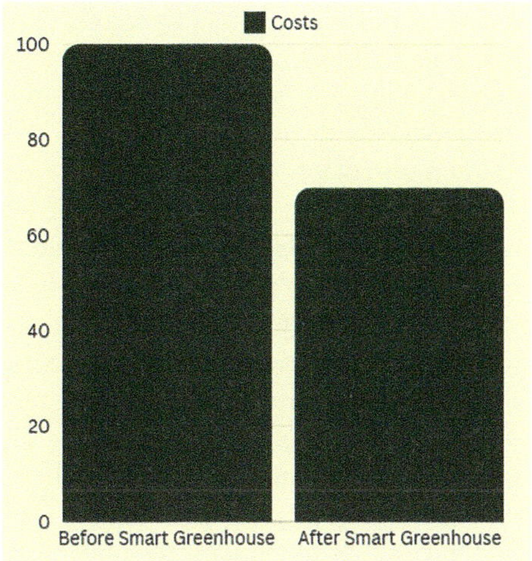

Fig. 4 Smart green house versus manual green maintenance cost

humidity level using a humidifier to get back to the humidity level needed. Similarly, temperature changes are managed automatically, ensuring that plants are never exposed to harmful conditions. This accuracy is very important because even small changes in temperature and humidity can effect on the plant's health, affecting everything from leaf color to fruit size and flavor.

Automated control also helps farmers to grow compatible crops together, ensuring that all plants receive the same care, by making sure you select the right crop that requires the same treatment. For example, tomatoes, peppers, and cucumbers require similar water and light levels, making them ideal to grow together in controlled environments. In comparison, incompatible crops with differing environmental needs suffer when grown together in traditional settings. Planting plants that don't need much water like rosemary alongside crops like lettuce, that need a lot of water, for example, often results in inconsistent watering and humidity, compromising the health and yield of both crops.

Research from the University of Wageningen found that pairing compatible crops in smart greenhouses boosted yields by 25–35% compared to traditional greenhouses. Farmers also noticed a big increase in produce quality, with reports of larger, tastier tomatoes and lettuce that stayed consistently crisp. The ability to precisely control temperature and humidity at every growth stage doesn't just increase yields it also ensures the produce meets top market standards. The result showed better quality fruits and vegetables for consumers, both in taste and appearance.

4.5 Sustainability: A Greener Future for Farming

Sustainability isn't just a nice to have anymore it's a must for modern agriculture. With climate change increasing and weather conditions becoming inconsistent and resources becoming harder to access, farming needs to shift toward methods that are both efficient and eco-friendly. Smart greenhouses are becoming one of the most important things in farming as they started making farming fast and very sustainable and producing the best quality of yield possible.

A big part of this improvement comes from automated systems that control temperature and humidity in real time. These systems adjust based on weather conditions detected from the sensors and what's happening inside the greenhouse and the specific temperature need of these plants. In traditional greenhouses, manual thermostats and ventilation systems usually mean inconsistent conditions because workers must constantly monitor and adjust them manually which usually makes it inconsistent. Smart systems, on the other hand, maintain ideal growing conditions all the time because they change depending on the plants' need. By monitoring and responding to changes in humidity and temperature, they save energy and prevent crop losses caused by unsuitable environments.

Better water usage isn't the only sustainability benefit. Smart greenhouses also focus on energy efficient practices. Automated heating and cooling systems require less energy, and that happens because they only work when needed, unlike the traditional manual way.

4.6 Crop Management and Compatibility: Key to Maximizing Yield

Better crop management is important for maximizing the efficiency of a smart greenhouse. Different plants have different needs for water, light, and growing incompatible crops together can lead to bad crop quality. For instance:

- **Tomatoes, peppers, and cucumbers**: These plants require warm weather conditions with consistent watering. They can be planted together and effectively grow in a smart greenhouse without the risk of over or underwatering.
- **Lettuce and spinach**: Both cool weather crops, they require lower temperatures compared to other plants and higher humidity. Planting them together ensures the best growth conditions.

However, problems start when incompatible crops are grown together. Growing plants that need water with plants that don't need as much water can result in poor growth or crop loss. In traditional greenhouses, farmers often struggle to manage these ideas of growing only compatible plants, leading to inconsistent yields and wasted resources.

Smart greenhouses solve this issue by using sensors to monitor the specific needs of each crop or group of crops that are compatible. This allows for better control over the plants, helping farmers to grow multiple types of plants together without treating them all differently. As a result, plants are healthier, waste is minimized, and produce quality improves.

4.7 Challenges in Transitioning to Smart Greenhouse

While the benefits of smart greenhouses are clear, making the switch from traditional manual systems to automated ones isn't without its challenges. The challenge is the initial setup cost. Installing sensors, automated irrigation systems, and climate control technology requires a significant upfront investment. For farmers in the beginning this might sound expensive but eventually it will pay off.

That said, the long-term benefits often pay off the initial expense. Savings on labor, water, and energy combined with higher crop yields usually allow smart greenhouses to pay for themselves within just a few growing seasons. To make the transition easier, many governments and agricultural organizations are stepping up with subsidies and giving money to help farmers get these more sustainable and efficient systems.

Another challenge is learning how to use them that comes with using new technology. Farmers who are used to hands-on methods might be hesitant to trust automated systems. However, with proper training and support, most farmers learn quickly. Once they do, they can spend less time on repetitive tasks and focus more on strategic crop planning, which helps them get the most out of their greenhouses.

5 Implementation

In the implementation section, this project demonstrates the design and functionality of an automated greenhouse environment utilizing IoT devices, modeled through the simulation tool Packet Tracer. The first step involves designing the greenhouse structure and integrating essential IoT devices, all connected to a central home gateway. This gateway interfaces with a smartphone, allowing full automation of greenhouse conditions.

To establish connectivity, each IoT device is linked to the home gateway via the IoT server. The lighting system, for example, is programmed to operate for one hour daily to meet the plants' specific light requirements. Access to the IoT monitoring application on the smartphone is secured with a unique username and password, allowing controlled adjustments to the plants best interest, including water levels and humidity, which are set through conditional "if" functions.

Further customization is applied to specific devices, such as the water sprinkler and humidifier. Initial default codes for these devices are changed to deliver lower water volumes and controlled humidity levels, as the original settings were excessive for the greenhouse's plant needs. For this research model, it is assumed that all plants share similar needs like similar water, temperature, and humidity requirements, meaning those levels should be measured and customized based on the plant's requirement (see Fig. 5).

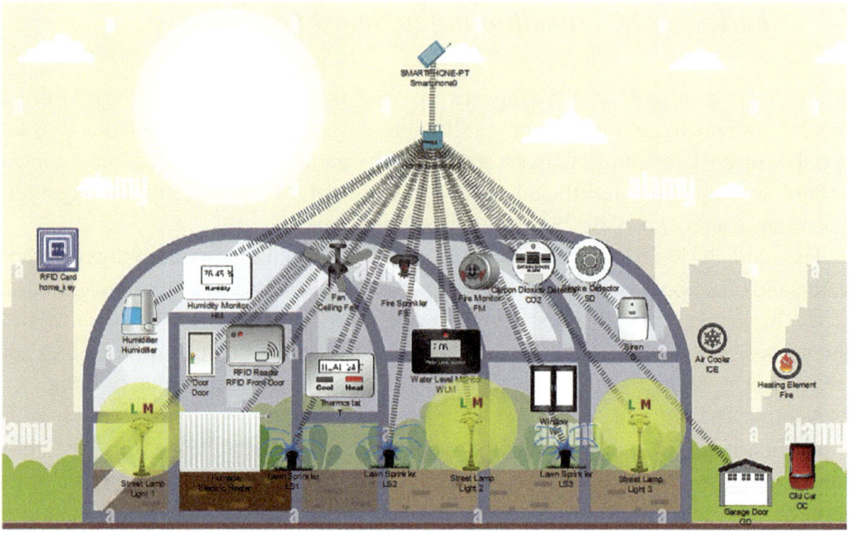

Fig. 5 Simulation of greenhouse in packet tracer

5.1 Discussion and Implications

The findings from this research suggest that smart greenhouses are not only more efficient and cost-effective but also offer significant scalability potential. By reducing the need for manual labor, smart greenhouses lower operational costs, making it easier for farmers to manage larger facilities. This scalability is crucial for meeting the food demands of a growing global population.

Smart greenhouses started really improving food security by minimizing the number of people required to enter the greenhouse and improving the overall productivity of each crop. In areas with very bad or changing weather conditions, the ability to precisely manage water, temperature, and humidity allows crops to be grown consistently, no matter the condition. This level of control makes sure the production regardless of any challenges.

Also, smart greenhouses could reshape how labor is used in agriculture. With fewer workers needed for daily operations, these systems help solve the labor shortages many farming regions are struggling with. Instead of spending time on boring repetitive manual tasks, workers can change their focus to more strategic, higher-value activities. It's a win–win for efficiency and productivity.

5.2 Challenges in Scaling and Implementation

While smart greenhouses have plenty of advantages, starting them up comes with challenges. The biggest problem is the initial setup costs. Installing IoT devices, automation software, and advanced climate control systems can be too expensive for smaller farms to afford. To make smart greenhouses more accessible, governments and agricultural organizations could step in with subsidies or low-interest loans to help farmers cover these costs. This solution will help small farmers and the country in general.

Another problem is learning how to use the system that comes with using new technology. Farmers who have spent years relying on traditional methods might be hesitant or scared to switch to automation, worrying it could ruin their old routines. To ease this transition, agricultural extension programs could provide training to help farmers get comfortable with smart greenhouse systems and understand the long-term benefits they offer. With the right support, these problems can be solved, making it the new way for greenhouses to operate.

5.3 Limitations and Future Research

This research mainly focused on how smart greenhouses improve water management, crop yields, and worker safety. However, the study was only for certain regions, which means it might not cover the other challenges or conditions faced in other parts of the world. Future research should look at how smart greenhouses perform in a variety of geographic and climatic settings to provide a more comprehensive understanding.

Another promising area for future study is the development of next-generation IoT devices. These could be designed to support a wider variety of crops, each with unique needs. Advances in sensor technology could make it possible to manage mixed-crop greenhouses more precisely, allowing farmers to grow different plants side by side more effectively. There's also a pressing need to explore cost-effective smart solutions for smaller farms. Making automation affordable and accessible could help more farmers adopt these systems and benefit from their advantage.

6 Conclusion

Smart greenhouses represent the future of sustainable agriculture. By integrating automation, they offer a reliable solution to the inefficiencies and risks associated with traditional greenhouses. Our findings demonstrate that smart greenhouses improve worker safety from 56 to 90%, reduce water waste from 30–40% to 5–10%, and increase crop yields by 25–35%. These systems enable precise environmental control, leading to healthier, higher-quality crops and lowering operational costs.

Transitioning to smart greenhouses will be essential in addressing the challenges of food security, labor shortages, and resource conservation. As climate change continues to threaten traditional farming methods, smart greenhouses provide an adaptable, efficient, and sustainable solution. By investing in this technology, the agricultural sector can move toward a more resilient future that meets global food demands while preserving natural resources.

References

1. Zhang, H., Qian, X.: Comparative analysis of traditional and smart greenhouses in resource utilization. Sustain. Agric. Resour. Effi. **19**(6), 214–225 (2019)
2. Gonzalez, R., Ruiz, L.: Water management efficiency in arid region greenhouses. J. Agric. Water Manage. **45**(3), 367–374 (2020)
3. Smith, J., Wang, Y., Kim, T.: Health and safety risks in traditional greenhouse environments. Occup. Health Saf. Agric. **22**(5), 110–118 (2021)
4. Sharma, M., Patel, R.: Reducing water waste with smart irrigation systems. Agric Sci. Technol. J. **15**(4), 123–130 (2020)
5. Lee, S., Choi, J.: Automation in agriculture: the role of IoT in smart greenhouses. Int. J. Agric. Technol. **12**(2), 89–105 (2018)

Exploring Consumers Perceptions and Preferences for Takaful Products in Oman: Insights for Market Growth and Innovation

Umar Ahmed, Ibrahim Gambo, Bashir Ahmad Fida, and Yousuf Mohammed Al Siyabi

Abstract The purpose of the study to explore Consumer Perceptions and Preferences for Takaful Products in Oman. Questionnaires were distributed to the 300 respondents across Muscat using convenient sampling technique. Descriptive statistics are used in data analysis. The research revealed that a significant proportion of respondents were familiar with both Takaful and conventional insurance. However, many of the respondents perceived no significant differences between Takaful and insurance products. Consequently, they express no preferences between Takaful and conventional insurance. However, when the respondents are given a choice, they identified several factors influencing their decisions to adopt Takaful. These factors include affordability of Takaful products, excellent services quality, Shariah compliant, simplicity and ease of the claims process and the availability of Takaful branches. To increase public awareness and market penetration, Takaful sector should adopt aggressive marketing strategies. Leveraging social media and other innovative marketing strategies can effectively promote Takaful products to general masses. Additionally, the sector should emphasize the distinctiveness of Takaful products compared to conventional insurance, particularly highlighting their Shariah compliance. Tailoring these products to meet specific needs of customers is vital. The sector can concentrate on areas such as health, life and other socio-economic products to improve on social welfare of the society.

Keywords Exploring · Consumers perceptions and preferences · Takaful products · Oman · Market growth · Innovation

U. Ahmed · I. Gambo (✉) · B. A. Fida · Y. M. Al Siyabi
Department of Business and Economics, Modern College of Business and Science, Muscat, Oman
e-mail: Ibrahim.Gambo@mcbs.edu.om

B. Awwad (ed.), *AI and IoT: Driving Business Success and Sustainability in the Digital Age*, Studies in Systems, Decision and Control 590,
https://doi.org/10.1007/978-3-031-88874-8_68

1 Introduction

Oman was the last country in the Gulf Cooperation Council (GCC) to embrace Islamic finance. In 2014, Takaful Oman and Al Madina Takaful become the first two Takaful companies established in the country. The transformation of Al Madina Takaful from a conventional insurance to Islamic insurance company is a significant milestone in the history of Takaful in Oman [1].

The entry of Islamic insurance companies into the industry has become a competitive force, encouraging competitive environment where traditional insurance providers must compete for market share. Since their inception, these companies have adopted aggressive strategies to expand their market base, achieving remarkable success, especially in their Initial Public Offerings (IPOs). The IPOs of both Al Madina Takaful and Takaful Oman were highly oversubscribed nearly four times for Al Madina, and 5.5 times for Takaful Oman. Collectively, these were able to raise almost OMR 59 million (USD 153.2 million) from the market [1].

It is noteworthy to mention that the sector performed well over the years achieving a market share of 14%. In 2022, the sector gross direct premiums increased by 7.7%, reaching RO 76.256 million compared to RO 70.779 million in 2021. However, gross premiums of general Takaful business slightly decreased to RO 43.800 million in 2022 from RO 44.160 million in 2021, accounting for 16% of the total general insurance market. On the contrary, the family Takaful experience significant growth, with gross direct premiums increased by 37.4% to RO 18.505 million in 2022 up from 13.471 million in 2021. This segment now represents 28% of the total life/family insurance market. Similarly, the health Takaful segment saw a 6.1% increase in gross direct premiums of Takaful companies for health insurance increased by 6.1%, rising to RO 13.951 million in 2022 from RO 13.148 million in 2021, constitutes about 7% of the total health insurance market [2]. Therefore, the purpose of the study to explore Consumer Perceptions and Preferences for Takaful Products in Oman. The chapter is organized into five sections, beginning with the introduction. Section Two provides a review of the literature on Takaful. Section Three outlines the research methodology, while Section Four presents the study's findings. The last section provides conclusion and implications of the study.

2 Literature Review

2.1 Concept of Takaful

Takaful derived from Arabic word *Kafala* which means suretyship, joint responsibility, guarantee, etc. [3]. Accounting and Auditing Organization for Islamic Financial Institutions (AAOIFI) defines *Takaful* as "a process of agreement among a group of persons to handle the injuries resulting from specific risks to which all of them are vulnerable. Thus, a process involves payment of contributions as donations and leads to the establishment of an insurance fund that enjoys the status of a legal entity and has independent financial liability. This fund's resources are used to indemnify any participant who encounters injury, subject to a specific set of rules and a given process of documentation. The fund is managed either a selected group of policy-holders or a joint-stock company that manages the insurance operations and invests the assets of the fund, against a specific fee." [4].

Takaful is an alternative to conventional insurance and the system is based on Islamic principles which prohibits *Riba* (interest), M*aysir* (gamblimg) and *Gharar* (uncertainty) which formed the basis of the conventional insurance. *Takaful* is a system based on mutual help (*ta'awun*), aimed at promoting virtue and supporting those who are in need or facing hardship [5]. It operates on foundational principles of cooperation, shared and collective protection. These principles foster the spirit of solidarity and ensure participants contribute to the social welfare of the community [3].

The principles of mutual responsibility are explained in the following hadiths as recorded in Sahih Al Bukhari and Sahih Al Muslim:

> The relationship between one believer and another is like that of a building where one part of the building strengthens the other parts.
>
> A believer to another believer is like a building where each part enforces the others.
>
> Each of you has a responsibility and each of you is responsible towards those under your responsibility.

The principle of cooperation is established through the Quran and Sunnah. For example, surah (Qur'an Surah Al-maidah:2). Abu Dawood narrated that Prophet (P.B.U.H) said:

> Whosoever removes the hardship of his brother, Allah will remove his hardship on the day of Judgment.

2.2 Perceptions Studies About Takaful

Reference [6] explored public awareness and knowledge of *Takāful* in Malaysia, revealing mixed levels of understanding. While 75% of respondents were aware of *Takāful*, a notable 25% were entirely unfamiliar with the concept. Further analysis showed significant gaps in knowledge about core *Takāful* principles. Specifically, 45% of respondents were unaware of its risk-sharing nature, 33.2% did not know about its prohibition of gambling (*qimar*), and 38.8% and 37.2% were unaware of its prohibition of uncertainty (*gharar*) and interest (*riba*), respectively. Nevertheless, approximately 70% of respondents identified *Takāful* as a Shariah-compliant product.

Similarly, [7] study in Kuwait underscored deficiencies in public awareness. Over half of the participants were unfamiliar with the concept of *Takāful*. Of the 48% who were aware, most had learned about it through word-of-mouth, highlighting the need for more robust marketing and educational initiatives to enhance understanding.

In Afghanistan, [8] research indicated a relatively high level of public knowledge about *Takāful*, although awareness levels were comparatively lower. Encouragingly, most respondents demonstrated a willingness to contribute to and subscribe to *Takāful* products, reflecting a positive attitude towards its adoption.

Reference [9] study in Somalia examined factors influencing the adoption of *Takāful* products. The findings identified perceived relative advantage, compatibility, cultural alignment, awareness, and trust as significant drivers of adoption. In contrast, perceived complexity was found to have an insignificant impact, suggesting that complexity may not pose a substantial barrier to *Takāful* adoption in the Somali context.

3 Research Method

The study adopted a quantitative survey methodology, consistent with [10] argument that surveys are effective for collecting data on attitudes and opinions, facilitating quantitative analysis, hypothesis testing, and generalization of results. A Likert scale questionnaire was developed, enabling respondents to indicate their level of awareness or unawareness with statements related to the research topic [11].

4 Findings and Discussions

4.1 Demographic Profile of the Participants

The results in Table 1 indicate that 90% of the respondents were male, while only 10% were female. In terms of age distribution, 40% of the respondents were aged between 20 and 30 years, 42% were between 31 and 40 years, 15% were between 41 and 50 years, and only 3% were above 50 years. Most of the respondents (92%) were Omani nationals, with expatriates accounting for the remaining 8%. Regarding marital status, 67% of the respondents were married, 30% were single, and 3% were divorced. In terms of educational attainment, 57% of the respondents held a bachelor's degree, 20% held a diploma, and 23% possessed a postgraduate degree. Employment-wise, 65% were working in the public sector, while the remaining 35% were employed in the private sector. The respondents' monthly income levels also varied: 40% earned between OMR 600 and 1000, 35% earned between OMR 1000 and 1500, and only 7% earned above OMR 1500. Most respondents (85%) have experience with insurance products and have been using them for over five years, while only 9% have been using them for less than three years.

Table 1 Socio-economic background of the participants

Gender	Frequency	Percentage
Male	225	90
Female	25	10
Age		
21–30	100	40
31–40	105	42
41–50	38	15
Above 50 years old	7	3
Nationality		
Omanis	230	92
Expatriates	20	8
Marital status		
Married	167	67
Single	75	30
Divorce	8	3
Education		
Diploma	50	20
Bachelor's degree	142	57
Postgraduate	58	23
Occupation		
Public sector	163	65
Private sector	87	35
Monthly income		
Below 325	5	2
325–600	40	16
600–1000	100	40
1000–1500	88	35
Above 1500	17	7
Experience		
Less than 3 years	23	9
3–5 years	40	16
Above 5 years	187	85

4.2 Consumers Awareness of the Differences Between Takaful and Conventional Insurance

Table 2 results show that most respondents (75%) are aware that conventional insurance is not Shariah-compliant due to the presence of *riba* (usury), *gharar* (excessive uncertainty), and *maysir* (gambling). However, this awareness does not significantly influence their decision-making when purchasing insurance. Sixty-six percent of respondents are aware that in conventional insurance, risk is transferred from policyholders to the insurance company. However, thirty-four percent of the respondents were not aware. More than half (57%) of the respondents are aware that in

Table 2 Consumers awareness of the differences between Takaful and conventional insurance

Factors	Aware	Not aware
Conventional insurance is considered non-compliant with Shariah principles due to the presence of *riba* (usury), *gharar* (excessive uncertainty), and *maysir* (gambling) elements	75	25
In conventional insurance, risks are transferred from the policyholders to insurance company	66	34
In conventional insurance, profits are maximized by minimizing policyholders' claims	57	43
Takaful is form of mutual help and cooperation	70	30
In Takaful, risk is shared between the policyholders and the insurance company	40	60
In Takaful, most products are like those offered by conventional insurance	62	38

conventional insurance, profits are maximized by minimizing policyholders' claims. However, forty-three percent of the respondents were not aware. Most respondents (70%) are aware that Takaful operates as a form of mutual assistance. However, this understanding does not significantly influence their decision-making when purchasing insurance. Sixty-two percent of the respondents are aware that Takaful offers products like those in conventional insurance. However, Majority of the respondents (60%) are unaware that in Takaful, risk is shared between the policyholders and the insurance company.

4.3 Consumers Preferences Between Takaful and Insurance

In relation to consumers preferences between Takaful and insurance, results in Table 3 indicate that most respondents (78%) continue to use conventional insurance despite being aware of its prohibition under Shariah principles. Only thirty-two percent opt for Islamic insurance. Insurance expenses vary among respondents: 42% spend less than OMR 100, 31% spend between OMR 100 and 200, and only 10% spend more than OMR 300. Nearly half of the respondents are unsure about their general preference for a specific type of insurance when given a choice. Takaful providers should capitalize on this market potential by enhancing awareness and educating consumers about their offerings.

4.4 Factors Influencing Consumers Preferences

Table 4 highlights the key factors respondents consider when choosing between Takaful and conventional insurance. These factors include Shariah compliance, ease

Table 3 Consumers preferences between Takaful and insurance

Statements	Percentage
Insurance subscription	
Conventional	78
Takaful (Islamic insurance)	28
Annual insurance expenses	
Less than OR 100	42
O.R. 100–200	31
O.R. 200–300	17
Above 300	10
Preferences	
Takaful	28
Insurance	26
Not sure	46

Table 4 Factors influencing consumers preferences

Statements	Important (Percentage)	Not important (Percentage)
Shariah compliant	80	20
Easy and faster to claim	83	17
Availability of insurance branches	71	29
Reasonable price	95	5
Excellent quality services	93	7

of claims processing, availability of insurance offices, affordability, and service quality. Eighty percent of respondents indicated that they would choose Takaful due to its Shariah compliance, while 20% did not view this as a significant factor. Additionally, 83% of respondents identified ease of claims processing after an accident as an important consideration, and 71% emphasized the importance of the availability of insurance offices. Affordability and service quality were overwhelmingly considered the most critical factors, with 95% and 93% of respondents, respectively, highlighting them as decisive when choosing between Takaful and conventional insurance.

5 Conclusion and Implications

The research revealed that a significant proportion of respondents were familiar with both Takaful and conventional insurance. However, many perceived little distinctions between the two. Overall, customers did not express a strong preference for either Takaful or conventional insurance.

The factors influencing customer choices included affordability, service quality, Shariah compliance, ease of claims, and accessibility. The study also identified untapped market segments, such as life, health, home, fire, and travel insurance, with most respondents not having subscribed to these products.

To enhance Takaful's market penetration, the industry should leverage social media and innovative marketing strategies to raise awareness and encourage customers to opt for Takaful products. These efforts should be accompanied by improvements in affordability, service quality, and claims processing.

To differentiate Takaful from conventional insurance, the industry should emphasize its unique selling points, particularly Shariah compliance, and tailor products to meet the specific needs and concerns of sceptical customers. Given the untapped potential in life, health, home, and fire insurance, the industry should focus on developing targeted marketing campaigns and product offerings to attract a broader customer base.

References

1. Oxford Business Group.: Oman's nascent Takaful segment now benefits from a legal framework (2017). https://oxfordbusinessgroup.com/reports/oman/2017-report/economy/getting-ahead-the-nascent-takaful-segment-now-benefits-from-a-legal-framework. Accessed 5 Nov 2024
2. Capital Market Authority (CMA).: Annual Report (2022). https://fsa.gov.om/documents/CMA-AR-2022-ENGLISH.pdf. Accessed 5 Nov 2024
3. Malik, A., Karimullah.. Introduction to Takaful theory and practice. Springer Nature (2019)
4. AAOIFI.: Takaful or Islamic Insurance, Accounting and Auditing Organization for Islamic Financial Institutions (2015)
5. Yusoff, Y.M., Latif, R.A., Yahaya, A.S., Malom, M.M., Zaharum, Z.: Factors that affect the investment performance of Takaful industry: Malaysia evidence. Int. J. Acad. Res. Bus. Soc. Sci. **12**(12), 2755–2765 (2022)
6. Hassan, R., Salman, S.A., Kassim, S., Majdi, H.: Awareness and knowledge of Takaful in Malaysia: a survey of Malaysian consumers. Int. J. Bus. Soc. Sci. **9**(11), 45–53 (2018)
7. Shammari, Y.S.: Indicators of Takaful awareness among Kuwaitis. J. Islamic Bank. Finance. (2015). Published by American Research Institute for Policy Development
8. Ahmada, G.S., Mahadia, N.F.: Public awareness, knowledge, and perception towards Takāful in Afghanistan. J. Islamic Soc. Financ. **1**(2) (2023)
9. Mukhtar, Y.A., Barre, G.M.: Determinants of customer perception on the adoption of Takaful in Somalia. J. Islamic Acc. Bus. Res. **15**(7), 1053–1071 (2024). https://doi.org/10.1108/JIABR-03-2022-0093
10. Burns, A.C., Bush, R.F.: Marketing Research (5th ed.). Upper Saddle River, NJ Prentice Hall (2006)
11. Sekaran, U., Bougie, R.: Research Methods for Business: A Skill-Building Approach, 7th edn. Wiley, West Sussex (2016)

The Impact of Community Leadership in Enhancing Volunteer and Development Initiatives

Khalid Alhosni and Mohamed Elgeddawy

Abstract This study aimed to shed light on the impact of community leadership in enhancing volunteer and development initiatives through the three theories of leadership: servant, transformational, and participatory and the dimensions of each theory through three dimensions: community building, trust for individuals, and cooperation in local community institutions in the Sultanate of Oman. We concluded that the variables for this study are community leadership as an independent variable and the promotion of volunteer and development initiatives as a dependent variable as a result of the influence of leadership. The dimensions of the three theories, community building, trust for individuals, and cooperation, came as the main dimensions for improving and activating volunteer and development initiatives. Based on the above, it is clear to us that the conceptual framework of community leadership, through the three theories, leads community leadership to strengthen voluntary and development initiatives. The variables interact with each other to achieve sustainable development and build an integrated and cooperative society. Proving the direct relationship between these variables will solve the study's problem and fill the research gap. This is what the researcher will seek to study and apply to all segments of society.

Keywords Community leaders · Volunteer initiatives · Development initiatives

1 Introduction

Community leadership is a type of leadership that aims primarily to serve, develop, and sustain society through the efforts of individuals to achieve the goals of society in general. In the Sultanate of Oman, Oman Vision 2040 received wide attention at all levels of all different sectors and community participation from people of experience,

K. Alhosni (✉) · M. Elgeddawy
Modern College of Business and Science, Muscat, Oman
e-mail: 20246772@mcbs.edu.om

M. Elgeddawy
e-mail: Mohamed.Elgeddawy@mcbs.edu.om

B. Awwad (ed.), *AI and IoT: Driving Business Success and Sustainability in the Digital Age*, Studies in Systems, Decision and Control 590,
https://doi.org/10.1007/978-3-031-88874-8_69

knowledge, and expertise. It is among the national priorities to have a clear role for leadership with the presence of voluntary social responsibility programs, as is the case in the Ministry of Social Development, which seeks to achieve the goals of Oman Vision 2040 related to strengthening volunteer and development initiatives, such as establishing a donation platform (charitable work portal) and increasing the number of members affiliated with associations. Depending on their classification, During the year 2023, the "Jood" platform was launched to serve volunteer charitable associations and teams in the Sultanate of Oman (Oman Vision 2023/2024 report), but these efforts require community leadership, and this leadership usually begins at the level of international policies, then regional policies, then local policies.

At the global level, the United Nations Sustainable Development Group (https://unsdg.un.org/ar) acts as a high-level forum for joint policy formation and decision-making. It also directs, supports, tracks, and supervises the coordination of development processes in 162 countries and territories. The United Nations Sustainable Development Group also has a set of seventeen goals to achieve a better and more sustainable future for all. Among the United Nations initiatives such as the International Monetary Fund and the World Bank are community initiatives such as the "Leave No One Behind" initiative to achieve sustainable development for the United Nations agenda. 2023 AD and the United Nations 2024 report indicated that the world is 17% away from achieving sustainable development goals to reach the right track.

At the regional level, there are many organizations, such as the League of Arab States and African countries, that promote the path of social leadership within their development goals, such as initiatives related to education, health, water, and other initiatives. This reinforces the necessity of enabling local communities to participate in sustainable development and achieve its goals, and this appears through the cohesion of individuals within the community, the presence of local councils, the management of various types of crises, and the development of projects that help individuals within the community.

In addition, when asking stakeholders about the challenges faced by community leaders related to promoting volunteer and development initiatives, by asking one of them, we found the lack of funding to be one of the main reasons for strengthening community initiatives. When asked by the other, he answered that among the challenges is the lack of volunteers in promoting community initiatives, and another added that there is a lack of awareness among many people in promoting volunteer and development initiatives.

The researcher analyzed the opinions of stakeholders and found that concerning the challenges related to funding, they do exist, as many people, because they do not find anything that touches on reality, abstain from funding many parties. By analyzing the other opinions related to the shortage of volunteers, we found that the culture of volunteering is not available in the country. Many residential neighborhoods even need community initiatives to enhance this awareness.

We can tell a success story of community leadership and its impact on promoting volunteer and development initiatives. In the sixth village of Al Khoudh in the Seeb State of the Sultanate of Oman, there was one of the local community's problems, which was the lack of a volunteer school serving the people of the neighborhood. However, thanks to local leadership, a community group was created that took it upon itself to solve this problem by establishing a charitable fund and communicating with individuals, groups, and official bodies to build an endowment school for the neighborhood's children that serves all different sects. Thank God, this school was established, which now serves the young before the old and serves women and men, but it has become a beacon of science and knowledge.

2 Problem Statement

We are in a rapidly developing world that all societies are witnessing, so it has become necessary for us to provide community leadership that works to direct individuals and groups toward achieving sustainable goals that achieve well-being and growth for society. From this standpoint, community leadership works to strengthen and motivate individuals, and this is the researcher's idea of having community leadership that works to improve volunteer and development initiatives, starting from the community itself under study to all other communities. We will review the impact of community leadership on enhancing volunteer and development initiatives, and we will try to find the vital role of these community leaders.

3 Research Gap

Although some research focuses on the role of leadership in general and its role in volunteer and development work in international, regional, and local communities, there is still an urgent need to understand how to understand community leadership that contributes to strengthening volunteer and development efforts to achieve tangible results in line with Oman Vision 2040. After we returned to the literature [1], we found that this book focuses on the development of local communities and the role of local leaders in developing these communities, whether urban or rural communities. In another study [2], this study aimed to identify the relationship between leadership skills and their impact. On managing volunteer work teams and mechanisms for activating leadership skills to enable social responsibility, In one of the studies in the Journal of the Faculty of Education at Benha University [3], the study focused on the most prominent social manifestations that show the extent of the cohesion of community members, which enhances the social value of the individual and society. Thus, it became clear to the researcher that this problem has previous roots. Through it, it is possible to understand the past and explore the future to find solutions for society and enhance community initiatives. Community leadership was not addressed separately in all previous studies, and there was a scarcity of scientific research that addressed this issue. From this standpoint, the researcher found that there is a rooting of this problem not only at the level of gray literature. And the opinions of the interest, but from the published social manifestations that show the extent of the cohesion of community members, which enhances the social value of the individual and society. Thus, it became clear to the researcher that this problem has previous roots. Through it, it is possible to understand the past and explore the future to find solutions for society and enhance community initiatives. Community leadership was not addressed separately in all previous studies, and there was a scarcity of scientific research that addresses this issue. From this standpoint, the researcher found that there is a rooting of this problem not only at the level of gray literature. And the opinions of the interest, but from the published literature.

4 Theoretical Framework

Leadership in general has linguistic and terminological connotations. Ibn Manzur [4, p. 370] believes that leadership is the triple verb (led), and the origin of the letter alif in the aforementioned verb is the letter waw, to become (drive). Leading is the opposite of the market. It is said that he leads the animal in front of it and drives it. Who is behind it? In terminology, [5, p. 124] sees that leadership is knowing what must be done and making others do it at their will, as defined it as the process of moving people towards the goal. Community leadership is also considered to have a major role in achieving social responsibility because the leader has the qualities and skills that enable him to direct individuals in the right direction for the benefit of all. Thus, he bears responsibility for the success of the company or place he is leading to achieve sustainable development [6].

Volunteer work is also defined as work that is chosen freely, without compensation, to benefit society. Researchers identify four dimensions that must be included in the definition of volunteer work, which are the voluntary nature of the work, the nature of the reward, and the context in which the volunteer work is carried out for the beneficiary [7], and the definition [8]. Volunteering means making an administrative effort based on a specific skill or experience out of desire and choice to perform a voluntary social service duty that has a return on individuals without compensation.

Here we discuss some theories related to community leadership, which represent the problem of the study's research and its impact on enhancing volunteer and development initiatives as follows.

4.1 Servant Leadership Theory

Servant leadership is defined as: a leadership approach in which the main task of the leader is to help and enable workers in their demands to satisfy their needs and ambitions and fulfill their interests. The leader places a high value to serving others at the expense of his own interests [9, p. 245]. It is also known as: leadership is based on developing employees, developing them and meeting their needs, which contributes to increasing their efficiency and creativity, It is the ability to achieve organizational goals by working to develop individuals, meet the needs of employees in the organization, and assume greater responsibilities at work [10, p. 73]. Through the two concepts, it becomes clear to us that servant leadership seeks to serve others, meet their needs, and develop individuals in a way that achieves sustainability. For society, servant leadership has several principles, as pointed out by [11, p. 619], including serving others, which we will discuss in the study.

4.2 Transformational Leadership Theory

Transformational leadership emerged in the late 1970s at the hands of political sociologist James Burns in his book Leadership in 1978, and was defined as "a process in which subordinates are encouraged to participate in setting the organization's vision and setting clear and realistic goals". Transformational leadership has a set of dimensions, but we will address one dimension that is related to the study's research, which is individual considerations, which have a significant and positive impact on changing individuals and groups in society and achieving the organization's goals.

4.3 Participatory Leadership Theory

The concepts of participatory leadership, also known as democratic leadership, have varied. Reference [12, p. 9] defined it as leadership that seeks for the leader to participate with workers in making administrative decisions in an actual and not formal way. Many definitions agreed that participatory leadership is based primarily on participation between workers and leaders. Participatory leadership also has several dimensions, including cooperation and teamwork to achieve the goals of organizations and the sustainability of societies. Through the three theories and their different dimensions, it becomes clear to us that some of these dimensions are related to the research problem.

The theory of servant leadership has the dimension of building society and strengthening social ties and collective spirit among members, and the theory of transformational leadership has the dimension of individual consideration and interest in the needs and aspirations of individuals, and providing personal support and guidance. As for the theory of participatory leadership, it has the dimension of cooperation and teamwork to achieve common goals. From this, we conclude that the three theories share in strengthening community building, trust for individuals, and collective cooperation, which makes these tools effective in achieving community leadership, and leaders can achieve a positive and sustainable impact on their societies.

5 Research Objectives

The aim of this study is to shed light on the impact of community leadership in enhancing voluntary and development initiatives through the three theories of servant, transformational and participatory leadership, and the dimensions of each theory in building society, trust for individuals and collective cooperation in local community institutions in the Sultanate of Oman. This is done through:

1. The impact of the leader's community leadership on the level of community building.
2. The impact of the leader's community leadership on the level of trust for individuals.
3. The impact of the leader's community leadership on the level of collective cooperation.

6 Research Questions

In this section, the researcher will try to present the research questions related to the study objectives by asking the following questions:

1. What does community leadership affect social relations and building and sustaining society?
2. What are the factors that cause building trust between leaders and society?
3. How can a leader contribute to creating an environment that motivates individuals to cooperate?

Through the questions raised for the research, the researcher will try to shed light on community leadership as a fundamental axis and the extent of its impact on society in a way that achieves growth and sustainability.

7 Literature Review and Previous Studies Related to the Study

The literature review is part of scientific research, where the researcher works on studying previous studies related to the research problem identifying similarities and differences between previous studies, and writing a summary of the details of the literature related to the research topic in terms of its objectives, methodology, recommendations and results, extracting criticism and identifying research gaps and strengths and weaknesses. Among this previous literature is the study [2], which aimed to know the correlation between leadership skills and the extent of their impact on the ability of female students to manage volunteer work teams and provide mechanisms for practicing leadership skills to enable community responsibility. This study used the quantitative method using the descriptive correlation method and the qualitative method using the documentary method. The researcher used the questionnaire tool to measure leadership skills and the scale for managing volunteer work teams.

The study [13] also aimed to identify the dimensions of transformational leadership represented in (ideal influence, intellectual consultation, inspirational motivation, and individual consideration) of social responsibility among workers in the Health Affairs Directorate in Asir. The study relied on the descriptive analytical approach through designing and distributing questionnaire forms to know the role of transformational leadership in enhancing the social responsibility of workers. The study [14] also aimed to identify the manifestations of the social specialist's practices of transformational leadership behaviors with members of the civil society association to develop their community participation, such as identifying the manifestations of the behavior of the spirit of challenge and the behavior of communicating effectively, as the importance of identifying the manifestations of the social specialist's practices of transformational leadership behaviors with members of the association lies. All of these studies emphasized a fundamental point, which is the necessity of developing social participation and the role of leaders in influencing initiatives. We also conclude that there is no specialized study on community leadership and its role in enhancing voluntary development initiatives. This is what we will try to search for during the current study and link the three leadership theories (servant—transformational—participatory) as dependent variables through dimensions or hypotheses and link them to the independent variable, which is community leadership.

8 Conceptual Framework

Community leadership here includes community leaders and organizations and is considered the main component of the study to know the extent of its impact on other components of community members. Thus, it appears to us that community leadership through different types of leadership affects the promotion of voluntary and development initiatives. The conceptual framework has three main components that affect each other, the first of which is the independent variable, which is community leadership. The independent variable affects the dependent variable, which is the promotion of voluntary and development initiatives, which is the second component of the conceptual framework. Through the three leadership theories (servant—participatory—analytical), we found that there are three dimensions to the promotion of voluntary and development initiatives (community building—trust for individuals—group cooperation). We can summarize them in the following points:

- The first component (independent): community leadership, which includes the three leadership theories.
- The second component (dependent): promoting voluntary and development initiatives, as a result of the influence of community leadership.

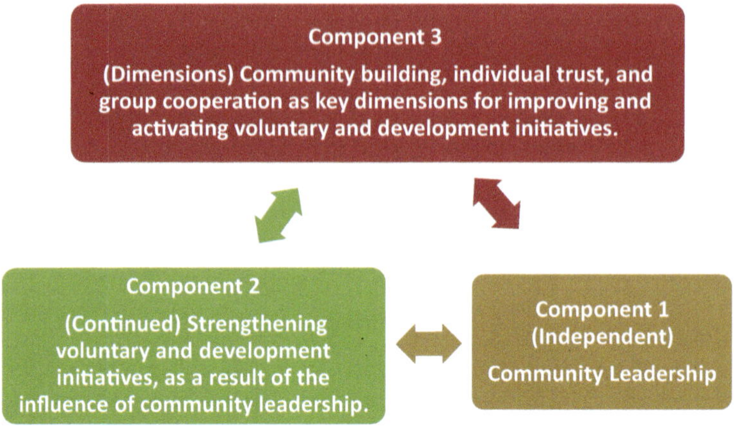

Fig. 1 Conceptual framework

- The third component (dimensions): community building, individual trust, and group cooperation, as key dimensions for improving and activating volunteer and development initiatives.

Figure 1 shows the conceptual framework for the study as follows.

It can be concluded from the above, that conceptual framework of community leadership, through the three theories, leads community leadership to strengthen voluntary and development initiatives, and the variables interact with each other to achieve sustainable development and build an integrated and cooperative society.

Proving the direct relationship between these variables is as stated in Fig. 1, which links the independent variable and the dependent variable. This is what the researcher will seek to study in the coming chapters by solving the study problem and filling the research gap.

References

1. Aleeq, A.M.Y.: Leadership and Community Development. Al-Mutanabbi Library (2013)
2. Al-Qahtani, F.B.M.S.: Leadership Skills in Managing Volunteer Work Teams and Mechanisms for Activating Them to Enable Social Responsibility in Light of the Objectives of the Kingdom's Vision 2030 Program (2023)
3. Saqr, A.A.M.A., Al-Omari, I.B.M.B.I.: The role of school leadership in activating volunteer work in Al-Kharj governorate secondary schools in light of national transformation initiatives. Coll. Educ. J. **30**(1), 406–426 (2019)
4. Ibn Manzur, J.A.M.B.M.: Tongue of the Arabs. Dar Al-Ma'rif for Publishing and Distribution, Cairo (2009)
5. Tarawneh, S.A.: The impact of organizational culture on the job satisfaction of employees. In: Jordanian Commercial Banks. Omdurman Islamic University (2012)
6. Al Khaleej Training and Education.: Community Leadership (2024). Retrieved from https://alkhaleej.com.sa/

7. Schugurensky, D., Mundel, K.: Volunteer work and learning: hidden dimensions of labor force training. International Handbook of Educational Policy, vol. 12. Manchester, UK (2005)
8. Obaid, M.A.R.: Volunteering and Giving. Ministry of Information (General Information Service), Cairo (2006)
9. Najm, N.A.: Administrative Leadership in the Twenty-first Century, 1st edn. Dar Al-Safaa for Publishing and Distribution, Amman (2011)
10. Al-Saleh, A.A.W.: Servant leadership of school principals as perceived by teachers in public education schools in the State of Kuwait and its relationship to their job satisfaction. Coll. Educ. J. **30**(4), 65–89 (2020)
11. Al-Kandari, A.A.R.A.A.T., Mahmoud, J.A.S.: Requirements for activating servant leadership in public education schools in the State of Kuwait. Coll. Educ. J. **36**, 611–628 (2021)
12. Abi, K.: Participatory Leadership and its Relationship to Managing Organizational Conflict in the Organization. Master's thesis, Mohamed Boudiaf University, M'sila (2020)
13. Al-Khadran, A.: The Role of Transformational Leadership in Promoting Social Responsibility. Arab Foundation for Science and Research Publishing (2023)
14. Abu Saleh, M.N., Al-Gamal, M.H.M.: A future vision for developing community participation in the transformative light association of NGOs. J. Hum. Stud. Sect. (2022)

Enhancing Vocational Skills in Adults with Autism: Evaluating the Effectiveness of ICT-Embedded PECS Training Through the Avaz Digital App in Alignment with SDG-4, 2030

Shahala Nassim, M. Prabavathy, Alex Natarajan, and Mona Abdelfattah

Abstract This quasi-experimental study investigated the effectiveness of ICT-embedded Picture Exchange Communication System (PECS) training in enhancing vocational skills among adults with autism spectrum disorder (ASD) aged 14 to 16 years, in alignment with the Sustainable Development Goal 4 (SDG 4) for inclusive and equitable education by 2030. Four participants used a purposive sampling technique in a three-month intervention that combined traditional physical prompting with updated PECS picture videos demonstrated through the Avaz app. The program equips participants to acquire and develop vocational skills autonomously, particularly in block-printing on sarees and necklace making, thus promoting self-sufficiency and job preparedness. This research examined the impact of ICT in enabling skill development, emphasizing the significance of cutting-edge educational approaches in enhancing independence and learning outcomes in adults with ASD. The study led to the achievement of SDG 4 by providing inclusive educational opportunities by supporting ASD adults in developing vocational skills, leading to more focus on digital interventions.

Keywords Vocational skills · Adults with autism · ICT-embedded · PECS training · Avaz digital app · SDG-4

S. Nassim (✉) · M. Abdelfattah
Modern College of Business and Science, Muscat, Oman
e-mail: Shahala.Nasim@mcbs.edu.om

M. Abdelfattah
e-mail: mona.abdelfattah@mcbs.edu.om

M. Prabavathy · A. Natarajan
Centre for Differently Abled Persons, Bharathidasan University, Trichy, India

B. Awwad (ed.), *AI and IoT: Driving Business Success and Sustainability in the Digital Age*, Studies in Systems, Decision and Control 590,
https://doi.org/10.1007/978-3-031-88874-8_70

1 Introduction

The difficulties of adolescents with Autism Spectrum Disorder (ASD) increase when they embark on their maturity period which demands a high intensity of emotional and stress management. It's a transitional period for them to be skilled and trained for a vocational life [1]. This transitional period is an important phase for autistic people and thus education needs to administer various sustainable programs to support them in achieving their career goals. It is essential to identify the unique abilities of autistic people and develop their job skills through innovative teaching methods and programs [2]. Education technology has recently been viewed as a way of meeting diverse academic needs in learning. The Picture Exchange Communication System (PECS), used mainly to enhance communication skills among autistic students, was later adapted for solving learning problems among autistic students [3]. Using PECS as a learning strategy helped in developing autistic students' communicative and behavioural problems [4]. The 2030 SDG 4 aims to provide high-quality education globally for a stable community by expanding learning opportunities and social tolerance [5]. Better learning opportunities for autistic people are fewer and modifications are required to enhance inclusive education for autistic students at classroom levels [6].

The study aims to enhance self-efficacy and job skills among adolescents with Autism Spectrum Disorder (ASD) by embedding Information and Communication Technology ICT with PECS. ICT is widely used among autism for different learning purposes and digital apps are recommended for better learning and communication processes, enhancing social and emotional regulations [7]. This research attempts to analyse the role of technology in the vocational training of autistic teenagers and to develop sustainable education by integrating inclusive and equity education. This research seeks to demonstrate how technology makes learning possible and how teachers, policymakers and other stakeholders can help people with autism achieve success. It is argued that the Avaz Digital App, designed to enhance the communication and social skills of people with autism, is likely to yield better results when embedded with the standard PECS. The Avaz app and PECS can address communication problems differently through engaging training that can yield employability outcomes, which is a new intervention concept.

1.1 Inclusive Education for ASD Using Technology

The term autism integration means not only that children with autism attend school with typical peers but also that learning environments are created that maximise the accommodation of all developmental needs from early childhood to school age. Inclusive education for autism is a wide call to give equal opportunities for all children despite their disabilities. The undertakings of such initiatives are, however, reliant upon several factors such as educator training and modification of practices in classroom instruction [8]. Integrating innovative technical tools is to bring an educational experience that caters to social inclusiveness by incorporating children with autism into regular social settings with the society in which they are involved [9].

Personalized learning experiences can be facilitated through mobile applications that adapt to individual educational requirements and accommodate diverse proficiency levels. Educators can develop targeted strategies for students with Autism Spectrum Disorders (ASD) by using various mobile application apps. Comprehensive digital platforms offer the ability to monitor student advancement and deliver insights to teachers and families, thus ensuring continuous assistance and intervention [10].

Evidence indicates that advanced technologies, like virtual reality (VR), avatar-based pictures, and web-based applications can improve the interest and cognitive skills of children with autism [11–13]. Technology with human interventions can enhance understanding and communication among autistic students which can be extended to their houses with their parents' help [14].

1.2 Effectiveness of the Avaz Digital App for ASD

The Avaz app is a digital application invented in India as an alternative tool for communication that allows children with ASD to express themselves more effectively, overcoming barriers to functional speech. The app integrates pictures and cards, making it suitable for children with ASD, Cerebral palsy, Down's syndrome and other disabilities. It is available on Android and iPhone in different languages and thus can be applied with multi-linguistic children. The app is less expensive, and minimal data is required than other apps. The app proved to be a helping tool to develop social interaction among ASD students [15]. The app can also be used for children with communication difficulties associated with Cerebral palsy. The tool proved its efficiency with Cerebral palsy children in improving their communication intention and thus a recommended tool for children with complex language difficulties and motor functions [16]. Children preferred using the Avaz app over traditional methods, highlighting its engaging nature and highly recommended by educators as an educational tool for ASD children as it helped to cater for learners through its multiple versions of text and images (17).

2 Literature Review

2.1 *Assistive Techniques Used for Autism Spectrum Disorder (ASD) Individuals*

Autism Spectrum Disorder (ASD) is a developmental disability that impacts perception, communication, behaviour, and social interaction. Repetitive behaviours and restricted activities characterize it. The condition varies widely in severity, making everyone's experience unique. ASD affects people from all backgrounds and can have long-term effects, but with proper support, individuals can lead fulfilling lives and exercise their rights, avoiding marginalization [18]. Weitlauf et al. [19] reviewed studies focused on behavioural interventions for children with ASD and found that high-intensity Applied Behavior Analysis (ABA) notably improved cognitive and language skills compared to community controls and other treatments. Early Intensive Behavioral Intervention (EIBI), a prominent ABA-based approach, was especially effective in enhancing communication skills. Many studies have shown positive results with various training strategies for autism, such as ABA, TEACCH, LEAP, ESDM, and DHSCP. Among these, ABA is the only method that has consistently demonstrated evidence-based improvements in the core challenges associated with autism. However, this field requires further investigation as there is no specific method to approach autism [20]. The appropriate integration of various methods according to the need will succeed in addressing ASD challenges.

2.2 *Effectiveness of PECS Training in Adults*

Picture Exchange Communication System (PECS) is a communication tool used to help children with autism express needs and ideas through pictures. Research suggests that PECS is effective in the early stages of language instruction, especially for teaching English vocabulary and basic conversational skills [21]. The National Institute for Health and Care Excellence (NICE) recommends group or individual social learning programs to support children and young people with autism, aiming to improve social interaction deficits. These programs use behavioural therapy techniques such as video modelling, peer feedback, imitation, and reinforcement to teach appropriate social behaviours [22]. Evidence from observational studies indicates that social skills groups can be effective in enhancing social interaction in adults with autism [23].

Picture Exchange Communication System (PECS) proved to be an effective tool for communication initiation among adults with developmental disabilities. The findings also showed that PECS improved their self-dependency and reduced problem behaviours [24]. Jaffal [25] analysed 15 studies on Video-Based Interventions (VBIs) with ASD students and found development in vocational, life, and academic skills respectively.

2.3 Impact of Technological Interventions on ASD Employment Opportunities

Stoner et al. [26] implemented PECS training with five non-speaking adults with intellectual disabilities and found three participants developed functional communication skills. However, two patients showed limited progress. This research highlights the importance of teaching functional communication to nonspeaking adults and suggests areas for future research to improve training effectiveness and outcomes. The Virtual Reality for Vocational Rehabilitation (VR4VR) system [27], developed at the University of South Florida's CARRT, uses virtual reality to assess and train individuals with severe cognitive and physical disabilities. The system is affordable, safe, and adaptable to individual needs, with customizable environments and real-time intervention. It includes a VR training area, optical motion tracking, curved screen, server computer, tablet interface for job coaches, and virtual assistive robot. The system focuses on training underserved individuals with cognitive disabilities such as autism and traumatic brain injury through six transferable skill modules aimed at improving job readiness, independence, and life skills. This scoping review identified the most used features of evidence-based digital interventions in mobile technologies for autism spectrum disorder (ASD), offering valuable insights for digital health human–computer interaction designers [28]. The review analyzed 42 studies on mobile apps designed to teach life skills to individuals with autism spectrum disorder (ASD). These apps address ASD learners' needs by personalizing content and breaking tasks into smaller steps using multimedia elements. Mobile technologies help to build routines and extend school interventions at home. Future research should aim to improve the accessibility of evidence-based interventions for learners, caregivers, and educators to support the well-being of individuals with ASD.

A study conducted by Van Larrhoven et al. [29] aimed to compare mobile technologies and universally designed prompting systems to enhance independent vocational performance in adolescents with autism spectrum disorder (ASD) and/or intellectual disabilities (ID). The results showed that both devices increased independent responses for three out of the four participants. The study also explored video modelling for improving social-communication skills in adults with autism and found variable effects. This research highlights the challenges ASD individuals face with high unemployment rates and offers insights into effective evidence-based interventions [30], making it exceptionally challenging to secure and maintain employment. The severe unemployment rates for adults with ASD, coupled with the high costs of services, underscore the importance of identifying best practices for employment support. This review examined 20 studies on vocational training for adolescents and adults with ASD, focusing on pre-employment, vocational skill development, and job retention. This highlights emerging trends and offers recommendations for future research.

The study conducted by Gollash et al. [31] examined the challenges faced by autistic individuals in vocational training and their transition into the labour market in Germany, focusing on how assistive technologies (AT) can improve employment opportunities. It identifies obstacles in workplace settings, particularly office environments, and highlights how tailored AT can enhance inclusion and support better employment outcomes. Babb et al. [32] found the effectiveness of video-based visual scene displays (VSDs) through a tablet-based AAC app to help adolescents with autism or Down syndrome complete volunteer activities. The findings suggest that video VSDs enhance participation by teaching new skills and supporting communication, thus improving involvement for individuals with complex communication needs. Smith et al. [33] reviewed evidence-based interventions to increase employment opportunities for individuals with disabilities by analyzing 46 articles. The study found strong evidence supporting ongoing support and social skills training for individuals with mental health disabilities. Moderate evidence has shown the effectiveness of simulation techniques, assistive technologies, and apps for work participation in individuals with various disabilities.

3 Hypothesis of This Study

a. ICT-embedded PECS training through the Avaz Digital App significantly enhanced vocational skills in adults with autism.

4 Or

b. ICT-embedded PECS training through the Avaz Digital App does not significantly enhance vocational skills in autistic adults.

5 Methodology of This Study

5.1 Study Design

This quasi-experimental study utilized a mixed-methods pre-test/post-test design to evaluate the effectiveness of ICT-embedded Picture Exchange Communication System (PECS) training in enhancing vocational skills in adolescents with Autism Spectrum Disorder (ASD). This approach allowed for the comparison of participants' skills before and after the intervention, providing insights into the impact of training. Four adolescents aged 14 to 16 years with autism spectrum disorder were purposively selected based on their willingness to participate, individual learning needs, and prior exposure to digital tools. Initial assessments evaluated participants' vocational skills and familiarity with technology, involving practical demonstrations of tasks such as block printing and necklace-making. The intervention spanned three months and combined traditional physical prompting with ICT-embedded PECS training using the Avaz app.

In the present study, a modified ABAB single-subject design [26] was employed to examine the changes in functional communication skills. Four participants went through all four PECS training steps easily and successfully mastered their skills in the home and the community. However, two individuals showed limited progress, and the PECS training did not significantly improve their communicative competence. The findings highlight the potential of PECS to enhance communication for some non-speaking adults while suggesting that individualized approaches may be necessary for those who struggle with this method. This study discusses implications for teaching functional communication and recommends further research to optimize PECS training for diverse needs.

5.2 *Training Components*

a. Video Modelling: Participants learned to identify and select video prompts for specific vocational tasks.
b. Task Execution: Each session included watching instructional videos and hands-on practice to complete the tasks, referencing the video for guidance.
c. Gradual Complexity: Training began with simple tasks and progressively increased in complexity as the participants mastered each level.

5.3 *Training Process*

a. Introduction to ICT Tools: Participants were familiarized with the Avaz app and instructed to select the PECS picture videos.
b. Video Selection: Participants learned to identify and click on the correct video for each task.
c. Video Watching: Participants observed the videos, following the instructions provided.
d. Task Execution: Participants practised the tasks with support, following the sequence in the video.
e. Increasing Complexity: Tasks started with one- or two-step instructions and became more complex as participants mastered more straightforward tasks.

5.4 *Post-Intervention Assessment*

Post-test: After three months, a post-test assessment evaluated participants' progress in vocational skills using the same tools as the pre-test for consistency. The results

Table 1 Quantitative analysis of the study

Score	test	Sample	Mean	SD	t-value	**Remark**
ICT-embedded PECS training using the Avaz app	Pre-test	4	11.3	5.2	4.3	**Significant at 0.05 level**
	Post-test		24.1	9.01		

from the pre-test and post-test were compared to measure advancements in skill acquisition and independence in task completion.

6 Data Analysis

Quantitative Analysis was used to compare pre-and post-test scores, assessing the significance of skill improvements.

As shown in Table 1, this study examined the impact of ICT-embedded PECS training delivered through the Avaz Digital App on vocational skills in adults with autism. The research compares pre-and post-test scores to determine whether the app significantly enhances skill development. Initial testing yielded a mean score of 11.3 (SD = 5.2), while post-intervention scores increased to a mean of 24.1 (SD = 9.01). A t-test revealed a t-value of 4.3, significant at 0.05, indicating a meaningful improvement, likely due to the intervention. This substantial improvement in post-test scores, with slight differences among learners in the mental strategies they employed, is used to refute the Null hypothesis, demonstrating that Avaz-mediated PECS training effectively increases the vocational skills in this population. The research recommends integrating ICT solutions, e.g., Avaz, and application for vocational training and special education curricula to enhance skill development among autistic adults.

7 Discussion and Conclusion

Research findings showed that the use of a digital app integrated with Picture Exchange Communication System (PECS) training can develop vocational competencies in teenagers with Autism Spectrum Disorder (ASD). The experimental group acquired the job skills for block printing and necklace making through video observations. This approach enhanced their self-efficacy in handling technology for their learning processes. These results align with the contributions made in previous research [31, 32] where video modelling is an inevitable intervention to develop vocational skills independently and research on video modelling is highly recommended.

The study highlights the benefits of evidence-based interventions by using digital cues to improve participants' motivation and involvement conforming to the results of Smith et al. [33]. This approach promotes the main aim of SDG Goal 4, which abides by the notion of inclusive education for all irrespective of their disabilities [5, 6]. The customization of the learning process according to individual skills and job demands enhanced the effectiveness of digital education with PECS. Such tailored learning is essential to develop vocational skills according to the needs of both individuals and organizations [2].

The study showed the prospects of conducting research with digital technologies to transform vocational training for people with Autism Spectrum Disorder (ASD) at low cost and effort. However, due to the smaller sample size and short period, more comprehensive studies to verify the study results with long-term effects of ICT training are required. The efficiency of technology in developing vocational skills among ASD adults can be analyzed qualitatively from the educators' and parents' perceptions as it can help researchers develop structured programs tailored according to the learners' needs [34].

The study recommends expanding the research on cultivating vocational skills in different job domains. More studies can be conducted to compare ICT-integrated programs and traditional training modalities for supporting adolescents with Autism Spectrum Disorder (ASD) to reach their vocational goals. This paper contributes to the ongoing discussion around the inclusion of others in the curriculum and the value of new teaching styles in achieving self-sufficiency in adolescents on the autistic spectrum.

References

1. White, S.W., Smith, I.C., Miyazaki, Y., Conner, C.M., Elias, R., Capriola-Hall, N.N.: Improving Transition to adulthood for students with autism: a randomized controlled trial of STEPS. J. Clin. Child Adolesc. Psychol. **50**, 187–201 (2019). https://doi.org/10.1080/15374416.2019.1669157
2. Zhou, K., Alam, B., Bani-Fatemi, A., Howe, A., Chattu, V.K., Nowrouzi-Kia, B.: Autism spectrum disorder in the workplace: a position paper to support an inclusive and neurodivergent approach to work participation and engagement. Discov. Psychol. 4 (2024). https://doi.org/10.1007/s44202-024-00150-5
3. Khoiriyah, K.: Picture exchange communication system (PECS): a strategy to improve children's speaking ability. J. Early Child. Care Educ. **3**, 111 (2021). https://doi.org/10.26555/jecce.v3i2.2589
4. Malhotra, S., Rajender, G., Bhatia, M.S., Singh, T.B.: Effects of picture exchange communication system on communication and behavioral anomalies in autism. Indian J. Psychol. Med. **32**, 141–143 (2010). https://doi.org/10.4103/0253-7176.78513
5. Adipat, S., Chotikapanich, R.: Sustainable development goal 4: an education goal to achieve equitable quality education. Acad. J. Interdisc. Stud. **11**, 174 (2022). https://doi.org/10.36941/ajis-2022-0159
6. Petersson-Bloom, L., Holmqvist, M.: Strategies in supporting inclusive education for autistic students—a systematic review of qualitative research results. Autism Dev. Lang. Impairments. **7**, 239694152211234 (2022). https://doi.org/10.1177/23969415221123429
7. López-Díaz, J.M.: Analysis of the impact of ICT through apps on students with autism spectrum disorder: a systematic review of the literature. Int. J. Inform. Educ. Technol. **14**, 785–790 (2024). https://doi.org/10.18178/ijiet.2024.14.6.2103

8. Johnson, A.M., Williams, S., Wilson, J.E.: The impact of teacher training programs on the implementation of inclusive education for children with autism in urban public schools in the United States. Res. Adv. Educ. **3**, 16–20 (2024). https://doi.org/10.56397/rae.2024.04.03
9. Konstantinos, N.: Ways of social inclusion and integration of children with autism in secondary education, inside and outside the school context, in the region Thessaly. J. Ecohumanism. **3**, 1484–1499 (2024). https://doi.org/10.62754/joe.v3i7.4311
10. Moraiti, I., Fotoglou, A., Drigas, A.: Digital and mobile applications for autism inclusion. Int. J. Online Biomed. Eng. (iJOE). **19** (2023). https://doi.org/10.3991/ijoe.v19i11.37895
11. Tsiopela, D., Jimoyiannis, A.: Using a web-based environment to enhance vocational skills of students with autism spectrum disorder. In: Springer eBooks, pp. 381–395 (2018). https://doi.org/10.1007/978-3-319-95059-4_23
12. Zhang, M., Ding, H., Naumceska, M., Zhang, Y.: Virtual Reality Technology as an Educational and Intervention Tool for Children with Autism Spectrum Disorder: Current Perspectives and Future Directions (2022)
13. Ren, Y., Liu, R., Sang, H., Yu, X.: Avatar-based picture exchange communication system enhancing joint attention training for children with autism. IEEE J. Biomed. Health Inform. 2024 Oct 29. https://doi.org/10.1109/JBHI.2024.3487589. Epub ahead of print. PMID: 39471110
14. Maseri, M., Mamat, M., Yew, H.T., Chekima, A.: The implementation of application software to improve verbal communication in children with autism spectrum disorder: a review. Children **8**, 1001 (2021). https://doi.org/10.3390/children8111001
15. Sonawane, J.V., Varshneya, H.: AVAZ application (trial version)—a voice for the nonverbal children with autism spectrum disorder: a pilot study. Indian J. Occup. Ther. **52**, 8 (2020). https://doi.org/10.4103/ijoth.ijoth_2_20
16. Sreekumar, S.: Using AVAZ to enhance communicative abilities of a child with cerebral palsy. Disabil. CBR Inclusive Dev. **25**, 95 (2014). https://doi.org/10.5463/dcid.v25i1.289
17. Sankardas, S.A., Rajanahally, J.: IPad: efficacy of electronic devices to help children with autism spectrum disorder to communicate in the classroom. Support Learn. **32**, 144–157 (2017). https://doi.org/10.1111/1467-9604.12160
18. Martolia, D., Gupta, R.: Assistive technology for autistic children: a review. Int. J. Curr. Microbiol. Appl. Sci. **9**, 2170–2175 (2020). https://doi.org/10.20546/ijcmas.2020.907.253
19. Weitlauf, A.S., McPheeters, M.L., Peters, B., Sathe, N., Travis, R., Aiello, R., Williamson, E., Veenstra-VanderWeele, J., Krishnaswami, S., Jerome, R., Warren, Z.: Therapies for Children With Autism Spectrum Disorder: Behavioral Interventions Update [Internet]. Rockville (MD): Agency for Healthcare Research and Quality (US); 2014 Aug. Report No.: 14-EHC036-EF. PMID: 25210724
20. Revena, F.-P.: Department of Physical Education and Sports Science, University of Athens, GREECE: Teaching strategies for children with autism (2015). https://doi.org/10.7752/jpes.2015.01024
21. Bondy, A.S., Frost, L.A.: The picture exchange communication system. Focus Autistic Behavior. **9**, 1–19 (1994). https://doi.org/10.1177/108835769400900301
22. National Collaborating Centre for Mental Health (UK).: Autism: Recognition, Referral, Diagnosis and Management of Adults on the Autism Spectrum. Leicester (UK): British Psychological Society; 2012. PMID: 26065067
23. Hillier, A., Fish, T., Cloppert, P., Beversdorf, D.Q.: Outcomes of a social and vocational skills support group for adolescents and young adults on the autism spectrum. Focus Autism Other Dev. Disabil. **22**, 107–115 (2007). https://doi.org/10.1177/10883576070220020201
24. Conklin, C.G., Mayer, G.R.: Effects of implementing the picture exchange communication system (PECS) with adults with developmental disabilities and severe communication deficits. Remedial Spec. Educ. **32**, 155–166 (2010). https://doi.org/10.1177/0741932510361268
25. Jaffal, M.A.: Video-based interventions for adolescents and young adults with autism spectrum disorder: a systematic review. Int. J. Mental Health Promot. **25**, 881–890 (2023). https://doi.org/10.32604/ijmhp.2023.028982

26. Stoner, J.B., Beck, A.R., Bock, S.J., Hickey, K., Kosuwan, K., Thompson, J.R.: The effectiveness of the picture exchange communication system with nonspeaking adults. Remedial Spec. Educ. **27**, 154–165 (2006). https://doi.org/10.1177/07419325060270030401
27. Bozgeyikli, L., Bozgeyikli, E., Clevenger, M., Gong, S., Raij, A., Alqasemi, R., Sundarrao, S., Dubey, R.: VR4VR: towards vocational rehabilitation of individuals with disabilities in immersive virtual reality environments. IEEE **42**, 29–34 (2014). https://doi.org/10.1109/vaat.2014.6799466
28. Villamin, G., Luppicini, R.: Digital interventions using mobile technologies for life skills development of learners with autism spectrum disorder: a scoping review. Int. J. Dev. Disabil. 1–14 (2024). https://doi.org/10.1080/20473869.2024.2380947
29. Van Laarhoven, T., Carreon, A., Bonneau, W., Lagerhausen, A.: Comparing mobile technologies for teaching vocational skills to individuals with autism spectrum disorders and/or intellectual disabilities using universally-designed prompting systems. J. Autism Dev. Disord. **48**, 2516–2529 (2018). https://doi.org/10.1007/s10803-018-3512-2
30. Seaman, R.L., Cannella-Malone, H.I.: Vocational skills interventions for adults with autism spectrum disorder: a review of the literature. J. Dev. Phys. Disabil. **28**, 479–494 (2016). https://doi.org/10.1007/s10882-016-9479-z
31. Gollasch, D., Engel, C., Branig, M., Schmalfuß-Schwarz, J., Weber, G.: Unlocking opportunities: empowering autistic adults in vocational training and employment through assistive technologies. In: Lecture Notes in Computer Science, pp 32–46 (2024). https://doi.org/10.1007/978-3-031-60884-1_3
32. Babb, S., McNaughton, D., Light, J., Caron, J., Wydner, K., Jung, S.: Using AAC video visual scene displays to increase participation and communication within a volunteer activity for adolescents with complex communication needs. Augment. Altern. Commun. **36**, 31–42 (2020). https://doi.org/10.1080/07434618.2020.1737966
33. Smith, D.L., Atmatzidis, K., Capogreco, M., Lloyd-Randolfi, D., Seman, V.: Evidence-based interventions for increasing work participation for persons with various disabilities. OTJR Occup. Ther. J. Res. **37**, 3S-13S (2016). https://doi.org/10.1177/1539449216681276
34. Li, G., Zarei, M.A., Alibakhshi, G., et al.: Teachers and educators experiences and perceptions of artificial-powered interventions for autism groups. BMC Psychol. **12**, 199 (2024). https://doi.org/10.1186/s40359-024-01664-2

Vulnerability Detection: Dynamic Analysis of Web Applications and Assessment of Penetration Testing Tools

Rahil Al Majizi, Hothefa Shaker, Basant Kumar, and Zeyad T. Sharef

Abstract This research investigates strategies for expediting the discovery of vulnerabilities in web applications through dynamic analysis, focusing specifically on penetration testing. Recognizing the imperative for the swift identification of security weaknesses in the dynamic web environment, the study explores the efficiency of penetration testing methodologies. By simulating real-world attack scenarios, this research aims to provide insights that enhance the effectiveness of security assessments. The findings offer valuable recommendations for fortifying web applications against emerging cyber threats, aligning to accelerate vulnerability discovery.

Keywords Web application · Injection attacks · Penetration testing · Sqlmap · Zaproxy

1 Introduction

A web application, often called a web app, is a program that operates on web servers and is accessed by users through a web browser over a network, usually the Internet. Unlike traditional desktop applications, web apps must be accessed through a browser and provide a user interface for interaction, typically using a client–server architecture. In the dynamic field of web applications, security faces continuous challenges from evolving cyber threats. This research focuses on tailored penetration testing strategies for web applications in overloaded server environments, addressing vulnerabilities intensified by excessive loads. Penetration testing, a proactive security approach, simulates real-world attacks to uncover and mitigate potential risks. Web application security and penetration testing uncover a steady

R. Al Majizi · H. Shaker (✉)
Modern College of Business and Science, Muscat, Oman
e-mail: hothefa.shaker@mcbs.edu.om

B. Kumar · Z. T. Sharef
College of Electronics Engineering, Ninevah University, Mosul, Iraq
e-mail: zeyad.tariq@unoninevah.edu.iq

B. Awwad (ed.), *AI and IoT: Driving Business Success and Sustainability in the Digital Age*, Studies in Systems, Decision and Control 590,
https://doi.org/10.1007/978-3-031-88874-8_71

center on the significance of defending advanced resources against advancing cyber dangers [1]. Analysts commonly advocate for testing as a proactive and basic methodology for distinguishing and tending to vulnerabilities in web applications. Whereas various ponders emphasize the need to recreate real-world assault scenarios, there's a discernible hole within the writing concerning the particular challenges displayed by over-burden servers. A subset of inquiries recognizes the increased helplessness of web applications beneath intemperate server loads, emphasizing the requirement for custom-fitted testing strategies. These considerations emphasize the centrality of understanding how server over-burden conditions can worsen security dangers, requiring a nuanced approach to entrance testing. The agreement is that the versatility of security conventions is foremost, considering the energetic scene of cyber dangers [2]. The current considers points to bridge this crevice by delving into the complexities of overloaded web servers due to some factors. A common vulnerability and issue associated with web applications is Debugging, which constitutes a routine task in both the development and maintenance phases of Single Page Applications (SPAs). The study aims to enhance the resilience of web applications by providing insights and practical recommendations for optimizing security protocols in dynamic server conditions. This paper is organized as follows: Sect. 2 provides a literature review on the process of penetration testing, followed bycommon related work (Sect. 3). Section 3 discusses the methodology by using different tools. Additionally, give the comparison of the result and analysis in Sect. 4.

2 Literature Review

2.1 Common Attacks and Threats with Web Applications

A Denial-of-Service (DoS) attack on a web application with a noxious endeavor to disturb its accessibility by inundating the focused-on server with an enormous volume of activity [2]. This may be shown through different procedures, counting Volumetric, Convention, and Application Layer threats, pointing to overpowering organization or application assets. Common strategies such as Ping requests, HTTP requests, Slowloris attacks, and SYN/ACK floods are conveyed to deplete server capabilities and hinder ordinary operations. Moderation procedures include activity sifting, rate constraining, stack adjusting, and the utilization of Web Application Firewalls (WAF) and Substance Conveyance Systems (CDN) to identify and relieve malevolent activity, guaranteeing the maintained accessibility and versatility of the net application against potential disturbances.

SQL injection (SQLi) could be a cyber threat that misuses vulnerabilities in a web application dealing with SQL language, empowering unauthorized get to a database. In this attack, SQL injection is filled into client inputs, such as shape areas or URL parameters, deceiving the application into executing unintended database inquiries. On the off chance that fruitful, aggressors can control, recover, or indeed erase delicate information. For example, a login page defenseless to SQLi may be controlled with an input like ' OR '1' = '1'–, bypassing confirmation and giving unauthorized get to. Preventative inquiries andorite input approval, parameterized inquiries [3].

Cross-Site Scripting (XSS) could be a predominant web application threat wherein scripts are injected into web pages and executed within the browsers of clueless clients. By misusing uncertain input areas, attackers can inject scripts that take delicate data, such as session treats, or control the substance displayed to clients. For occasion, a powerless look bar might permit the input, driving to the execution of self-assertive code. Relieving XSS includes input approval, yield encoding, and utilizing security instruments like the Substance Security Approach (CSP) to confine script execution, subsequently defending clients from potential bad script injections.

2.2 *Web Application Penetration Testing*

It is a pivotal feature of moral hacking, including efficiently surveying the security of online stages to distinguish and correct potential vulnerabilities that can be misused by some people. Penetration testing (PT) could be a strategy for evaluating and assessing the security of computerized resources by arranging, creating, and executing conceivable assaults that point to finding and abusing vulnerabilities [4]. Figure 1 shows the Phases of the presentation test.

The insightful investigation into quickening the revelation of vulnerabilities through energetic examination in web applications utilizing infiltration testing

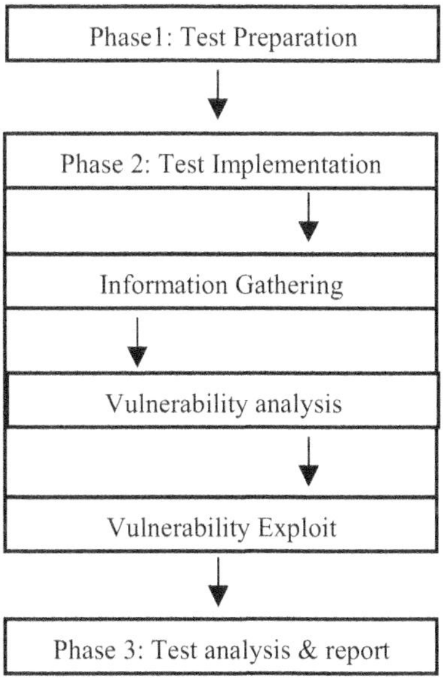

Fig. 1 Penetration testing phases

uncovers a nuanced and comprehensive scene of inquiry about and built-up hones. A huge number of scholarly requests underscore the fundamental significance of proactive security measures, with a particular center on the techniques inherent to entrance testing. This accentuation positions entrance testing as an essential approach for the speedy recognizable proof of vulnerabilities, contributing altogether to the continuous talk inside the scholastic community.

Ibarra-Fiallos presented web applications continue to face a high incidence of injection attacks, with recognized organizations such as OWASP consistently ranking them among the top ten security risks for web applications [5].

Wang et al. [6] discussed the address issues encountered during runtime, developers frequently engage in the process of bug reproduction as an initial step to comprehend and ultimately rectify observed bugs. Web applications face a spectrum of common vulnerabilities that pose significant threats to their security.

Gandikota et al. [7] examined that SQL Injection (SQLi) exploits weaknesses in database layers, allowing unauthorized access or data manipulation. Cross-site scripting (XSS) enables attackers to inject malicious scripts, compromising user data or manipulating content. Cross-site request Forgery (CSRF) tricks authenticated users into unintended actions, exploiting fake requests. Security misconfigurations and Insecure.

Chorás et al. [8] studied the list of beat 10 most basic dangers related to web applications security, given by OWASP (Open Web Application Security Venture) demonstrates Injection (counting Organized Inquiry Dialect (SQL), Working Framework (OS) and Lightweight Registry Get to Convention (LDAP) infusions) as a major powerlessness.

3 Methodology

3.1 Requirements and Parameters Identification

This section is based on exploring the common types of vulnerabilities in web applications. This section aims to show the different tools used by penetration testers. In addition to analyze the testing results from the tools to examine the effectiveness of each.

Target system: Damn Vulnerable Web Application (DVWA) is a room in TryHackMe website. It is a vulnerable machine for demo testing and scanning for web application vulnerabilities.

Testing Environment: the testing is done on the penetration testing platform (*kali*) is running on the Linux operating system. The classification of the tools needed are illustrated in Table 1 to ensure the success of the implementation.

Table 1 Requirements needed for implementation

Software	Operating system & Version
VirtualBox	Oracle VM VirtualBox 7.0.8
Kali Linux	Kali GNU 2023.1
Vulnerable machine	DVWA on TryHackMe
Testing tools and others	SQLmap, Zaproxy, nessuss, whatweb, Draw oi

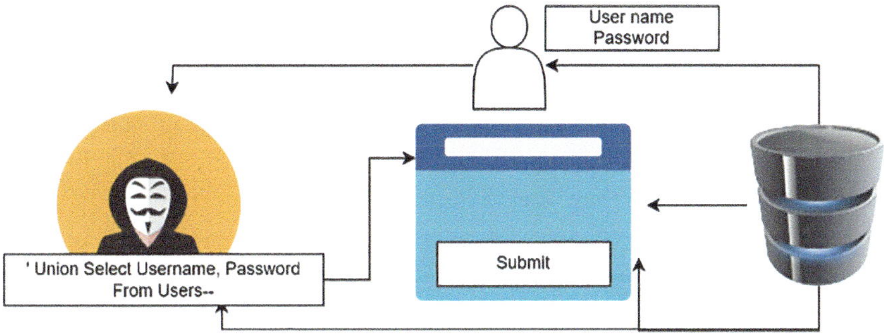

Fig. 2 SQL injection attack

Recently, a cluster of threats has risen, focusing on frameworks and computers inside web applications to pick up unauthorized get to and misuse delicate information. At the same time, other assaults influence organizes assets by actuating busyness, rendering them inaccessible to veritable clients. This troublesome strategy, frequently coming about in refusal of benefit, includes the injection of malicious code and other malevolent exercises. Figure 2 explains one of the common attacks in web applications (SQL injection attack).

Web application vulnerabilities pose security risks by exposing vulnerabilities that malicious actors can exploit. One of the risky examples of web application vulnerability is SQL injection, where attackers manipulate input fields to inject malicious SQL code into a database query. This unauthorized access can lead to unauthorized data retrieval, manipulation, or even deletion, compromising the confidentiality and integrity of sensitive information.

Dynamic Analysis Parameters

- **Types of Vulnerabilities**: Through this paper, we highlight the common web application vulnerabilities.
- **Testing Techniques**: Outline the dynamic analysis techniques to be utilized, including automated scanning, manual testing, and fuzzing.

Tool Configuration: Detail the configuration settings for each tool, including:

- **SQLMap (Manual)**: need to set the DVWA IP, Scan the number of databases and types, and injection techniques using cookies.

- **WhatWeb (Manual)**: recognize the level, information about user-agent, and proxy settings.
- **ZAPROXY (Automated)**: Proxy configuration, set automated scan, identify the target IP.
- **Nessus (Automated)**: Select the advanced scan and test, and set the IP and description of the target machine.

Manual Testing Using Sqlmap Tool

To run sqlmap, it needs to identify and understand the available databases, tables, columns, and cookies. The use of sqlmap is to test the website from sql injection attack.

To check all the scripts or flags used with this tool by run*: man sqlmap or sqlmap –help.*

With the help of this tool, we generate the table name and dumb all users to get their credentials.

Command: *python sqlmap.py -u < target url > -D dvwa -T users –column.*

Flags:

- -D: Specify the name of database
- -T: Specify the name of table
- -columns: Specify the columns that SQLMap should retrieve from the database

Manual Testing Using Whatweb: Whatweb tool helps to identify websites. It recognizes web technologies including content management systems (CMS), blogging platforms, statistic/analytics packages, JavaScript libraries, web servers' status and numbers, and embedded devices.

Automated Testing Using Zaproxy: The OWASP Zed Attack Proxy (ZAP) is an open-source tool created by the Open Web Application Security Project (OWASP) for web application security testing. It assists developers, security testers, and penetration testers in identifying vulnerabilities and security flaws in web applications. After the testing is completed, there will be different sections that can help the investigator. History, search, alerts, output, WebSockets, spider and active scan.

Automatic Implementation Using Nessus Essentials: Nessus, developed by Tenable, is a platform that scans for security vulnerabilities across devices, web applications, operating systems, cloud services, and other network resources. It is a remote security scanning tool that scans a web application and alerts you if it finds any vulnerabilities that malicious hackers could exploit to access any network-connected computer or server.

Performance Metrics

- **Detection Rate**: Identify the number of vulnerabilities detected by each tool. Table 2 illustrates the type of vulnerabilities.

Table 2 Type of vulnerabilities

Tool	Vulnerability or alert
SQLMAP	SQL injection
Whatweb	0
Zaproxy	XSS, SQL, CSRF
Nessus	Weak encryption algorithms, Cipher Block Chaining, ICMP timestamp request, enable MD5 and 96-bit MAC algorithms

4 Result and Discussion

Table 3 summarizes the output of each tool from the four security testing tools: SQLmap, WhatWeb, ZAP (Zed Attack Proxy), and Nessus. Each tool focuses on a specific aspect of security assessment. SQLmap focuses on identifying SQL injection vulnerabilities and extracting database information, while WhatWeb gives details about web servers, CMS identification, and security mechanisms. ZAP identifies various security vulnerabilities like XSS and CSRF, and provides web server details and HTTP request/response information. Nessus offers comprehensive vulnerability scan results, including CVE identifiers and risk ratings, along with remediation recommendations. Together, these tools provide a holistic view of potential security weaknesses in web applications and networks.

SQLmap is a powerful tool that automates the detection and exploitation of SQL injection vulnerabilities. With the –cookie flag, it can incorporate specific cookies, such as PHPSESSID, in its requests. This feature is especially beneficial for web applications that use session cookies to manage user sessions. By including these cookies, SQLmap can authenticate as a specific user, enabling access to restricted sections of the application and extracting hash values and database results.

WhatWeb tool identifies technologies used by websites and provides details about cookies like PHPSESSID and their security attributes, and gives a comprehensive understanding of a web application's security posture. WhatWeb might reveal that the PHPSESSID cookie lacks the Secure and HTTP Only flags similar to the result on DVWA, indicating potential vulnerabilities. This information can be used

Table 3 Tool's output

Tools	Output result
SQLmap	Identified SQL injection vulnerabilities, Database names, tables, and columns, Extracted data from databases, Database user and version information and hash value of the password
Whatweb	Web server details (e.g., Apache, Nginx), CMS identification, Frameworks and libraries used Security mechanisms
Zaproxy	XSS, CSRF, alert, risk level, HTTP request and response
Nessus	Vulnerabilities, risk levels, recommendation

by SQLMAP to craft more effective attacks, exploiting these weaknesses to gain unauthorized access to sensitive data.

SQLmap and WhatWeb are essential tools in web security and penetration testing, offering several common features. Both are open-source and operate primarily through command-line interfaces, which allows for easy scripting and integration into automated workflows. They streamline the process of identifying vulnerabilities and collecting crucial information, significantly improving the efficiency of security assessments. While SQLmap focuses on detecting SQL injection vulnerabilities and extracting detailed database information, WhatWeb excels in providing a comprehensive analysis of web technologies, CMS platforms, and security mechanisms. Despite their different focuses, both tools are valuable for reconnaissance, offering unique insights that contribute to a thorough understanding of a web application's security posture. Together, they form a robust toolkit for identifying and mitigating potential security threats. Despite their different focuses, SQLmap and WhatWeb share some similarities.

ZAP (Zaproxy) and Nessus are both widely used tools in security testing, but they serve different purposes and excel in distinct areas. ZAP focuses on web application security, offering powerful fuzzing capabilities and an open-source platform ideal for identifying vulnerabilities in web applications. In contrast, Nessus provides comprehensive vulnerability scanning across various systems, including networks, servers, and applications. Its integration with enterprise security frameworks makes it a strong option for holistic security assessments. Key differences between the two include:

- **Scope**: ZAP focuses on web applications, while Nessus covers a broader range of systems.
- **Cost**: ZAP is free and open source, whereas Nessus is a commercial tool with a free version available.
- **Community Support**: ZAP benefits from a large open-source community, while Nessus has dedicated support from Tenable.

Zaproxy generates a report with all the web application vulnerabilities with the level of impact of each vulnerability as shown in Fig. 3.

Alert of web application vulnerabilities with the level of risk as illustrated in Fig. 4 can be generated from the report provided in Fig. 3.

Through Nessus Essential tool in Fig. 5, it helps to identify the weaknesses associated with DVWA machine. Besides that, the report generated explain also the possible solutions and level of severity with scores.

The total number of vulnerabilities is 29 (web application vulnerability and non-web-app) classified based on their severity. Moreover, this tool provides more details about the vulnerability through the plugin section. The remote SSH server is susceptible to a man-in-the-middle prefix truncation vulnerability known as Terrapin. This flaw can enable a remote attacker, positioned as a man-in-the-middle, to bypass integrity checks and weaken the connection's security. Nessus excels in web application testing, delivering comprehensive results that are particularly valuable in enterprise environments where integration with other security systems and detailed

		Confidence				
		User Confirmed	High	Medium	Low	Total
Risk	High	0 (0.0%)	0 (0.0%)	0 (0.0%)	0 (0.0%)	0 (0.0%)
	Medium	0 (0.0%)	2 (11.8%)	2 (11.8%)	1 (5.9%)	5 (29.4%)
	Low	0 (0.0%)	1 (5.9%)	4 (23.5%)	0 (0.0%)	5 (29.4%)
	Informational	0 (0.0%)	1 (5.9%)	4 (23.5%)	2 (11.8%)	7 (41.2%)
	Total	0 (0.0%)	4 (23.5%)	10 (58.8%)	3 (17.6%)	17 (100%)

Fig. 3 Zaproxy report

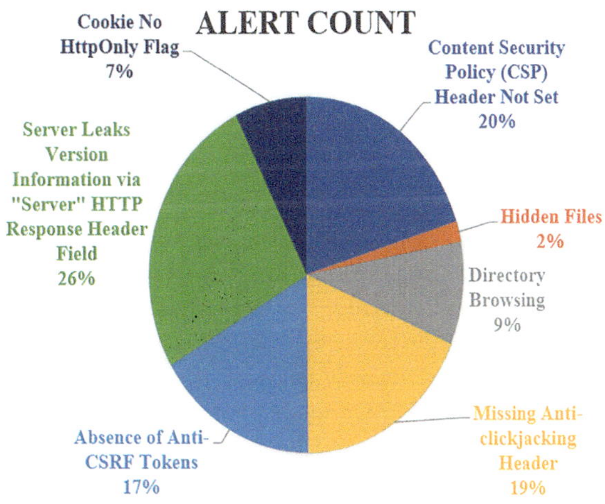

Fig. 4 Alert of web application vulnerabilities

reporting are essential. If you require a tool that addresses a broad spectrum of vulnerabilities across various systems, Nessus is an outstanding option. However, for a solution specifically focused on web application security with extensive fuzzing capabilities, Zaproxy (ZAP) may be more appropriate.

Vulnerabilities Total: 29

SEVERITY	CVSS V3.0	VPR SCORE	EPSS SCORE	PLUGIN	NAME
MEDIUM	5.9	6.1	0.9653	187315	SSH Terrapin Prefix Truncation Weakness (CVE-2023-48795)
MEDIUM	4.3*	-	-	90317	SSH Weak Algorithms Supported
LOW	3.7	3.6	0.5961	70658	SSH Server CBC Mode Ciphers Enabled
LOW	3.7	-	-	153953	SSH Weak Key Exchange Algorithms Enabled
LOW	2.1*	4.2	0.8808	10114	ICMP Timestamp Request Remote Date Disclosure
LOW	2.6*	-	-	71049	SSH Weak MAC Algorithms Enabled
INFO	N/A	-	-	18261	Apache Banner Linux Distribution Disclosure
INFO	N/A	-	-	48204	Apache HTTP Server Version
INFO	N/A	-	-	39520	Backported Security Patch Detection (SSH)

Fig. 5 Nessus report

5 Conclusion

In conclusion, web applications, or web apps, are programs that run on web servers and are accessed via web browsers over networks, typically the Internet. Unlike traditional desktop applications, web apps require a browser for access and utilize a client–server architecture for user interaction. The dynamic nature of web applications presents ongoing security challenges due to evolving cyber threats. This research highlights the importance of tailored penetration testing strategies for web applications, particularly in overloaded server environments where vulnerabilities are exacerbated by high loads. By simulating real-world attacks, penetration testing proactively identifies and mitigates potential risks. The study aims to bolster the resilience of web applications by offering insights and practical recommendations to optimize security protocols in dynamic server conditions.

References

1. Wen, S.F., et al.: A quantitative security evaluation and analysis model for web applications based on OWASP application security verification standard. Comput. Secur. **135** (2023)
2. Praseed, A., et al.: DDoS attacks at the application layer: challenges and research perspectives for safeguarding web applications. IEEE Commun. Surv. Tutor. **21**(Inc.), 661–685 (2019)
3. P. A. Sonewar et al. (2016), "Detection of SQL injection and XSS attacks in three-tier web applications. In *2016 International Conference on Computing Communication Control and automation (ICCUBEA)* IEEE (pp. 1–4).

4. Ghanem, M.C., et al.: Hierarchical reinforcement learning for efficient and effective automated penetration testing of large networks. J. Intell. Inf. Syst. **60**(2), 281–303 (2023)
5. Ibarra-Fiallos, S., et al.: Effective filter for common injection attacks in online web applications. IEEE Access **9**, 10378–10391 (2021)
6. Wang, D., et al.: Application monitoring for bug reproduction in web-based applications. J. Syst. Softw. **207** (2023)
7. Gandikota, P.S.S.K., et al.: Web application security through comprehensive vulnerability assessment. Procedia Comput. Sci. **230**, 168–182 (2023)
8. Chorás, M., et al.: Machine learning techniques applied to detect cyber attacks on web applications. Log. J. IGPL **23**(1), 45–56 (2014)

Software Defect Detection and Machine Learning: Bibliometric Approach

John Philipose Villoth, Saramma John Villoth, and Joseph Mani

Abstract The world is eagerly awaiting the development of machine learning technologies since there has been extensive research on the potential applications of machine learning for software defect identification. Examining publications that emphasize the bibliometric features of machine learning-based software defect detection is the goal of the study. The IEEE database was searched using a variety of keywords, yielding 219 papers about the subject between 2007 and 2024. Analysis revealed more publications in the subject domain between 2007 and 2024. Machine learning, convolutional neural networks, curriculum, and machine learning systems that co-occur with AI are some of the emerging themes. More international collaboration founded on common democratic values is required for the development of machine learning technologies for software defect detection.

Keywords Machine learning · Software defect detection · IEEE · VOS viewer · Artificial intelligence · Bibliometric analysis

1 Introduction

Machine learning, which draws attention and changes reality with computers that perform better than they can and imitate human intelligence, is still essential to reveal a wide range of developments in the modern world [1, 2]. Our marketplace, educational system, way of life, and other aspects are being redesigned by machine learning technologies. It is currently touted as a tool to support the development and advancement of the machine learning concept. With the development of robotics and computing technology, there has been a noticeable shift toward goods and services

J. P. Villoth (✉) · J. Mani
Modern College of Business and Science, Muscat, Oman
e-mail: John.villoth@mcbs.edu.om

S. J. Villoth
University of Technology and Applied Sciences, Nizwa, Oman
e-mail: John.villoth@mcbs.edu.om

B. Awwad (ed.), *AI and IoT: Driving Business Success and Sustainability in the Digital Age*, Studies in Systems, Decision and Control 590,
https://doi.org/10.1007/978-3-031-88874-8_72

enabled by intelligence. Because machine learning is versatile and long-lasting, it has been widely used in various fields, from robotics to aircraft flight control.

Since its inception, machine learning technologies have developed, leading to a more expansive and dynamic definition. Machine learning includes the study and creation of intelligent gadgets and complex computer programs. It is not restricted to biological methods of observation, even though it seeks to mimic human intelligence [3]. It can also be thought of as the system's ability to accurately evaluate and understand vast volumes of data and utilize them to create formulas for independent learning and judgment.

Machine learning-powered technologies have a significant impact on our daily lives due to their amazing capacity to change people's thoughts, behaviors, and interactions [4]. Machine learning is having an increasing impact on a variety of industries, such as engineering, agriculture, and health. Machine learning has the potential to greatly aid in the creation of adaptive learning environments by leveraging learner data to personalize instructional materials, assessments, and prompt feedback. Numerous administrative tasks, like registration, course scheduling, admissions processes, and providing evidence to support educational decisions, can be automated by machine learning-driven algorithms. This minimizes paperwork and the chance of human error. Machine learning-driven intelligent tutoring systems are revolutionizing education with their ability to provide personalized and flexible learning opportunities.

This also provides tailored feedback to fit the individual needs of each learner, evaluates preferences, reinforces learning objectives, boosts learner engagement, enhances time management, reduces costs, fosters ICT proficiency, and offers unbiased development suggestions. Machine learning, which allows researchers to examine large datasets and uncover hidden patterns and relationships that would be missed by conventional investigators, also makes software flaw identification possible [5]. Furthermore, by integrating research methodologies for complex themes, machine learning maximizes the use of cross-source searching, topic exploration, cross-field technique application, and interdisciplinary, multidisciplinary, and transdisciplinary research.

Machine learning systems can identify learning trends and provide targeted advice for academic success. The application of machine learning presents significant prospects for the future since machine learning-enabled educational systems can also be used to assess student participation and the classroom environment [6]. Knowledge analysis and summarization are essential given the increasing amount of scientific research examining the potential application of machine learning. Although there are many papers reviewing scientific research on the potential of machine learning, as far as the researchers are aware, there are very few bibliometric studies.

For example, the common issue of education was the focus of a thorough bibliometric investigation of the larger landscape of big data and machine learning applications [7]. Another study looked at educational research on machine learning. It did not, however, specifically engage in the bibliometric analysis of the body of research in the field of machine learning. This is why the current work intends to conduct a thorough bibliometric analysis of the IEEE database's publications on potential machine learning applications in software defect detection that were published between 2007 and 2024. The following research questions will be addressed here:

2 Literature Review

A careful examination of its implications in this sector has been spurred by the growing worldwide augmentation of software defect detection within machine learning. Through many targeted studies and systematic reviews, this has prompted academics to investigate the various areas of machine learning use in software defect detection, offering insightful information. Furthermore, it seemed that machine learning technologies were more integrated into the fields of computer science and engineering.

Nevertheless, the study's regional focus restricts how broadly such models can be applied in various educational contexts. Numerous institutions highlighted the wide range of machine learning applications and the significant impact of machine vision in fostering entrepreneurial abilities. This useful data shows that to maximize learning outcomes, machine learning systems must be tailored to certain educational situations [8]. The versatility of machine learning for improving educational experiences is further highlighted by studies focusing on its various applications in academic domains, including validating academic credits, facilitating online proctored examinations, improving language instruction, and its effects on learners' achievement [9].

While specific research and systematic reviews offer important insights into the subject of machine learning's integration into software defect detection, a comprehensive bibliometric study could provide a robust framework for merging these numerous contributions. Such a study would allow for a thorough view of the academic landscape by exposing future trends, thematic focus, and the evolving trajectory of machine language applications in software defect detection [10]. Their research indicates that the United States is the top publisher of proceedings papers, which have emerged as the most popular publication format. This study demonstrated that the usage of machine learning is spreading beyond specialist domains, indicating that it may have potential multidisciplinary applications that could impact the next phase of machine learning integration in software defect detection.

However, the bibliometric study conducted between 2007 and 2024 found a serious gap despite these noteworthy efforts and encouraging transdisciplinary improvements [11]. Despite the general interest, there is still room for improvement in the thorough integration of software fault detection with machine learning. In addition to generating interest in particular fields or levels, this shows that there aren't many thorough bibliometric studies addressing machine learning's holistic integration in software fault detection. Additionally, this makes it more difficult to fully comprehend the advancements and patterns in the sector, pointing out areas that need more research. Additionally, it hinders the ability to make well-informed decisions on the use of machine learning in software fault detection and hinders the advancement of knowledge regarding machine learning's effects on software defect detection [12].

3 Methodology

3.1 Data Source and Search Strategy

IEEE, a renowned and comprehensive database for bibliometric analysis, was used to search for articles and collect data in November 2024. This method was carried out in three stages: defining the topic, eligibility conditions, and parameters; screening the articles; and incorporating them. The first search string concentrated on papers about Software Defect Detection Using Machine Learning using the relevant keywords [13]. Strict inclusion and exclusion criteria were used, and the document type was defined as "Article" published in "English" in academic "Journals" and "Conferences" between 2007 and 2024. The original search strategy yielded 219 articles.

3.2 Search Strategy Flow Diagram

The first search approach found several software defect detection subcategories, with machine learning being the focus (Fig. 1). The researcher examined 219 papers' abstracts and titles, evaluating those that particularly discussed the possibility of employing machine learning to identify software flaws. 182 publications were found to be appropriate for additional bibliometric analysis after this phase. 182 IEEE articles made up the final data collection, which was extracted in CSV format for Excel. This data was then imported into VOS viewer 1.6.19, a free software program with an easy-to-use graphical user interface for network visualization and text analysis.

For the data analysis, the researcher examined the distribution of publications by year and country, journal metrics, and citation counts. This exploration utilized the complete data set collected from IEEE, which was exported into both Excel and VOS viewer for more in-depth insights.

3.3 Study Selection and Data Management

The articles identified by the initial search technique included numerous subcategories of specific AI concepts and machine learning programs that were largely focused on studying AI rather than the applications of machine learning for software defect detection [12]. In this way, following the initial screening of the search

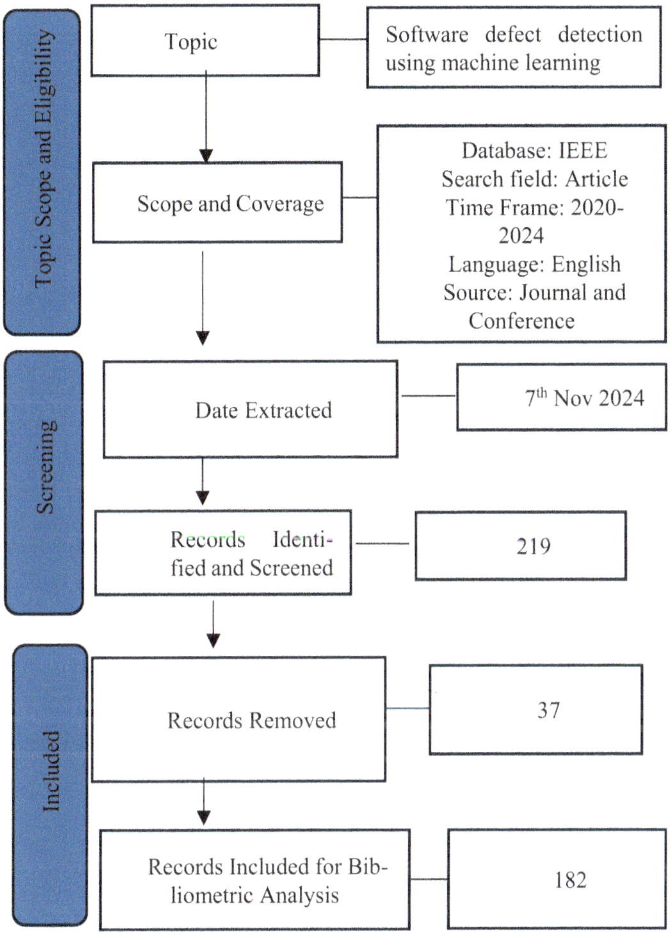

Fig. 1 Search strategy flow diagram

results, the researcher manually examined the papers to determine their suitability for inclusion in the included phase. The researcher looked at the abstracts and titles of 219 studies, assessing those that specifically address machine learning's potential. Following this stage, 182 publications were identified as suitable for further bibliometric examination. The final 182 IEEE data was then extracted and saved in CSV Excel format. Keywords, citations, abstracts, bibliographic information, and funding details. Subsequently, this information was transferred to VOS viewer 1.6.19, a free software application with a user-friendly graphical interface for text analysis and network visualization.

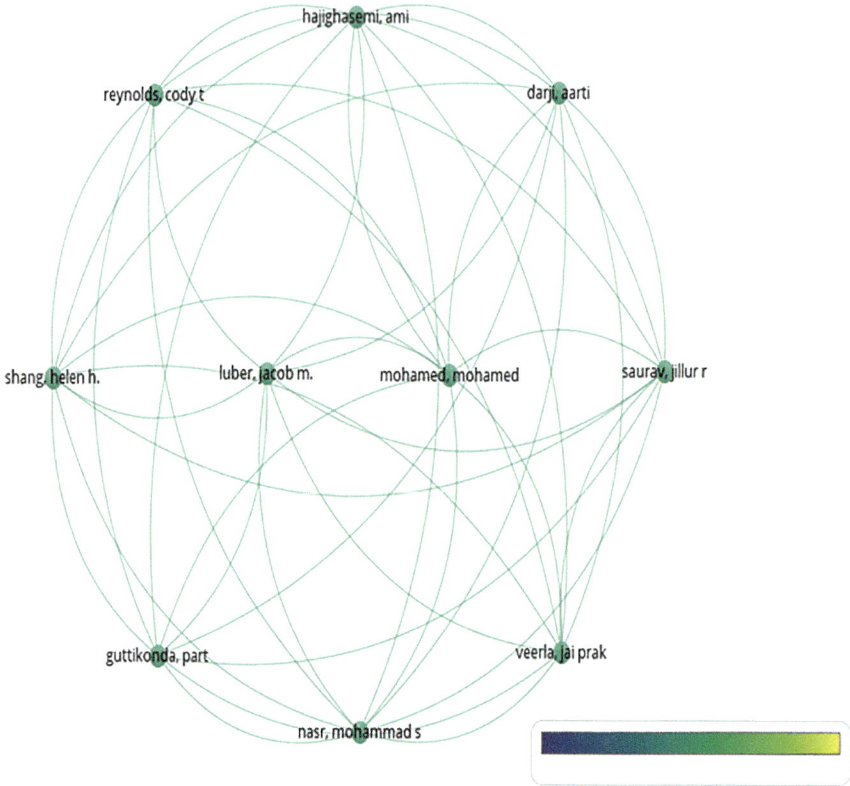

Fig. 2 Network visualization of Collaboration among authors

3.4 Data Analysis

First, the distribution of publications across years and countries was analyzed, along with journal metrics and publication citation counts. To obtain deep insights, all the IEEE data collected for this study was exported to Excel and VOS Viewer. Network visualizations of co-authorship (authors, countries, institutions), co-occurrence keywords, and co-citation (authors, references) networks were then produced using VOS Viewer. The following conclusions are predicated on the study's objectives as shown in Fig. 2.

3.5 Distribution of Publications Over the years

The study looked at how studies published in the IEEE database between 2007 and 2024 were distributed. The results of this investigation are shown in Fig. 3. It shows

Fig. 3 Distribution of publication years

that in the 2020s and beyond, academics began to focus more on machine learning. The frequency of scholarly articles about its various uses in machine learning-based software fault detection has increased exponentially during the review period.

3.6 *Co-authorship Network Analysis*

Nearly every facet of scientific cooperation networks may be precisely identified using bibliometric tools for co-authorship network analysis. Co-authorship is the term used when numerous authors or organizations formally recognize their contributions to a scientific article. Countries, organizations, and writers participating in collaborative authorship are represented by nodes in the co-authorship network (Fig. 4). First, the distribution of publications across years and countries was analyzed, along with journal metrics and publication citation counts. To obtain deep insights, all the IEEE data collected for this study was exported to Excel and VOS Viewer. Network visualizations of co-authorship (authors, countries, institutions), co-occurrence keywords, and co-citation (authors, references) networks were then produced using VOS Viewer.

3.7 *Co-authorship Analysis*

The co-authorship analysis of the study's authors in articles published in the IEEE database between 2007 and 2024 is displayed in Fig. 5. Two authors' collaboration is indicated by a line (link) joining their names. The degree of author participation is indicated by the link thickness and node distance.

In bibliometric analysis, a network of collaboration among authors refers to the patterns of co-authorship relationships that develop when multiple researchers work

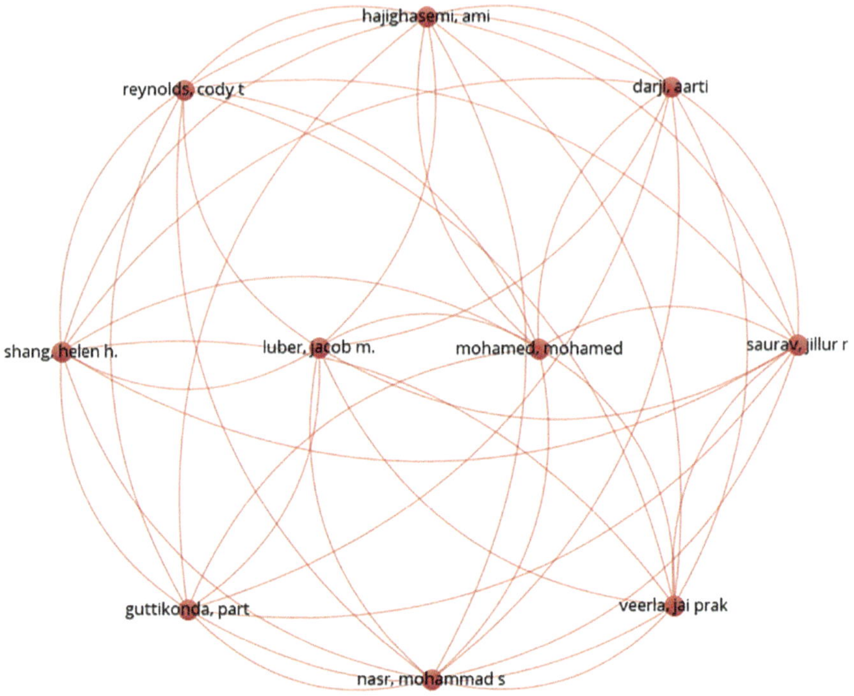

Fig. 4 Co-authorship analysis

together on academic publications. This network visualizes how authors collaborate within a specific field and highlights key players, research groups, or institutions that drive knowledge creation in various areas.

In summary, a network of collaboration in bibliometric analysis serves to visualize and understand the relationships between authors, their collaborative behaviors, and the collective efforts that drive scientific progress in specific fields.

3.8 Co-authorship Analysis

The examination of author co-authorship networks is shown on Fig. 5. The researcher established a threshold level for this analysis, needing at least one document and five citations. Just 182 of the 219 authors that were taken into consideration met this requirement. It is clear from examining the co-authorship of research on the topic that writers who collaborate often publish together in smaller groups with three linkages. Furthermore, a few of the writers released their works on their own. The co-authorship analysis of the study's authors in articles published in the IEEE database between 2007 and 2024 is displayed in Fig. 5. Two authors' collaboration is indicated by a line (link) joining their names.

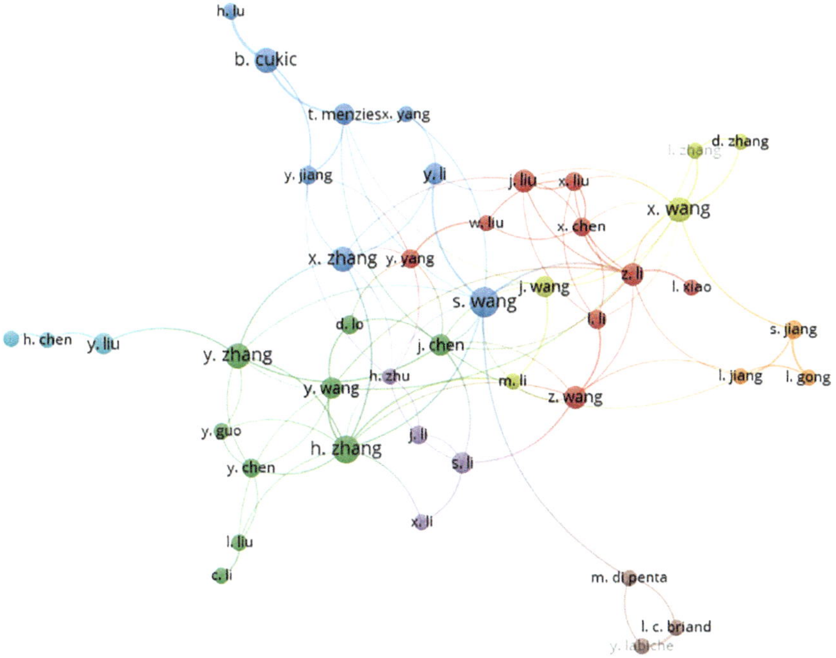

Fig. 5 A co-authorship network visualization

3.9 Co-citation Network Analysis

In bibliometric analysis, a co-citation network refers to the pattern where two or more authors, papers, or sources are frequently cited together in academic literature as shown in Fig. 6. This suggests that if two authors (or works) are often referenced in the same citations by other publications, they may share related topics, ideas, or academic contributions, even if they do not collaborate directly.

In summary, co-citation networks serve as a valuable tool for understanding the interrelationships between scholarly works and tracking the evolution of knowledge within a given field.

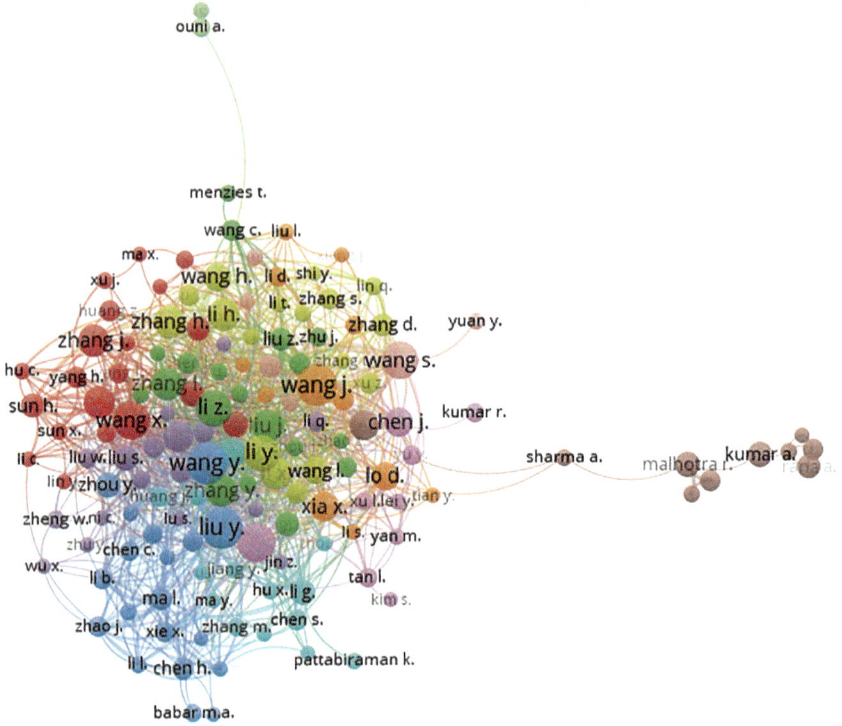

Fig. 6 Co-citation network analysis

4 Conclusion

This study focused on the potential for software defect detection in machine learning from 2007 to 2024 by performing a bibliometric analysis of publications in international journals from the IEEE database. For visualization, VOS Viewer 1.6.19 software was used. The number of studies on machine learning in software defect detection has increased rapidly during the past 20 years. The study's results also showed that, between 2007 and 2024, the number of publications in the field increased significantly.

For innovations and advantages that benefit everyone, the paper also highlights the need for greater international cooperation in the sector. By identifying new topics, the profile results of publications, and important countries in the field of software defect detection in machine learning, the study also establishes the foundation for future research. In this sense, it encourages scholars to investigate complex connections and open-ended inquiries, ultimately creating new paths in this emerging area. The study's overall focus is on the emerging domains of software defect detection and covers the connections between the emerging novel machinery and e-commerce flows that change its marketability, as software defect detection has been perceived as a market term rather than a social term with surplus funding like trade, machinery, and e-commerce flows which change its primary prospects from classrooms to virtual media.

References

1. 2022 Crypto Crime Report. BlockBR Blog (2022). https://blog.blockbr.com.br/wp-content/uploads/2022/06/2022-crypto-crime-report.pdf. Accessed: 5 Nov 2024
2. Yadav, A., Yadav, T., Kumar, R.: Defect prediction in imbalanced code repositories using deep neural networks. J. Syst. Softw. **152**, 124–142 (2019)
3. Meher, S., Satapathy, S.C.: A Comparative Study of Deep Learning Models for Software Defect Prediction (2020)
4. Fadlisyah, M., Thanoon, M., Budi, A.: A survey on deep learning in software engineering: past, present, and future directions. J. Syst. Soft. **173**, 110892
5. Singh, A., Chaturvedi, A.: A Hybrid Approach for Software Defect Prediction Using Machine Learning with Ensemble of Filtering Methods (2018)
6. Shihab, E., Jiang, Z., Adams, B.: Efficiently Detecting Defective Files in Large-Scale Software Systems Using Deep Learning (2019)
7. Markovic, V., Njegus, A., Bulaja, D., Zivkovic, T., Zivkovic, M., Mani, J.P., Bacanin, N.: Employee reviews sentiment classification using BERT encoding and AdaBoost classifier tuned by modified PSO algorithm. In: Proceedings of the 2nd International Conference on Innovation in Information Technology and Business (ICIITB 2024), Muscat, Oman, pp. 22–37. Atlantis Press, Dordrecht, Netherlands (2024)
8. Zhang, H., Zhang, Z., Ernst, M.D.: A comprehensive survey on deep learning applications to software engineering. J. Syst. Softw. **157**, 110396 (2019)
9. Goodfellow, I., Bengio, Y., Courville, A.: Deep Learning. MIT Press (2016)
10. Amershi, S., Cakmak, M., Knox, W.B., Kulesza, T.: Power to the people: the role of humans in interactive machine learning. IEEE Trans. Hum.-Mach. Syst. **46**(4), 460–473 (2014)
11. Xia, X., Lo, D.: Heterogeneous defect prediction. IEEE Trans. Softw. Eng. **45**(10), 1078–1099 (2018)
12. Jovanovic, L., et al.: Optimizing machine learning for space weather forecasting and event classification using modified metaheuristics. Soft. Comput. **28**, 6383–6402 (2024)
13. Milutinovic, N. et al.: Tuning Artificial Neural Network for Healthcare 4.0. by Sine Cosine Algorithm. 2023 International Conference on Intelligent Data Communication Technologies and Internet of Things (IDCIoT). 510–513 (2023).

Sustainable Digital Learning Readiness in Higher Education: Emotional Intelligence, Psychological Security, and Technological Trust in Oman

Fadi Abdelfattah, Rami Oweis, Safoura Navari, and Ahmed Al Mata'ni

Abstract This research investigates the role of Emotional Intelligence (EI), Psychological Security (PS), and Technological Trust (TT) in mitigating the fears of learning digitally with emerging technologies to induce sustainable readiness for digital learning (SDLR) among higher education. This paper employs AI-augmented analysis in conjunction with traditional empirical approaches and investigates the interaction of these spheres within Oman. Key findings revealed that 87.7% of participants had prior experience with digital learning technologies, with 26.2% using electronic devices for 5–7 h daily. AI-based projections highlighted how behavioral shifts in EI, PS, and TT can predict SDLR, providing valuable insights for designing interventions to enhance digital learning readiness. Results indicate that EI can act as a building block for learner adaptability towards digital learning, and PS counteracts resilience and engagement in the virtual context. TT also boosts preparedness by instilling confidence in digital platforms. The current study adds to the developing body of research on sustainable education and provides some insight into the psychological and technological preconditions for optimizing digital learning ecosystems. Our findings highlight the importance of developing emotional and psychological skills among learners in digital education, which are essential for educators and policymakers to create sustainable and long-term strategies in higher education.

Keywords Sustainable digital learning readiness · Emotional intelligence (EI) · Psychological security (PS) · Technological trust (TT) · AI in education

F. Abdelfattah (✉) · R. Oweis · S. Navari · A. Al Mata'ni
Modern College of Business and Science, Muscat, Oman
e-mail: fadi.abdelfattah@mcbs.edu.om

R. Oweis
At a sabbatical leave from Jordan University of Science and Technology, Irbid, Jordan

B. Awwad (ed.), *AI and IoT: Driving Business Success and Sustainability in the Digital Age*, Studies in Systems, Decision and Control 590,
https://doi.org/10.1007/978-3-031-88874-8_73

1 Introduction

The unprecedented transition from traditional classroom settings to online learning environments, accelerated by the global COVID-19 pandemic, has transformed higher education systems worldwide. This shift underscored the need for more than just access to technological tools; it highlighted significant gaps in learners' readiness for digital education. While often presumed that "digital natives" would seamlessly adapt to online learning, evidence revealed that many learners struggled to engage effectively, indicating a more profound, multidimensional construct underpinning readiness for digital learning [1, 2].

Digital learning readiness extends beyond technical proficiency; it encompasses emotional, psychological, and cognitive preparedness that collectively influence learners' ability to thrive in online environments [3]. Emotional Intelligence (EI), defined as the ability to perceive, understand, and regulate emotions, plays a pivotal role in fostering resilience and motivation, which is critical for navigating the challenges of digital learning [4, 5]. Similarly, psychological security (PS), which is the belief in one's ability to cope with emotional and academic stressors, provides a supportive foundation for engagement in virtual learning [6, 7]. Trust in technology (TT) and confidence in the reliability and functionality of digital platforms further reinforce this readiness by enabling students to utilize technological tools effectively [8].

While previous studies have examined these factors independently, their interrelated impact on readiness for digital learning remains underexplored, particularly in the context of sustainable education systems. This gap is particularly relevant in Oman, where higher education institutions increasingly integrate digital and hybrid learning models. Understanding how EI, PS, and TT interact can inform strategies for developing sustainable, inclusive digital learning ecosystems tailored to local contexts.

This study employs an AI-augmented analytical framework to investigate the relationships between EI, PS, and TT in shaping readiness for digital learning among university learners in Oman. By combining traditional empirical research with AI-driven predictive modeling, this research aims to provide actionable insights for educators and policymakers, advancing the discourse on sustainability in higher education. The findings contribute to the growing knowledge of digital readiness, emphasizing the critical role of emotional and psychological factors in fostering effective and resilient learning environments.

2 Literature Review

Amid the COVID-19 pandemic, we have seen a frequent transition from traditional teaching to digital learning platforms, which has led many to examine what makes learners ready for online learning. Apart from the technical tools we need to facilitate digital learning, emotional and psychological factors have also been recognized as vital to success. Literature provides evidence for describing Ei = (PS + TT) as intrinsic elements in learners who confront their ability to face collaborative learning spaces having undergone digital transformation.

2.1 Emotional Intelligence (EI) and Sustainable Digital Learning Readiness (SDLR)

The term Emotional Intelligence (EI) was first introduced by Salovey and Mayer [1] to refer to the individual ability to perceive, understand, control, and regulate emotions in oneself and others. Since then, this construction has been expanded, and Goleman [2] presented a more inclusive model containing personal and social competencies. The importance of EI as a skill complementary to academic skills has risen even more as learners in digital learning environments are confronted with emotional burdens like isolation, stress, and difficulties in self-regulation.

EI has gained even more importance in a digital learning format, where physical interactions are reduced, and learners must work on their self-motivation to engage with the material instructed. Empirical research has shown that learners with higher EI can cope with the psychological strains of online learning, such as Pipers et al. AI and ML] [6] found that learners positively affect learners. EIs activate emotional coping strategies better, which causes academic achievement in digital environments. In addition, learners with a degree in EI are better equipped to manage the challenges associated with self-directed learning, which is the hallmark feature of online learning [4].

Alenezi [5] found direct effects of EI on learners' online learning readiness in motivation, resilience, and adaptive behaviors. Managing emotions, mainly when times get tough, ensures learners stay engaged with their academic work and learn to navigate challenges in online classrooms. According to [6], learners with solid emotional competencies can better adjust to online education's demands and thus achieve better achievements.

2.2 Psychological Security (PS) in Digital Learning

Psychological security (PS), or the belief that one can cope with emotional difficulties, is essential to learners' readiness for online learning. In the case of digital education, where DGBLs offer new ways to absorb content but also entail new risks, PS is defined as the perception that one can successfully navigate a virtual learning environment in ways that promote his or her emotional and psychological well-being. Evidence shows that psychologically secure learners engage with the materials meaningfully and continue to study even when digital education becomes isolated and stressful [7].

It also improves learners' self-regulation and emotion management, essential for achieving their goals in online learning environments. According to [8], PS is a defense against anxiety and uncertainty that may be generated in the digital learning context. The study showed that learners with higher PS could cope with the pressures of online learning and feel more satisfied with the process [3]. Additionally, there is a positive association of PS with academic performance in online learning [9]. Another essential part of the psychological safety in learning through internet channels is freedom from emotions that make learners feel vulnerable and thrown into a dilemma to handle the uncertainty, sometimes making online classes feel more exhilarating than actual classrooms. According to [10], PS allows learners to adopt adaptive coping mechanisms, pushing them toward better learning outcomes by asking for help managing stress effectively. This highlights that promoting a psychologically safe space for the learners participating in digital learning environments is crucial.

2.3 Technological Trust (TT) and Sustainable Digital Learning Readiness (SDLR)

Digital trust (DT) is the extent to which learners trust the digital platform used in online education, which is reliable, safe, and effective. Learners must trust technology so that they can make use of technological tools and resources. Trust in technology is vital for the practical learning experience, especially in online learning environments where learners access course materials, submit assignments, and communicate with classmates and instructors via digital platforms [11].

Studies have found that learners who feel confident in the technology they use to learn are less inclined to disengage from digital learning environments. According to [12], learners' engagement and participation in online learning depend on their trust in technology. Learners tend to be more invested in putting time and effort into utilizing technological tools when they have access to support, leading to better learning outcomes. On the other hand, distrust in technology can result in disconnection and grouchiness, making learners less inclined to engage with digital learning [13].

Trust in technology can likewise correlate with learners' feelings about their digital environment's security and privacy. A study by [14], secure and user-friendly platforms are emphasized to develop trust in learners. According to the research, learners who felt technological tools were safe spaces, reliable, and secure tend to be more confident in their use, thus helping them with key, self-directed learning behaviors essential for success in an online education environment. The study further revealed that learners familiarized with the previous tools found it easier to surf new ones [7].

2.4 EI, PS, and TT in Sustainable Digital Learning Readiness (SDLR)

As per the literature, EI has a direct positive mediated by PS and TT with the intent of continuing digital learning among learners. These factors do not function separately; instead, they influence one another to create the circumstances in which learners participate in e-learning. For example, greater emotional regulation among learners with higher EI results in relatively lower psychological insecurity. Consequently, this PS provides a foundation for learners to trust the technology they utilize, resulting in increased efficacy of engagement with online learning [15].

ScienceDirect Research by García-Ruiz et al. Learners with dual high EI and PS are likelier to adopt technology and learn online [16]. Their study revealed that EI and PS predict learners' SDLR, and both factors jointly enhance trust in technology. This indicates a need to understand emotional versus psychological differences as learners navigate emotion and learning digitally.

In addition, research indicates that improving EI and psychological safety can nurture a sense of trust in technology, resulting in higher levels of SDLR. However, when online learning environments are here to stay, knowing how these constructions relate with others to designing educational interventions preparing learners for future success while becoming active adults is more important than ever.

3 Methodology

We used a quantitative method approach to survey-based data collection, where traditional survey instruments were analyzed using AI-powered analysis and causal modeling to explore the relationships between Emotional Intelligence (EI), Psychological Security (PS), and Technological Trust (TT) for predicting digital readiness.

3.1 Data Collection

The research design was quantitative, and the data were collected from university learners in Oman. We adopted random sampling to include a sample containing learners with varying levels of academic backgrounds.

Constructs The survey instrument was developed to measure three main characteristic features: Emotional Intelligence (EI), Psychological safety (PS), and Trust in Technology (TT). Validated scales were used to assess each of the constructs. EI was measured with the Mayer-Salovey-Caruso Emotional Intelligence Test (MSCEIT) [1], which assesses an individual's capacity to perceive, understand, and regulate emotions. A modified scale version established by Zhou et al. was used for PS assessment. This is [2] where you assess your confidence in dealing with emotionally challenged issues. TT was assessed as described in [3] and referred to learners' perceptions of the reliability and effectiveness of digital learning platforms.

3.2 Data Analysis and Results

The data were evaluated through Partial Least Squares Structural Equation Modeling (PLS-SEM) to test relationships between EI, PS, TT, and SDLR. The model estimation was done using Smart PLS 3.3 software, and bootstrapping was used to test the hypothesized paths' significance. PLS-SEM was chosen because it enables sophisticated modeling of complex relationships involving multiple constructs [3] and is often well-suited for predictive purposes in educational research [4].

Alongside PLS-SEM, AI-driven predictive Analytics were jointly applied to analyze the interaction behaviors and predict the three psychological factors (EI, PS, and TT) on SDLR. The study then projected possible future scenarios of SDLR using AI algorithms relevant to current data on how EI, PS, and TT behavior changes can be used as factors or behavioral indicators to predict learners' readiness for an online environment.

Whenever possible, this research design was guided by ethical considerations. All participants provided informed consent, and data was collected anonymously, stored in secure locations, and identified, when possible, to maintain confidentiality. Participants were informed of the purpose of the study and their right to withdraw at any time.

The sample comprised 130 university learners, 71.5% females, and 28.5% males. A significant portion of participants (78.5%) were undergraduate learners, while 16.9% held a diploma and 2.3% were postgraduate learners. Regarding daily electronic device usage, 26.2% of participants spent 5–7 h per day on electronic devices. 87.7% of the respondents had prior experience with digital learning technologies. The demographic details are presented in Table 1.

Assessment of the validity and reliability of measurement scales for Emotional Intelligence (EI), Psychological Security (PS), and Sustainable Digital Learning Readiness (SDLR). Internal consistency (Cronbach's Alpha, Composite Reliability CR) and Convergent validity (AVE) were assessed for each construct. Results showed that EI and DLR reported acceptable reliability, but PS has a very low internal consistency (Cronbach's Alpha = 0.345), which the small number of items can explain to measure this construct. Table 2 summarizes the reliability and validity of all constructs.

Table 1 The demographic details

Demographics	Number	Percentage (%)
Male	37	28.5
Female	93	71.5
Diploma	22	16.9
Bachelor's degree	102	78.5
Postgraduate	5	2.3
< 1 h	5	3.8
1–3 h	29	22.3
3–5 h	27	20.8
5–7 h	34	26.2
7–9 h	21	16.2
> 9 h	14	10.8
Yes (experience with digital learning)	114	87.7
No	16	12.3

Table 2 The reliability and validity of the study variables

Variables	Cronbach's Alpha	rho_A	CR	AVE
Emotional intelligence (EI)	0.635	0.723	0.788	0.501
Psychological security (PS)	0.345	0.537	0.683	0.463
Sustainable digital Learning readiness (SDLR)	0.814	0.817	0.878	0.643

3.3 Hypothesis Testing

Bootstrapping-based hypothesis testing showed that EI positively predicted digital learning readiness ($\beta = 0.369$, $p < 0.001$) and that it also significantly predicted PS ($\beta = 0.490$, $p < 0.001$), while the latter was similarly related to digital learning readiness ($\beta = 0.284$, $p < 0.01$). PS also significantly mediated EI in "sustainable digital learning readiness" ($\beta = 0.139$, $p < 0.01$), suggesting its role in improving learners' digital education preparedness. Table 3 is where we present the results from our hypothesis testing.

Table 3 Hypothesis testing results

Relationships	Sample (O)	Sample mean (M)	Standard deviation	T value	*P* values
EI → SDLR	0.369	0.372	0.096	3.834	0.0
EI → PS	0.49	0.503	0.074	6.658	0.0
PS → SDLR	0.284	0.291	0.092	3.092	0.002

Table 4 The indirect effects (mediation) results

Mediation path	Sample (O)	Mean (M)	Standard deviation	T value	*P* values
EI → PS → SDLR	0.139	0.146	0.051	2.707	0.007

3.4 Indirect Effects (Mediation)

The mediation analysis revealed that psychological security (PS) partially mediates the relationship between EI and SDLR.

The standardized indirect effect (EI → PS → SDLR) was significant, with a path coefficient of 0.139 (t-value = 2.707, $p = 0.007$). The indirect effects (mediation) results are presented in Table 4.

4 Discussion and Conclusion

The findings of this study emphasize the contribution of EI, PS, and TT in predicting readiness for digital learning among learners. The results reinforce a nascent research program that recognizes the significance of digital education's psychological and emotional aspects, as opposed to its technical components. Emotional Intelligence (EI) and Psychological Security (PS) are related to personal and social well-being and how we work with each other, all of which matter in how learners approach assessment structures and performance within online learning settings.

The positive impact of EI on SDLR is notably in line with prior studies that have found EI to predict better academic performance and responsiveness to online learning environments [17, 18]. High-EI learners can better manage their effect and remain motivated and engaged with that content. These skills are essential for e-learning, where you do not have as much face-to-face time with learners, and they need to be self-directed learners.

In addition, PS was an essential mediator between EI and SDLR. The findings suggest that learners who believe they can effectively navigate emotional and academic challenges online are at least somewhat more likely to flourish. Such a result is consistent with studies where the emotional reinforcement dimension helps learners to be more able to adapt and survive in online settings 1.

Although TT has not been tested independently, it may significantly influence teachers' and learners' SDLR. The efficiency of learners using digital platforms and technologies is affected by trust. With the speed of digital adoption in education only increasing, trust in technology is one of the most critical aspects on which we can operate to help foster engagement and participation. As a result of the increasing complexity in digital learning environments, future studies could further investigate the role of PS and EI as predictors of TT or vice versa.

5 Limitations and Future Research

Although this analysis offers critical perspectives, it is not without limitations. This study has several limitations, the most significant being that it relies on a convenience sample of only 130 university learners from Oman who answered self-report questionnaires. For example, self-reports are subject to biases such as social desirability and recall bias that may compromise data accuracy. Many more geographically diverse and varied academic background sample characteristics will undoubtedly help in the generalizability of outcomes for the findings of this study.

In addition to this, psychological security is measured. The PS was measured with a limited number of items, which could explain why the scale produced Cronbach's Alpha of 0.345 was considered low [18]. Future work needs to triangulate PS with more diverse items to refine the measure and further examine its reliability/validity. Furthermore, the study explored learners only from psychological and emotional perspectives without considering other external factors like family support for the study, availability of institutional resources, or teaching quality, which all can influence student readiness for digital learning.

Finally, the cross-sectional nature of this study precludes making any causal inferences. While the relationships between EI, PS, and SDLR are significant, longitudinal studies should be conducted to see whether these will lead to academic success for learners in a digital learning environment.

Many potential directions for future research may further contribute to advancing the literature on readiness for digital learning. First, it may be valuable to examine the role of forms of technological trust, which can serve as both an antecedent and a consequence of EI and PS. It may be valuable to examine the development of learners' trust in technology as they progress through courses online to gain insights into how these constructs co-travel.

Second, future research should investigate improving learners' EI and PS. The promising impact of this study's findings warrants testing the application of targeted interventions like EI training or psychological resilience programs to facilitate learners' enhancement of digital learning readiness and academic success. Such interventions may be implemented as part of online learning curricula or available as additional support programs.

The next avenue of future inquiry may encapsulate the significance of digital literacy in terms of EI, perception of PS, and readiness for learning in a digital environment. Digital skills—i.e., knowing how to use online platforms to explore resources and assess digital information—are essential for effective online learning. Exploring the relationships among digital literacy, EI, and PS may thus deliver a richer picture of what it means to be ready for a digital learning environment.

In addition, research designs incorporating quantitative data could be taken in future research, as highlighted by qualitative work from focus groups or interviews. This strategy could enable researchers to access the emotional, cognitive, and developing contexts that define student challenges in digital settings and their assets of being high on EI and PS as they learn.

References

1. Salovey, P., Mayer, J.D.: Emotional intelligence. Imagin. Cogn. Pers. **9**(3), 185–211 (1990)
2. Goleman, D.: Emotional Intelligence: Why It Can Matter More Than IQ, 1st edn. Bantam Books, New York, NY, USA (1995)
3. Pijpers, M., Bastiaens, T., Van der Meijden, A.: The role of emotional intelligence in online education: a study of students' emotional coping and academic achievement. J. Educ. Psychol. **112**(2), 231–243 (2020)
4. Engin, C.: Analysis of students' online learning readiness based on their emotional intelligence level. Univ. J. Educ. Res. **5**(12A), 32–40 (2017)
5. Alenezi, A.: The relationship of student's emotional intelligence and the level of their readiness for online education: a contextual study in the example of university training in Saudi Arabia. Образование и Наука **22**(4), 89–109 (2020)
6. Zhou, Z., Zhang, M., Zhong, X.: Psychological security and students' adaptation to online learning: the mediating role of emotional intelligence. J. Educ. Psychol. **113**(5), 923–935 (2021)
7. Babore, N., et al.: Psychological effects of the COVID-2019 pandemic: perceived stress and coping strategies among healthcare professionals. Psychiatry Res. **293** (2020)
8. Cejudo, J., García-Fernández, J.M., Trigueros, R.: Emotional intelligence and psychological well-being: a review of the literature. Int. J. Environ. Res. Public Health **15**(11), 2396 (2018)
9. Extremera, N., Sánchez-Álvarez, N., Rey, L.: Pathways between ability emotional intelligence and subjective well-being: Bridging links through cognitive emotion regulation strategies. Sustainability **12**(5), 2111 (2020)
10. Falloon, G.: From digital literacy to digital competence: the teacher digital competency (TDC) framework. Educ. Tech. Res. Dev. **68**(5), 2449–2472 (2020)
11. Blayone, T., et al.: Profiling the digital readiness of higher education students for transformative online learning in the post-Soviet nations of Georgia and Ukraine. Int. J. Educ. Technol. High. Educ. **15**(1), 119 (2018)
12. Hair, F., Risher, J.F., Sarstedt, M., Ringle, C.M.: When to use and how to report the results of PLS-SEM. Eur. Bus. Rev. **31**(1), 2–24 (2019)
13. López-Bonilla, G., García-Ruiz, C.: Psychological security and students' adaptation to online learning: the role of emotional intelligence and digital learning readiness. J. Educ. Psychol. **113**(5), 907–921 (2021)
14. Goleman, M.: Emotional Intelligence 2.0, 1st edn. TalentSmart, San Francisco, CA, USA (2009)
15. Abdelfattah, F., Al Alawi, A.M., Dahleez, K.A., El Saleh, A.: Reviewing the critical challenges that influence the adoption of the e-learning system in higher educational institutions in the era of the COVID-19 pandemic. Online Inf. Rev. (2023). https://doi.org/10.1108/OIR-02-2022-0085
16. Al Halbusi, H., Abdelfattah, F., Ferasso, M., Alshallaqi, M., Hassani, A.: Fear of failure for entrepreneurs in emerging economies: stress, risk, finances, hard work, and social support. J. Small Bus. Enterp. Dev. (2023). https://doi.org/10.1108/JSBED-01-2023-0035
17. Rahman, M.S., Hossain, M.A., Abdelfattah, F., Ibne Mokter, A.M.: Avoidance behavior towards using pirated software: testing a seven-component model on SME employees. Inf. Technol. People (2021). https://doi.org/10.1108/ITP-12-2019-0621
18. Dahleez, K.A., El-Saleh, A.A., Al Alawi, A.M., Abdel Fattah, F.A.M.: Student learning outcomes and online engagement in a time of crisis: the role of e-learning system usability and teacher behavior. Int. J. Inf. Learn. Technol. **38**(5), 473–492 (2021). https://doi.org/10.1108/IJILT-04-2021-0057
19. Hassan, M.S., Al Halbusi, H., Ahmad, A.B., Abdelfattah, F., Thamir, Z., Raja Ariffin, R.N.: Discretion and its effects: analyzing the role of street-level bureaucrats' enforcement styles. Int. Rev. Public Adm. **28**(4), 480–502 (2023). https://doi.org/10.1080/12294659.2023.2286671

Exploring the Impact of Social Media on Personal Communication, Business Engagement, and Civic Participation in Jordan

Mahmoud Abouraia, Ayman Bassam Nassoura, Hothefa Shaker, and Fadi Abdelfattah

Abstract The study investigates the role that social media plays in enriching people's interpersonal relationships and business connections, as well as in sharing information, thus influencing the cultural and social life of a nation. Further, it investigates how social media can assist in brand awareness and customer attraction to Jordan's business sector, particularly the Small and Medium-sized enterprises (SMEs) that leverage social media to widen market access. In this sense, social media has become a more powerful tool for raising brand awareness but seems to have a smaller effect on sales and revenues. It presents the results of a survey completed by 400 participants, which elicits information about Jordanian users' activities, perceptions of the benefits, and protagonists' obstacles associated with social media. It adds insights proffering social media for social and economic improvement in Jordan's technological advancement.

Keywords Social media · Digital communication · Business engagement · Civic participation · Jordan · Social media impact

M. Abouraia · H. Shaker (✉) · F. Abdelfattah
Modern College of Business and Science, Muscat, Oman
e-mail: hothefa.shaker@mcbs.edu.om

M. Abouraia
e-mail: Mahmoud.abouraia@mcbs.edu.om

A. B. Nassoura
Department of MIS, City University Ajman, Ajman, UAE
e-mail: a.bassam@cu.ae.ae; nassuora@gamil.com

B. Awwad (ed.), *AI and IoT: Driving Business Success and Sustainability in the Digital Age*, Studies in Systems, Decision and Control 590,
https://doi.org/10.1007/978-3-031-88874-8_74

1 Introduction

The advanced techniques of technology as well as the emergence and widespread of the internet in the running of everyday lives and operations of business have reached unprecedented levels [1]. Social networks which are part of Web 2.0 technology [2], have also assumed a central role in interpersonal communication, commerce, and politics [3, 4]. Recently, the number of Jordanians using social media has risen tremendously, especially among the youth and the college-educated [5, 6]. These platforms have not only reinvented social encounters but have also gained immense importance in restructuring the business as well as the political arena of the country. Indeed, the use of social media has increased among Jordanians; currently, 77% of Jordanians use social media, primarily Facebook, Instagram, and Twitter [7]. The present paper traces the shift in the use of social media in Jordan, and its effect in the spheres of interpersonal relationships and interaction, business functions and involvement in civil society. In a personal capacity, social media has created improved communication and prospects of networking among the Jordanian community through reaching friends and families and other members in real-time dynamic [8, 9]. In the case of business sighted, social media has been noted to be an economic strategy of marketing as evidenced among the SMEs in Jordan, where most of them have a constraint of financial capital rather than competitiveness marketing and brand visibility [10]. SMEs can expand the reach of their products and services and engage with customers on social media sites such as Facebook and Instagram [11]. Yet, it has been a matter of controversy whether these platforms can contribute to direct customers' purchases because many companies experience difficulties in defining ROI on social media advertisements [12]. Moreover, Jordan's social media mainly consists of civil engagement that involves citizens in social, political and cultural issues. Politics and political activities have been enhanced by social media platforms through availing a platform where citizens can express by their political mileage and engage in demonstrations should there be one [13]. Nevertheless, drawbacks, including misinformation, privacy issues, and the digital divide, are present and have sparked debates on the usability and permissibility of social networks in Jordan [3]. Election seasons and political events have however been identified as having fake news and misinformation spread notably especially from Facebook. Such cases demonstrate the necessity for improved rules and regulations of social networks, as well as for increasing the rate of conscious use of social media by individuals. This paper will unfold these various aspects of the social media in Jordan by establishing the impact of these platforms across personal, business, and political realms. It will give information on the advantages and disadvantages of using social media in Jordan that is; it will suggest the business and government on how they can make good use of social media opportunities while dealing with social media risks. The study enhances the understanding of the role of social media in interaction in Jordan and provides important implications for the rest of the Middle East and North Africa (MENA) context.

2 Literature Review

In this literature review, several pertinent sections for analyzing the role of social media in Jordan will be discussed: the effect of social media on personal communication, business interaction and civil participation. It will also discuss the implications of social media use and analysis focusing on SMEs in Jordan. Facebook for instance has changed the way people interact; users can telephone or chat with their friends without regard to the distance between the two of them. This has led to the formation of more complex and advanced social networks. Sadiq et al. [8] established that social apps, including Facebook, Twitter, and Instagram, have merged with interpersonal communication in the course of interaction, especially among the youthful population in Jordan [8]. Social media mostly creates both formal and informal communication where through such platforms people get to share their experiences, interactions, and even relationships. Based on the study by Abu-Shanab et al. [5], social media networks in Jordan reveal the functionality of online communities that act as both social forums and arenas for sharing knowledge. Not only has it brought about relative ease in keeping in touch with family and friends, but these platforms have brought about new ways of social networking where people can find friends who have similar issues. Nevertheless, social media has many advantages for carrying out personal messaging services, but at the same time, it poses a threat to close, physical contact forms of communication. Hani et al. argue that some of these experts have concluded that people may suffer from social isolation resulting from excessive social media use especially for younger generations who rely much on social media to communicate [11]. This is because technology is also a cause of weakened relationships since some interactions occur online and not face to face hence, it may not bring out the quality relationships than the physical ones. Nevertheless, social media, as it has been mentioned before, plays a significant role in daily interpersonal communication broadly in Jordan and particularly among the youth on the increase.

Playing the global undercurrent, social media has become an influential force in the business marketing segment. Business people of Jordan, especially SMEs have considered social media platforms such as Facebook, Instagram, and LinkedIn mainly because of the large, targeted audience that is easily accessible compared, to other traditional methods of marketing [10]. To SMEs, social media platforms are affordable means of communicating with the customer, creating brand familiarity and direct contact with the specifier or buyer [11]. Many Jordanian organizations including such, especially the small and medium ones, have also benefited from these platforms to enhance their public presence and access more markets. Both the paid ads and organic post allows the companies to interact with a wide range of target customers hence improving the competitive advantage.

Besides, politics, social media has enhanced civic participation where people can spread knowledge on social issues and equally call for support on group projects [3]. In humanitarian crises/disasters or public health concern activities, social media has been adopted for sharing information, relief organizational activities and to support affected communities. As alluded earlier, social networking services have expanded the horizon of participation, but there are worries of the spread of fake news particularly in polarized settings. It can indeed have serious impacts on people's perception and may actually demoralize people to vote from the social platforms that are social media.

Alalwan et al. [10] noted that the inexpensive characteristic of social media marketing helps small businesses to go beyond conventional marketing platforms [3]. However, some issues which are discussed below need to be resolved to get the best of these platforms. One of the key risks is that many entrepreneurs and employees, particularly in SME businesses have rather limited experience in the use of digital technologies. Also, due to a high frequency of change in the algorithms used in social networks and the growing competition for consumers' attention, SMEs can no longer constantly advertise their products on social media [11].

Another major concern is the problem of data protection and confidentiality or security. This research indicates that many people in Jordan today are still very worried about their privacy especially when using social media. There are some points that SMEs should consider and guarantee that they do not violate any legislation by using customers information. The legal governing social media platforms in Jordan is in its infancy and there is a need for related policies regarding the protection of data, fake news, and hazard [3].

However, there are increasing prospects for the future of social media in Jordan, mainly concerning SMEs. The emerging trends of social media for business marketing along with the rising trend of digital technology in Jordan offer a huge prospect for small business to compete effectively in the market [8]. In addition, e-commerce integration into social media has opened up more prospects of getting to the customers and making the sales by SMEs [13]. The use of social software in Jordan and its assimilation is preparing Jordan to be more of a connected society. And as more people become computer literate and more get access to the internet social media is bound to remain a core component of communication and business as well as civil society in the future in Jordan. It requires effort by policymakers, businesses, and citizens for those platforms to benefit society, as well as confront their problems. As mentioned by Zoughb et al. [7], digital literacy and the responsibility for ethical use of social media will be vital to guarantee these platforms will have a positive impact on the development of Jordan's socio-economy.

3 Methodology

The present research uses a quantitative method to explore the significance and effects of social media in Jordan. The use of self-administered survey questionnaires was appropriate to study the behaviors, perceived uses and perceived barriers among social media users in the country. Having identified several dimensions the survey instrument was designed to capture; the frequency of one's social media use, what platforms the respondent frequents, the reasons for use, effects observed on business interaction, and one's level of participation in civil activities.

The target population for the survey embraced all the Jordanians in the society irrespective of age, education level or occupation. Stratified random sampling was used to identify 400 participants to complete the survey as this represented the diverse population of individuals in Jordan. Thus, the recruitment policy captured individuals who lived in urban areas as well as the rural sector, and in addition, covered workers of different ages and professions. Samples were recruited through social networks and personal email accounts only, which crisscrossed the globe. The survey was conducted in two languages, Arabic and English to suit the different language compliances. Participants were told the objectives of the study and could choose to participate and all responses submitted were kept anonymous to enhance the privacy and security of respondents. Consequently, statistical software was used to estimate and perform descriptive and inferential statistics on the collected data. For the pattern of social media usage, descriptive statistics were used.

4 Result

The results of the survey shall be used to underscore the extent of our connectedness to social media in Jordan in all aspects of individual, workplace, and community. The demographic profile of the respondents (Table 1) shows a high concentration within the 25–34 age group (35%), with the majority having at least a bachelor's degree (50%). Male respondents represent 60% of the sample. This demographic distribution suggests that the survey primarily reflects the perspectives of young, educated individuals, a group that may actively use social media.

Table 1 Demographics of survey respondents

Age group	Percentage (%)
18–24	25
25–34	35
35–44	20
45–54	15
55+	5
Gender	
Male	60
Female	40
Education level	
High school	10
Bachelor's degree	50
Master's degree	30
Ph.D.	5
Other	5

Table 2 Frequency and purpose of social media usage

Usage frequency	Percentage (%)
Daily	80
Weekly	15
Monthly	3
Rarely	2
Purpose of use	
Connecting with friends/family	35
Professional networking	20
Staying informed	20
Entertainment	15
Shopping/Brand engagement	10

The data indicates that social media is primarily used daily (80%) among respondents, emphasizing its integral role in their lives (Table 2). Connecting with friends and family (35%) and professional networking (20%) are the main purposes, followed by staying informed (20%). This highlights the significance of social media as a multi-functional tool for personal connection, professional growth, and information access.

WhatsApp is the most widely used platform (90%), followed by Facebook (70%) and Instagram (60%) (Table 3). This distribution reflects the popularity of messaging apps for personal communication in Jordan, while visual and social media platforms like Instagram also hold substantial engagement.

For those using social media for business, the primary benefit reported is increased brand awareness (40%), followed by improved customer engagement (25%) (Table 4). This suggests that businesses view social media as a valuable channel for building visibility and strengthening customer relationships rather than direct sales.

A majority (55%) of respondents use social media for political or social engagement, with news and current events being the most common topics (40%) (Table 5). This finding illustrates social media's influence as a platform for public dialogue and awareness of social and political issues.

The primary concern regarding social media usage is privacy (35%), followed by exposure to misinformation (25%) (Table 6). These findings highlight the critical

Table 3 Social media platforms preferred by users

Platform	Percentage (%)
WhatsApp	90
Facebook	70
Instagram	60
Twitter	35
LinkedIn	20

Table 4 Impact of social media on business use

Business benefit	Percentage (%)
Brand awareness	40
Customer engagement	25
Sales generation	20
Market research	15

Table 5 Social media's role in political and social engagement

Use for engagement	Percentage (%)
News and current events	40
Social causes/activism	30
Political discussions	20
None of the above	10

Table 6 Concerns about social media usage

Concern	Percentage (%)
Privacy	35
Exposure to misinformation	25
Data security	20
Addiction/Overuse	15
Cyberbullying	5

need for improved privacy measures and fact-checking to protect users and maintain trust in social media platforms.

A significant majority (60%) of respondents view digital literacy as "Very Important" for safe and effective social media usage (Table 7). Additionally, respondents believe social media will be crucial for business growth (30%) and civic engagement (25%) in Jordan's future, underscoring the role of social media in shaping the country's economic and social development.

5 Discussion

The survey highlights the deep integration of social media in Jordanian society, serving social, economic, and political purposes. Nearly two-thirds of respondents are active users, with 71% being young adults and 63% holding higher education degrees, using social media daily for information, sharing personal updates, and job-seeking. Social media primarily strengthens interpersonal relationships and societal bonding, beyond entertainment. From a business perspective, Jordanian companies leverage social media for brand promotion and customer relations, though its impact on direct sales remains limited. This suggests that while social media enhances

Table 7 Digital literacy and future prospects of social media in Jordan

Importance of digital literacy	Percentage (%)
Very important	60
Important	25
Somewhat important	10
Not important	5
Future prospects for social media	
Business growth	30
Civic engagement	25
Education and learning	20
Entertainment	15
Political engagement	10

visibility, its commercial potential is still evolving as digital marketing practices mature.

The study also reveals that 70% of respondents use social media for information and social causes, reflecting its role in civic engagement and accountability. This aligns with global trends, emphasizing its importance in raising awareness and fostering political and social discussions in Jordan.

Challenges such as privacy concerns and misinformation were significant, with users expressing fears about data security and unverified information. Addressing these issues requires improving digital literacy and enhancing platform accountability to ensure a safer social media environment.

Finally, the respondents value digital literacy, recognizing it as an essential skill for networking, opinion management, and business. Social media has become integral to daily life in Jordan, with the potential to drive societal and economic growth by maximizing its benefits and mitigating its drawbacks.

6 Conclusion

In this paper, the positive changes that stem from the use of social media in Jordan are discussed in terms of personal and social change processes. Public usage of social media has gotten central in the lives of Jordanians especially among the young educated individuals, hence it is used as a means of communication, as professional reference and information resource. This demonstrated usage shows that social media gifts the society, especially those who use it a platform to foster interpersonal relationship and to become digitally connected, with a voice.

Speaking of business opportunities, social media is one of the most effective ways to build brand awareness and engage customers, though its benefits drastically increase the lead's conversion. Since Jordanian companies are now shifting to online business models there are still opportunities for social media to help increase economic impact. The targeted approaches used in correlation with social media and other marketing tools may further strengthen its position in conversion and broadening market niche, providing SMEs with opportunities to establish stronger presence in the sphere of digital economy.

Also, the use of the social media is another aspect in civil activities of an individual where such people use social media in making their contributions to the society, and demonstrating participation in matters touching on politics as well as coordinating events within the society. This has made social media as a potent force for change for the voices of the people in Jordan as it expands the means through which citizens engagement in public forums can be supported. However, to realize this potential, it is imperative that two main issues of disinformation and privacy be solved. Awareness raising and development of legislation and policies may enhance user's ability to use those platforms safely and responsibly and create a credible and stable digital space.

Finally, the significance of social media in Jordan is enormous; it affects the social, economic and civil aspects of Jordanians' life. If these problems are solved and the mentioned advantages utilized, the Jordanian society will be able to develop the use of social media as a communication platform, economic recovery tool, and active engagement in building the strong democratic future where the digital world is involved and implemented.

References

1. Kumar, A.: How technology changed our lives: has it improved life today? (2024). Retrieved from https://www.simplilearn.com/how-has-technology-improved-our-lives-article
2. SyamimiSaid, N., et al.: Review on web performance. J. Eng. Appl. Sci. **9**(1), 18–23 (2014)
3. Khashman, et al.: The impact of social media misinformation in Jordan: challenges and regulatory frameworks. Middle Eastern J. Soc. Media **3**(2), 211–225 (2022)
4. Martin-Gomez, de Luna, M.: The role of social networks in communication in the scientific research community. J. Technol. Sci. Educ. **14**(2), 291–305 (2023)
5. Abu-Shanab: Social media, trust, and civic engagement in Jordan. J. Soc. Polit. Stud. **18**(4), 210–223 (2021)
6. Abdelsalam, et al.: Social media platforms and political participation: a study of Jordanian youth engagement. Soc. Sci. **12**(7), 1–18 (2023)
7. Zoughb, et al.: The influence of social media on public opinion in Jordan. Jordanian J. Commun. **19**(1), 45–59 (2022)
8. Sadiq, et al.: The role of social media in interpersonal communication in Jordan. Arab J. Commun. Stud. **6**(2), 99–110 (2021)
9. Majed, et al.: The role of social media in shaping public opinion among Jordanian university students. J. Infrastruct. Policy Dev. **8**(8), 1–25 (2024)
10. Alalwan, et al.: The impact of social media on SMEs in Jordan: opportunities and challenges. Int. J. Inf. Manag. **37**(6), 1193–1204 (2027)
11. Hani, Basyuni: Social media marketing in Jordan: the role of Facebook and Instagram in SME brand promotion. J. Bus. Econ. **10**(4), 54–67 (2021)
12. Rosenberg, et al.: Social media in political engagement: the Jordanian case. Int. J. Polit. Soc. Media **15**(1), 65–78 (2021)
13. Shafiyah, N., et al.: Review on electronic commerce. Middle-East J. Sci. Res. **18**(9), 1357–1365 (2013)

Leveraging Artificial Intelligence in Marketing: A Survey of Effectiveness, Challenges, and Future Prospects

Mahmoud Abouraia, Ayman Bassam Nassoura, and Hothefa Shaker

Abstract This research aims to advance the knowledge on the use of artificial intelligence (AI) in predictive marketing with its use, efficiency, and problems encountered during its implementation in different sectors. The ability of artificial intelligence in vast consumption of analytical data and clienteles makes predictive marketing a powerful contact that can enrich personalization, improved focusing as well as elevated marketing ROI. The current research employs a survey design to gather data from 170 marketing professionals with different organizational affiliations. It is revealed that AI has been taken up in the context of predictive marketing with a relatively high prevalence, an insight that was found most dominant within fields including customer segmentation and recommendation. Customers also revealed increased customer sales and better campaign results through using artificial intelligence. Nevertheless, it is vital to understand that several problems remain—data protection, legal requirements, and existing skills shortage in the people. In summary, this work established that AI has the possibility to change the nature of predictive marketing but highlighted that it could only be realized once these research questions are addressed. That is why the findings of this study are essential for organizations interested in using AI in marketing, as well as enhance understanding of the strategic uses of AI and practical challenges faced in actual applications.

Keywords Predictive marketing · Artificial intelligence (AI) · Customer insights · Data privacy · Personalization · Marketing ROI

M. Abouraia · H. Shaker (✉)
Modern College of Business and Science, Muscat, Oman
e-mail: hothefa.shaker@mcbs.edu.om

M. Abouraia
e-mail: Mahmoud.abouraia@mcbs.edu.om

A. B. Nassoura
Department of MIS, City University Ajman, Ajman, UAE
e-mail: a.bassam@cu.ae.ae; nassuora@gamil.com

B. Awwad (ed.), *AI and IoT: Driving Business Success and Sustainability in the Digital Age*, Studies in Systems, Decision and Control 590,
https://doi.org/10.1007/978-3-031-88874-8_75

1 Introduction

Artificial Intelligence (AI) development progresses astonishingly in different fields, including marketing where data play the crucial role of providing insights into marketing strategies. As the volume of customer data increases daily through customer interactions and behavior, AI assists in analyzing patterns, forecasting future behavior and offering relevant service delivery in a way that will improve customer satisfaction and business outcomes [1]. This shift towards data-centric marketing allows firms to gain competitive advantage by achieving better insight of the distribution of their resources, targeting the exact clientele that will provide a good ROI as well as using predictive marketing analytics as noted by Jarek and Mazurek [2].

Predictive marketing, being an application of machine learning models, helps the marketer to predict what the customer will do next and the movement in the market. From this, the approach offers better opportunities in enabling proactive rather than reactive marketing activities, improvement of customer engagement and customer loyalty [3, 4]. For instance, purchasing patterns, churn rates or likely consumer interests can be forecasted in organizations using predictions models so that the campaigns that are being run, are in tune with the likely interests of the consumers. Moreover, the mentioned capabilities also enable such companies to accurately fine tune the marketing strategies in real time based on changes in customer behaviors and the markets. Nevertheless, the application of AI in predictive marketing is done with some difficulties. The greatest challenges of organizations when it comes to deploying AI technologies include; Data privacy, accuracy of the models and ethical questions. It is crucial to adhere to legal standards of privacy as in GDPR and be transparent regarding AI decisions to gain customer trust and to minimize possible biases [5]. Overcoming all these challenges is basic to the development of AI practices that protect the rights of consumers and help maximize marketing returns.

This paper explores AI in predictive marketing highlighting how organizations can appropriately harness data for insights for marketing while also discussing the ethical, technical and regulatory concerns of doing so. In its current form, this research aims at providing insight into the practical applications, the advantages and of course the drawbacks of using predictive marketing alongside other essential insights regarding what it would take to apply the concept with leveraged success in the corporate world. With 170 questionnaires completed, this study offers practical understanding of perceptual attitudes, advantages, and challenges to AI-based predictive marketing. Finally, this study is expected to make meaningful input to the emerging literature on AI in marketing with specific reference to the use of such tools in formulating and implementing customer-focused policies that are aware of the need to safely use customers' data.

2 Literature Review

Predictive marketing is defined as a methodology that applies various models to estimate future customers' behavior and trends, which in turn help businesses to adapt to potential future needs for improving the marketing strategies [6]. In contextualizing predictive marketing, the key enabler is coined as AI, majorly the ML whose function is to analyze vast data sets and predict subsequent actions to proactively tailor marketing efforts. Thomas [7] recognizes that through the application of artificial intelligence such as the creation of predictive models, companies can advance in customer targeting and optimizing ROI for their endeavors alongside the ability to change quickly [7]. In today's world where more consumer data is produced on daily basis, AI helps organizations capture even more about consumers' behavior thus giving a competitive advantage over another firm in the market [1].

Classically applied in predictive marketing tasks comprise of; customer segmentation, demand forecasting, churn prediction, and use in personalized recommendation engines [8]. For instance, AI can group customers according to the likelihood that they will make a purchase then market items to these groups with higher chances of a purchase [9]. It also enables analysis of future trends in the demands for products and services so that companies can stick to a proper resource use strategy [10]. Further, it can also guess a customer's tendency to leave by using data which guides companies in employing measures that can better retain the customer [11]. Some of the models used in the current study have included collaborative filtering and decision trees as these have been used to generate customer recommendations hence improving sales [12].

As evidenced in the paper, AI incorporation in the marketing function can improve decision making at different hierarchies. Over time, as AI algorithms continue to learn from the data provided, they get better, giving marketers better all-encompassing insights for decision making [3]. Marketing using AI results in high performing campaign messages that reflect customer needs and tendencies boosting the conversion percentages together with customer satisfaction [2]. Lim et al. [13] pointed out that AI can help businesses achieve the most of their marketing dollar through identifying which channel or campaign to invest in. It also makes it easier for an organization to make decisions on the basis of market situation and consumers' trends hence increases on flexibility [13].

However, AI, when integrated with predictive marketing, has its challenges as discussed below. An important issue is the quality of data that is collected. AI models depend on large amounts of data complimented by quality to deliver effective predictions. This implies that one may end up developing not only incorrect models for the marketing issue at hand but also wrong marketing strategies [5]. Moreover, AI models refer to different complexities, which may be challenging for individuals not so familiar with technical frameworks to understand or even trust the models [14]. Another issue is compatibility, or the difficulty some marketing organizations face to ensure new, enhanced AI tools interconnect with existing marketing systems; this is likely to necessitate more capital expenditure on structured ensuring that corporate staff members are adequately trained to optimally exploit new innovations in AI tools [1]. The other essential matter is the several consequences of AI usage, including data privacy problems and the problem of AI bias. Customers are getting privacy conscious and hence firms must be honest of how they use AI, to avoid backlash [5].

With the development of AI predictive marketing decreasing consumer data privacy is a key issue of ethical implication. Current rules such as the General Data Protection Regulation (GDPR) in Europe have outlined the way data can be collected and used by businesses [15]. Businesses need to be certain that they have not violated these stipulations, and that they are not misleading to customers regarding how AI models used inside of organizations are concluding about specific data pertaining to the customer. Furthermore, there has been an issue of algorithmic bias where an AI model is created based on an unfair set of data, and this leads to stereotyping or exclusion of specific groups [16]. Promoting fairness and non-bias in AI decision-making structures is very important in making sure that predictive marketing is a noble practice that will not in any way harm the consumers.

Different AI tools and technologies are used in the implementation of predictive marketing techniques. Using multiple algorithms by regression analysis, decision tree, or clustering, segmentation, forecasting, and trend analysis are now used in business intelligence [17]. The machine learning subfield known as deep learning is widely used for image and voice recognition, and various natural language processing uses, as pointed out by LeCun et al. [18]. Salesforce Einstein, IBM Watson, and Google AI are examples of the AI that companies can leverage to have AI templates suitable for a given marketing strategy [1]. They allow adapting AI insights into CRM and marketing automation systems to grow from a small-scale business to become larger and more efficient.

AI used in the context of predictive marketing is a field that has tremendous prospects. Over time, with advances in AI the predictive models will improve in efficiency resulting in business and organizations enjoying more detailed insight into consumers' behavior. The adoption of AI will not only make marketing results better, but also make it possible for organizations to provide customers with more singular and interesting experiences [6]. Also, improvements will be observed in AI over the years, and this will enable marketers to future understand the intent of the customer and how to improve the customer journey when interacting with the product. The synergy of AI with IoTs and AR will offer marketers fresh ways to engage with customers providing added value [3].

Predictive marketing is also known as a new trend in artificial intelligence that is driving changes in business decision-making and the actions of companies targeting customers [19]. In the case of the consumer database, AI offers businesses significant ways of identifying future behavior, refining marketing campaigns, and enhancing the way consumers deal with a specific business [20]. But there are problems like quality issues, and ethical issues, which the stakeholders must pay attention to implement successfully. AI will remain very central to the future of marketing, with massive opportunities for organizations to learn how to communicate much better with their customers.

3 Methodology

This research employed a quantitative research paradigm to assess the AI adoption success and obstacles in predictive marketing. The present study employed a structured online survey of marketing professionals, and the survey was conducted for two months amassing a total of 170 responses. In selecting these participants, the focus was on levels of experience and positions held particularly in marketing or related industries thus the participants included suited the study objectives. The survey was comprised of four major parts. The first section obtained demographic details about the respondents and their levels of experience and company industry sectors. The second part scrutinized how fundamental AI is ingrained in marketing plans and how customer segmentation and other related applications like predictive analytics and personalization were implemented. In the third part, I examined how organizations view the capability of AI in enhancing user interactions, optimizing the targeting of campaigns, and speaking to the marketing ROI. Lastly, regarding concerns and barriers to the implementation of AI that respondent reported was the problem encountered with data quality, privacy, regulation, and skills of workforce.

The majority of questions were posed in the form of a Likert scale, multiple choice and open ended that allowed for both quantitative and further qualitative results. Quantitative data analysis used descriptive statistics to measure the extent of AI use, success, and issues related to the marketing application of predictive marketing. To investigate the interconnection between measures like industry type and the extent of AI implementation or company size and perceived issues, cross-tabulations were applied. Moreover, open-ended questions were to identify patterns Why and What which provided better understanding of the advantages and risks of introducing AI in marketing. The following are the limitations of this research. Someone reading this study may wish to ask the following questions: One of the study's limitations is based on self-records; respondents may not give a thorough view of AI's potential or constraints in their organizations. In addition, it indicates that the overall sample totaling 170 respondents gave important results though more diverse and more significant sample could reveal more peculiarities related to specific industry. Altogether, this methodology provides a broad picture of AI's application in marketing and provides quantitative data along with qualitative feedback of AI and its opportunities as well as threats in predictive marketing.

4 Results and Findings

Table 1 provides an overview of the demographic distribution among the 170 survey respondents. Most respondents identified as marketing professionals (35%) or data scientists (25%), reflecting a strong representation from the operational and analytical side of marketing. Respondents' experience levels ranged widely, though a substantial portion (40%) reported having between 4 and 7 years of experience. In terms

Table 1 Respondent demographics

Role	Percentage (%)
Marketing professional	35
Data scientist	25
Marketing strategist	20
C-level executive	10
Researcher/Academic	10
Experience level	
1–3 years	20
4–7 years	40
8–10 years	25
Over 10 years	15
Organization type	
Small/Medium enterprise (SME)	45
Corporate enterprise	30
Marketing agency	15
Research institution	10
Industry	
Retail	30
Finance	20
Healthcare	15
Technology	25
Other	10

of organization type, 45% worked in small-to-medium enterprises (SMEs), indicating a notable representation from smaller business entities. Participants were from diverse industries, with retail (30%) and technology (25%) being the most common, providing a balanced insight across different market sectors.

Table 2 reveals the usage trends of AI in marketing among respondents. A substantial majority (75%) reported that their organizations leverage AI for predictive marketing activities. The primary application of AI was customer segmentation (40%), followed closely by personalization (30%), and churn prediction (20%), indicating a strong focus on customer-centric strategies. Machine learning emerged as the most widely used AI tool, adopted by 60% of respondents, while predictive analytics platforms and natural language processing were also common. Notably, 40% of respondents used AI insights on a weekly basis, suggesting regular integration of AI-driven data into marketing decisions.

Table 3 assesses the perceived effectiveness of AI in enhancing marketing outcomes. About half of the respondents rated AI's predictive accuracy as "Moderately Accurate," with 30% finding it "Very Accurate," demonstrating confidence in AI's potential to provide valuable insights. The effectiveness of AI in enhancing marketing strategy was rated highly, with an average score of 4 out of 5. Moreover,

Table 2 Usage of AI in marketing

Current AI Use in Marketing	Percentage (%)
Yes	75
No	25
Primary uses of AI	
Customer segmentation	40
Personalization	30
Churn prediction	20
Demand forecasting	10
Content recommendation	15
AI tools used	
Machine learning	60
Predictive analytics platforms	30
Natural language processing	25
Frequency of AI-driven decisions	
Weekly	40
Monthly	30
Daily	20

60% of respondents indicated that AI offers a significant competitive advantage, underlining its perceived value in gaining market differentiation.

Table 4 highlights the main challenges faced by organizations when implementing AI in marketing. Data privacy concerns (40%) and high costs (30%) were identified as the primary obstacles, reflecting common issues in AI deployment. Transparency in AI-generated predictions remained an area for improvement, with 35% rating AI outputs as "somewhat opaque." To mitigate these issues, many organizations focused

Table 3 Effectiveness of AI in marketing

Predictive accuracy rating	Percentage (%)
Very accurate	30
Moderately accurate	50
Somewhat accurate	15
Not accurate	5
Effectiveness in enhancing strategy (1–5)	*Average rating*
Strategy enhancement	4.0
Competitive advantage	
Significant advantage	60
Somewhat advantage	25
Minimal advantage	10
No advantage	5

on increasing staff training (30%) and enhancing data security (25%), aiming to improve both the technical and operational dimensions of AI integration.

Table 5 examines the anticipated future role of AI in marketing, with a majority (55%) of respondents expressing a high likelihood of expanding AI applications in the next 1–2 years. Respondents pointed out predictive customer insights (35%) and campaign optimization (25%) as the most promising areas for AI development. Open-ended feedback suggested a strong belief in AI's ability to drive personalized marketing and enhance decision-making frameworks, indicating that professionals see AI as integral to future marketing innovations.

Table 4 Challenges in Implementing AI

Primary challenges	Percentage (%)
Data privacy concerns	40
High costs	30
Lack of skilled personnel	25
Integration difficulties	20
Transparency in AI predictions	
Very opaque	25
Somewhat opaque	35
Transparent	25
Very transparent	15
Solutions implemented	
Increased staff training	30
Invested in data security	25
Improved data integration	20

Table 5 Future prospects of AI in marketing

Likelihood of expanding AI use	Percentage (%)
Very likely	55
Somewhat likely	25
Not likely	10
Beneficial areas for AI in marketing	
Predictive customer insights	35
Campaign optimization	25
Real-time personalization	20
Customer journey mapping	15

5 Discussion

The results in this paper highlight trends in AI application and its growing role in modern marketing. More than half of the respondents reported using AI to predict marketing insights, confirming its integration as a standard tool in marketing strategies. AI is primarily applied in customer segmentation, personalization, and churn prediction, with machine learning being the dominant technology. These findings align with literature emphasizing AI's role in analyzing large datasets to derive customer insights. High adoption rates and positive results for prediction accuracy and strategy improvement suggest marketers view AI as both effective and advantageous, with 60% believing AI provides a competitive edge. However, challenges such as data privacy, high costs, and a lack of transparency (the "black box" issue) hinder trust and broader adoption. Additionally, a shortage of skilled personnel underscores the need for human capital development.

The study also indicates that marketers plan to increase AI adoption, particularly in customer prediction and campaign efficiency. Future AI innovations are expected to focus on customer journey mapping and real-time personalization, emphasizing AI's potential in creating highly customized marketing communications. For organizations, key recommendations include prioritizing data privacy, choosing transparent AI tools, and investing in workforce training to address skill shortages. Finally, organizations expanding their use of AI should seek cost-effective solutions and strategies. This paper offers insights into AI's current and potential impact on marketing, raising questions for future research on transparency regulations, adoption processes, and the competencies required for effectively leveraging AI-driven insights.

6 Conclusion

This study demonstrates how AI can revolutionize marketing by improving prediction of strategies, customer categorization, and interactions. The survey findings show that AI is not only used by many companies but also viewed as a key competitive advantage that helps to improve targeting and client's retention. However, the study also identifies significant problems which include data privacy concerns, the lack of public transparency and scarcity of qualified staff. These barriers clearly call for organizations to go for explainable AI and improve the workforce to make the best use of AI. The outcome indicates that reliance on AI increases and it will become more important for marketers as it creates new solutions for improving customer engagement and perfecting ad campaigns and promotion strategies. There is nothing as critical to the success of AI in practice as respecting the technological aspect on one side, and the ethical, transparent, and best-practice-deed data management on the other side. Finally, it is suggested that future research should address how the emerging legislation in AI will affect marketing and investigate what new competencies the increasing sophistication of AI is going to require. To sum up, there are some unrevealed problems in AI-marketing, but generally, further development is possible with a big potential of AI tools in the companies who decided to use it for better understanding the needs of the customers and interact with them more effectively.

References

1. Marr: Artificial Intelligence in Practice: How 50 Successful Companies Used AI and Machine Learning to Solve Problems (2020)
2. Jarek, Mazurek: Marketing and artificial intelligence. Cent. Eur. Bus. Rev. **8**(2), 46–55 (2019)
3. Davenport, Ronanki: Artificial Intelligence for the Real World. Harv. Bus. Rev. **96**(1), 108–116 (2018)
4. Rahman, M., et al.: The new wave of AI-powered luxury brands online shopping experience: the role of digital multisensory cues and customers' engagement. J. Retail. Consum. Serv. **72**, 103273 (2023)
5. Binns, et al.: It's reducing a human being to a percentage': Perceptions of justice in algorithmic decisions. In: The 2018 CHI Conference on Human Factors in Computing Systems, pp. 1–14 (2018)
6. Chaffey: Digital Marketing: Strategy, Implementation, and Practice. Pearson Education (2020)
7. Thomas: Using AI to drive marketing decisions: a look at the future of predictive marketing. J. Mark. Trends **20**(2), 19–32 (2018)
8. Baker, Dellaert: AI in marketing: practical applications and opportunities. J. Mark. **83**(3), 45–67 (2019)
9. Gao: The role of artificial intelligence in marketing. J. Bus. Res. **117**, 25–35 (2020)
10. Chakraborty, et al.: Predictive analytics in marketing: review and future directions. Mark. Intell. Plann. **36**(6), 763–778 (2018)
11. Kim: Predicting customer churn with machine learning models. J. Bus. Analytics **9**(1), 45–57 (2019)
12. Hernández, Pérez: Predictive analytics in marketing: the future of customer behavior analysis. Int. J. Mark. Stud. **12**(2), 56–72 (2020)
13. Lim, et al.: AI in marketing: benefits, challenges, and the role of human judgment. Mark. Rev. **18**(3), 157–174 (2017)
14. O'Reilly: The challenges of AI implementation in marketing: a review of literature. J. Mark. Technol. **8**(2), 35–42 (2020)
15. Wright, De-Hert: Privacy Impact Assessment. Springer (2018)
16. Dastin. (2018). Amazon scrapped secret AI recruiting tool that showed bias against women. Reuters. Retrieved from https://www.reuters.com
17. Sterne: Artificial Intelligence for Marketing: Practical Applications. Wiley (2019)
18. LeCun, et al.: Deep learning. Nature **521**(7553), 436–444 (2015)
19. Bacanin, N., et al.: Quasi-reflection learning arithmetic optimization algorithm firefly search for feature selection. *Heliyon* **9**(4) (2023)
20. Shafiyah, et al.: Review on electronic commerce. Middle-East J. Sci. Res. **18**(9), 1357–1365 (2013)

Leveraging Green Technologies for Sustainable Business Ventures in the Middle East: Opportunities, Challenges, and Future Trends

Mahmoud Abouraia, Ayman Bassam Nassoura, and Hothefa Shaker

Abstract Green technologies are increasingly recognized as essential tools for boosting sustainable business ventures in the Middle East. This article aims to analyze the potential of green technologies, including renewable energy sources, energy-saving technologies, eco-materials and water-saving technologies, to make business environmentally sustainable in the region. It explores the opportunities arising from green technologies from environmental, economic and social perspectives and examines the gaps companies see in their implementation, including costs, ignorance and policies. It is in this context that this article seeks to contribute to the existing literature on sustainability and innovation by providing an exploration of green technologies and their implications for business opportunities in the field.

Keywords Green technologies · Sustainable business ventures · Renewable energy · Middle East · Circular economy · Environmental impact · Economic benefits · Sustainability

1 Introduction

Awareness of environmental sustainability has recently increased in the Middle East due to ecological problems and rapid economic growth in this region. These issues have allowed green technologies to be part of business growth programs. Green nature

M. Abouraia · H. Shaker (✉)
Modern College of Business and Science, Muscat, Oman
e-mail: hothefa.shaker@mcbs.edu.om

M. Abouraia
e-mail: Mahmoud.abouraia@mcbs.edu.om

A. B. Nassoura
Department of MIS, City University Ajman, Ajman, UAE
e-mail: a.bassam@cu.ae.ae; nassuora@gamil.com

B. Awwad (ed.), *AI and IoT: Driving Business Success and Sustainability in the Digital Age*, Studies in Systems, Decision and Control 590,
https://doi.org/10.1007/978-3-031-88874-8_76

technologies identify renewable energy, eco-friendly materials, energy-saving solutions and waste management as remedies that Middle Eastern countries can use to solve environmental problems while promoting economic growth [1]. With the diversification processes in all countries in the region and the emphasis on reducing dependence on fossil fuels, green technologies are highlighted as strategic priorities for sustainable business development [2]. There are also other factors driving the Catalan landscape to adopt green technologies in the Middle East, such as the country's regulatory frameworks and the country's national vision for sustainable development. For example, Saudi Arabia's Vision 2030 and the United Arab Emirates' Vision 2021 both set out targets for environmental protection, reducing "greenhouse gas" emissions and boosting energy. Renewable [3]. Acceptance of green technologies by stakeholders is also admitted due to awareness by the consumer and cooperation by international partners in the Middle East. This realization has ensured that the different eco-consumer segments have forced businesses to enhance sustenance in their activities while partnership with IGOs meaning that they have a chance to access the green technology advancement and investment [3]. For example, the IRENA based in the UAE has played a significant role of advocating for renewable energy technology as well as building capacities within the Middle East region [4].

Renewable technologies provide the best keys to unlock the opportunities of sustainable economic activities within the region. With resource conservation and the reduction of waste being more or less implemented into organizational models on an ongoing basis, they have migrated towards utilizing renewable sources of energy including solar power which would be more suitable for the climatic conditions. For instance, the Mohammed Bin Rashid Al Maktum Solar Park in Dubai is one of the largest solar systems in the world with the goal of producing solar power at 5000 MW by 2030 and adding solar infrastructure of sustainable energy [5]. These kinds of activities exemplify how green technologies can contribute to sustainable business models, as they increase resource utilization, decrease business costs, and invite inventions [6]. There are several factors that limit new organizations and especially the small firms, these being; limited operating capital, lack of technology, and inexperienced workers in ecofriendly process [7]. Big organizations can afford to procure green solutions due to their availability of capital; SMEs, on the other hand, face challenges of high procurement costs for these technologies [8]. Solving these challenges calls for the collaboration of different governments, business and industry, and international organizations to offer incentives, policies and practices, and training for green technology [9].

2 Literature Review

Green technology is then understood as innovation and processes that reduce the effects that humans impose on nature with the aim of achieving sustainable use of resources, minimizing emissions of carbon and other gases into the atmosphere and reduce the generation of waste. Some of the important green technologies are electricity generation from solar, wind and hydroelectric systems which provide solutions in place of conventional fossil fuel systems with low CO_2 emission rates respectively [10]. Sustainable technologies like; LEDs, smart grids, and efficient heating and ventilation systems aim to reduce energy consumption in homes and businesses [11]. Renewable energy is gradually advancing in the Middle East owing to access to solar energy with large-scale solar installation in the UAE, Saudi Arabia and other nations forming a part of national environmental and efficiency initiatives [12]. Recycling and waste to energy is another activity that has also received improvement in the region given the rise in waste production especially in the urban area and the new technologies associated with it. They are green technologies which are fundamental in meeting environmental problems while nurturing the growth of the economy.

Sustainable business ventures might therefore be described as those business ventures which bear equal responsibility to those of environmental conservation and management and those of profitability and growth [13]. The goal is sustainable, sustainable value for shareholders and stakeholders, sustainable value for society and the environment. Sustainability has recently emerged as an important factor in the Middle Eastern strategic plans of businesses due to the calls from the populace, authorities, as well as global standards. For example, the companies in UAE and Saudi Arabia have integrated sustainability reporting standard for their concern of minimizing the impact of environment while on the other hand adding value to the economy [14].

Geissdoerfer et al. [13] found that business models that implement the concept of circular economy enjoy less dependency on resources, lower operating costs, and more resistance to changes in regulation [13]. It has also been established that, Middle Eastern firms that have embraced green technologies have found it effective in practicing sustainable development objectives as firms that integrate renewable energy and waste management techniques can record better performances, policy compliance [15]. For example, solar power is well-applied in the commercial-use, its technology costs are decreasing and the converted efficiency of the photovoltaic panels is rising [16, 17].

Effective energy conservation technologies allow companies to achieve the envisaged goals and objectives of sustainable business performance by utilizing energy resources efficiently, reducing costs, and minimizing the adverse effects on the environment. Some of these are LED lighting that considerably reduces energy demand when compared to general lighting and smart controlling systems that track and regulate energy consumption in functional buildings in real-time [18]. In production function, energy efficient equipment and techniques in production reduce the amount of energy used in completing the production process. They create possibilities for businesses to achieve sustainability objectives without necessarily having to pay the price of reduced efficiency; therefore, they are ideal in industries that require tremendous energy.

The use of green technologies is very instrumental in enhancing sustainable business strategies as it has potential for bringing numerous environmental benefits. Measures like recycling and using of sustainable materials lower pollution as it will manage to cut on degrading ecosystems and limiting out of helpful emissions. Combined, these practices make lead to a more sustainable and ecology friendly model of business [19]. Green technologies provide numerous economics opportunities for businesses to propel sustainable business ventures as lucrative. Green certification and eco-friendly practices also give companies an opportunity to access green financing in that many financial institutions give concessional finance to green committed companies [20]. The use of information communication technology from an environmental perspective impacts a firm's social responsibility hence improves social responsibility since consumers associate that firm with environmentally friendly products. The following is more motivation is derived and employees are proud pushing retention and talent attraction to companies with environmental goals [20].

It is very clear that the initiation cost stands to be a major challenge arresting the green technologies implementation portfolio. While green technology may give cost savings in the long run, the cost barrier is still a challenge mainly because green financing is limited or nonexistent in some areas [21]. Furthermore, businesses may be unaware of options for sustainable solutions or may as yet be unaware of possible improvements to efficiency that would lead them do consider green solutions. Those are important for increasing the awareness of the businesses that have limited understanding of green technologies and closing the gap of the knowledge necessary for the improvement of the green technologies 'take-up' [22]. Governments therefore need to put forward clear policies and incentives to create greater opportunities for the selection of green technology [23]. The innovations of green technologies also have major technological constraints, especially to the less developed regions in sustainable technology. World Bank [24] found that technology infrastructure and compatibility are critical factors in improving green technologies' diffusion [24].

Green computing technologies are being adopted as more organizations strive to reduce their power consumption in IT systems. Carbon capture and storage CCS is another new technique in the Middle East which involves the efforts of cutting greenhouse gases particularly carbon dioxide from industrial processes; it is use by oil rich nations that are seek ways of reducing emission [25]. As for visually distinct innovations, recyclable packaging, and focusing on biodegradable materials, there is a growing trend in response to customer appeals for environmentally friendly products [4]. AI and big data analytics are fast becoming key enablers to making green technologies work much more effectively and to have a much bigger impact in the Middle East. Sustainability as a consumer preference is sneaking into the Middle Eastern markets much more, as individuals demand more earth-friendly goods. Such change is promoting the use of green technologies by organizations and incorporating sustainability issues in their offerings. It is already emerging in segments such as hospitality, retail and real estate where firms are deploying green technologies to align their offerings with customer demands and hence gain customers' loyalty [26].

2.1 Types of Green Technologies in Sustainable Ventures

Renewable energy is a preeminent element of green technology for industries and provides efficient solutions to conventional fuel sources. So important sources are solar energy, wind energy, hydroelectric energy power, and biomass energy which can substantially minimize the greenhouse effect and consumption of non-renewable energy. For example, solar power is well-applied in the commercial use, its technology costs are decreasing and the converted efficiency of the photovoltaic panels is rising [16]. Wind energy is also picking up due to the growth of large wind project portfolios because industries require significant amounts of power. Hydropower and biomass energy augment these sources with always available energy production solutions and opportunities for conversion to energy, in line with circular economy thinking [17]. Organizations adopting these renewable resources show organizational commitment to the environment and corporate societal responsibilities, besides enjoying long-term cost advantages.

2.2 Energy Efficiency and Conservation

Effective energy conservation technologies allow companies to achieve the envisaged goals and objectives of sustainable business performance by utilizing energy resources efficiently, reducing costs, and minimizing the adverse effects on the environment. Some of these are LED lighting that considerably reduces energy demand when compared to general lighting and smart controlling systems that track and regulate energy consumption in functional buildings in real-time [18]. In production function, energy efficient equipment and techniques in production reduce the amount of energy used in completing the production process. They create possibilities for businesses to achieve sustainability objectives without necessarily having to pay the price of reduced efficiency; therefore, they are ideal in industries that require tremendous energy.

2.3 Sustainable Materials and Waste Reduction

Green products and management of wastes form a significant component of green business initiatives. Most companies, nowadays, minimize landfill waste by using biodegradable and recyclable material as much as possible. Recycling technologies and up cycling encompass operational processes to transform waste products into new form of products, therefore establishing a closed loop that follows circular economy principles proposed by [19]. For instance, it is common to find firms in the consumer goods sector incorporating sustainable packaging made from recycled plastics or biodegradable material as the environmental pollution effects, coupled by increased consumer demand for green products.

2.4 *Impact of Green Technologies on Sustainable Business Ventures*

Environmental Benefits: The use of green technologies is very instrumental in enhancing sustainable business strategies as it has potential for bringing numerous environmental benefits. Incorporation of energy from natural sources such as the solar and wind energy will enable firms to alleviate the use of fossil-based energy hence reducing greenhouse gases emissions thus combating climate change. Similarly, resource conservation technology that involves installation of water efficient gadgets and energy efficient equipment reduces wastage of important natural resources. Measures like recycling and using of sustainable materials lower pollution as it will manage to cut on degrading ecosystems and limiting out of helpful emissions. Combined, these practices make lead to a more sustainable and ecology friendly model of business [24].

Economic Benefits: Green technologies provide numerous economics opportunities for businesses to propel sustainable business ventures as lucrative. Companies have less operational expenses, mostly due to optimization on energy consumption because long term energy saving systems and green energy causes less money to be spent on utilities. Moreover, when the public becomes more sensitive to the issues of the environment it will prove to be a plus for firms that apply environmentally friendly practices to retain their consumers and gain more clients with a similar nature. Green certification and eco-friendly practices also give companies an opportunity to access green financing in that many financial institutions give concessional finance to green committed companies [24].

Social Benefits: The use of information communication technology from an environmental perspective impacts a firm's social responsibility and hence improves social responsibility since consumers associate that firm with environmentally friendly products. A need for consumers to buy from companies that are socially responsible and concerned with environmental issues gives an organization better brand image and customer interaction. Moreover, they also improve the rate of satisfaction since employees get to embrace company values and responsibilities to a sustainable planet. The following is more motivation is derived and employees are proud pushing retention and talent attraction to companies with environmental goals [20].

Challenges and Limitations: However, those businesses interested in green technologies must consider the opportunities together with their disadvantages and constraints. The attractiveness of green technologies can be offset by the high initial investments required for the projects like renewable energy systems or a waste management facility particularly for enterprises that fall in the small to medium bracket. Firstly, inequalities in the ability to obtain technology reduce the use of green technology in some areas where facilities of renewable energy or recycling are not sufficiently advanced. Furthermore, it will be critical to have regulatory backing as well as policy support for many of these green ventures as many of them require prescriptive frameworks through which change can be efficiently spurred particularly in emerging economy markets [23].

3 Discussion

Interestingly, the Middle East is currently standing at an important crossroads of its development towards green business models and new technologies are working to green the region's economy. The incorporation of renewable energy, energy efficiency, environmentally friendly materials, and water saving technologies in business investment not only has considerable and numerous environmental impacts but also has great economic and social value. However, getting to this point of routine usage of these technologies is not without hitches. Another advantage that industries are likely going to benefit from when adopting green technologies is the possibilities available for diminishing the impact of the industries on the environment. Although the Middle East is culturally notorious for its overreliance on fossil fuels, the move towards solar and wind power is a good one.

Furthermore, aspects of making use of efficient technologies such as LED in lighting, smart grids and efficient appliances will help in reducing the operational costs of organizations across the manufacturing and selling sector, as well as in food and accommodation. By doing this they assist in save on energy and thus there are increased savings on bills. In addition, as consumers consider being environmentally conscious globally, companies that invest in sustainability are likely to enjoy a loyal market base and better markets and customers' retention. As it is with the social benefits of adopting green technologies which are relevant also to the Middle Eastern scenario. Growing awareness of environmental-consciousness makes consumers expect organizations to implement environment-friendly policies. This shift is particularly more and more prominent among the generations Y and Z, the latter of which are willing to work for or buy from companies that share their values. The successes in implementing green technologies enable businesses not only to improve their business image and thus attract the expanding band of environmentally suspicious consumers. In addition, many corporate business entities are eager to promote sustainability and the organizations that practice sustainability could enjoy improved worker morale due to awareness for sustainability goals. Such positive effect on employee engagement and retention is this critical in the employment of motivated and efficient employees.

However, there are several challenges which limit the implementation of green technology in the Middle East. However, one of the biggest challenges is high capital costs associated with implementation of these technologies. The costs of implementing green technologies thus remain high owing to the few options of financing green ventures. However, a common problem for green investments is a general and strategic ignorance of green technology by eminent business people and entrepreneurs of the region. While there is a good demand for green solutions, many companies either know little about the technologies they can buy or they do not see that green technologies could be profitable for them in the long run. This lack of knowledge mandates further teaching and training, public awareness, and information dissemination to make business leaders understand the necessity of sustainability and how green technologies can help in realization of sustainability. Regulatory issues also help to explain why the introduction of green technologies and products is progressing so slowly. Although the governments in the Middle East countries have shown increasing support towards sustainability through reliefs such as tax incentives, subsidies, and funding, the legal structures in place are relatively complex and varied.

4 Conclusion

Harnessing green technologies is a profound chance for organizations operating in the Middle Eastern region to adopt green practices in the course of promoting sustainable economic development. For this area and many others in the world, which used to rely on fossil fuels and their products, the opportunity to create, save money and protect the environment is huge. Green power technologies, proficient energy usage technologies, environmentally friendly materials, and water saving technologies impact the decrease of environmental effects on industries and at the same time increase their productivity and profitability. Besides environmental impact like low carbon emission and resource conservation, green technologies provide several commercial benefits like cost reduction, enhanced marketability, and avenues to seek green credits. Also, there is an outstanding social component embracing sustainability as a result of better brand image, improved employee engagement, and improved customer relations—all of which are crucial for successful business operations in the long run.

But there are issues regarding wide implementation—cost effectiveness, the absence of a unified approach to problem solving, and general lack of awareness. Eliminating these barriers needs the government and private sectors to work actively. To that end, governments can be instrumental by providing well-coordinated and stable policies, and favorable conditions for green technology uptake. Similarly, organizations require educational support and training to enhance the understanding of the sustainable activities among businessmen and managers. Looking toward the future, emerging innovations in green technologies, such as energy storage systems, carbon capture, and AI-driven sustainability solutions, are poised to drive further advancements. As demand for eco-friendly products and services grows, businesses that embrace green technologies will be better positioned to meet consumer expectations and contribute to global sustainability efforts. The Middle East is on the cusp of a green revolution, with green technologies offering a powerful pathway to a more sustainable, resilient, and economically viable future. By overcoming current barriers and capitalizing on the opportunities ahead, businesses in the region can play a critical role in shaping a sustainable future for both the economy and the environment.

References

1. Al-Saidi, Elagib: Ecological challenges and sustainable solutions in the Middle East. Ecol. Ind. **80**, 50–60 (2017)
2. Mogielnicki: Green finance and renewable energy development in the Gulf. Middle East Inst. **14**(2), 156–170 (2021)
3. Al-Fattah: Vision 2030 and sustainability in Saudi Arabia: a roadmap for green innovation. Environ. Sci. Policy **111**, 137–144 (2020)
4. IRENA: The Middle East's Progress in Green Innovation. International Renewable Energy Agency (2023)
5. DEWA: Mohammed bin Rashid Al Maktoum Solar Park. Dubai Electricity and Water Authority. DEWA (2022)
6. Abdou, Raneem: Sustainable business models in the Middle East: the role of green technologies. J. Clean. Prod. **235**, 233–245 (2019)
7. Salem, Baqer: Challenges and opportunities for sustainable business in the Middle East. Arab Econ. J. **56**(3), 78–89 (2018)
8. Jaber: The adoption of sustainable practices in Middle Eastern SMEs. Environ. Manag. Sustain. J. **65**(1), 11–24 (2020)
9. Al-Sari, et al.: Green technology and sustainable practices for Middle Eastern SMEs. Sustain. J. **13**(4), 2010 (2021)
10. Ellabban, et al.: Renewable energy resources: current status, future prospects, and their enabling technology. Renew. Sustain. Energy Rev. **39**, 748–764 (2014)
11. Dincer: Renewable energy and sustainable development: a crucial review. Renew. Sustain. Energy Rev. **4**, 157–175 (2020)
12. Wiedmann, et al.: Environmental policy and sustainable development in the UAE. J. Environ. Plann. Manag. **63**(8), 1317–1332 (2020)
13. Geissdoerfer, et al.: The circular economy—a new sustainability paradigm? J. Clean. Prod. **143**, 757–768 (2017)
14. El-Sayegh, Mohammed: Sustainability in the construction industry in the Middle East. Energy Rep. **7**, 57–63 (2021)
15. IEA: Renewables 2020: Analysis and forecast to 2025. International Energy Agency (2020)
16. REN21: Renewables Global Status Report. Renewable Energy Policy Network for the 21st Century (2021)
17. Aloulou: Renewable energy and sustainability in the Arab world. Energy Policy J. **95**, 275–284 (2016)
18. Lawrence Berkeley National Laboratory: Energy-Efficient Building Technologies (2019)
19. UNEP: Sustainable Fashion Practices in Emerging Economies. UNEP (2022)
20. Harvard Business Review: The Impact of Sustainability on Employee Satisfaction and Customer Loyalty. Harvard Business Review (2021)
21. International Renewable Energy Agency: Financial Challenges in Renewable Energy Adoption. International Renewable Energy Agency (2022)
22. UN Environment Program: Awareness and Knowledge Gaps in Sustainable Technology. UN Environment Program (2021)
23. OECD: Green Technology Adoption and Policy Support. OECD (2023)
24. World Bank: Green Financing and Economic Benefits for Sustainable Ventures. World Bank (2022)

25. World Economic Forum: Governmental Support for Green Technology in the Middle East. World Economic Forum (2023)
26. PwC: Eco-Conscious Consumerism and Market Trends in the Middle East. PwC (2023)

AI-Driven Marketing in Oman: Opportunities, Challenges, and Industry Perspectives

Mahmoud Abouraia, Ayman Bassam Nassoura, Hothefa Shaker, Bandar Al Harbi, and Fadi Abdelfattah

Abstract The paper discusses the prospects of AI in Oman market environment through the lens of industry experts. AI adoption remains a burning issue today in different parts of the globe as it acts as a game-changer in numerous industries; so is the case with the marketing departments. The research was carried out by the administration of questionnaires among 150 marketing practitioners to identify current AI status, its advantages, limitations and further prospects in the Omani context. The findings presented here highlight the fact awareness of AI in marketing context is high, but there are issues regarding data quality and availability of talent, as well as ethical issues around data privacy and AI bias. However, there exists a number of barriers and challenges which has impeded the adoption of AI solutions in Oman, no doubt there are many businesses that understand the potential of AI and are seeking to incorporate such into their marketing strategies. The paper argues that when correctly applied AI can prompt radical improvements in marketing in Oman, delivering benefits to the applying organization while enhancing customer experience and returns.

Keywords AI-driven marketing · Artificial intelligence (AI) · Marketing automation · Customer insights · Data quality

M. Abouraia · H. Shaker (✉) · F. Abdelfattah
Modern College of Business and Science, Muscat, Oman
e-mail: hothefa.shaker@mcbs.edu.om

M. Abouraia
e-mail: Mahmoud.abouraia@mcbs.edu.om

F. Abdelfattah
e-mail: fadi.abdelfattah@mcbs.edu.om

A. B. Nassoura
Department of MIS, City University Ajman, Ajman, UAE
e-mail: a.bassam@cu.ae.ae; nassuora@gamil.com

B. Al Harbi
King Abdulaziz University—KSA, Jeddah, Saudi Arabia
e-mail: Balharbi0513@stu.kau.edu.sa

B. Awwad (ed.), *AI and IoT: Driving Business Success and Sustainability in the Digital Age*, Studies in Systems, Decision and Control 590,
https://doi.org/10.1007/978-3-031-88874-8_77

1 Introduction

Artificial Intelligence (AI) has become dominant throughout all fields, including marketing [1, 2]. The use of AI in marketing gives business the possibility of achieving better communication with customers, better decision making about all kinds of marketing and enhancing the effectiveness of marketing itself [3, 4]. Machine learning and natural language processing, supplemented with other AI technologies and predictive analytics, open up an opportunity to transform the ways companies approach consumer analysis, audience segmentation, content customization, and marketing strategy implementation.

In Oman, AI has only recently started featuring as a part of marketing management as companies are keen to adapt to the digital and data environment [5]. A report by the Oman Ministry of Transport, Communications, and Information Technology [6] revealed that the Oman has improved on the use of digital solutions and technologies [7], AI inclusive of various sectors. The Oman Digital Economy Strategy [8] indicated that while most Omani firms indicated interest in using AI for marketing purposes [9], few firms have applied the technology in a way that can transform business. Al-Kaabi and Al-Mazroei [4] had identified that marketing functions in many Oman businesses are plagued by the shortage of qualified professionals specializing in AI and machine learning [10]. Furthermore, Al-Busaidi [1] established that the top two critical issues regarding the marketing application of AI have increased with time and include data privacy and ethicality of processing consumer data and fairness of algorithms used [11]. For Oman, marketing is on the cusp of freshly new opportunities through leveraging AI when the country Invests right in infrastructure, education and ethical consideration. However, as reported by Sultan Qaboos University [12], AI implementation in marketing could mean big boost for Omani companies competing in the new digital economy [8].

2 Literature Review

AI marketing offers unprecedented opportunities and unique risks at the same time [6, 13]. Al-Jabri and Sohail [3] mentioned that machine learning techniques for segmentation, customer behavior prediction analytics, and natural language processing for marketing content in real-time communication are crucial in the current strategies. These technologies assist firms to have insight analysis of the customers and possible changes in the market before they occur. In Oman currently, the application of AI in marketing is not yet very developed but the possibility exists that it can be applied. Al-Kaabi and Al-Mazroei [4] explained that Omani firms are using AI technologies to improve consumers' interactions when doing business, particularly within Oman's retail and service industries [14]. In a recent survey conducted by Oman Chamber of Commerce and Industry [11] revealed that Oman firms have not exploited AI much, the reason being poor awareness and lack of qualified personnel to control such systems. According to Al-Jabri and Sohail [3], Katie [7], AI can assist companies in developing long-term customer relationships by providing customers with suitable products & service offers and relevant contents [15, 16]. One of those that survey participants pointed out as a major advantage of marketing with the help of AI is customer satisfaction and retention. According to Al-Busaidi [1], challenges and opportunities of big data for Oman businesses include data quality and data availability since many firms lack the resources for data storage and accumulation. Al-Busaidi [1] examines some of the ethical issues that people have with AI-based marketing, especially in getting data about consumers. Al-Kaabi and Al-Mazroei [4] noted that one of the main challenges to AI in Oman is skill absence for employment and handling of AI options and tools fill by all organizations [12]. Sultan Qaboos University [12] opines that, Universities and training institutions in Oman have a task of responding to the growing demand of skills n AI and data analytics by presenting the courses and certifications [8]. Al-Busaidi [1] discussed the ethical issues that come with the use of AI in targeted marketing department stating that such organizations have to abide by such principles of data protection act and protect consumer trust [11].

3 Methodology

This research adopted a quantitative research perspective to analyze the prospects and the issues that attend the application of AI in marketing in Oman. The study instrument of the current research was developed as a structured questionnaire for accessing industry experts from the retail, finance, health care, and technology industries. This approach allowed the researchers to get the big picture of the current state of implementation of AI in marketing in the country. This study used cross-sectional survey research design to collect data from the participants at one given time. For this study, a purposive sampling technique of 150 professionals from different fields were selected. The researchers aimed this technique at the active user of marketing or AI technologies including the marketing managers, data analysts, specialists in IT, or business executives. Recruitment was through emails which informed the participants of the study's purpose, the questionnaire, an undertaking of confidentiality and anonymity. Data collection took place over four weeks to provide enough time for the respondent to complete it. Descriptive and inferential methods of statistics were applied in the analysis.

4 Results and Findings

The collected survey data hold valuable information for understanding the trends about the AI marketing in Oman. The demographic information of the respondents is shown in Table 1. It proves that a significant part of the sample, which is 40%,

Table 1 Demographic information of survey respondents

Demographic variable	Category	Percentage (%)
Age	18–30 years	40
	31–45 years	35
	46–60 years	20
	60+ years	5
Gender	Male	60
	Female	40
Education level	Bachelor's degree	50
	Master's degree	30
	Doctorate	5
	Other	15

belongs to the 18–30 years' group, which presupposes that AI marketing technologies are employed by an active, young workforce. While 60% of respondents are male, 50% of respondents possess a bachelor's degree; the sample is, therefore, relatively young and professionally active with 30% of respondents having Master's degrees.

Regarding the awareness of AI-driven marketing, Table 2 highlights that most respondents (85%) are familiar with AI in marketing, with 45% being "Very Aware" and 40% "Somewhat Aware." This suggests that AI is a recognized concept in the industry, though there is room for more awareness and education, as 15% of respondents are either not aware or unsure about it.

When exploring the perceived benefits of AI in marketing, Table 3 indicates that the major advantages of AI adoption are seen in increasing efficiency (75%) and gaining better customer insights (70%). While cost reduction (50%) and improved customer targeting (65%) are also significant benefits, they are perceived to a lesser extent.

Table 2 Awareness of AI-driven marketing among respondents

Level of awareness	Percentage (%)
Not aware	10
Somewhat aware	40
Very aware	45
Not sure	5

Table 3 Perceived benefits of AI-driven marketing by respondents

Benefit	Percentage (%)
Increased efficiency	75
Better customer insights	70
Cost reduction	50
Improved customer targeting	65

Table 4 Challenges faced in implementing AI-driven marketing in Oman

Challenge	Percentage (%)
Data quality and availability	60
Budget constraints	50
Lack of skilled professionals	55
Integration with existing systems	40

The challenges associated with adoption of AI are significant, as shown in Table 4. The largest barriers identified are data quality and availability (60%) and the lack of skilled professionals (55%). These challenges point to the importance of addressing data management and talent development as part of any AI strategy. Budget constraints (50%) and the integration with existing systems (40%) are also concerns, though they appear less pressing than the issues of data and talent.

In terms of readiness for AI adoption, Table 5 reveals that 70% of respondents are at least partially ready for AI integration, with 25% being "Fully Ready." This suggests a favorable outlook for AI adoption, but the 30% who are either not ready or unsure indicate that challenges remain in fully embracing AI-driven marketing.

Table 6 highlights the expected impact of AI on marketing effectiveness. A significant 60% of respondents believe that AI will greatly improve customer engagement, followed by 50% expecting better campaign performance and 55% anticipating an improved return on investment.

When examining AI usage across industries, Table 7 shows that the retail (60%) and finance (50%) industries lead in AI adoption. These sectors are more data-driven and customer-centric, making them ideal candidates for AI-driven marketing strategies.

The future outlook for AI in Oman is highly positive, as Table 8 demonstrates. A majority (65%) believe that AI will become integral to all marketing strategies, with

Table 5 Readiness for AI Adoption in marketing among respondents

Readiness level	Percentage (%)
Fully ready	25
Partially ready	45
Not ready	20
Not sure	10

Table 6 Expected impact of AI on marketing effectiveness

Expected impact	Percentage (%)
Significant improvement in customer engagement	60
Moderate improvement in campaign performance	50
Improvement in return on investment	55
No impact or negative impact	5

Table 7 AI usage in marketing across different industries

Industry	Percentage (%)
Retail	60
Finance	50
Healthcare	30
Technology	40

Table 8 Future outlook on AI-driven marketing in Oman

Outlook	Percentage (%)
AI will become integral to all marketing strategies	65
AI will be a niche tool for certain segments	20
Uncertain or don't know	15

Table 9 Perceived ethical concerns of AI-driven marketing among respondents

Ethical concern	Percentage (%)
Privacy concerns	70
Data security	60
Algorithmic bias	50
Lack of transparency	45

only 15% uncertain about its future role. This reflects a strong belief that AI will play a significant role in shaping the future of marketing in Oman.

Lastly, ethical concerns regarding AI in marketing are highlighted in Table 9. The most significant concerns are related to privacy (70%) and data security (60%), indicating that respondents are cautious about how AI handles personal data.

5 Discussion

The survey findings are beneficial for understanding the current/Omani position and future of AI-based marketing. Considering the demographic characteristics of the respondents, it can be stated that the implementation of AI in marketing is realized by young, educated and innovative employees. Most participants are from the younger age groups, with a significant provision having completed at least a bachelor's degree. This means that the professionals in the marketing sector also have better knowledge regarding the tools and ideas allied to AI solutions implementation. It can also be inferred from the data collected evaluating the level of awareness of marketing personnel about AI that the vast population of participants have at least a fair idea about how they can use AI. This suggests that the marketing industry has been infiltrated by AI to a certain extent because there is that part of the market that still needs to be convinced. To help such employees to understand the possibilities of AI, the authorities of their companies could expose them to more examples of these technologies, perhaps by inviting trainers, offering workshops, seminars and more examples and cases of AI in their work.

The survey also reveals the main advantages of AI applications in marketing, based on which efficiency and customer knowledge stand out as major applications of machine learning. The survey shows that there are substantial barriers impeditive to the implementation of AI. Data quality and availability are identified as potential issues that reveal the fact that to support AI technologies, structured and reliable data must be used. However, the survey shows that a good percentage of the organizations in Oman are ready or parts ready for AI oriented marketing. This positive outlook makes one deduce that business entities may be slowly embracing the use of AI and trying to integrate it into their operations.

6 Conclusion

The results presented in this work prove the increasing demand and the possibilities for AI in marketing in Oman, while showing the strengths and weaknesses of organizations in implementing AI innovations in their campaigns and promotion activities. Since the commercial, administrative and service sectors are mostly staffed by educated individuals who are aware of what AI can do, there is a good starting point for marketing AI. But the research also outlines several challenges that should be solved, such as, data quality problems, shortage of qualified personnel, and ethical concerns regarding privacy and bias. Nevertheless, the survey reveals that majority of organizations that are affected by AI believe that this technology will enhance marketing effectiveness, enhance customer understanding and increase organizational performance. That so many managers are convinced of the positive impact of AI means that the businesses are ready to adopt AI solutions to cause a change. Some industries, like retail and finance, are already ramping up their use of AI, while the healthcare sector struggles with implementing it more because of regulatory issues and because AI is not easily integrated into the marketing plans of a healthcare company.

Further, the study establishes that majority of organizations are prepared for some level of AI integration but there is a need for organizations to ensure that they have educated and trained all stakeholders about practical use of AI as well as its benefits. If businesses are to harness AI as a transformative means of innovating and improving their customer interface, then, greater attention must be paid to the development of the so-called data intensity and AI readiness in human capital. No data and legal concerns will be principles aside when it comes to using AI in marketing. With AI being already involved in the operations of the marketing field at deeper levels, it will be important to remain as transparent as possible, solve the questions of privacy, and become aware of the harmful effects of the AI algorithms' biases. AI has gained popularity in business organizations, and therefore, firms need to implement ethical standards for the right implementation of AI. Although there are some obstacles that can be encountered in the way to achieving common use of AI in marketing in Oman, there are some benefits that can be obtained. With addressable data quality, skill shortage, and ethical problems in the way, AI will change the face of the marketing landscape to benefit businesses and customers alike. If education continues to be invested and infrastructure, along with ethical practices adopted, Oman's marketing sector can open itself up and in a way that can enable it to lead the self-some way in utilizing AI in marketing.

References

1. Al-Busaidi: Data privacy and ethics in AI-driven marketing: perspectives from the Omani business landscape. J. Bus. Ethics Digit. Innov. **12**(4), 203–218 (2021)
2. Al-Harthy, Al-Hadhrami: The role of artificial intelligence in enhancing marketing strategies in Oman's retail sector. J. Bus. Res. **45**(2), 75–88 (2021)
3. Al-Jabri, Sohail: Artificial intelligence in marketing: a case of emerging economies. Int. J. Inf. Manag. **53**(102106) (2020)
4. Al-Kaabi, Al-Mazroei: Artificial intelligence and digital marketing: exploring challenges and opportunities in Oman. Middle East J. Bus. Technol. **17**(3), 125–138 (2022)
5. ATG: Artificial Intelligence (AI) has Emerged as a Powerful and Transformative Force (2023)
6. Oman Ministry of Transport Communications and Information Technology: Digital Oman Strategy 2020–2025. Oman Ministry of Transport, Communications, and Information Technology (2020)
7. Katie: Unlocking the Power of Hyper-Personalization: How AI-Driven Marketing is Redefining Customer Engagement (2024). Retrieved from https://useinsider.com/hyper-personalization/
8. Oman Digital Economy Strategy: Oman Digital Economy Survey: AI Adoption in Business and Marketing. Sultanate of Oman. Oman Digital Economy Strategy (2023)
9. Lukan, E.: 11 Powerful Uses of AI in Marketing (2024)
10. OHI Leo Burnett: AI in Marketing and Advertising—What Omani Brands Need to Know (2024)
11. Oman Chamber of Commerce and Industry: The State of Artificial Intelligence in Oman: Challenges and Opportunities. Oman Chamber of Commerce and Industry (2022)
12. Sultan Qaboos University: AI Adoption in Oman's Small and Medium Enterprises: Insights into Marketing Transformation. Sultan Qaboos University Press (2023)
13. OSOS: The Future of Technology in Oman: Opportunities and Challenges (2024)
14. Rashid, A.B., Kausik, A.K.: AI revolutionizing industries worldwide: A comprehensive overview of its diverse applications. Hybrid Adv. **7** (2024)
15. Rachel, et al.: An evaluation on the impact of artificial intelligence (AI) in facilitating digital marketing in telecommunication industry Oman. GSJ **11**(6), 1647–1677 (2023)
16. Bacanin, N., et al.: Quasi-reflection learning arithmetic optimization algorithm firefly search for feature selection. Heliyon **9**(4) (2023)

Exploring Algorithmic Trading: Technology-Driven Market Transformation

Amalknath Anilkumar and Mani Joseph

Abstract Algorithmic trading or algo trading involves the utilization of computer programs to conduct trades, without intervention of humans. It's for increasing the efficiency, reducing the cost of transaction, It has grown from early systems that just automating orders to today's high-frequency trading (HFT) and predictive models that powered by the machine learning. Even though algo trading has optimized the market efficiency and liquidity it has also developed issues such as high volatility, systemic risks and flash crashes. Due to the development of machine learning and technology in general, trading strategies are enhancing the ability to adapt; as such, algorithmic trading is now the primary mode of financial trading.

Keywords Algorithmic trading · High-frequency trading (HFT) · Machine learning · Market volatility

1 Introduction

Algorithmic trading, is the process of executing decision made with the help of mathematical models and Advanced Computer Programs in the process of buying and selling securities in the financial markets. Such algorithms are known to execute predefined strategies such as, arbitrage, Market making or high frequency trading to exploit some opportunity on the market. The significance of algorithmic trading lies in its ability to rapidly analyze vast quantities of data, execute large orders, and achieve precision and speed beyond the capability of human traders [1]. It enhances market efficiency, reduces transaction costs, and facilitates liquidity, thus playing an essential role in the functioning of modern financial markets.

A. Anilkumar (✉) · M. Joseph
Modern College of Business and Science, Muscat, Oman
e-mail: Amalkanth.anilkumar@mcbs.edu.om

B. Awwad (ed.), *AI and IoT: Driving Business Success and Sustainability in the Digital Age*, Studies in Systems, Decision and Control 590,
https://doi.org/10.1007/978-3-031-88874-8_78

2 Purpose of the Survey

This survey aims to give a overview of algorithmic trading, its evolution, main methodologies, technology, market impact, and the corresponding risks and regulatory response. Specifically, this survey addresses several questions: What are the various types of algorithmic trading strategies used in modern markets? What impact has the advancement in machine learning and artificial intelligence had on algorithmic trading techniques? What does algo trading mean for market efficiency, liquidity, and systemic stability? The survey also seeks to learn about regulatory measures adopted to mitigate the risks faced by these systems. By addressing these questions, the survey attempts to examine how algorithmic trading has altered the financial markets and the associated risks it represents for market participants and regulators. Understanding these dynamics is important not only to market practitioners but also to policymakers and researchers who are interested in how technology, financial stability, and the regulatory framework work together.

3 Foundations of Algorithmic Trading

3.1 Basic Principles and Components of Algorithmic Trading Systems

Algorithmic trading systems operate on several key principles: automation, optimization, and systematic decision-making. These systems are designed to generate and execute trades based on pre-programmed strategies that can include rules for price, timing, and quantity, with minimal human intervention[2] The main components of an algorithmic trading system include:

Data Acquisition and Analysis

Historical prices, real-time market data other data's such as social media and news can be used to generate effective signals. Data acquisition and data cleaning processes ensure that algorithms have all the right information to make right decisions [3].

Signal Generation and Strategy Design

Market information is collected by the signal generation engine which identifies trading opportunities as a result. Historic data is used to design and evaluate trading strategies referred to as back-testing to establish their viability in real markets [4].

Order Management and Execution Algorithms

Orders are handled by an order management system (OMS), and execution algorithms such as VWAP and TWAP reduce transaction costs and make less impact to the market [5]. Smart Order Routing (SOR) is a mechanism, which takes orders and routes them to trading with the best price or liquidity available to minimize transaction costs and the impact on the market.

Risk Management

Prevention of financial losses is crucial to good risk management. The risk module is always watching exposure, leverage and that the trades stay within specified limits. In fact, this is crucial in high frequency trading environments, where even a few milliseconds can cause substantial market change [6].

3.2 Key Technologies Driving Algorithmic Trading

High Frequency Trading or HFT platforms are an important part of current day algorithmic trading and they make transactions within milliseconds with the help of certain algorithms. These platforms improve trade speed via co-location and direct market access to reduce latency and therefore capture transient market opportunity [7]. Thus, combining with the HFT, the cloud computing offers the required set of technically sophisticated computation-al capabilities suitable for generating vast high-frequency trading accounts or backtesting and handling substantial data traffic. This ecosystem is built further by Edge compu-ting which helps in decreasing latency that is important in initiating orders and exploit brief market failures.

Algo trading has seen enormous benefits from progressive enhancement of both Machine learning (ML) and Artificial Intelligence (AI) as they enhance the accuracy of trading opportunities and the modulation of trading strategies within the market. Corporate strategies are thus refined over time through the feedback received through the techniques such as reinforcement learning [8]. In addition, predictive analytics, neural networks, and Natural Language Processing (NLP) provide extended trading models with concepts of overall market sentiment and tendencies of the price change [9].

3.3 Types of Algorithmic Trading Strategies

Algorithmic trading techniques make use of computer programs to take advantage of one kind of market anomaly or another. Arbi-trage strategies for example, involve the exploitation of price differences of similar commodities in different markets or within one market. More crucially, statistical arbitrage strategies such as pairs trading are based on price history and the principle of mean reversion [10]. Market making techniques offer liquidity and facility for efficient market functioning; the firms who engage in market making simultaneously put both buying and selling orders to benefit from the differential between bid price and ask price and use sophisticated risk management in order to achieve the right mix of inventory [11].

Momentum trading strategies are used to track the movements in the market and buy assets that show bullish momentum while selling assets with bearish momentum, the application of machine learning technologies improves the efficiency of trend identification and the right time for executing the trades [12]. In statistical arbitrage, deep learning models are more used to discover complex, non-linear relationships among financial data that traditional statistical methods may overlook [13]. Market impact minimization methodologies like VWAP (Volume Weighted Average Price) and Implementation Shortfall are adopted by large investors that seek to spread their transactions over time to keep their effects on the market price minimal [14].

In addition, sentiment based trading algorithms analyze the market through news, reports, and social media using NLP that use the sentiment analysis to predict the market direction as a result of changing emotions of traders [15]. Adaptive and reinforcement learning strategies are two approaches that lied on the horizon of the algorithmic trading, where models are able learn new control rules depending on the current changes in market and improve the future's profits based on previous trading results [16].

4 Algorithmic Trading Strategies

4.1 Statistical Arbitrage

Statistical arbitrage (StatArb) is a mean reverting trading strategy that tries to exploit inefficiencies in between different asset prices. It also involves a familiar strategy known as pairs trading, which involves choosing two securities that are closely related to each other, purchasing the cheaper one while simultaneously short selling the more expensive one, mainly because the spread is felt to narrow over time [17]. This strategy mostly uses statistical models and machine learning to predict the overall relationship between the prices of assets.

4.2 Trend Following Strategies

Trend following operates under the assumption that an investment has an inherent path (upward or downward) which it will continue to move on for a certain period of time. This strategy aims at trying to find that a trend, whether bearish or bullish, has continued and trade based on that trend. Moving averages, Moving averages convergence divergence (MACD), and price abnormalities are few indicators employed for trading in trends. Over the last decade, trend -following approaches were improved through machine learning and artificial intelligence, which made such approaches more flexible for the current market condition [12].

4.3 Market Making

Market making algorithms continuously place buy and sell orders for a particular asset to profit from the spread between the bid and ask price. The goal of market makers is to provide liquidity to the market while profiting from small price differences, which is essential for creating an efficient market [11]. Market makers adjust their orders dynamically to maintain inventory balance and avoid significant exposure to market volatility.

4.4 Mean Reversion

Mean reversion strategies are based on the assumption that a particular set of assets' price will always oscillate and return to the mean value, accordingly. In this strategy traders search for security that has moved dramatically away from its average and then perform transactions to cover gains as soon as prices revert. This approach is normally used in combination with the Bollinger Bands or Relative Strength Index (RSI) to determine if it is overbought or oversold [18]. latest advancement has integrated machine learning utilities to improve the detection of the mean reversion trade opportunities.

5 Case Studies or Examples of Successful Strategies

5.1 Renaissance Technologies - Medallion Fund

Renaissance Technologies' Medallion Fund is one of the best performing examples of trading by algorithmic means. Employing statistical arbitrage and quantitative model, the fund posted continuously high returns that were superior to the conventional market benchmarks [19]. The Medallion Fund does not employ fundamental analysis to identify securities to buy but used proprietary algorithms to decipher patterns in market data and thereby generate trades across all classes of assets, with high-frequency trading and pattern recognition.

5.2 Goldman Sachs—Trend Following in Commodity Markets

Goldman Sachs, used trend-following techniques in commodity markets, leveraging across the different commodities such as crude oil & metals. This approach was especially effective during market events with longer trends, like the world wide commodities boom at the start of this decade [20].

6 Technological Advances in Algorithmic Trading

6.1 Role of AI and Machine Learning in Algorithmic Trading

Machine learning along with Artificial intelligence have greatly influenced algorithmic trading and brought into existence, smart, dynamic and autonomous trading systems. ML models differ from conventional rule-based algorithms in that they parse and draw inference from massive data, make accurate predictions, and adapt in real time, thereby enhancing trade efficiency. Supervised learning is used frequently in the process of predicting assets; on the other hand, unsupervised learning helps in classifying the assets depending on their patterns of behavior [9].

In particular, deep learning models have been successful in the analysis of financial time series data, to elaborate long term dependencies, and for future price movements forecasting's in particular Long Short-Term Memory (LSTM) [9]. These models enhance decision making since they are easier to estimate and capture temporal relationship in prices of more assets. However, with the emergence of trade automation, Natural Language Processing (NLP) has enabled algorithmic trading systems to interpret unstructured information like articles, reports, and social media activity in order to determine the mood of investors. With help of structured and unstructured data, AI-based trading strategies can respond effectively to the occurrence of vital news affecting the prices and gain a deeper view on trading and its methodology [8].

6.2 Impact of Big Data Analytics in Strategy Development

Big data analytics plays a fundamental role in the development of algorithmic trading strategies. The availability of information from a variety of structured data (price, volume) and unstructured data (articles, social media posts) has provided traders with extraordinary understanding of market conditions. By utilizing tools such Hadoop and Apache Spark, trading systems can effectively manage huge amounts of data as well as make prompt analyses that can help in arriving at better decisions [21].

Machine learning integration with big data analysis enables the trader to recognize the interdependencies, and trends that are not easily recognizable with other common analysis. Big data analysis improves risks management since it covers market behavior, and the traders can devise strategies that would work regardless of the type of market risks that occur [21]. Furthermore, this approach ensures continues learning as MLL algorithms become more accurate in their predictive models with each new batch of data which results in improved trading strategies in the future.

6.3 *Advances in Computing Power and Their Effects on Trading Speeds and Complexity*

Development in computational processing has been central to the development of algorithmic trading, especially in areas of high frequency trading (HFT). The current computing platforms, such as graphics processing units (GPUs) and Field programmable gate arrays (FPGAs), offer sufficient processing power to execute algorithms with microseconds latency for executing trades. These technologies have the capability of performing high-speed analysis of market data and the real time execution of orders and these features become very useful when operating in a competitive trading environment [22].

Further, the increasing use of cloud computing has shifted the environment of trading by making computational resources on the cloud without investing in huge structures. Thanks to the cloud technologies, many small businesses can employ refined trading algorithms, which have been too costly to be integrated earlier [23]. This has in turn lead to increased competition between players within the market arena.

7 Market Impact and Risks of Algorithmic Trading

7.1 *Impact on Market Liquidity and Volatility*

Algorithmic trading affects market liquidity and volatility in a twofold manner. On the positive note, algorithmic trading's market making strategies widens the market by continually providing the market with buying and selling opportunities. These factors are useful in helping reduce the bid-ask spreads and make the trading more efficient for all stakeholders in a market. High-frequency trading (HFT) has been proven to enhance liquidity and the stability of markets [7]

However, algorithmic trading has been pointed out as prone to increased market fluctuations especially during high-risk situations. When numerous algorithms respond to the same market signals, the speed of their response leads to large fluctuations in price. Given that HFTs constantly open and close positions, which leads to liquidity gaps in volatile times, this will lead to a lot of shifting of asset prices. For instance, when algorithms withdraw liquidity during risk events, they exacerbate negative feedback that causes steep volatility.

7.2 *Analysis of Flash Crashes and Market Anomalies*

The first and foremost risk of algorithmic trading is that the possible flash crashes, which means that the market price drops for a short period and then rebounds very quickly. Some of them are: The flash crash of May 6, 2010 The Dow Jones Industrial Average fell by about 1000 points in the space of a few minutes before recovering shortly after. The actual cause of this event was the simultaneous operation of several algorithms that started to sell assets as soon as they noticed that there was an increased trading volume, which, in turn, led to sudden asset depreciation [17].

There are market anomalies resulting from algorithmic trading when multiple algorithms respond to the same inputs in a way that induces a chain of buying or selling that goes beyond the market's capacity. Inaccurate pricing may occur when algorithms misunderstand the state of the market and issue wrong decisions that spread through interconnected systems. Such risks have been addressed through the introduction of regulatory measures such as circuit breakers designed to halt trading during dramatic price action to allow for human intervention and the restoration of order to the market [24].

8 Regulatory and Ethical Considerations in Algorithmic Trading

8.1 Overview of the Global Regulatory Landscape for Algorithmic Trading

The rules and directives of algorithmic trading is changing globally as the authorities seek to harness the efficiency gain while avoiding the negative impact on financial markets. Leaders of the world's stock markets have imposed certain rules to prevent algorithmic trading from distorting market integrity. For example, the MiFID II trading in the EU has put into force strict guidelines that regulate the functioning of trading algorithms encompassing procedural guidelines of testing and risk management, and real-time monitoring [25]. These rules are intended to reduce some of the problems that have arisen with regards to the use of automated trading tools, such as 'flash' crashes and the manipulation of the marketplace.

In the United States, regulation has been developed through the regulations set by the Securities and Exchange Commission (SEC) and the Commodity Futures Trading Commission (CFTC); mainly because of the main principles of transparency, surveillance, and accountability of trading. An example of such rule is the Regulation SCI (System Compliance and Integrity) whereby trading firms have to guarantee that their system are sound and capable of meeting the volume and speed of trading in today environment. This regulation is designed to mitigate technically caused failure scenarios which can impact a whole market [26].

9 Specific Regulations and Guidelines from Major Financial Markets

United States (SEC and CFTC Regulations): The SEC has therefore shifted to increasing the transparent-ness of its processes and demanding that trading algorithms only adhere to standards that act in the interest of market integrity. Algorithmic traders involved in the implementation of trades under Regulation ATS (Alternative Trading Systems) have to provide certain detailed information about trading platforms and strategies, hence, making the market significantly more transparent. However, further protection to investors is offered by the Regulation NMS (National Market System) that looks to make the orders pass through and execute in the best possible price regardless of the market [27].

The Commodity Futures Trading Commission (CFTC) also addresses algorithmic trading under Regulation AT (Automated Trading), which mandates pre-trade risk controls, monitoring, and requirements for the registration of certain trading systems to prevent market abuse and ensure stability [28]. In other words, these rules ensure that market anomalies are averted and instead assure that surveillance and testing are routinely conducted on systems.

United Kingdom (Financial Conduct Authority, FCA): The Financial Conduct Authority has published guidelines for proper conduct of algorithmic trading within the UK. Among other obligations, firms must have adequate risk management systems, test trading algorithms adequately, and monitor properly the live trading activity [29]. Additionally, detailed documentation of the algorithmic trading system and changes made to it, including the effect of such changes, needs to be maintained by the firm, available for inspection by the regulators.

By the FCA, in coordination with the Bank of England, other measures have been introduced to introduce circuit breakers, which halt trading temporarily during extreme price swings, to prevent flash crashes. These will help stabilize market activity and are a part of larger efforts regarding market stability and fairness especially due to the increased reliance on automation and algorithmic systems for trading.

10 Ethical Considerations in the Deployment of Trading Algorithms

The ethical implications of algorithmic trading are of utmost importance, particularly in light of relatively high reliance on AI-based strategies that are getting only more complex. One very urgent issue is market manipulation. Misuses of the program trade flow are through manipulative activities like spoofing (placing orders that are not intended ever to be executed just to give a signal that others misinterpret) and quote stuffing (overloading the market with large volumes of orders to slow down competition). The SEC and other regulatory bodies already have been vigilant in such practices, with hefty penalties for violators [9].

Another ethical issue is again connected to justice and equal opportunity in the market place. Algorithmic trading is beneficial to firms with better technology and this has created debates on whether such trading systems provide equal chance to traders in global financial markets. For instance, HFT firms themselves are capable of executing the trades faster than traditional traders, which leaves the later at a disadvantage. Worldwide ethical standards are being deliberated to make trading fair including for all trading participants to be of same latency as to access market information.

However, use of artificial intelligence and machine learning in practicing algorithmic trading poses debates over issues to do with both transparency and accountability. It needs to be noted that many AI based trading strategies are 'black box' in nature; which indicates that it could be even challenging for firms to comprehend or justify the rationale for any particular transaction. This lack of transparency presents great difficulties for regulators to determine whether such systems are legal and moral. This principle has therefore been advanced in financial regulations, implying that firms should have ways of explaining AI outcomes to the regulators and stakeholders.

11 Future Trends and Challenges in Algorithmic Trading

11.1 Emerging Trends in Algorithmic Trading Technologies and Strategies

This area of algorithmic trading is expanding over the years with new technologies and strategies that reshaped the context of financial markets. Another of the trends that is becoming evident is the rising usage of artificial intelligence (AI), deep reinforcement learning to design self-optimizing trading algorithms specific to real-time market data. These algorithms are capable of freely searching for potential profitable deals and adjusting trade execution techniques by themselves [30]. Another trend includes the application of Generative Adversarial Networks (GANs) for simulating markets to analyze trading strategies performance and refine signals for uncertainty in market conditions.

The other emerging trend is the increases application of quantum computing in the markets. While it is still very much in its infancy, clients will increasingly be able to use this new form of computing to solve complex problems related to the particularities of algorithmic trading mechanisms at incomparably higher speeds. This could cut the time spent on marketing simulation exercises by a significantly large margin and enhance the efficiency of forecasting models [31]. Furthermore, the use of NLP is going beyond consumer sentiment analysis and entering the sphere of event trading where an application is capable of trading on news events in split seconds, a Substantial competitive advantage.

The use of such additional data is also being developed progressively. Traders are now using other classes of data, including satellite imagery, geolocation data, and weather reports, to build up new trading perspectives that other types of data cannot offer [32]. This trend can be attributed to a larger phenomenon of using big data to gain informative advantage in the market.

12 Challenges Faced by Traders and Regulators in Keeping up with Rapid Technological Advancements

The quick unfolding of the technologies in algorithmic trading gives rise to several issues for traders and the regulators. The first major problem that hurts traders' performance is technology degradation. Trading firms are steadily investing in hardware modernization to sustain latency advantages as well as implement latest AI and machine learning algorithms. There is therefore a heavy cost implication for the procedure, especially for small firms that do not have the financial might as compared to some institutions [33].

From a regulatory point of view it becomes very challenging to be in tune with some of these technological developments in trading systems. Sadly, regulatory authorities are facing difficulties in comprehending and surveilling the relatively new and more complex and less transparent trading mechanisms that are now mounting, including deep learning driven or quantum computing driven. Due to the opaque nature of most AI applications, it becomes difficult for regulatory authorities to evaluate the level of compliance. As the examples of decentralized finance (DeFi), functioning based on blockchain technologies, show, it becomes even more challenging to regulate the activity of increasingly complex and widespread networks, as the existing tools of supervisory oversight are rarely sufficient for the decentralized and global environment [34].

13 Conclusion

This survey describes how algorithmic trading permanently changes the appearance of the financial market by making it more efficient, liquid and accurate through trading strategies such as market making, statistical arbitrage, and trend following. These strategies have been improved by technologies like Artificial Intelligence, machine learning and big data to provided adaptive as well as improved models. But they also pose some threats, such as volatility, flash crashes, and systemic threats, which further assert the need for credible regulation and risk control.

The future of algorithmic trading will be based deep learning, quantum computing, and further use of alternative data and will result in a further evolution towards full automation and self-adaptation of algorithmic trading strategies allowing for fast market responses. These advancements come with problems such as Technological risks, transparency threats, and fairness problems. Current and emerging regulation may include global standard, AI transparency requirement.

There is a need to balance innovation and regulation in order to sustain growth. As is seen, the key values include ethical advancement, transparency as well as efficient scrutiny needed to harness technological innovation to stabilize the market with fairness. This will ensure a resilient financial ecosystem,

References

1. Hendershott, T., Jones, C.M., Menkveld, A.J.: Does algorithmic trading improve liquidity? J. Financ. **66**(1), 1–33 (2011)
2. Gomber, P., Arndt, B., Lutat, M., Uhle, T.: High-frequency trading. J. Inf. Technol. **26**(4), 312–324 (2011)
3. Kolanovic, M., Krishnamachari, R.: Big Data and AI Strategies: Machine Learning and Alternative Data in the Financial Industry. J.P. Morgan (2017)
4. Harris, L.: Trading and Exchanges: Market Microstructure for Practitioners. Oxford University Press (2003)
5. Hasbrouck, J., Saar, G.: Low-latency trading. J. Financ. Markets **16**(4), 646–679 (2013)
6. Fung, W., Hsieh, D.A.: Hedge fund benchmarks: a risk-based approach. Financ. Anal. J. **60**(5), 65–80 (2004)
7. Brogaard, J., Hendershott, T., Riordan, R.: High-frequency trading and the execution costs of institutional investors. Financ. Rev. **52**(1), 37–68 (2017)
8. Boni, F., Brownlees, C.: Reinforcement learning for algorithmic trading in financial markets: a review and practical implementation. J. Financ. Data Sci. **3**(2), 25–45 (2021)
9. Zhang, L.: AI in financial trading: the rise of deep learning techniques. J. Financ. Markets Anal. **14**(3), 230–245 (2022)
10. Gatev, E.G., Goetzmann, W.N., Rouwenhorst, K.G.: Pairs trading: performance of a relative value arbitrage rule. Rev. Financ. Stud. **19**(3), 797–827 (2006)
11. Menkveld, A.J.: High-frequency trading and the new-market makers. J. Financ. Markets **16**(4), 712–740 (2013)
12. Fischer, T., Krauss, C.: Deep learning with long short-term memory networks for financial market predictions. Eur. J. Oper. Res. **270**(2), 654–669 (2018)
13. Huck, N.: Pairs trading and outperformance: evidence from the European equity market. J. Bank. Finan. **55**, 1–14 (2015)
14. Bertsimas, D., Lo, A.W.: Optimal control of execution costs. J. Financ. Markets **1**(1), 1–50 (1998)
15. Tetlock, P.C.: Giving content to investor sentiment: the role of media in the stock market. J. Finan. **62**(3)
16. Deng, Y., Bao, F., Kong, Y., Ren, Z., Dai, Q.: Deep direct reinforcement learning for financial signal representation and trading. IEEE Trans. Neural Netw. Learn. Syst. **28**(3), 653–664 (2016)
17. Kirilenko, A.A., Kyle, A.S., Samadi, M., Tuzun, T.: The flash crash: the impact of high-frequency trading on an electronic market. J. Finan. **72**(3), 967–998 (2017)
18. Balvers, R., Wu, Y., Gilliland, E.: Mean reversion across national stock markets and parametric contrarian investment strategies. J. Financ. **55**(2), 745–772 (2000)
19. Zuckerman, G.: The Man Who Solved the Market: How Jim Simons Launched the Quant Revolution. Penguin Books (2019)
20. Jegadeesh, N.: Foundations of technical analysis: computational algorithms, statistical inference, and empirical implementation: discussion. J. Finance **55**(4), 1765-1770 (2000)

21. Smith, J., Allen, R.: Big data analytics in algorithmic trading: the impact of real-time processing on market efficiency. J. Financ. Data Sci. **4**(1) (2022)
22. Reis, A., Silva, M., Fernandes, P., Oliveira, L.: High-performance computing in high-frequency trading: the role of GPUs and FPGAs in ultra-low latency trading systems. J. Comput. Finan. **24**(3) (2021)
23. Takahashi, H., Wang, Y.: The impact of cloud computing on algorithmic trading: democratizing high-performance computing in financial markets. J. Financ. Technol. Innov. **7**(1) (2023)
24. Biais, B., Foucault, T., Moinas, S.: Equilibrium fast trading. J. Financ. Econ. **142**(3), 1055–1082 (2021)
25. Zagaglia, P.: Regulatory approaches to algorithmic trading: insights from MiFID II and global perspectives. J. Financ. Regul. Compliance **30**(2), 178–199 (2022)
26. Gomber, P., Haferkorn, M.: Regulatory requirements for algorithmic trading systems: analysis and implementation challenges. J. Financ. Markets **56**, 101310 (2021)
27. Chatterjee, S., Chiu, J.: U.S. regulatory measures for algorithmic trading: balancing innovation and market stability. J. Financ. Compliance **10**(1), 23–39 (2023)
28. Farmer, J.D., Sklaroff, L.: Regulation AT and the evolution of automated trading oversight. Rev. Futures Markets **19**(3), 275–290 (2021)
29. Khan, R., Turner, J.: Algorithmic trading regulations in the UK: FCA's approach to managing systematic risk. J. Financ. Regul. Compliance **31**(1), 48–65 (2023)
30. Zhang, Y., Huang, T., Li, X.: Reinforcement learning in algorithmic trading: emerging techniques and strategies. J. Financ. Data Sci. **5**(1), 34–49 (2023)
31. Das, S., Singh, R., Roy, A.: Quantum computing in algorithmic trading: opportunities and challenges. J. Comput. Finan. **18**(2), 87–102 (2023)
32. Kumar, A., Anderson, D.: Alternative data in financial markets: harnessing the power of big data in trading strategies. J. Financ. Innov. **7**(3), 120–135 (2022)
33. Huang, L., Wang, R.: The cost of technological advancement in algorithmic trading: challenges for small firms. J. Financ. Markets Technol. **9**(1), 23–39 (2023)
34. Morris, T., Lang, P.: The impact of decentralized finance on algorithmic trading regulation. J. Financ. Regul. Compliance **31**(4), 87–102 (2022)
35. Chinco, A., Clark-Joseph, A.D., Ye, M.: Sparse signals in the cross-section of returns. J. Finan. **74**(1), 449–492 (2019)

Building a Smart Economy: Exploring AI's Role in Oman's Economic Progress

Mahmoud Abouraia, Ayman Bassam Nassoura, Hothefa Shaker, and Bandar Al Harbi

Abstract This research examines the efficiency of Artificial Intelligence (AI) as a tool that can contribute to Oman's economic growth and influence different industries. As AI technologies continue to change the global economies, this study seeks to determine AI's existence, acceptance and implementation in the Omani market. A survey was conducted with 145 professional users who are involved in their professional lives and are leaders and shareholders of businesses from the financial, healthcare, educational, and oil and gas sectors. The research evidence shows that though AI is gradually getting acclaim for helping organizations to foster efficiency and creativity, its implementation is still facing certain constraints in Oman as data quality issues, lack of AI professionals, and regulation. This research work aids in the understanding of the impact of AI on the Oman economy and a guide to businesses and policymakers on the venue and opportunities to adopt AI for improved economic development. It also underlines the demand for raising education and training levels in AI-associated disciplines and creating permissive conditions for applying AI in the country regarding ethical standards.

Keywords Artificial intelligence (AI) · Economic progress · Smart economy · AI adoption · Industry innovation · Economic diversification

M. Abouraia · H. Shaker (✉)
Modern College of Business and Science, Muscat, Oman
e-mail: hothefa.shaker@mcbs.edu.om

M. Abouraia
e-mail: Mahmoud.abouraia@mcbs.edu.om

A. B. Nassoura
Department of MIS, City University Ajman, Ajman, UAE
e-mail: a.bassam@cu.ae.ae

B. Al Harbi
King Abdulaziz University, Jeddah, Kingdom of Saudi Arabia
e-mail: Balharbi0513@stu.kau.edu.sa

B. Awwad (ed.), *AI and IoT: Driving Business Success and Sustainability in the Digital Age*, Studies in Systems, Decision and Control 590,
https://doi.org/10.1007/978-3-031-88874-8_79

1 Introduction

Artificial Intelligence (AI) remains one of the groundbreaking phenomena in the global economy as it impacts several sectors including healthcare, finance, manufacturing retail. AI technologies are now planning to revolutionize efficiency and innovativeness, as well as economic growth. As Oman is keen on diversifying from a rentier economy, AI brings formidable opportunities to foster this change and enhance the competitiveness of domestic Industries [1].

The official development plan of Oman enshrined in Oman's Vision 2040 highlights the development of a knowledge-based economy and AI is seen as a key enabler for this goal. Because AI can analyze massive quantities of data to give decision support, improve business efficiency, and discover new opportunities for business advancement. Nevertheless, there are several barriers to the adoption of AI in Oman, which are as follows technological implementation, deficiency of AI expertise, data privacy and ethical issues [2]. In the Gulf Cooperation Council (GCC) countries like the United Arab Emirates and Saudi Arabia have already started adopting the use of AI policy interventions to enhance economic diversification and public service delivery [3]. Although Oman is increasingly embracing AI technologies, there is scarce literature that seeks to establish how those technologies are being deployed in the country's economy.

2 Literature Review

AI can significantly drive economic development in emerging economies including Oman by increasing production efficiency, encouraging innovation and rebalancing the economic structure. To Oman, AI became a paramount option for the development of a non-rentier economy by enhancing the formation of non-oil income [4], forthcoming [5]. AI contributes to the future improvements of agriculture, manufacturing, and service industries, where Oman Vision 2040 unveiled the importance of digital transformation for development [3, 4]. The use of AI will enable Oman to increase efficiency, create a sustainable economy and increase non-oil economic revenues [6]. There are key sectors in Oman that could experience a lot of benefits when implementing AI solutions. Some of the benefits that the financial industry can realize include fraud detection, risk management, and customer relationship management [7]. In antecedent sections of this paper, it has been presented that, in healthcare, AI for disease diagnosis, treatment planning, and institutional management are possible. Supply chain management and vehicle routing logistics can unlock cost efficiency in the supply chain and maintenance of vehicles to support diversification strategies [8]. These applications demonstrate the ability of AI to add new value and generate employment in various related high-need fields [9].

However, Oman like other developing countries has some constraints that hinder it from properly adopting AI Some of the constraints include a lack of qualified and limited and poor quality of data accumulated, Poor quality of infrastructural facilities required in the development and deployment of AI Oman can also hinder AI adoption through Insufficient numbers of AI professionals and data scientists are problematic for AI adoption, requiring capacity development interventions [10, 11]. Confidentiality and data quality also limit implementation depending on the type and effectiveness of regulatory policies [12]. It is, therefore crucial that Oman pays precise attention to the acceptance of technology and policies that give room for the growth of AI [3]. AI can also enhance the consideration of resources in Oman, regarding energy and water and the agricultural industry, for the sake of long-term economic growth and development without depleting valuable commodities [7, 13]. Solving problems like skill deficits and data problems, to go with investments in education, infrastructure, and policy, will allow Oman to harness the potential of AI and bid to become a leader of the AI-driven global economy [4, 14].

3 Methodology

This study used quantitative research to establish the role of AI in developing Oman's economy. To achieve the aim of the current study, a structured questionnaire was developed to capture professionals, business people and key stakeholders of different sectors operating in the Sultanate of Oman with specific reference to the current level of awareness, adoption and barriers/ opportunities of applying AI technology in the Omani economy. The study used a descriptive survey research design given its main function of compiling data to make a description of AI in Oman. In order to collect the data, an online questionnaire was administered to 145 professionals, business persons, and managers, mainly in the oil and gas sector, the financial and banking sector, the health sector, and the educational sector. To make it friendly to all the respondents, the survey was done in both English and Arabic. Data collection was conducted in two weeks of November 2024 allowing respondents adequate time to respond.

4 Results and Findings

Figure 1 presents the respondents' demographic data, based on a sample size of 145 individuals. The age distribution shows that 40% of respondents are between the ages of 21 and 30, while a smaller proportion, 10%, are over 50. This suggests that the sample is primarily composed of young professionals. Regarding education, a majority (60%) hold a bachelor's degree, and only 10% have a Ph.D., indicating a well-educated sample. Employment-wise, 70% of respondents are employed, with most working in the private sector (45%), followed by the government sector (30%). This diversity in employment status and sector provides a comprehensive perspective on AI's role across different industries in Oman.

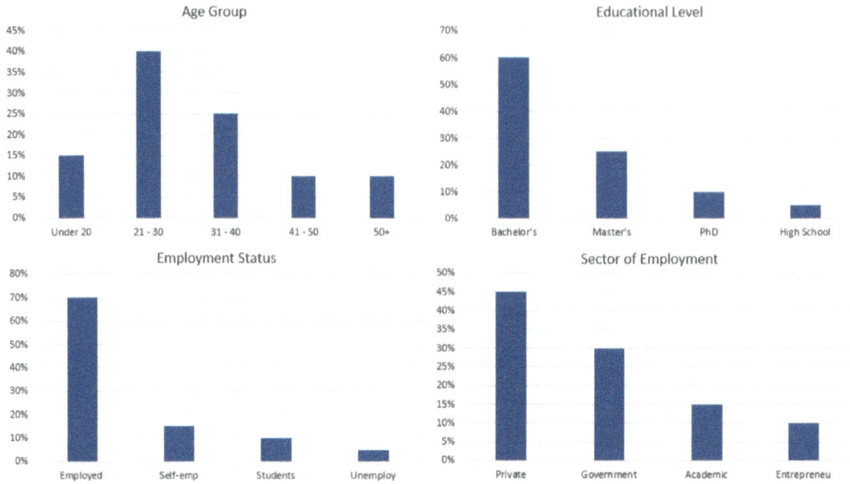

Fig. 1 Demographic information of respondents

Table 1 presents the respondents' demographic data, based on a sample size of 145 individuals. The age distribution shows that 40% of respondents are between the ages of 21 and 30, while a smaller proportion, 10%, are over 50. This suggests that the sample is primarily composed of young professionals. Regarding education, a majority (60%) hold a bachelor's degree, and only 10% have a Ph.D., indicating a well-educated sample. Employment-wise, 70% of respondents are employed, with most working in the private sector (45%), followed by the government sector (30%). This diversity in employment status and sector provides a comprehensive perspective on AI's role across different industries in Oman.

Table 2 highlights the level of awareness and understanding of AI among respondents. Around 25% of respondents rated their understanding as very high, and 35% rated it as high, showing that a majority have a solid grasp of AI concepts. However, 25% reported a moderate understanding, and 10% rated their awareness as low. When asked about familiarity with AI's potential across industries, 30% were very familiar, while 40% were somewhat familiar, suggesting that while there is general awareness of AI's potential, there is still room for growth in deepening this knowledge.

Table 3 examines the extent of AI adoption across Oman's key industries. 40% of respondents believe AI is moderately integrated, while 10% feel it has been extensively integrated. However, 50% believe that AI is either not much integrated or not integrated at all, suggesting that there are still barriers to widespread adoption. In terms of which sectors benefit most from AI, 35% of respondents identified the oil and gas sector as the leader, followed by finance and banking (30%). These sectors are likely the first to adopt AI due to their capital-intensive and data-heavy operations, while sectors like healthcare and manufacturing have room for further AI integration.

Table 1 Demographic information of respondents

Demographic factor	Frequency (%)
Age group	
Under 20	15
21–30	40
31–40	25
41–50	10
51+	10
Education level	
High school	5
Bachelor's degree	60
Master's degree	25
Ph.D.	10
Employment status	
Employed	70
Self-employed	15
Student	10
Unemployed	5
Sector	
Government	30
Private sector	45
Academia/research	15
Entrepreneurship	10

Table 2 Awareness and understanding of AI

	Frequency (%)
AI awareness level	
Very high	25
High	35
Moderate	25
Low	10
Very low	5
Familiarity with AI's potential in industries	
Very familiar	30
Somewhat familiar	40
Neutral	20
Unfamiliar	10

Table 3 AI adoption in Oman's economy

	Frequency (%)
AI integrated into Oman's key industries	
Yes, extensively	10
Yes, moderately	40
Not much	30
Not at all	20
Sectors benefiting the most from AI	
Oil and gas	35
Finance and banking	30
Healthcare	15
Manufacturing	10

Table 4 identifies the key challenges to AI adoption in Oman. The most frequently cited issue is the lack of skilled professionals (45%), followed by concerns about data privacy and security (40%) and high implementation costs (35%). These challenges suggest that to facilitate AI adoption, Oman will need to focus on talent development, ensure robust data security protocols, and address the financial barriers to entry. Issues like lack of infrastructure (20%) and resistance to change (15%) were seen as less critical, but still worthy of attention in the longer term.

Table 5 reveals that most respondents (40%) believe that the government is somewhat active in promoting AI adoption, while 30% feel that the government is very active and supportive. This indicates that while the government has made efforts in this area, there is still a perceived need for greater support and clearer policies. Only a small percentage (5%) believe that the government is not very active or not active at all, suggesting that most respondents recognize the government's role but feel there is room for improvement.

Table 6 examines respondents' views on the impact of AI on employment and the skills needed for an AI-driven economy. Forty percent believe AI will lead to job creation, while 30% are concerned about job displacement. The remaining 30% felt that AI would have no impact or were uncertain. When it comes to the skills required, data science and analytics (40%) and AI programming and machine learning (35%) are seen as the most valuable, highlighting the need for technical expertise in the

Table 4 Challenges to AI adoption in Oman

Challenges to AI adoption	Frequency (%)
Lack of skilled professionals	45
Data privacy and security concerns	40
High implementation costs	35
Lack of infrastructure	20
Regulatory and policy limitations	10
Resistance to change	15

Table 5 Government's role in AI adoption

Government role in promoting AI adoption	Frequency (%)
Very active and supportive	30
Somewhat active	40
Neutral	20
Not very active	5
Not active at all	5

Table 6 Impact of AI on employment and skills

	Frequency (%)
Impact of AI on employment	
Job creation	40
Job displacement	30
No impact	20
Uncertain impact	10
Most valuable skills for an AI-driven economy	
Data science and analytics	40
AI programming and machine learning	35
Cybersecurity	15
Business strategy and AI integration	10

workforce. Skills like cybersecurity (15%) and business strategy (10%) are also recognized as important, but to a lesser extent.

Table 7 provides insight into respondents' outlook on the future of AI in Oman's economy. A majority (50%) are optimistic, while 20% are very optimistic, indicating a positive outlook toward AI's role in the country's economic development. Regarding government focus areas, 45% of respondents believe that the government should prioritize accelerating AI integration in key industries, followed by ensuring ethical AI use and data privacy (30%). This emphasizes the need for a balanced approach that fosters AI growth while addressing social and ethical concerns.

5 Discussion

The results of the survey can serve as a practical source of information on the situation with Oman's economy and its relation to AI. The respondents seemed to provide dependable knowledge of AI and its opportunities, and that can be attributed to their diverse educational backgrounds and employment. This is an indication that the surveyed group is not ignorant of technology in their working environment they are able to use modern tools. However, it shows that the general public seems not to possess sufficient knowledge about AI beyond the common hype. Moreover, given

Table 7 Future AI prospects in Oman

	Frequency (%)
Future AI integration in Oman's economy	
Very optimistic	20
Optimistic	50
Neutral	20
Pessimistic	10
Government focus areas for AI policies	
Accelerating AI integration in key industries	45
Ensuring ethical AI use and data privacy	30
Training and education in AI	15
Building AI infrastructure	10

the fact that the understanding of the risks and benefits of applying AI is relatively present among organizations, the current level of AI implementation in Oman's strategic industries seems to be at a nascent stage. This means that sectors like Oil & Gas and Finance are singled out as the main beneficiaries of AI, as is the case around the world. On the other hand, such sectors as healthcare and manufacturing have better conditions for AI implementation, but they have lower growth rates, which means that AI in these spheres has more potential that needs further investment. The lack of qualified human resources to handle the demand arising from AI projects is even much more of a concern to Oman than low awareness and infrastructural limitations. Other challenges relate to data protection, security measures and the high cost of implementing AI solutions. Most of these challenges, though, are not peculiar to Oman but are seen in the rest of the world more specifically in countries with relatively dormant regulatory environments and weak IT facilities. Secure data should also be adopted to enhance the security of the data shared in AI. The Omani government's engagement is expected to play a strategic role in shaping policies, rewards and a favorable setup that supports AI development.

To fully benefit from the integration of AI in Oman, the nation needs to address human capital, creating local talent that will be needed for a nation with an enhanced use of Artificial Intelligence. Analyses of respondents' attitudes toward the subject demonstrate that there is optimism about the AI future and its positive impact on Oman's economy. However, this optimism should not leave us with no plan for addressing current issues that face students. It is crucial for the government as well as private companies to address problems concerning professional jurisdiction, data confidentiality, and requisite funding. The following are considered priorities to promote the use of AI: Promoting AI adoption in critical industries, encouraging proper use of AI technology, and grants for education and training on AI technology. Therefore, while the awareness and adoption of AI in Oman remain commendable, much more needs to be done to unlock the barriers to enhanced adoption of AI. The government's intervention in providing a favorable environment both in form and content, in terms of policy, incentives, and other supporting structures, is important.

If managed properly with an effective strategy and serious capital investments, artificial intelligence holds the key position in Oman for economic diversification and new opportunities for prosperous innovations.

6 Conclusion

The survey shows an increase in concern about technology and its use in Oman's economy. However, this is coupled with the problems about its use. Although people recognize the positive effects of adopting AI and remain optimistic about these innovations, the gulf between knowledge of AI solutions and their application in the markets is quite vast. These include a lack of skilled professionals, data privacy issues and the relatively high costs of AI implementation, which are the key challenges that companies experience. However, there is an exciting and demonstrable promise of how applications of AI might give rise to growth and innovation amidst oil and gas organizations, as well as the finance sector as a whole, where AI applications can shed new light on opportunities for optimization. However, much effort should be applied by the central and local governments as they are charged with the mandate of ensuring the adoption of AI by supporting policies and infrastructure and investing in education and training. Through the outlined approach to the solutions to the concerns regarding workforce readiness and the use of AI in Oman, it is possible to realize the future impact of AI on the country's economy. AI is expected to have a positive outlook in the future for the Oman economy and a lot of provisions can be made to attain its capability to revolutionize certain critical sectors and create more employment opportunities besides improving the status of Oman in the global economy. Thus, it can be concluded that to allow the growth of an efficient AI ecosystem, The public and private sectors must work together to decrease barriers that are currently present.

References

1. Al-Youbi: Overcoming barriers to AI adoption in Oman: challenges and opportunities. J. Middle East. Econ. Dev. **15**(2), 45–63 (2023)
2. Al-Maashari, Al-Dhuhli: Artificial intelligence applications in healthcare: opportunities and challenges in Oman. J. Healthc. Technol. **9**(2), 45–60 (2021)
3. Alshammari: AI in the GCC: transforming economies and industries. Gulf Stud. Rev. **18**(4), 120–134 (2022)
4. Al-Hashimi: Overcoming data challenges in AI adoption in Oman. Oman Bus. Technol. J. **4**(3), 30–40 (2022)
5. World Economic Forum: The Future of Jobs Report 2021. World Economic Forum. (2021)
6. Al-Kindi: Economic implications of artificial intelligence in the Omani economy. Oman J. Econ. **6**(4), 110–120 (2021)
7. Alshammari, Al-Khushayni: Leveraging AI for economic growth: lessons from the GCC region. Middle East. Econ. Rev. **22**(1), 45–59 (2023)

8. Al-Ghassani, Al-Maashari: Building AI infrastructure for Oman's digital economy. J. Digit. Innov. **11**(2), 112–123 (2022)
9. Al-Hashimi, Al-Mahrouqi: The role of AI in the economic diversification of Oman. Oman Econ. Rev. (2022)
10. Al-Riyami, et al.: The role of AI in Oman's logistics and transportation sectors. Oman J. Eng. Technol. **7**(3), 134–145 (2022)
11. Al-Khushayni: Integrating AI for sustainable development in Oman. Oman Environ. Sci. Rev. **6**(3), 145–157 (2021)
12. Al-Rawahi: The impact of artificial intelligence in Oman's logistics industry. J. Logist. Supply Chain Manage. **8**(2), 90–102 (2022)
13. Al-Suleimani, Al-Belushi: Enhancing financial services through artificial intelligence in Oman. Int. J. Fin. Technol. **10**(1), 50–65 (2021)
14. Al-Mahrouqi, et al.: Challenges and opportunities for AI-driven economic growth in Oman. Gulf Econ. Rev. **5**(3), 78–92 (2020)

The Impact of Ethical Leadership on Job Performance in Small and Medium Enterprises in the Sultanate of Oman from the Point of View of Employees, a Case Study on an Omani Small Enterprise: A Concept Paper

Nasser ALbadi and Fadi Abdelfattah

Abstract This study aimed to reveal the impact of practicing ethical leadership behaviors on job performance from the point of view of employees in small and medium enterprises in the Sultanate of Oman through a case study on an Omani company, where the study sought to know the reality of practicing ethical leadership behaviors represented in (fairness, credibility, participation, and empathy) and its impact and role in the performance of employees, and the study assumes a positive impact of ethical leadership on job performance, and this study was conducted on the case community of 20 individuals, working In an Omani SME company, through a personal interview with each individual, then these interviews underwent deductive qualitative analysis to test the hypothesis and reach conclusions. The study found a positive impact of practicing ethical leadership behaviors (fairness, credibility, participation, and empathy) on the performance of employees in the company. This study recommends designing training programs to develop ethical leadership skills and behaviors among the company's leaders—under study—based on real needs, and recommends developing regulations and policies in the company—under study—to support the application of ethical leadership behaviors, and provide an organizational climate that supports fairness, credibility, participation in decisions and empathy with employees.

Keywords Ethical leadership · Fairness · Credibility · Engagement · Empathy · Job performance

---

N. ALbadi (✉) · F. Abdelfattah
Modern College of Business and Science, Muscat, Sultanate of Oman
e-mail: 20247203@mcbs.edu.om

F. Abdelfattah
e-mail: fadi.abdelfattah@mcbs.edu.om

N. ALbadi
ALbadi Investment Group LLC, Saham, Oman

B. Awwad (ed.), *AI and IoT: Driving Business Success and Sustainability in the Digital Age*, Studies in Systems, Decision and Control 590,
https://doi.org/10.1007/978-3-031-88874-8_80

1 Introduction

The human element is the most important resource for any institution, as it is the one who relies on it to increase productivity, raise market share, and improve the competitive position of the institution in front of other institutions, and therefore taking care of this resource, developing it and helping it improve its performance was one of the greatest tasks of institutional leaders.

Great leaders and influential leaders throughout history were able to inspire followers, influence them and extract their best, to achieve the goals of the nation, state or institution, and one of the most important reasons for this inspiration and the impact that these leaders have was their ethics, such as integrity, justice, empathy, service to others, credibility, reliability and participation, which led to the followers trusting them, and preparing to serve the projects of those leaders, even if it cost them their lives and themselves sometimes, hence the thinking about how Taking advantage of the model of great and influential leaders in ethical leadership, the principle of ethical leadership was introduced in the business and corporate sector.

The issue of moral leadership is an issue as old as man, and the prophets and messengers are the best example of moral leadership since the dawn of humanity, and philosophers since ancient times have dealt with the principle of moral leadership, such as Aristotle, who spoke about the moral person who occupies the position of leadership, and embodies the virtues of courage, moderation, generosity, self-control, honesty, social communication, humility, fairness and justice, Confucius, who spoke about wisdom and courage, and the philosopher Al-Farabi, who provided insights on ethical leadership from an Islamic perspective [1].

Various institutions in our time have begun to pay attention to the ethical behavior of leaders, in light of the increase in unethical issues in business environments and the failure of many organizational leaders to practice ethical behavior, and the negative impact this has had on the behavior and performance of employees [2], and the first to address the theory of ethical leadership was the political sociologist James Francis Burns, in 1978 [3], and then studies on the subject followed.

In the Sultanate of Oman, for example, a study [4] recommended the importance of designing awareness and training programs for educational leaders to adopt ethical leadership practices, and a study [5] recommended promoting ethical leadership practices and principles in school management in the Sultanate of Oman, so that ethical standards are adopted as a basic criterion in the selection, appointment and promotion of principals, and some government agencies such as the Ministry of Education participated in conferences related to professional work ethics, with the aim of Developing the capabilities of senior leaders in the Ministry in ethical paths that develop performance and raise job satisfaction [6].

In this context, the researcher in the company—under study—noticed the existence of a research gap related to ethical leadership practices, based on tracking the positions of the company's leadership through the daily dealing of leadership with employees, reading and analyzing the minutes of the company's meetings and decisions, and monitoring the behavior of employees, while comparing this with the institution's regulations and policies and what it declares, as well as by holding some interviews with employees and stakeholders who are related to that company, so the researcher found a research problem that can be studied, and recommendations made on it, and it was This problem is that there is a defect in the practice of ethical leadership behaviors in the company—under study—and a defect in the performance of its employees accordingly.

The importance of this study comes from its originality, as it is the first study conducted on this company—under study—and that it can provide results that can be studied to be generalized to small and medium enterprises and that it provided recommendations that can be used in the application of ethical leadership practices in similar companies in the Sultanate of Oman.

The importance of this study is also confirmed due to the scarcity of studies on ethical leadership in small and medium enterprises in the Sultanate, despite the presence of (140,068) companies between small, small and medium—at the time of preparing this research—registered on the website of the Small and Medium Enterprises Development Authority [7], but the site does not contain any study on ethical leadership practices in those companies, and the researcher did not find—according to his knowledge—any study published on the Internet in this aspect, despite finding studies Many address ethical leadership in various government institutions, especially on ethical leadership in the Ministry of Education in the Sultanate of Oman, its directorates and schools.

This study aims to achieve several objectives, the most important of which is to raise awareness of the importance of practicing ethical leadership behaviors among leaders and employees of small and medium enterprises in the Sultanate of Oman, and to provide these companies with studies that enable them to focus on the basic dimensions of ethical leadership (fairness, credibility, participation, and empathy), and to establish regulations, policies and procedures that help them achieve them on the ground, and enable them to prepare and design training and qualification programs to prepare ethical leaders.

2 Previous Literature and Studies

Based on the above presented in the introduction of the challenges related to ethical leadership practices in the company—under study—here comes the importance of reviewing the previous literature on similar challenges faced by other institutions, to clarify the impact of these challenges and the factors associated with them on ethical leadership practices in the case company, which enables the researcher to study the problem and its dimensions and reach the results of the research.

The study of Koudouda and Salami [8], which aimed to explore the impact of ethical leadership practices on job commitment through the intermediate variable "intention to leave work", and was conducted on 75 individuals working in the Directorate of Distribution of Gas and Electricity in the Batna region in Algeria to answer the questionnaire, and the study found a positive impact of ethical leadership practices on organizational commitment, and a negative impact on the intention to leave work, and the intention to leave work negatively affected organizational commitment, and the intention to leave work had a role Partial mediation in the relationship between ethical leadership and organizational commitment.

The study of Al-Salami et al. [9], which aimed to identify the reality of practicing ethical leadership behaviors among secondary school leaders in Jeddah Governorate, and its relationship to the level of trust in the leader among male and female teachers, was conducted on a random sample of (439) individuals, and the study found that the practice of ethical leadership among the study community was high from the point of view of teachers, and that the level of trust in school leaders was also high, and concluded that there was a statistically significant positive correlation At the level of significance (0.01) between the degree of ethical leadership practice and the level of trust in the leader among the study sample, the study recommended the importance of working on professional development to disseminate the contents of ethical leadership.

Mabrouki and Nazli study [10], this study aimed to identify the impact of ethical leadership in enhancing the organizational loyalty of employees in the Electricity and Gas Distribution Corporation, in the Ouargla region of Algeria, and it was conducted on a random sample consisting of (60) individuals, through questionnaires and statistical analysis, and the study concluded that there is a high application of ethical leadership and a high level of organizational loyalty in Sonelgaz, and the existence of a statistically significant impact between ethical leadership and organizational loyalty, and the existence of a positive and strong relationship between them.

Therefore, previous studies, after linking them to the problem of the study and the opinions of stakeholders, indicate that the application of ethical leadership practices in their dimensions (fairness, credibility, participation, and empathy), will positively affect job performance in the company—under study—and this study comes to solve the problem of deficiencies in the application of ethical leadership practices, and the associated job performance, and address a research gap that has not been addressed before in that company.

3 Research Question

After the problem of the study is clarified and proved, and after determining its objectives, the researcher has reached the main question that summarizes this research gap and determines its features accurately, so that question is as follows: What is the reality of ethical leadership practices in small and medium enterprises in the Sultanate of Oman from the point of view of employees, and what is their impact on their job performance?

From this main question specific to the research gap, the researcher has set out to branch that question into two branches, each branch of which covers a variable of the study variables, and therefore these sub-questions of the study are as follows.

1. What is the reality of ethical leadership practices from the point of view of employees in the company—under study?
2. What is the impact of ethical leadership in its four dimensions (fairness, credibility, participation, and empathy) on job performance?

4 Research Methodology

This study will be based on qualitative research (descriptive, analytical, deductive), where the researcher will use qualitative research to arrive at the results of the research.

As for the collection of information and data related to the study, the study will depend on collecting data through interviews with stakeholders in the company—under study—and employees in it, (3) stakeholders were interviewed, and (20) members of the company's employees—under study. After that, these interviews were analyzed to reach knowledge of the reality of ethical leadership in the company, and to reach the results of testing the hypotheses of the study through the descriptive and deductive analytical approach.

5 Conceptual Framework

Based on the research methodology and the questions and hypotheses of the study, the researcher has developed this conceptual framework, in order to clarify the relationship between the variables of the study, within the framework of ethical leadership practices in the company—under study—and this framework shows the impact of the main factors identified by the literature and previous studies on job performance.

5.1 A Subsection Sample

The researcher has summarized the conceptual framework of this study in a mental map that shows the independent variables and their role in the dependent variable, as shown in Fig. 1, where the figure shows the impact of ethical leadership practices represented in (fairness, credibility, participation, and empathy) on the performance of employees in each of (organizational commitment, achieving goals, and creativity) in the company—under study—as shown in Fig. 1.

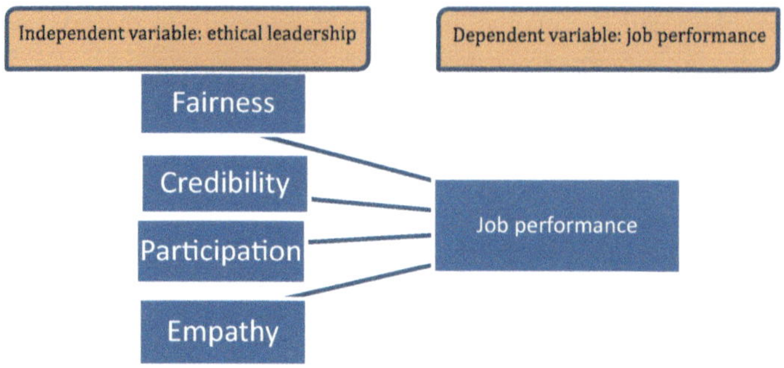

Fig. 1 Study variables

5.2 Ethical Leadership

Ethical leadership means spreading fairness and showing respect for the personal qualities of employees, sincerity, honesty, reliability, empathy, kindness and employee participation in decision-making [11, p. 3950], and it is a work environment to prepare the rationale in dealing within professional relationships, in which the leader is an example to be followed through his actions and ethical behaviors at work, in a manner commensurate with the rules of work and career [12, p. 4].

When we say that a leader exercises ethical leadership, it means that the leader is committed to acting through the values, beliefs, and ethical standards he believes in without responding to external pressures [13, p. 1086].

5.3 Fairness

A fair leader works to make fair decisions and provide support and assistance to subordinates, and is characterized by the ability to set clear standards [14, p. 137], and the ethical leader deals with everyone in a correct and equal manner, makes fair and equitable choices for all, and does not practice favoritism [8, p. 97].

5.4 Credibility

The honest moral leader is honest and honest in dealing with others, if the leader is honest, he gains reliability and reliability from others, earns their respect, and thus can influence them [10, p. 5], and the moral leader is characterized by the conformity of actions with words, and the fulfillment of promises and non-breaking them [8, p. 97].

5.5 Participation

Participation means the ethical leader's keenness to enhance employees' self-confidence and abilities, develop the spirit of responsibility and commitment, enhance the spirit of cooperation and collective decision, appreciate different opinions and diverse ideas, and attribute the successes achieved by the organization to the cooperation and diligence of all [15, p. 3], and this dimension reflects the ethical leader's ability to build a climate that stimulates teamwork among employees, and create a suitable environment for communication and interaction between individuals [14, p. 137].

5.6 Empathy

A compassionate ethical leader who practices respect for others, values their differences and ideas, encourages them to highlight their creativity and ambitions, puts the well-being and interests of his employees at the forefront, and acts in a way that achieves the interests of all parties and benefits all [10, p. 5].

5.7 Job Performance

The term job performance refers to the employees' performance of the job duties and responsibilities required of them and specified in the job description in full, and their commitment to achieving the company's objectives and performance indicators specified therein [16, p. 1207].

5.8 Features of This Frame

This framework is characterized by simplicity and clarity, as it provides a clear concept of the impact of independent variables on job performance, which facilitates understanding the relationship between the independent variables and the dependent variable, and one of its advantages is also based on the relevant previous literature, which studied problems similar to the problem of this research, where the researcher built the study variables based on those previous literature, and this enhances the accuracy and credibility of the research hypotheses. One of the advantages of this framework is the flexibility through which this framework can be expanded to suit other variables, allowing it to be applied in other environments that require ethical leadership.

5.9 *Disadvantages of This Framework*

The study finds that one of the disadvantages of this conceptual framework is its neglect of some other factors affecting job performance, such as the conviction of the company's leadership—under study—of the feasibility of ethical leadership practices, internal factors such as policy and procedures regulations, external factors such as laws governing SMEs in the Sultanate of Oman, and the level of culture about ethical leadership practices among employees.

5.9.1 How to Overcome the Problems of This Framework

The disadvantages of this conceptual framework can be overcome by adding new factors and variables according to the company's need—under study—or according to the updated data and information that can be provided later, and providing measurement tools to measure the impact of each independent variable, and some intermediate variables such as employee satisfaction, and measuring its impact on the efficiency of job performance.

5.9.2 The Environment in Which This Framework Can Be Applied

This conceptual framework can be applied to SMEs in the Sultanate of Oman by training SMEs leaders on ethical leadership principles and practices, developing behaviors that support fairness, credibility, participation and empathy in the work environment, and creating policies and a general organizational culture that contribute to collaboration, teamwork, justice and respect for others.

5.9.3 Field-Based Research

A set of field research can be conducted to study the research variables and factors affecting job performance in the case institution, by distributing questionnaires to employees, conducting field interviews with them to know the challenges they face due to the poor application of ethical leadership practices, and conducting interviews with the organization's management and stakeholders, to know the efforts made to develop ethical leadership practices in the status institution and the existing obstacles.

Acknowledgements This Paper Was Supported and Funding from the Modern college of Business and science and was Supervised by Dr. Fadi Abdelfattah

References

1. United Nations Office on Drugs and Crime website. https://www.unodc.org/e4j/ar/integrity-ethics/module-4/key-issues.html (2018)
2. Jandali, B.: The impact of ethical leadership on organizational commitment among employees: a case study at Algeria Telecom, Operational Directorate of Ouargla, Kasdi Merbah University, Ouargla, Algeria, Master's Thesis (2018)
3. Magdy, A.: https://www.maktabtk.com/blog/post/154 (2018)
4. Al-Rawas, A.b.S.b.A.T,: The impact of ethical leadership on organizational commitment: a field study in the Directorate General of Education in Dhofar Governorate. J. Faculty Educ. Tanta Univ. **89**(1), 42–63 (2023)
5. Al-Alawi, K.b.A.b.N,: The Degree of Ethical Leadership Practice for Public School Principals in Al Dakhiliya Governorate, Sultanate of Oman and its Relationship to Organizational Justice from the Teachers' Perspective. National Center for Educational Research and Development, Cairo (2021)
6. Educational Portal website: https://home.moe.gov.om/topics/1/show/2147 (2013)
7. Website of the Authority for Small and Medium Enterprises Development (https://www.sme.gov.om). Sultanate of Oman (2024)
8. Koudouda, N., Salami, M.: The impact of ethical leadership on organizational compliance: the role of mediation of the intention to leave work: a case study of the Directorate of Gas and Electricity Distribution in Batna. J. Strat. Dev. **10**(4), 91–110 (2020)
9. Al-Salami, A.R.H.H., Mahrizi, A.A.D., Al-Sulami, H.S.H.: The reality of practicing ethical leadership behaviors among secondary school leaders and its relationship to trust in the leader from the point of view of teachers in Jeddah Governorate. J. Educ. Al-Azhar Univ. **190**(Part (4)), 296–336 (2021)
10. Mabrouki, I., Nazli, S.: The impact of ethical leadership on enhancing organizational loyalty to employees: a field study at Sonelgaz Distribution Excellence Ouargla. Master's Thesis, Kasdi Merbah University, Ouargla, Algeria (2020)
11. Yalmaz, E.: The analysis of organizational creativity in schools regarding principals ethical leadership characteristics. Procedia Soc. Behav. Sci. **10**(02), 3949–3953 (2010)
12. Chichoun, F.: The impact of ethical leadership on the quality of career life. Master's Thesis, Kasdi Merbah University—Ouargla, Algeria (2023)
13. Moussa, A.F., El-Din, H., Fahmy, A.A.: The impact of trustworthy leadership on organizational proficiency by application to Egypt Post, Raya. Int. J. Commercial Sci. **3**(10), 1082–1154 (2024)
14. Sewify, A.M.: The effect of ethical leadership on empowerment: field study Al-Azhar University in Assuit—Arab Republic of Egypt. J. Econ. Adm. Legal Sci. **4**(11), 132–157 (2020)
15. Al-Rumaidi, B.S., Al-Asraj, A.S.A.: The impact of ethical leadership on the intention of employees to leave Egyptian tourism companies: organizational trust as an intermediary variable. J. Tour. Hotels **5**(1/1) (2021)
16. Abdel Hadi, A.H.M.M.: The impact of ethical leadership on improving job performance efficiency, a field study. Sci. J. Bus. Environ. Stud. **11**(2/2), 1186–1216 (2020)

The Role of Integrating Sustainable Development Goals in Education to Enhance Students' Well-Being

Muna Mohammed Al Mukhaini, Rami Oweis, and Safoura Navari

Abstract This conceptual paper aims at establishing the impact of Sustainable Development Goals on notions of SDG awareness within the curricula context. By focusing on how the students' well-being is affected by the curricula. Implementation of the SDGs at universities helps students develop personal relevant knowledge concerning sustainable practices and their significance. This integration necessitates decisions to include sustainable practices in educational curricula. This research thus proposes an exploratory research design to examine the effects of implementing the SDGs into the curricula of HEIs (Higher Education Institutes) to understand the extent of the effect of implementing the SDGs on students' level of awareness and commitment to sustainable practices, and the improvement of partnerships among HEIs. This topic has not been extensively covered in the existing literature and thus, the study seeks to contribute new insights.

Keywords Sustainability · Sustainable development goals (SDGs) · Well-being · Curriculum · Education

M. M. Al Mukhaini (✉) · R. Oweis · S. Navari
Modern College of Business and Science (MCBS), Muscat, Oman
e-mail: muna.almukhaini@mcbs.edu.om

R. Oweis
e-mail: rami.oweis@mcbs.edu.om

S. Navari
e-mail: safoura.Navarifarmad@mcbs.edu.om

R. Oweis
Jordan University of Science and Technology, Irbid, Jordan

B. Awwad (ed.), *AI and IoT: Driving Business Success and Sustainability in the Digital Age*, Studies in Systems, Decision and Control 590,
https://doi.org/10.1007/978-3-031-88874-8_81

1 Introduction

Due to the shift towards implementing Sustainable Development Goals (SDGs) across various sectors globally, it is important to achieve these goals within the higher education sector. This integration makes students understand the need to contribute in sustainable practices not only in higher education institutions but in all other settings. Higher education institutions have been identified as key in nurturing of capacities for sustainable practices, which are crucial for transforming local, regional, and global visions. The mission of these institutions is to promote community well-being, which can be accomplished by providing high-quality education and fostering partnerships with various organizations. Any society with a vision of enhancing the contribution of each individual to help with the achievement of sustainable development goals should have courses relative to the goals [1]. In regards to education, many researchers emphasize the inclusion of the SDGs in social education [2]. Besides, there is the need for greater collaboration among the institutions in the higher education sector in order to address the gaps in the educational experiences that conforms to the set SDGs [3]. Therefore, there is need to facilitate environments and resources through which knowledge is imparted to students as this acts a core tool in the well-being of the students. In addition, the cultivation of relationship and relational context particularly in the context of education plays an important role [4] for maintaining emotional stability among students [5]. Therefore, a much emphasis has been directed towards creating indices to assess students' quality of life in academic, financial, physical and psychological domains [6]. This paper discusses how higher education institutions should consider these factors to enhance students' well-being and the overall learning environment. Thus, it is imperative for these institutions to create an environment that supports personal growth and enhances overall student well-being [7].

With the establishment of the SDGs, institutions understood that working towards these goals improves the well-being of individuals within educational settings by promoting sustainable practices. A third of the SDGs focuses on good health and well-being that includes education as one of its goals. This aim ensures that students are prepared to live healthy lifestyles and maintain psychological well-being while embracing sustainable practices. However, institutions may face a several challenges affecting students' well-being, including increased workload, which may hinder their ability to engage in sustainable activities [8]. Moreover, incorporating well-being into management education assists in developing leaders capable of navigating various social contexts, aligning with the third goal of the SDGs [9].

Therefore, this study will focus on the role of integrating SDGs into the curriculum and engaging students in sustainable activities to enhance students' values and awareness of SDGs, thereby impacting their well-being as shown in Fig. 1.

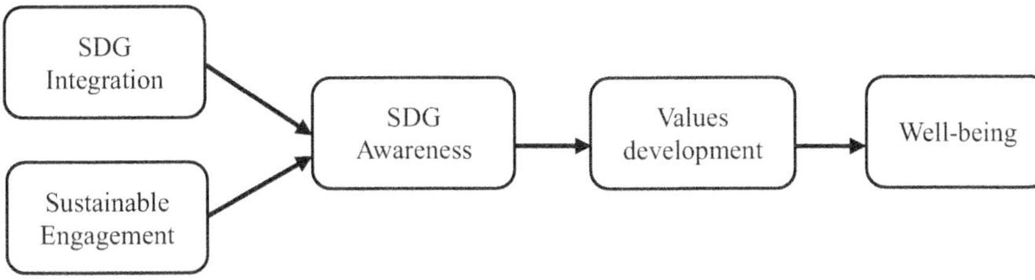

Fig. 1 Conceptual framework for SDG's based education impact on well being

2 Literature Review and Constructs Development

2.1 Integrating SDGs into Curriculum

Education acts as a crucial foundation that sharpens students' consciousness in various disciplines. Different techniques are used to impart the principles required for the accomplishment of educational goals. In the paradigm of sustainable development, improving practices requires a comprehensive approach that focuses on educating both teachers and students about its significance. Among various educational methods, problem-based learning has been singled out for its emphasis since it allows learners to get acquainted with various global issues through their involvement in several sustainable projects [9].

However, the incorporation of sustainable development goals into learning outcomes is crucial to the support of sustainability in all disciplines. Still, it is crucial to implement such goals into educational curricula effectively in order to increase students' awareness, as such. This integration process must also be assessed in terms of its efficacy, particularly as universities strive to enhance their global positioning, which could be enhanced by environmental contributions to sustainable developmental goals. Hence, the use of quantitative system can help measure the effort of universities in achieving these goals and locate where improvement is needed [10].

2.2 Student Engagement on Sustainable Activities

With universities desiring to improve their global rankings and promote sustainable development, it is important to highlight the role of students in sustainability initiatives. Implementing sustainable practices in the curriculum helps the students participate in the achievement of sustained goals, but this is not the only method available. Allowing the students to participate in sustainable projects enables them to develop essential skills for addressing environmental issues and reaching sustainability targets

[2]. Furthermore, the infrastructure of educational institutions, including living laboratories and scientific experiments focused on environmental sustainability, is critical in improving student competencies [11]. By developing such infrastructure, HEIs nurture future generations that are both aware and responsive to environmental issues. Involving the students in environmental activities positively affects their attitude and behaviour towards sustainability [12]. Therefore, educating students about the SDGs promotes their active engagement in higher education institutions [13], making the integration of these goals into curricula essential.

2.3 The Role of SDG Integration into Curriculum and Sustainable Engagement to SDG Awareness

Higher education institutions aiming to enhance students' awareness and participation in green activities must recognize the important outcomes associated with implementing the SDGs in their curricula. Many researchers have stressed on the need to effectively develop and improve curricula for teaching in congruence with these sustainable imaginaries. When students engage with courses across various disciplines, they gain a broader understanding of sustainability, which helps them better evaluate environmental issues [14]. According to [15], conducting workshops can successfully raise students' awareness of environmental challenges and leads to the development of proposals and solutions for sustainable projects. In addition, through the engagement in sustainability activities, students facilitate the formation of positive attitudes and bring increased motivation to practice the successful accomplishment of sustainable goals [16]. However, as there is a shift towards the implementation of activities promoting the establishment of principles of the sustainable development agenda as well as incorporation of sustainable development goals into curricula's, it is important to embrace the commitment of students to engage in the sustainable initiatives.

Proposition 1 Integrating SDGs into the curriculum and engagement in sustainable activities enhances student awareness, fostering proactive advocacy for sustainability challenges.

2.4 The Role of SDG Awareness on Values Development

Awareness of the SDGs is vital ingredient in shaping and instilling environmental values in students, ultimately influencing their actions toward achieving sustainable objectives. Sustainable development is associated with various significant values, such as environmental responsibility, social justice, and the preservation of natural resources. It is possible to improve understanding of the environmentally friendly practices by incorporating such values into curricula for education. Currently, many institutions have invested in teacher preparation programs that aim to develop sustainable development standards, and this intervention has shown that social values increase people's awareness of sustainable development [17]. Moreover, environmental awareness is essential for garnering support for the SDGs, as it highlights how values can enhance individual participation in addressing environmental challenges [18]. Therefore, it is a necessity for HEIs to continue with their effort in raising student awareness of the SDGs to enhance the success of green practices. Nevertheless, if students and other stakeholders of an institution do not have systematic knowledge of the SDGs, they may encounter major challenges in increasing their environmental impacts. This underscores the need to focus efforts on increasing students' awareness of these important goals [19].

Proposition 2 Awareness of SDGs shapes and instills environmental values in students, driving actions toward sustainable objectives.

2.5 The Impact of Values Development on Students' Well-Being

Possessing an optimistic attitude and such emotional, physical, and psychological aspects is linked to sustainable values. Some of these include values such as environmental responsibility, belongingness, and equality, as all these values contribute to the prosperity of various aspects of life. Education has a crucial role in nurturing sustainable values like citizenship and personal development [20]. Moral values have a role in facilitating a positive educational environment, as they help build strong relationships between teachers and students, which in turn enhances student well-being [21]. Thus, it is very difficult to isolate the role of values in combating some environmental issues affecting the world. Being aware of this relationship prepares the students to address environmental issues [22]. Also, understanding cultural diversity and individual differences among students affects their awareness of values. It is essential to develop and adapt these values to align with SDGs and promote student well-being.

Proposition 3 Values development positively impacts students' well-being by instilling environmental responsibility, a sense of belonging, and equality, contributing to their overall prosperity.

2.6 *The Role of SDG Integration and Sustainable Engagement on Students' Well-Being*

Previous literature suggests that the integration of SDGs into curriculum improves the students' awareness of these goals [14] and fosters their engagement in sustainable activities [15]. Engaging students in addressing sustainability issues fosters a sense of responsibility towards solving these problems [22]. In addition, an educational environment that promotes sustainable practices fosters feelings of belongingness and fairness [23]. Hence, educational institutions cannot underestimate significance of adding elements of sustainability into the educational processes. This integration not only increases students' awareness but also helps cultivate values that contribute to their overall well-being, as illustrated in Fig. 1. Additionally, the use of different educational strategies aimed at sustainable goals contributes to the transformation of students' behaviours regarding sustainable well-being, promoting students' mental health and environmental health interaction [24].

Proposition 4 SDG integration and sustainable engagement enhance students' well-being through the mediating roles of SDG awareness and values development.

3 Methodology

This paper provides a conceptual framework towards assessing the implementation of the integration of SDGs and engaging students in sustainable activities to improve student wellbeing. Key factors mediating this relationship include students' awareness of the SDGs and the development of sustainable values. The methods used in this research rely on the analysis of literature concerning the integration of SDGs into curricula, sustainable student activities, and their impact on well-being. It is the intention of this review to then build a comprehensive conceptual framework. This research is also concerned with the interaction between the implementation of the SDGs and how this would enhance students' awareness of these goals. This would ensure the development of sustainable values such as environmental conservation, equity, and efficient resource utilization, which are vital in enhancing students' well-being. In addition, the proposed conceptual framework underscores the importance of including the SDGs into curriculum and engaging student in various green activities as this can shape student behaviour by promoting positive values and subsequently improving well-being. This conceptual framework serves as a foundation for future empirical research, providing hypotheses to test how to integrate the SDGs to raise awareness, develop values, and improve well-being.

4 Conclusion

This research affirms the need to implement the SDGs into curricula as well as, the activities of learner for improvement of their well-being. In this regard it is important for educational institutions to improve this by bringing awareness of the SDGs and supporting sustainable values. According to the studies used for this research, this integration concerns awareness and forms values to enhance well-being. Some research emphasises the need for educating students regarding the SDGs given that they facilitate a shift towards the development of an environmentally conscious identity by individuals. When higher education institutions introduce these goals to students, they cultivate a generation that understands sustainable practices better, which enhances their contributions to resource preservation and environmental protection. Therefore, this paper suggests that leaders in higher learning institutions support the integration of the SDGs in the course curricula and promote sustainable activities among students and develop resources to support the same. Thus, the inclusion of SDGs into curricula and practice of green activities by students helps students to develop positive attitudes to sustainable practices and usage of their knowledge, as well as enhance their wellbeing. Future research needs to develop empirical studies to test the conceptual framework presented in this research and examine how the adoption of SDGs can enhance the quality of life. This study provides valuable insights for educational institutions seeking to improve their sustainable practices and enhance student well-being.

5 Theoretical Implications

The current paper contains an overview of the articles describing the incorporation of the SDGs into curricula, students' participation in environmentally friendly activities and their impact on well-being. It reveals that the perceived awareness of SDGs and the development of values mediate the relationship between integrating SDGs and improving student well-being. Implementing the SDGs in the curriculum enables students to learn about sustainable goals and green initiatives, as well as understand the exact degree of sustainability on and off the university campus. Furthermore, this study contributes to the existing theories by demonstrating how the integration of SDGs can influence the students' values and enhances their commitment and responsibility for sustainable development. Students play a crucial role in committing to green practices and conserving environmental resources, which can be achieved through a better understanding of sustainable practices and their implications. The study also reveals different strategies for integrating SDGs and provides theoretical insights into how the involvement of students in achieving sustainable goals can be strengthened.

References

1. Prior, D.D., Mysore Seshadrinath, S., Zhang, M., McCormack, M.: Measuring sustainable development goals (SDGs) in higher education through semantic matching. Stud. High. Educ. 1–14 (2024)
2. Monzó-Martínez, A., Ortiz-Cermeño, E., Martínez-Agut, M.P.: Global citizenship education and its role in sustainability at the university level. Educ. Sci. **14**(8), 847 (2024)
3. Roll, K., MacLeod, M., Agbodjah, S., Sánchez Siller, I.M.: Learning in action: embedding the SDGs through the reach alliance. Sustain. Earth Rev. **7**(1), 15 (2024)
4. Jiang, W., Saito, E., Zhang, H., Waterhouse, P.: Being-well-in-relationships': re-conceptualising students' wellbeing in secondary education. Discourse Stud. Cult. Polit. Educ. 1–14 (2024)
5. Vasileva, I.V., Chumakov, M.V., Chumakova, D.M.: Students' perceptions of well-being. Ekol. Cheloveka (Hum. Ecol.) **30**(12), 909–918 (2023)
6. Khatri, P., Duggal, H.K., Lim, W.M., Thomas, A., Shiva, A.: Student well-being in higher education: scale development and validation with implications for management education. Int. J. Manag. Educ. **22**(1), 100933 (2024)
7. Baik, C., Larcombe, W.: Student wellbeing and students' experiences in higher education. In: Research Handbook on the Student Experience in Higher Education, pp. 74–88. Edward Elgar Publishing (2023)
8. Abdelrahim, Y., Zafer, A.: How students' well-being, education for sustainability, and sustainable development relate: a theoretical model (2024)
9. Nguyen, L.T.V., Cleveland, D., Nguyen, C.T.M., Joyce, C.: Problem-based learning and the integration of sustainable development goals. J. Work.-Appl. Manag. **16**(2), 218–234 (2024)
10. Bertone, E., Bischeri, C., Boddy, J.: Integration of the sustainable development Goals (SDGs) in architecture curriculum–quantifying the extent of SDG contributions. Environ. Educ. Res. 1–20 (2024)
11. Nansen, C.: Active learning, living laboratories, student empowerment, and urban sustainability. Sustainability **16**(10), 3902 (2024)
12. Thompson, K., Herriges, M.: Embedding environmentally sustainable practices in learning and teaching: a student perspective. Open Sch. Teach. Learn. **3**(2) (2024)
13. Filho, W.L., Trevisan, L.V., Dinis, M.A.P., Ulmer, N., Paço, A., Borsari, B., Sierra, J., Salvia, A.: Fostering students' participation in the implementation of the sustainable development goals at higher education institutions. Discov. Sustain. **5**(1), 22 (2024)
14. Martínez-Borreguero, G., Maestre-Jiménez, J., Mateos-Núñez, M., Naranjo-Correa, F.L.: Integrating energy and sustainability into the educational curriculum: a pathway to achieving SDGs. Sustainability **16**(10), 4100 (2024)
15. Baron, J., Lawrence, S.A., Nevins-Bennett, C.: Sustainable development goals: a case of the community college student initiative project. Eur. J. Educ. Pedagog. **5**(3), 24–35 (2024)
16. Hdeib, R., Aouad, M., Awawdeh, S., Mahfoodh, S.A.: Assessing students' awareness, attitudes, and actions towards sustainable development goals (SDGs). In: 2024 ASU International Conference in Emerging Technologies for Sustainability and Intelligent Systems (ICETSIS), pp. 188–194. IEEE (2024)
17. Momani, H.I.: Social values and their relationship to teachers' awareness of sustainable development standards. J. Educ. Soc. Res. **13** (2023)
18. Krsnik, S., Erjavec, K., Klopčič, M.: Impact of citizens' personal values, knowledge, awareness, informing, advertising, and truth of environmental and climate challenges in support of the sustainable development goals. Sustainability **14**(12), 7333 (2022)
19. Selvam, V., Ashok, D., Rajalakshmi, V.: Assessment of students awareness, knowledge and accessibility about SDGS. SPAST Abstr. **1**(01) (2021)
20. Broadbent, C., Boyle, M.: Promoting positive education, resilience and student wellbeing through values education. Eur. J. Soc. Behav. Sci. (2014)

21. Aziz, R., Mulyadi, M., Hadi, M.S., Wahyuni, E.N., Rubaidi, R.: Students' well-being development in the classroom: a mixed-method study. Int. J. Eval. Res. Educ. (IJERE) **13**(3), 1666–1672 (2024)
22. Tudball, L.: Developing student wellbeing through education for sustainability. Second. Int. Res. Handb. Values Educ. Stud. Wellbeing **955** (2023)
23. Tripon, C., Gonța, I., Bulgac, A.: Nurturing minds and sustainability: an exploration of educational interactions and their impact on student well-being and assessment in a sustainable university. Sustainability **15**(12), 9349 (2023)
24. Vatovec, C., Ferrer, H.: Sustainable well-being challenge: a student-centered pedagogical tool linking human well-being to ecological flourishing. Sustainability **11**(24), 7178 (2019)

Climate Change and Sustainable Development Goals. A Systematic Literature Review

Rami Oweis, Sanjeet Singh, and R. Jayaram

Abstract The climate changes have severe social, economic, and ecological impacts. The Sustainable Development Goal 13 (SDG 13) aims for climate action by focusing on climate resilience; climate education and awareness; climate adaptation; climate funding and technology support; and climate planning and capacity enhancement. This paper aims to consolidate the literature on climate change and SDG 13, by focusing on challenges, and solutions for attaining targets of SDG 13. A systematic literature review of document publications from the resources of the Web of Science is conducted by using PRISMA guidelines. Targets of SDG 13 are used for thematic analysis. This review includes a discussion about promising themes for further research on climate change and SDG 13.

Keywords Climate change · Climate action · Climate finance · Climate education · Climate mitigation · Climate adaptation · Climate awareness · Natural disaster · SDG sustainable development goal

1 Introduction

Climate and environment have strong mutual bondage. The degradation of the environment can create climatic changes and may result in severe threats to life on Earth [44]. Climate change has a significant influence on public health too. In an Anthropocene, the major questions are how can we stop the climate catastrophes? Can we remain silent on climate change? What climate actions are needed for a sustainable and safe future? What is holding us back from climate actions? These questions are

R. Oweis (✉)
Modern College of Business and Science, Muscat, Oman
e-mail: rami.oweis@mcbs.edu.om

Jordan University of Science and Technology, Irbid, Jordan

S. Singh · R. Jayaram
Faculty of Management Studies, Marwadi University, Rajkot, Gujarat, Jordan

B. Awwad (ed.), *AI and IoT: Driving Business Success and Sustainability in the Digital Age*, Studies in Systems, Decision and Control 590,
https://doi.org/10.1007/978-3-031-88874-8_82

related to sustainable development goals, especially SDG 13. The Sustainable Development Goal 13 (SDG 13) aims to fight climate change with its ambitious goals and targets for 2030.

The targets of SDG 13 include strengthening resilience and adaptive capacity to climate-related hazards and natural disasters; integrating climate change measures into national policies, and strategies; improving education, and awareness on climate change mitigation, adaptation, and impact reduction; generation of early warning on climate change issues; building human and institutional and human capacity for managing climate-related issues; efforts for raising green climate fund of 100 billion dollars annually for combating with climate change; and promoting capacity building initiatives for better climate planning and management in least developed countries and vulnerable and marginalized communities.

Intergovernmental Panel on Climate Change has pointed out the urgency for deep reduction of emissions to avoid climate calamities. The major challenges for this objective include the mounting fossil fuel consumption; constraints in raising climate finance; emissions and climate damage from road constructions [50]; challenges and resistance from farmers, investors, and politicians [8]; financial development, urbanization, and agriculture in emission [2]; economic complexity [37]; and climate incompatible policies and poor implementation.

Several measures are adopted to fight against climate challenges and the prominent among them include sustainability and climate orientation in technology decisions, institutional quality and energy choices [4]; the carbon capture and utilization [17]; reducing greenhouse gas emission through the technology integration; promotion of renewables [53]; taxation on energy consumption [12]; practicing of climate-smart agricultural practices [41, 43]; community-based adaptations [20]; climate orientation in major economic and non-economic policies [18, 33, 36, 37]. Green information systems [10] and green product innovations [1] can also play a huge role in climate protection through sustainability.

Very few reviews have addressed the targets, challenges, and achievements related to SDG 13 and this gap in the literature is addressed in this paper. This review of SDG 13 has six chapters. This section of the introduction deals with SDG 13, targets, challenges, popular measures, and common indicators tracked in respect of targets of SDG 13; followed by a discussion of the research methodology used in this paper. A detailed thematic discussion on targets of SDG 13 is the third chapter, The fourth chapter discusses the research gaps and themes for further research. The conclusion of the results and findings of this paper is the fifth section.

1.1 Research Objectives

1. To discuss the targets of SDG 13, by focusing on the targets, challenges, and achievements
2. To sort out research gaps in literature related to SDG 13 and to spot the promising themes.

1.2 Research Questions

1. What are the climate-related targets of SDG 13?
2. What are the leading issues associated with the targets of SDG 13?
3. What are the research gaps and themes for further research related to SDG 13?

2 Review Methodology

This review of climate-related targets SDG 13 is based on a single source. This paper adopted the single-source research model followed in the works. The database of the Web of Science is used for preparing this review, which is a comprehensive database of scholarly publications on diversified streams of knowledge. Web of Science database deals with more than two hundred and fifty disciplines and is rich in scientific documents on climate actions and SDG 13.

The Booleans used for this work are selected with due care by the chances of the wrong set of papers, as warned in the works [3], related to the choice of search terms, combinations, and query structure on the relevance of search results. This paper used the Boolean TS = ("sustainable development goal 13 " OR "SDG 13") and target-specific Booleans TS = ("climate adaptation") AND sustainable development goal); TS = (("national disaster risk") and "sustainable development goal"); TS = "local disaster risk" TS = ("climate change policies OR strategies or planning) TS = ("education or awareness) and (climate change mitigation adaptation) TS = (("United Nations Framework Convention on Climate Change) TS = (("Raising capacity AND Climate Change) was used on 17/10/2023. Figure 1, visualises the record screening and record selection process.

The query in Web of Science obtained, 376 papers. The search is made on topic-based research. This paper followed PRISMA guidelines 2020 for paper selection. After the removal of duplicates and anonymous articles by reading the title and abstract the 370 papers are used for thematic discussion, inspired and motivated by the structure and model adopted in the works [5, 16, 19, 47]. Among 370 papers,

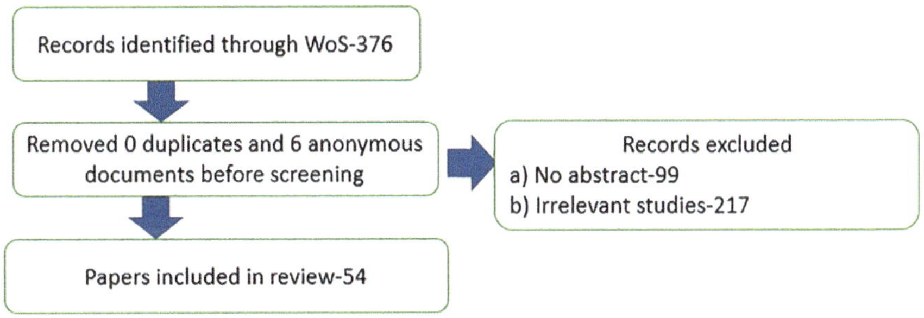

Fig. 1 Document screening and selection process

99 papers without abstract were excluded. Similarly, 217 records were irrelevant to the themes discussed in this review. The final paper is based on 54 records (Refer to Fig. 1).

3 Thematic Analysis

Thematic analysis of this review is based on the targets associated with SDG 13. The research themes and sub-themes are developed from the targets of SDG 13. Four major research themes are discussed in this paper. The theme for discussing the developments of SDG 13 in this review is visualized in Fig. 2,

3.1 Education and Awareness on Climate Change Mitigation, Adaptation, Impact Reduction, and Early Warning

Environmental education and climate change awareness have a significant role in building climate change adaptation skills. The study among university students in Turkey identified that a one-unit increase in awareness and sense of responsibility leads to a 0.638-unit increase in perceived risk and a 0.557-unit increase in environmental behaviors respectively [52]. Climate change awareness and education related to mitigation, adaptation, impact reduction, and early warnings are crucial for attaining targets of SDG 13 [14, 21, 30, 31, 40]; and for energy efficiency, and disaster risk reduction [40].

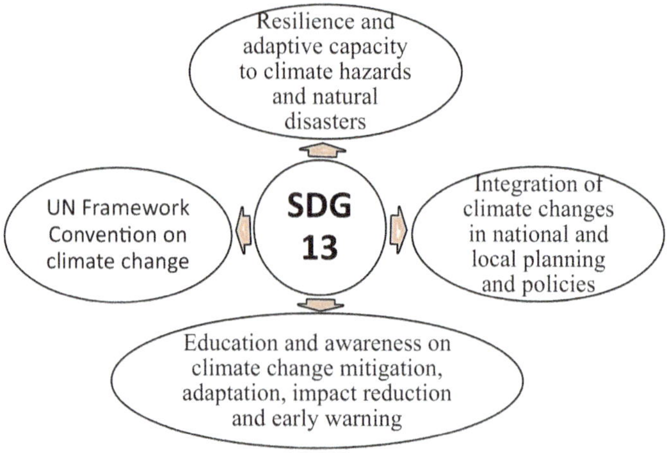

Fig. 2 Themes under discussion

Climate change has both physical and mental impacts on children and proper education and awareness of the same is inevitable [23, 26]. However, climate change awareness and education are confronted with a diversified set of challenges [6, 11]. Moreover, mere climate change awareness and education may not be sufficient for effective adaptation to climate change [34]. The participation level of society is another crucial factor for attaining the targets of SDG 13.

Agriculture is one of the vulnerable areas of climate change and can enhance the risk and uncertainty in this sector [45]. Climate change education and training are very effective in agriculture and allied activities in the background of climate change [28].

Climate change education can also be through the inculcation of climate change concepts in co-curricular and extracurricular opportunities for students [7]; through botanical garden visits [39]; and science-based role-play simulation exercises [35]. In the formal education route, climate change awareness can be inculcated by introducing the various dimensions of the climate change concept in the formal education curriculum (Solis-Espallargas and [46]). The climate change adaptation curriculum should be well-planned to meet professional and social issues related to climate change [13]. Moreover, educational intervention can remove much of the confusion and misguidance related to climate change [15, 38] among the student community and motivate them for better climate actions.

The Indigenous knowledge systems can also be used to enrich climate change education and can also explored for strengthening the knowledge base of climate education and awareness [27]. The Geographical Information System (GIS) can also deployed for enhancing climate change awareness [42]. The exhibition of climate change impact photographs [24]; and video clips [49] can be successful tools to enhance climate change awareness and education. The Citizen participation model is successfully used in Singapore for climate change education among society and adopted a multi-stage grassroots approach [51].

3.2 Resilience and Adaptive Capacity to Climate Hazards and Natural Disasters

Climate change awareness level and education level are important determinants affecting the adaptive capacity to climate hazards and natural disasters. Natural disasters demand to cope with several stress and emotional challenges including the role of supporting families and communities. In short, resilience after climate hazards and natural disasters can be at both individual and community levels [25]. Several communities have been victims of repeated natural disasters, and the stress management, adaptive capacity, and resilience of this section is a challenging task to recover, especially at the psychological level [48]. Children and women are highly vulnerable to climate hazards and natural disasters.

The culture of the communities [32] is an important factor affecting resilience and adaptive capacity to climate hazards and natural disasters. Poor climate change adaptation can be due to many reasons including poor action plans, initiatives, and execution; knowledge gaps in adaptation, weak institutional structure and weak government actions, poor funding levels and lack of general awareness [29]. In short, resilience after climate hazards and natural disasters can be at both individual and community levels [25].

3.3 Integration of Climate Changes in National and Local Planning and Policies

Integration and institutionalization of climate change in national policies is an urgent matter to be implemented. The local policy measures on climate change are also equally important as the local government or agencies have a crucial role in climate change adaptation at local levels. The local policy actions for climate change have great importance and such policies are heavily dependent upon institutional implementation, and technology. One of the challenges in the implementation of climate change adaptation is related to the difficulties in reaching local policy-makers [9].

3.4 United Nations Framework Convention on Climate Change

The signatories to the United Nations Framework Convention on Climate Change (UNFCCC) are motivated and assisted to initiate climate change considerations in their national-level agendas, establish institutional systems to face climate change and help in better interconnections and coordinated efforts for climate change measures, The national policies on climate change should adhere to the United Nations Framework Convention on Climate Change (UNFCCC). The appraisal of the climate change mitigation and adaptation regulatory frameworks can be with the UN climate change convention. Many of the achievements of the United Nations Framework Convention on Climate are affected by the breakout and aftereffects of the COVID-19 pandemic [22].

4 Research Gap and Future Research Agenda

Future researchers can work upon the broad themes of measures and initiatives for building resilience and adaptive capacity for fighting against climate change actions; climate integrated national and local policy actions; climate education and climate awareness; climate change monitoring and generation of early signals; rehabilitation issues due to climate change; capacity building for managing climate change; and climate finance.

Several challenges are confronted due to climate change researchers can focus on recommended measures for solving those challenges. The rehabilitation of victims of climate change is another promising theme for further research. The existing literature on the national and local policy actions, measures for climate change early signals, rehabilitation, capacity building, and climate change challenges for defending climate change is limited. These segments have great importance in policy matters and for climate actions. Future researchers can also research technological innovations for faster and more accurate prediction and generation of early warning signals related to climate change.

Policy initiatives and cooperation and tie-ups for reducing climate finance constraints and easing up of availability of climate funding. Research can be for enhancing climate compatibility of policies and programs and efficiency in such policy implementation. Research can also be initiated for strengthening media and for overcoming the adverse impacts of pandemics like COVID-19 on climate change issues. Reviews and evaluation studies on policy implementations and individual measures for facing climate change issues can be meaningful and help in better documentation of climate change issues, designing of climate change management, rehabilitation, and funding.

Academicians and educational experts can work on the integration of climate change concepts in teaching modules and integration with the curriculum. Development of teaching materials educational videos and documentaries can also improve the general awareness related to the importance of climate protection. Research can be on disaster recovery initiatives due to climate change and technology integration for rescue and rehabilitation efforts. Society and community participation should also be improved for climate actions and researchers, academicians, policy experts, politicians, and administrators have an important role in this regard. Researchers should take the initiative to handle the stress and emotional challenges of vulnerable sections due to climate change, especially, children, old age people, women, the disabled, and the section of people who are economically and mentally suffering due to climate changes. Researchers can play a great role in rebuilding infrastructures, education, public health, water and sanitation, energy and communication facilities and transportation in locations suffered due to climatic changes or natural disasters.

5 Conclusions

SDG 13 is related to climate action, having the mission of taking urgent climate actions to combat extreme climate events and their impacts including food shortages, and economic, social, and biodiversity losses. SDG 13 has a strong association with several other SDGs including poverty goals, food security, energy, education, health, and many more. The 2030 targets for climate action focus on strengthening resilience and adaptive capacity; climate policies at national and local levels; climate education and awareness; climate change mitigation, impact reduction and generation of early warning; capacity building for climate issues; raising of climate funds; climate planning and assistance for least developed economies and for vulnerable communities.

The major challenges include the raising of climate funds for combating climate change-related issues; rebuilding of infrastructures; climate-incompatible policies and poor implementation practices. Poor awareness level of climate change and its impacts; and poor social and community level participation in vested interests and politicisation of events is a serious concern for climate actions. Similarly, the community adaptation to such change and, the competence of administrators, policymakers, and politicians are creating challenges for the efforts related to climate actions. Moreover, COVID-19 has eroded many of the achievements related to climate actions. The emission level data of countries point out that the emissions are high in high-income countries and less in low-income countries. But further empirical evidence is essential before concluding on such assumptions.

These ambitious targets can be achieved through emission reduction, promotion of renewables, climate education, and awareness; technological and institutional innovations; energy choices; carbon capturing; climate-smart agricultural practices; can strengthening climate actions. Climate change and awareness and education are major determinants for the development of climate change adaptation skills and for the achievement of targets of climate actions, energy efficiency, food security, and reducing casualties, losses, and risks related to climate change. The vulnerable communities including children, women, senior citizens, disabled people, and people outside financial security need support and climate awareness to cope with several stress and emotional challenges during climate-related disasters. Climate change awareness should be integrated with education through formal and informal routes, which can develop strong awareness in society related to climate change issues, impacts, and solutions and help in strengthening climate resilience and adaptation. Citizen-level participation and community awareness can strengthen climate actions. Curriculum should be better designed for this and innovative co-curriculum activities can be planned for such events. Climate-smart agriculture can strengthen agriculture in vulnerable areas and can minimize food security challenges and farmer income drainage during climate change catastrophes.

Further research can be on climate change resilience; adaptive capacity enhancement; better climate-oriented policy orientations at the national and local levels; climate education and awareness; climate funding; and rehabilitation and community support. Future researchers can focus on the challenges of climate change, its impact, measures, strengths, and weaknesses of such measures and requirements of policy reforms, reforming implementation plans, and technology integrations in plan execution. Research can be for getting better climate funding, database management, and tracking of fund utilization. The research in this sector can be promising for the academic community for imparting lessons about climate change among youth through curricular and non-curricular modules; documentaries and sponsored films. The leading funding agencies on this topic are the European Commission and the

National Natural Science Foundation of China with eight funded projects, followed by UK Research Innovation with seven funded projects.

References

1. Abdelfattah, F., et al.: The future of competitive advantage in Oman: integrating green product innovation, AI, and intellectual capital in business strategies. Int. J. Innov. Stud. **8**(2), 154–171 (2024). https://doi.org/10.1016/j.ijis.2024.02.001
2. Adebayo, T.S., et al.: Drivers of environmental degradation in Turkey: designing an SDG framework through advanced quantile approaches. Energy Rep. **8**, 2008–2021 (2022). https://doi.org/10.1016/j.egyr.2022.01.020
3. Armitage, C.S., Lorenz, M., Mikki, S.: Mapping scholarly publications related to the sustainable development goals: do independent bibliometric approaches get the same results? Quant. Sci. Stud. **1**(3), 1092–1108 (2020). https://doi.org/10.1162/qss_a_00071
4. Anwar, A., Chaudhary, A.R., Malik, S.: Modeling the macroeconomic determinants of environmental degradation in E-7 countries: the role of technological innovation and institutional quality. J. Public Aff. (2022). https://doi.org/10.1002/pa.2834
5. Bansal, S., Garg, I., Sharma, G.D.: Social entrepreneurship as a path for social change and driver of sustainable development: a systematic review and research agenda. Sustainability **11**(4) (2019). https://doi.org/10.3390/su11041091
6. Bonifacio, A.C., Takeuchi, Y., Shaw, R.: Mainstreaming climate change adaptation and disaster risk reduction through school education: perspectives and challenges. In: Shaw, R., Pulhin, J.M., Pereira, J.J. (Eds.) Climate Change Adaptation and Disaster Risk Reduction: Issues and Challenges, pp. 143–169. Emerald Group Publishing Ltd (Community Environment and Disaster Risk Management), England (2010). https://doi.org/10.1108/S2040-7262(2010)0000004013
7. Brugmann, R., et al.: Expanding student engagement in sustainability: using SDG- and CEL-focused inventories to transform curriculum at the University of Toronto. Sustainability (Switzerland) **11**(2) (2019). https://doi.org/10.3390/su11020530
8. Campbell, B.M., et al.: Urgent action to combat climate change and its impacts (SDG 13): transforming agriculture and food systems. Curr. Opin. Environ. Sustain. **34**, 13–20 (2018). https://doi.org/10.1016/j.cosust.2018.06.005
9. Clar, C., Steurer, R.: Why popular support tools on climate change adaptation have difficulties in reaching local policy-makers: qualitative insights from the UK and Germany. Environ. Policy Gov. **28**(3), 172–182 (2018). https://doi.org/10.1002/eet.1802
10. Dahleez, K. et al.: An investigation of government involvement and green information systems in driving management practices. In Alareeni, B., Hamdan, A. (Eds.) Navigating the Technological Tide: the Evolution and Challenges of Business Model Innovation. Springer Nature Switzerland, Cham, pp. 327–335 (2024)
11. Des Combes, H.J. et al.: Recognition of prior learning (RPL) in resilience (climate change adaptation and disaster risk reduction) in the Pacific: opportunities and challenges in climate change education. In Filho, W. L. and Hemstock, S. L. (Eds.) Climate Change and the Role of Education, pp. 363–370. Springer International Publishing AG (Climate Change Management), Switzerland (2019). https://doi.org/10.1007/978-3-030-32898-6_20
12. Dogan, B., et al.: How environmental taxes and carbon emissions are related in the G7 economies? Renewable Energy **187**, 645–656 (2022). https://doi.org/10.1016/j.renene.2022.01.077
13. Erlandsson, L., et al.: Combating climate change through the estate and curriculum—a whole-institution commitment at nottingham trent university', in Azeiteiro, U. M., Filho, W. L., and Davim, J. P. (Eds.) Higher Education Institutions in a Global Warming World: the Transition

of Higher Education Institutions to a Low Carbon Economy, pp. 1–22. River Publishers (River Publishers Series in Management Sciences and Engineering), Denmark (2017)
14. Feinstein, N.W., Mach, K.J.: Three roles for education in climate change adaptation. Climate Policy **20**(3), 317–322 (2020). https://doi.org/10.1080/14693062.2019.1701975
15. Franco, I.B., Tapia, R., Tracey, J.: SDG 13 climate action climate education: identifying challenges to climate action. In Franco, I B et al. (eds) Actioning the Global Goals for Local Impact: Towards Sustainability Science, Policy, Education and Practice, pp. 219–228. Springer Japan KK (Science for Sustainable Societies), Japan (2020). https://doi.org/10.1007/978-981-32-9927-6_14
16. Sharma, G.D., et al.: COVID-19 and environmental concerns: a rapid review. Renew. Sustain. Energy Rev. **148**(January) (2020)
17. Ioannou, I. et al.: Trade-offs between sustainable development goals in carbon capture and utilisation. Energy Environ. Sci. (2022). https://doi.org/10.1039/d2ee01153k
18. Iqbal, K.M.J., et al.: Mix-method modelling of actors' capacity for environmental sustainability and climate compatible development in energy sector. Environ. Sci. Pollut. Res. (2022). https://doi.org/10.1007/s11356-022-19399-1
19. Jain, M., Sharma, G.D., Mahendru, M.: Can I sustain my happiness? A review, critique and research agenda for economics of happiness. Sustainability (Switzerland) **11**(22) (2019). https://doi.org/10.3390/su11226375
20. Khan, A.S., et al.: Chennai city and coastal hazards: addressing community-based adaptation through the lens of climate change and sea-level rise (CBACCS). In Filho, W. L. et al. (eds) Climate Change, Hazards and Adaptation Options: Handling the Impacts of a Changing Climate, pp. 777–798. Springer International Publishing AG (Climate Change Management), Switzerland (2020). https://doi.org/10.1007/978-3-030-37425-9_39
21. Kolenaty, M., Kroufek, R., Cincera, J.: What triggers climate action: the impact of a climate change education program on students' climate literacy and their willingness to act. Sustainability **14**(16) (2022). https://doi.org/10.3390/su141610365
22. Leal Filho, W., Setti, A.F.F., et al.: An overview of the interactions between food production and climate change. Sci. Total. Environ. **838**(3) (2022). https://doi.org/10.1016/j.scitotenv.2022.156438
23. Leger-Goodes, T. et al.: Eco-anxiety in children: a scoping review of the mental health impacts of the awareness of climate change. Front. Psychol. **13** (2022). https://doi.org/10.3389/fpsyg.2022.872544
24. Liu, Y., Liu, J., Su, Y.: Low-carbon awareness and behaviors: effects of exposure to climate change impact photographs. Sage Open **11**(3) (2021). https://doi.org/10.1177/21582440211031487
25. Madsen, W., O'Mullan, C.: Perceptions of community resilience after natural disaster in a rural Australian town. J. Community Psychol. **44**(3), 277–292 (2016). https://doi.org/10.1002/jcop.21764
26. Martin, G., et al.: Review: the impact of climate change awareness on children's mental well-being and negative emotions—a scoping review. Child Adolesc. Ment. Health **27**(1, SI), 59–72 (2022). https://doi.org/10.1111/camh.12525
27. Mbah, M., Ajaps, S., Molthan-Hill, P.: A systematic review of the deployment of indigenous knowledge systems towards climate change adaptation in developing world contexts: implications for climate change education. Sustainability **13**(9) (2021). https://doi.org/10.3390/su13094811
28. Mkonda, M.Y.: Awareness and adaptations to climate change among the rural farmers in different agro-ecological zones of Tanzania. Manag. Environ. Qual. **33**(6), 1502–1527 (2022). https://doi.org/10.1108/MEQ-10-2021-0241
29. Mthembu, D.E., Nhamo, G.: Aligning SDG 13 with South Africa's development agenda: Adaptation policies and institutiona frameworks. Jamba-J. Disaster Risk Stud. **14** (2022). https://doi.org/10.4102/jamba.v14i1.1155
30. Obisesan, M., Chitakira, M.: Investigating climate change awareness and adaptation strategies among female farmers in the Lephalale municipal area in South Africa. Afr. J. Sci. Technol. Innov. Dev. **13**(6), 727–734 (2021). https://doi.org/10.1080/20421338.2020.1796005

31. Okaka, F.O., Odhiambo, B.D.O.: Urban residents' awareness of climate change and their autonomous adaptive behaviour and mitigation measures in the coastal city of Mombasa, Kenya. S. Afr. Geogr. J. **100**(3), 378–393 (2018). https://doi.org/10.1080/03736245.2018.1500302
32. Raich, M. et al.: Impact of culture on community resilience in natural disaster situations. Eur. J. Public Health **27**(3) (2017)
33. Rej, S., et al.: The asymmetric influence of environmental-related technological innovation on climate change mitigation: what role do FDI and renewable energy play? Environ. Sci. Pollut. Res. (2022). https://doi.org/10.1007/s11356-022-23182-7
34. Rodriguez-Cruz, L.A., Niles, M.T.: Awareness of climate change's impacts and motivation to adapt are not enough to drive action: a look of Puerto Rican farmers after Hurricane Maria. PLOS ONE **16**(1) (2021). https://doi.org/10.1371/journal.pone.0244512
35. Rumore, D., Schenk, T., Susskind, L.: Role-play simulations for climate change adaptation education and engagement. Nat. Clim. Chang. **6**(8), 745–750 (2016). https://doi.org/10.1038/NCLIMATE3084
36. Saba, C.S.: CO2 emissions-energy consumption-militarisation-growth nexus in South Africa: evidence from novel dynamic ARDL simulations. Environ. Sci. Pollut. Res. (2022). https://doi.org/10.1007/s11356-022-23069-7
37. Saint Akadiri, S. et al.: Testing the role of economic complexity on the ecological footprint in China: a nonparametric causality-in-quantiles approach Energy Environ. (2022). https://doi.org/10.1177/0958305X221094573
38. Schrot, O.G. et al.: Teenagers expand their conceptions of climate change adaptation through research-education cooperation. In: Filho, W.L., Hemstock, S.L. (Eds) Climate Change and the Role of Education, pp. 525–547. Springer International Publishing AG (Climate Change Management), Switzerland (2019). https://doi.org/10.1007/978-3-030-32898-6_29.
39. Sellmann, D., Bogner, F.X.: Climate change education: quantitatively assessing the impact of a botanical garden as an informal learning environment. Environ. Educ. Res. **19**(4), 415–429 (2013). https://doi.org/10.1080/13504622.2012.700696
40. Senevirathne, M., et al.: A common framework for MOOC curricular development in climate change education—findings and adaptations under the BECK project for higher education institutions in Europe and Asia. Prog. Disaster Sci. **14** (2022). https://doi.org/10.1016/j.pdisas.2022.100222
41. Shahbaz, P., et al.: Adoption of climate smart agricultural practices through women involvement in decision making process: exploring the role of empowerment and innovativeness. Agric.-Basel **12**(8) (2022). https://doi.org/10.3390/agriculture12081161
42. Shahid, Z., Piracha, A.: Awareness of climate change impacts and adaptation at local level in Punjab, Pakistan. In: Maheshwari, B., Singh, V.P., Thoradeniya, B. (Eds.) Balanced Urban Development: Options And Strategies For Liveable Cities, pp. 409–428. Springer International Publishing AG (Water Science and Technology Library), Switzerland (2016). https://doi.org/10.1007/978-3-319-28112-4_25
43. Shahmohamadloo, R.S., et al.: The sustainable agriculture imperative: a perspective on the need for an agrosystem approach to meet the United Nations sustainable development goals by 2030. Integr. Environ. Assess. Manag. **18**(5), 1199–1205 (2022). https://doi.org/10.1002/ieam.4558
44. Siwedza, S., Shava, S.: Insurance, increasing natural disaster risks and the SDGs: a focus on Southern Africa. In: Nhamo, G., Odularu, G.O.A., Mjimba, V. (Ed.) Scaling Up SDGs Implementation: Emerging Cases From State, Development and Private Sectors. Gewerbestrasse, vol. 11, pp. 129–138. Cham (Ch-6330, Switzerland: Springer International Publishing AG (Sustainable Development Goals Series). https://doi.org/10.1007/978-3-030-33216-7_9
45. Sohail, M.T. et al.: Determining farmers' awareness about climate change mitigation and wastewater irrigation: a pathway toward green and sustainable development. Front. Environ. Sci. **10** (2022). https://doi.org/10.3389/fenvs.2022.900193
46. Solis-Espallargas, C., Moron-Monge, H.: How to improve sustainability competences of teacher training? Inquiring the prior knowledge on climate change in primary school students. Sustainability **12**(16) (2020). https://doi.org/10.3390/su12166486

47. Srivastava, M., et al.: What's in the brain for us: a systematic literature review of neuroeconomics and neurofinance. Qual. Res. Financ. Mark. **12**(4), 413–435 (2020). https://doi.org/10.1108/QRFM-10-2019-0127
48. Stratta, P., et al.: Multiple natural disaster exposure effects on stress adjustment and resilience in clinical samples. Curr. Psychol. (2021). https://doi.org/10.1007/s12144-021-01844-y
49. Wang, X., Chen, J.: Fear emotion reduces reported mitigation behavior in adolescents subject to climate change education. Climatic Change **174**(1–2) (2022). https://doi.org/10.1007/s10584-022-03419-7
50. Wenz, L., et al.: Road to glory or highway to hell? Global road access and climate change mitigation. Environ. Res. Lett. **15**(7) (2020). https://doi.org/10.1088/1748-9326/ab858d
51. Wi, A.: Citizen participation as a key enabler for successful public education policies in climate change mitigation in Singapore. Int. Res. Geogr. Environ. Educ. **28**(1), 53–69 (2019). https://doi.org/10.1080/10382046.2018.1430789
52. Yilmaz, V., Gulec, P., Ari, E.: Impact of climate change information of university students in Turkey on responsibility and environmental behavior through awareness and perceived risk. Environ. Dev. Sustain. (2022). https://doi.org/10.1007/s10668-022-02319-1
53. Zhao, L., Saydaliev, H.B., Iqbal, S.: Energy financing, Covid-19 repercussions and climate change: implications for emerging economies. Clim. Chang. Econ. **13**(03) (2022). https://doi.org/10.1142/S2010007822400036.

Sustainable Development Goals of Equality. A Review and Future Agenda for Research

Rami Oweis, Sanjeet Singh, Naresh Dilawarsing Jadeja, and Meenu Bhardwaj

Abstract The sustainable development goals for equality can be achieved through economic growth promotion of all forms of equality; improved regulation; strong and clear migration policies and measures for development, and reducing migration costs. The major obstacles to these objectives are the existence of a diversified set of inequalities, migration challenges, and poor regulation and institutional quality. This review synthesizes the documents from the Web of Science database related to SDG 10 by a detailed thematic analysis. This paper recommends a detailed research agenda and policy recommendations for the attainment of sustainable development goals of equality. This research points out the key focus points to be addressed for sustainable migration policy and income equality. Future researchers can focus on research related to migration challenges, various forms of inequalities and solutions; regulation of international organizations, and the development of least-developed countries.

Keywords Equality · Migration · Regulation · Discrimination · Economic growth · SDG · SDG 10

R. Oweis (✉)
Modern College of Business and Science, Muscat, Oman
e-mail: rami.oweis@mcbs.edu.om

Jordan University of Science and Technology, Irbid, Jordan

S. Singh · N. D. Jadeja
Faculty of Management Studies, Marwadi University, Rajkot, Gujarat, India

M. Bhardwaj
Chandigarh University, Gharuan, Mohali, India

B. Awwad (ed.), *AI and IoT: Driving Business Success and Sustainability in the Digital Age*, Studies in Systems, Decision and Control 590,
https://doi.org/10.1007/978-3-031-88874-8_83

1 Introduction

Equality within and among countries is the tenth sustainable development goal (SDG 10). Equality helps to spread prosperity and helps to unleash the potential of everyone without discrimination. The inequality is not limited to economic inequality, but social, political, and many other. Inequality takes shape in many forms. Child labor, labor market inequality, social inequality, structural inequalities, and gender inequality are major threats faced by initiatives for equality, causing discrimination. Gender inequality can also lead to income inequality. Ethnic and religious discrimination is another roadblock to ensuring equality.

The sustainable development goals of equality are tracked by the targets for improving income growth; promoting social, political, and economic equality; promoting equal opportunity and discarding discriminating laws, policies, and practices; equality in fiscal, social protection, and wage policies; improved regulation and monitoring of global markets and financial institutions; improving the voice and participation of developing countries in global economic and financial institutions; designing and implementation of well-planned migration policies; measures for the development of the least developed countries; and reducing the migration remittance cost [46, 56]. However, the targets and indicators of SDG 10 do not cover all forms of inequalities and hence these goals of equality cannot be monitored and achieved comprehensively.

The Gini index is one of the strongest measures for tracking income inequality [13], on a scale of 0–100, where zero represents the highest equality in income distribution and vice versa [39]. Research among African countries has observed that internet penetration and enhancement of fixed broadband subscriptions can reduce the Gini coefficient [8]. Various limitation of using the Gini index is pointed out in the works [41]. The Gini index is one of the indicators used for tracking SDG 10 and the other is the Palma ratio. The Palma ratio is also used for tracking equality. Palma ratio is the proportion of all income received by the 10% of people with the highest disposable income and the share of all income received by the 40% of people with the lowest disposable income [15, 39]. Palma ratio is one of the important indicators for tracking SDG 10 and the study on the role of information and communication technology in sustainable development in the African region has concluded that mobile phone and internet penetration can reduce the Palma ratio and there the inequalities of income [8]. Several researchers have pointed out the limitations in monitoring the sustainable development goals of equality by focusing on the Gini index and Palma ratio, as they do not cover all forms of inequality [40]. SDG 10 has several targets for achieving the goals of equality, but the monitoring concentrates on the Palma ratio and Gini coefficient. These two indicators should be closely monitored to contain income inequality and efforts should be taken to contain the other inequalities too. The major focus of SDG 10 is related to migration policies, equality, development of least developed countries, and regulation of international organizations. The summary of Gini index performance and Palma ratio of leading regions is summarised in Table 1. Europe and America perform well in both the indicators and Asian and African countries lags in both the parameters.

Six chapters are included in this paper. The first is an introduction section of SDG 10, from an international perspective, followed by a discussion on the review methodology. The third section is a detailed thematic discussion of SDG 10 from an international perspective. The recommendations for future research are included in the fifth section. The final section of this paper is the conclusion.

Table 1 Summary of performance of leading regions on Gini Index and Palma ratio

Gini Index						
	Asia	Africa	Europe	North America	Oceanic	South America
Countries without Gini Index	9	3	3	Nil	3	Nil
Average Gini index	41.64	46.79	35.04	45.5	42.01	50.06
Countries below the Gini Index of 25	Zero	Zero	1	Nil	Nil	Nil
Countries falling between 25–50 in Gini index	28	34	All other	2	9	5
Countries falling between 50–75 in Gini index	4	17	Nil	1	2	6
Palma Ratio						
Countries without a Palma ratio	12	4	3	Nil	4	Nil
Average Palma ratio	1.57	2.44	1.18	1.8	1.66	2.51
Countries with a Palma ratio below 1.5	16	9	26	Nil	2	Nil
Countries with Palma ratio between 1.5–2	10	13	7	2	8	1
Countries with a Palma ratio above 2	3	38	Nil	1	2	10

Source Sustainable Development Report 2021, Cambridge University Press

1.1 Review Objectives

1. To summarize the initiatives and achievements for equality based on SDG 10 objectives
2. To consolidate the literature on SDG 10
3. To identify the themes for future research related to equality related (SDG 10)

1.2 Review Questions

1. What are the constraints for achieving the equality-related sustainable development goals (SDG 10)?
2. What are the achievements of equality-related sustainable development goals?
3. Which are niches for future research?

2 Review Methodology

Among various databases of scientific literature, the Web of Science has outstanding records in scientific publications on diversified topics. Web of Science is a collection of well-diversified disciplines. Web of Science resources include articles, conference proceedings, and book chapters, with a huge collection of literature related to SDG 10 and various dimensions of equality. This review used the web of science (single-source model) and there are several classic examples of reviews based on a single source [22, 28, 52].

The Booleans used for this work are "Sustainable Development Goal 10 " and "SDG 10" and indicator specific Booleans "migration policy sustainable development goal" "migration remittance transaction cost" "equal opportunities discrimination sustainable development goal" "social OR fiscal equality policy sustainable development goal" "income growth inequality sustainable development goal" "social-economic political exclusion sustainable development goal" "regulation global financial sustainable development goal" "development least developed country sustainable development goal" was used on 19/02/2022 and drew 202 papers were obtained for the review on topic-based research. This paper followed paper selection based on PRISMA guidelines as shown in Fig. 1. As the first step the duplicates and anonymous articles are eliminated, and by reading the title and abstract the one hundred and nine papers are used for thematic discussion. This paper was motivated by the structure and model adopted in the works [9, 16, 23, 50].

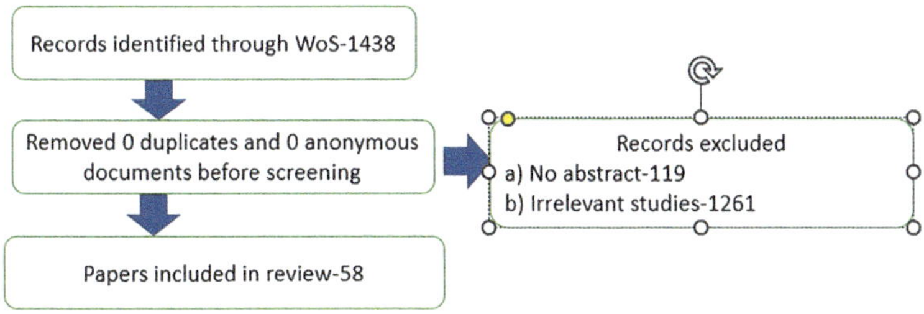

Fig. 1 Details of paper screening and selection

3 Thematic Analysis

The goal of reducing the inequalities among and within countries is the objective of sustainable development goal 10. This paper looked into the targets and developed the themes for the thematic analysis in this paper. The major themes identified are the migration policy and related issues; income inequality; and development of the least developed countries.

3.1 Migration Policy

Migration policies have specific country priorities and are influenced by a variety of factors. The immigration policies may change over the period. Smooth and safe migration policies and opportunities are essential for the achievement of SDG 10, for controlling borders and border-related tensions [43], and for controlling illegal immigration [45]. The immigration policy can also play a major role in the subjective well-being of non-immigrant populations [53]. Political interests and the motives for retaining power [33]; national security, and national problems [58]; anti-immigrant movements [30]; human rights [25, 44] risk of terrorism [11]; and employability [19] are also the dominant factors influencing the migration policies. Marriage migration, educational migration, and working visa permits; child migration [48]; social security of migrants [57], growth, structural change, and regional inequality [29], domestic priorities [3] and migrant rights are the other important points to be considered during the design and practice of the sustainable migration policies. The cost of remittance is very significant in remittance flow and thus a reduced remittance cost is one of the important targets of SDG 10 [4]. The cost of migrants also affects unemployment in economies [1]. The key points to be considered for sustainable migration policy after considering SDG 10 targets are visualized in Fig. 2.

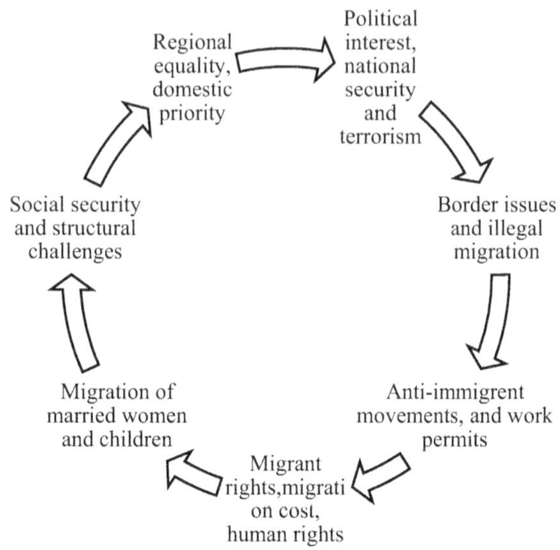

Fig. 2 Focus points for reform and development of sustainable migration policy, based on SDG 10

3.2 Income Equality

Income equality is the most researched concept associated with SDG 10. Income inequality is a big challenge and undesirable in the achievement of equality goals of sustainable development goals. The increased level of inequalities in income can lead to worsening poverty levels [49]; social unrest, and political instability [26]. Several factors influence income equality, especially labor, human development index [5]; development of the agriculture sector [5]; credit disbursal by small banks, and employment in SMEs and MSMEs [7]; globalization [20]; liberalization [27], government quality [21]; economic agglomeration [34]; infrastructure development [38]; sectoral growth and monetary policy [47]; economic growth and foreign direct investment [51]; and trade openness [59].

Income inequality and economic growth have different relations depending on the development of the country [54]. The inequality of income may exist in the initial periods of every economy [6]. Economic growth in the true spirit can improve income equality in the later stages of the economy, as income inequality can negatively affect economic growth [18]. However, the findings of [17] and Mukhlis et al. [34] contradict this. Very strong and concrete policies are essential for tackling this inequality, and for progressing for sustainability of the economy.

Human capital investments are very essential for improving the economic conditions and productivity of the middle-income population and thereby improving equality [2]. Contradictions exist in the use of taxation as a tool for promoting income equality. [12] are against the use of taxation tools to bring income equality, but contradictory findings related to the use of taxation tools for reducing inequality are by [14], and [31]. Financial inclusion [32, 42] and fiscal decentralization [35]

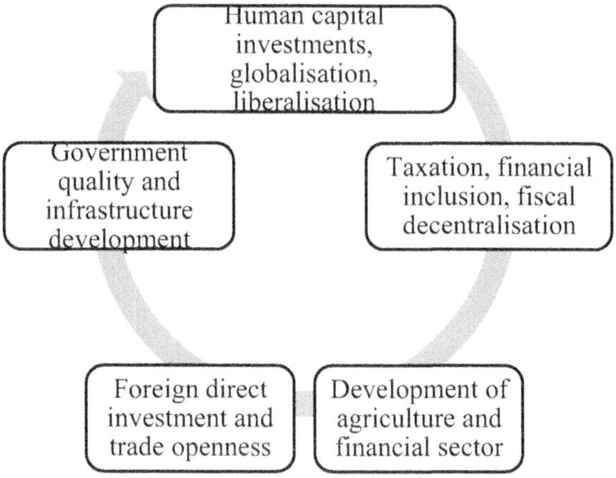

Fig. 3 Focus points for promoting income equality

can also be good measures to contain inequality and improve economic growth. The issues to be considered for reducing income inequality are visualized in Fig. 3.

3.3 Development of Least-Developed Countries

The sustainable development objectives of equality can be completed with the development of the least developed countries. The development of the least developed countries should not be limited only to economic development, but to climate protection and green technology integration too. The economic development among the least developed Asian countries during the period 2001 to 2015 has shown high economic growth, boosted performance by the manufacturing and service industries [36], similar result was also obtained by the study of the economic performance of least developed countries during 2002–2016 [37]; and less reliance with agriculture. However, the slow economic reforms [36], poor business climate [37], corruption, and poor governance are the major challenges in this region [37]. Another factor affecting the development of least developed economies is political systems and structure; and political stability [55]. Socio-economic development and tourism development can be a big boost for the development of the least developed countries [24]. Human rights development in the least developed countries is another important factor to be focused on for improving the performance of the least developed countries [10].

4 Future Research Agenda

The potential themes for research related to sustainable development goals of equality are discussed in the following sections of thematic, methodological, and policy recommendations. This section can be useful for future researchers to find solutions for bringing equality to all forms of society. A snapshot of of research agenda discussion is visualised in Fig. 4.

4.1 Thematic Recommendations

Future research related to SDG 10 can play a big role in the goals of equality and related research topics. The themes for future research related to the sustainable development goals related to all forms of equality should focus on the evaluation, development, and strengthening of migration policies and reducing migration costs. Research can be for ending all forms of inequality and discrimination in society. The promising research can be on the regulation of international organizations for enhancing equality and the development of least-developed countries to bring equality and stop all forms of discrimination.

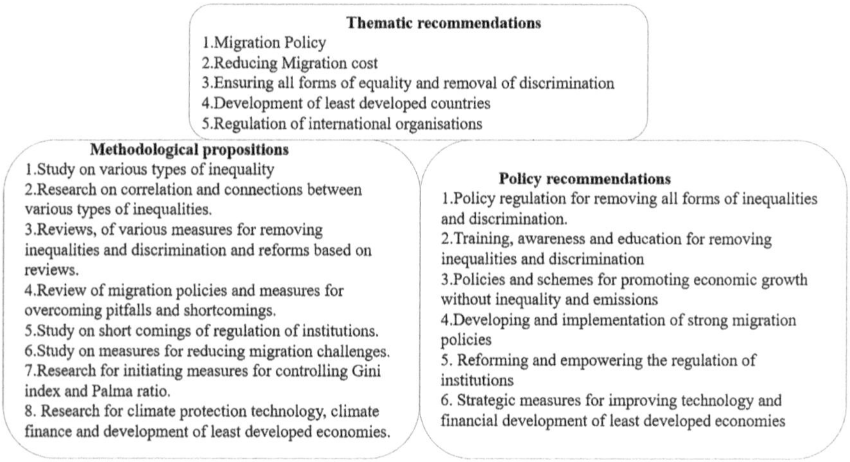

Fig. 4 Snapshot of research agenda discussion

4.2 Methodological Propositions

All forms of inequality in connection with the targets of SDG 10 are not deeply explored. Future researchers can conduct detailed studies of individual inequalities, the status, reasons, and challenges, and can develop solutions for removing the same. Empirical studies can be conducted related to correlation and connections between various types of inequalities in society. Such studies can be very meaningful for developing solutions for facing inequalities. Review studies including systematic reviews can be conducted to explore various policy measures adopted to remove inequalities in society.

A review of the existing migration policies and migration costs is promising for controlling and monitoring inequality. Clear-cut migration policies should be developed after a critical review of existing migration policies and the pitfalls and shortcomings in the policy, and execution should be reformed. Research can be for improving parameters associated with the Gini Index and Palma ratio. Further research can also be initiated on the regulation of the institutions and development of least-developed countries. Bibliometric analysis can also be planned on the individual types of inequalities and the associated challenges.

4.3 Policy Recommendations

Strong policy measures can be developed by the administrators and policymakers for ending economic, social, political, and gender equality. Young generations should be informed and trained related to various types of inequalities and should be encouraged to cooperate on the initiatives for reducing the evils of inequality and discrimination in society. The lessons of equality should be part of education and strong academic reforms are required for spreading the message of equality in classrooms. Policy reforms in the education sector can also be used as a measure for attracting more research funding for the initiatives for an equal world. This paper also recommends for reform and development of strong migration policies after considering both internal and external challenges. Reforms and empowerment of regulation of the international institutions and strategic policy development and reforms for the rapid overall development of the least developed economies.

5 Conclusions

The attainment of targets of SDG 10 depends on growth in income, improving all forms of equality, improved regulation, establishment, and execution of strong migration policies and measures for the development of least-developed countries and reducing migration costs. The SDG 10 aims for equality within and among countries. There are several forms of inequality in society, but the majority of the literature focuses on income inequality and other forms of equality are not explored. The other major types of inequalities to be urgently addressed are inequality related to gender, employment opportunities, recruitment; social structure, education, and religion. Similarly, proper management of parameters of the Gini Index and Palma ratio is also essential for better attainment of targets of SDG 10.

This paper recommends strong measures for ending economic, social, political, and gender equality and discrimination based on religion, caste, etc. Equality lessons should be included in all levels of education and the adults should be trained for the same. The equality awareness should be spread in the society and citizens should be encouraged to participate in the equality initiatives. This paper also recommends strengthening the migration policy with a holistic view. This paper can be very useful for researchers, academicians, practitioners, and administrators in the fields of reforms, equality, economic growth, and migration. Future researchers can focus on various research niches associated with the goals and targets of SDG 10. We recommend that future researchers on SDG 10 focus both on descriptive and empirical research. Promising research can be initiated in the segments related to migration policies, and migration costs; measures for ending inequality and discrimination; regulation of international organizations, and the development of least-developed countries. Future researchers can also focus on the individual determinants of the Gini index and Palma ratio for the better attainment of SDG 10. The untouched indicators and targets of SDG 10 also promise several themes for future research and funding. Future researchers can focus on projects related to SDG 10, the leading funding sponsors on this topic include the European Commission, UK Research Innovation, Austrian Science Fund, Economic Research Council, and Bill Melinda Gates Foundation.

The limitation of this paper is the availability of very few papers on individual targets of sustainable development goals of equality. This constraint limits the discussion to a few numbers of targets and indicators related to SDG 10. The reliance on secondary data is another limitation.

References

1. Abdelfattah, F.: Society, cost of migrants, and digital economy impact on unemployment of Omani Youth's : an empirical investigation of the gender moderate role. Res. Sq. 1–19 (2023)
2. Abrigo, M.R.M., Lee, S.-H., Park, D.: Human capital spending, inequality, and growth in middle-income Asia. Emerg. Mark. Financ. Trade **54**(6), 1285–1303 (2018). https://doi.org/10.1080/1540496X.2017.1422721
3. Adam, I., et al.: West African interests in (EU) migration policy. Balancing domestic priorities with external incentives. J. Ethn. Migr. Stud. **46**(15), 3101–3118 (2020). https://doi.org/10.1080/1369183X.2020.1750354
4. Ahmed, J., Mughal, M., Martinez-Zarzoso, I.: Sending money home: transaction cost and remittances to developing countries. World Econ. **44**(8), 2433–2459 (2021). https://doi.org/10.1111/twec.13110

5. Akbar, U.U., Aimon, H., Yulhendri: Analysis and prospect of economic growth and income inequality in West Sumatra Province. In Dwita, V., Handayani, D.F., Sari, Y.P., Sari, A.Y., Yeni, I. (Eds.) Proceedings of the 2nd Padang International Conference On Education, Economics, Business and Accounting (PICEEBA-2 2018), pp. 155–165. Atlantis Press (AEBMR-Advances in Economics Business and Management Research), France (2018)
6. Akinci, M., Yuce Akinci, G., Yilmaz, O.: What causes income inequality: economic growth or financial development? Ank. Avrupa Calismalari Derg.-Ank. Rev. Eur. Stud. **19**(2), 327–364 (2020)
7. Anwar, A.I., Mangilep, M.A.A.: The effect of economic growth, microfinance institution, and micro and small enterprises to income inequality in South Sulawesi. In: Reni, A., Munir, A.R., Munizu, M., Phuakkhong, C., Ahmad, N.H., Siddiqui, S.H., Kura, K.M. (Eds.) Proceedings of the 3rd International Conference on Accounting, Management and Economics 2018 (ICAME 2018), pp. 543–546. Atlantis Press (AEBMR-Advances in Economics Business and Management Research) France (2018)
8. Asongu, S.A., Odhiambo, N.M.: How enhancing information and communication technology has affected inequality in Africa for sustainable development: an empirical investigation. Sustain. Dev. (2019). https://doi.org/10.1002/sd.1929
9. Bansal, S., Garg, I., Sharma, G.D.: Social entrepreneurship as a path for social change and driver of sustainable development: a systematic review and research agenda. Sustainability **11**(4) (2019). https://doi.org/10.3390/su11041091
10. Bedrinana, K.G.A., Martin, J.A.R., Ananos, F.T.: Human rights in the least developed countries of Asia: an index for quantifying sustainable development goal 3 (good health and wellbeing). Int. J. Environ. Res. Public Health **18**(9) (2021). https://doi.org/10.3390/ijerph18094747
11. Bigo, D., Guild, E.: International law and european migration policy: where is the terrorism risk? LAWS **8**(4) (2019). https://doi.org/10.3390/laws8040030
12. Biswas, S., Chakraborty, I., Hai, R.: Income inequality, tax policy, and economic growth. Econ J **127**(601), 688–727 (2017). https://doi.org/10.1111/ecoj.12485
13. Chattopadhyay, B., De, S.K.: Estimation of Gini Index within pre-specified error bound. Econometrics **4**(3) (2016). https://doi.org/10.3390/econometrics4030030
14. Chen, S.-H.: Inequality-growth nexus under progressive income taxation. J. Macroecon. **65** (2020). https://doi.org/10.1016/j.jmacro.2020.103234
15. Cobham, A., Schloegl, L., Sumner, A.: Inequality and the tails: the Palma proposition and ratio. Global Pol. **7**(1), 25–36 (2016). https://doi.org/10.1111/1758-5899.12320
16. Sharma, G.D., et al.: COVID-19 and environmental concerns: a rapid review. Renew. Sustain. Energy Rev. **148**(January) (2020)
17. Gnangoin, Y.T.B. et al.: Public spending, income inequality and economic growth in asian countries: a panel GMM approach. Economies **7**(4) (2019). https://doi.org/10.3390/economies7040115
18. Hien, L.Q.: The impact of income inequality on economic growth: empirical evidence from Vietnam. J. Asian Financ. Econ. Bus. **9**(2), 305–312 (2022). https://doi.org/10.13106/jafeb.2022.vol9.no2.0305
19. Hoang, N.T.H., Hamid, M.O.: A fair go for all? Australia's language-in-migration policy. Discourse-Stud. Cult. Polit. Educ. **38**(6), 836–850 (2017). https://doi.org/10.1080/01596306.2016.1199527
20. Huh, H.-S., Park, C.-Y.: A new index of globalisation: Measuring impacts of integration on economic growth and income inequality. World Econ. **44**(2), 409–443 (2021). https://doi.org/10.1111/twec.12998
21. Hung, N.T., et al.: Relationship between government quality, economic growth and income inequality: evidence from Vietnam. Cogent Bus. Manag. **7**(1) (2020). https://doi.org/10.1080/23311975.2020.1736847.
22. Jabbour, C.J.C.: Environmental training in organisations: from a literature review to a framework for future research. Resour. Conserv. Recycl. **74**, 144–155 (2013). https://doi.org/10.1016/j.resconrec.2012.12.017

23. Jain, M., Sharma, G.D., Mahendru, M.: Can I sustain my happiness? A review, critique and research agenda for economics of happiness. Sustainability (Switzerland) **11**(22) (2019). https://doi.org/10.3390/su11226375
24. Jasinski, M.: Tourism and development aid and the socio-economic development level of geographically disadvantaged and least developed countries. Prace Kom. Geogr. Przemyslu Pol. Towar. Geogr.-Stud. Ind. Geogr. Comm. Pol. Geogr. Soc. **33**(3), 127–147 (2019). https://doi.org/10.24917/20801653.333.9
25. Jimenez-Alvarez, M.G., Espineira, K., Gazzotti, L.: Migration policy and international human rights frameworks in Morocco: tensions and contradictions. Journal of North African Studies, 26(5, SI), 893–911 (2021). https://doi.org/10.1080/13629387.2020.1800208
26. Koh, S.G.M., Lee, G.H.Y., Bomhoff, E.J.: The income inequality, financial depth and economic growth nexus in China. World Economy **43**(2), 412–427 (2020). https://doi.org/10.1111/twec.12825
27. Kwon, R.: Can we have our cake and eat it too? Liberalization, economic growth, and income inequality in advanced industrial societies. Soc. Forces **95**(2), 469–502 (2016). https://doi.org/10.1093/sf/sow077
28. Lage Junior, M., Godinho Filho, M.: Variations of the kanban system: literature review and classification. Int. J. Prod. Econ. **125**(1), 13–21 (2010). https://doi.org/10.1016/j.ijpe.2010.01.009
29. Leight, J.: Comment on the effect of migration policy on growth, structural change, and regional inequality in China{"}, by Hao, Sun, Tombe and Zhu. J. Monet. Econ. **113**(SI), 135–137 (2020). https://doi.org/10.1016/j.jmoneco.2020.03.007
30. Lutz, P.: Variation in policy success: radical right populism and migration policy. West Eur. Polit. **42**(3), 517–544 (2019). https://doi.org/10.1080/01402382.2018.1504509
31. Mao, C.-W.: Growth, income inequality, and capital income taxes: evidence from a seemingly unrelated regression model on panel data. Econ. Bull. **36**(3), 1463+ (2016)
32. Menyelim, C.M. et al.: Financial inclusion, income inequality and sustainable economic growth in sub-saharan african countries. Sustainability **13**(4) (2021). https://doi.org/10.3390/su13041780
33. Miller, M.K., Peters, M.E.: Restraining the huddled masses: migration policy and autocratic survival. Br. J. Polit. Sci. **50**(2), 403–433 (2020). https://doi.org/10.1017/S0007123417000680
34. Mukhlis, I., Hidayah, I., Sariyani: Economic agglomeration, economic growth and income inequality in regional economies. Econ. J. Emerg. Mark. **10**(2), 205–212 (2018). https://doi.org/10.20885/ejem.vol10.iss2.art9
35. Neubauerova, E., Tomcikova, M.: The importance of the method of income inequality measure in the relation to fiscal policy—case of V4 countries and Austria. In: Vankova, I. (Ed.) Proceedings of the 12th International Scientific Conference Public Economics and Administration 2017, pp. 233–239. VSB-Tech Univ Ostrava (Verejna Ekonomika a Sprava) Czech Republic (2017)
36. Nowak, W.: Economic development in Asian least developed countries. In: Cingula, M., Przygoda, M., Detelj, K. (Eds.) Economic and Social Development (ESD 2017), pp. 128–137. Varazdin Development and Entrepreneurship Agency (International Scientific Conference on Economic and Social Development), Croatia (2017)
37. Nowak, W.: Development trends in least developed countries. In: Grunwald, N., Zakrzewska, M. (Eds.) Proceedings of the 5th International Scientific Conference on Modern Economics, pp. 41–48. Univ Wismar, Germany (2018)
38. Nugraha, A.T., et al.: The role of infrastructure in economic growth and income inequality in Indonesia. Econ. Soc. **13**(1), 102–115 (2020). https://doi.org/10.14254/2071-789X.2020/13-1/7
39. OECD: Income inequality, OECD (2022). Available at: https://data.oecd.org/inequality/income-inequality.htm. Accessed: 13 Mar. 2022
40. Oestreich, J.E.: SDG 10: reduce inequality in and among countries. Soc. Altern. **37**(1), 34–41 (2018)
41. Osberg, L.: On the limitations of some current usages of the Gini Index. Rev. Income Wealth **63**(3), 574–584 (2017). https://doi.org/10.1111/roiw.12256

42. Ratnawati, K.: The impact of financial inclusion on economic growth, poverty, income inequality, and financial stability in Asia. J. Asian Financ. Econ. Bus. **7**(10), 73–85 (2020). https://doi.org/10.13106/jafeb.2020.vol7.no10.073
43. Rodriguez Guerra, J.J.: European integration and migration policy. Vegueta-Anuario De La Facultad De Geografia E Historia **16**, 461–480 (2016)
44. Santi, S.: Paraguay's new migration policy: human rights and security as pillars for the political treatment of immigration. Estudios de Derecho **77**(169), 213–242 (2020). https://doi.org/10.17533/udea.esde.v77n169a09
45. Schmoll, C.: Controlling irregular migration policy options and unwanted consequences. In: Triandafyllidou, A. (Ed.) Routledge Handbook of Immigration and Refugee Studies, pp. 361–370. Routledge (Routledge International Handbooks), England (2016)
46. Tracker, S.D.G.: Build resilient infrastructure, promote sustainable industrialization and foster innovation. SDG Tracker (2021). https://doi.org/10.1515/9783839441213-001
47. Siami-Namini, S., Hudson, D.: The impacts of sector growth and monetary policy on income inequality in developing countries. J. Econ. Stud. **46**(3), 591–610 (2019). https://doi.org/10.1108/JES-08-2017-0243
48. Singleton, A.: Data: creating the empirical base for development of child migration policy and protection. In: Bhabha, J., Kanics, J., Hernandez, D.S. (eds.) Research Handbook on Child Migration, pp. 334–344. Edward Elgar Publishing Ltd., England (2018)
49. de Souza, H.G.: Effects of economic growth and income inequality on poverty in Brazil. Economia Sociedad Y Territorio **19**(60), 25–45 (2019). https://doi.org/10.22136/est20191312
50. Srivastava, M., et al.: What's in the brain for us: a systematic literature review of neuroeconomics and neurofinance. Qual. Res. Financ. Mark. **12**(4), 413–435 (2020). https://doi.org/10.1108/QRFM-10-2019-0127
51. Suryahani, I., Susilowati, I., Nugroho, S.B.M.: Impact of economic growth per capita and foreign direct investment on income inequality in Indonesia. In: Hadiyanto, M., Warsito, B. (Eds.) 3rd International Conference on Energy, Environmental and Information System (ICENIS 2018). E D P Sciences (E3S Web of Conferences), France (2018). https://doi.org/10.1051/e3sconf/20187310013
52. Talan, G., Sharma, G.D.: Doing well by doing good: a systematic review and research agenda for sustainable investment. Sustainability (Switzerland) **11**(2) (2019). https://doi.org/10.3390/su11020353
53. Tatarko, A., Jurcik, T., Hadjar, A.: How migration policy shapes the subjective well-being of the non-immigrant population in European countries. J. Cross Cult. Psychol. **52**(3), 316–333 (2021). https://doi.org/10.1177/00220221211001531
54. Topuz, S.G.: The relationship between income inequality and economic growth: are transmission channels effective? Soc. Indic. Res. (2022). https://doi.org/10.1007/s11205-022-02882-0
55. Torun, T., Ilgun, M.F.: The effect of political factors on financial development: an analysis of less developed and developing countries. Eskiseh. Osman. Univ. Iibf Derg.-Eskiseh. Osman. Univ. J. Econ. Adm. Sci. **13**(3), 181–204 (2018)
56. United Nations: Build Resilient Infrastructure, Promote Inclusive and Sustainable Industrialization and Foster Innovation, Department of Economic and Social Affairs (2021). Available at: https://sdgs.un.org/goals/goal9. Accessed: 19 Dec. 2021
57. Vasyltsiv, T., Lupak, R., Kunytska-Iliash, M.: Social security of Ukraine and the EU: aspects of convergence and improvement of migration policy. Balt. J. Econ. Stud. **5**(4), 50–58 (2019). https://doi.org/10.30525/2256-0742/2019-5-4-50-58
58. Vedeneeva, V.T.: Poland's migration policy: formation of paradigm (1989–2019). Mirovaya Ekon. I Mezhdunarodnye Otnos. **64**(12), 105–112 (2020). https://doi.org/10.20542/0131-2227-2020-64-12-105-112
59. Yang, Y., Greaney, T.M.: Economic growth and income inequality in the Asia-Pacific region: a comparative study of China, Japan, South Korea, and the United States. J. Asian Econ. **48**, 6–22 (2017). https://doi.org/10.1016/j.asieco.2016.10.008

Sustainability of Risk Management Practices in the Mining Industry: The Case of Using Hedging Strategy in Oman

Ahmed Al Wahaibi

Abstract Mining Industry encounters many financial risks like volatile commodity prices, currency fluctuations, and operational costs variations. The mining sector in Oman is facing a challenge to mitigate the volatility in commodity prices. Previous literature has debated the benefit of hedging as a tool to preserve financial health. Thus, the aim of this research paper is is to investigate how Omani mining sector hedge and why it does so. This study employed a descriptive-deductive research approach (DDM). The research design for this study is a qualitative case study design, using Minerals Development Oman (MDO) as a case study. The empirical data collected through primary and secondary data. Two semi-structured interviews were conducted with the Chief Financial Officer (CFO), and Director of Treasury at MDO in order to get their expertise and opinions of hedging. The sampling population was selected because they have the most and relevant information and expertise that the researcher needs to analyse and answer the two research questions. The findings of this study indicated that Omani mining sector is currently working on hedging against future prices of the copper metal. In addition, the findings showed that the factors for heading at Omani mining sector are for financial, market, and investment reasons. The main theoretical contribution of this study is the new knowledge of how and why Omani mining sector hedge as part of its risk management practices. Future qualitative research may involve all related shareholders and stakeholders to obtain more in-depth information about hedging.

Keywords Sustainability · Risk management · Risk mitigation · Hedging strategy · Omani mining

A. Al Wahaibi (✉)
Modern College of Business and Science, Muscat, Oman
e-mail: 20246738@mcbs.edu.om

B. Awwad (ed.), *AI and IoT: Driving Business Success and Sustainability in the Digital Age*, Studies in Systems, Decision and Control 590,
https://doi.org/10.1007/978-3-031-88874-8_84

1 Introduction

Mining contributes to the global economic growth by providing raw materials for construction, manufacturing, technology and energy. However, mining industry encounters many financial risks such as volatile commodity prices, currency fluctuations, and operational costs variations. Mining operations are particularly prone to financial exposure requiring strong financial risk management procedures to protect the business and keep it profitable [14].

Most of the leading global organizations have risk management strategies that includes hedging as a method for financial risk protection [9]. Hedging is defined as an exposure reduction method that helps firms avoid adverse price action in financial markets. Organizations use different financial derivatives such as forwards, futures, swaps and options to trade risks of commodity price movements, currency fluctuations, and interest rate changes [6]. The rise of the use of technology and data analytics have made hedging even more effective as firms are able to make more informed decisions and react to the market dynamically [1].

The mining sector in Oman offers a unique context to study hedging strategies. Oman is a member of the Gulf Cooperation Council (GCC) and as such its economy has traditionally depended on hydrocarbon revenues. But in recent years, the government has tried to diversify its economic base by introducing initiatives like Oman Vision 2040, which views mining as a major growth industry for sustainable development [2].

MDO is a closed joint-stock company in the Sultanate of Oman owned by 76.8% to Oman Investment Authority, and the remaining 23.2% is owned by Oman National Investment Company. MDO generated 9.1 million Omani Rials as a revenue and reduced its loss to 2.4 million Omani Rials in 2023. MDO group net asset reached 136.4 million Oman Rials in 2023 [24].

In light of the above, this paper aims to explore how the Omani mining sector decide on best hedging. This study addresses a literature gap by examining hedging in the Middle East, which is an area with special economic, regulatory and geopolitical conditions [20]. Although hedging dynamics have been extensively studied in developed economies, there is relatively little research on hedging in developing markets [30]. The research study has practical implications for researchers and mining industry practitioners. The outcome of the study expands the current literature on heading in the mining industry and contribute to creating resilience and sustainability in an uncertain global commodity marketplace.

2 Literature Review

2.1 *Risk Management, Hedging and Sustainability in the Mining Industry*

Organizations are facing a more complex risks in the international markets such as differences in currency and legal policies [26]. Risk management is defined as 'coordinated efforts to manage and direct an organisation's risk' [16]. Hedging is a risk mitigation tool that transfer risks to a third party and through financial instruments [13]. Financial risks could be hedged, for example companies can reduce commodity prices risk, currency exchange risk, or interest rate risk [11]. The literature highlighted four costs that could be reduced to gain shareholder value through a risk management programme in a competitive market which are: "bankruptcy cost, taxes cost, stakeholders cost, and cost of raising funds" [29]. According to Smith and Stulz [27], organizations hedge mainly for three reasons that are related to taxes, financial hardship, and avoidance of risks.

The central logic behind hedging is to establish a position that compensates for all risks imposed by the market [19]. The "perfect hedge" is the hedge that can remove all market risks that could impact the value of an organization such as price risks or interest rate risk. The real-world practices have shown that a "perfect hedge" is almost impossible, and hence the purpose of hedging is to be in a close position to "perfect hedge" as possible. In addition, an organization has the option to take a short-term or long-term hedging position, or both. A short-term hedging occurs when an organization owns, or is going to own, an asset and would like to sell it at a future date. While the long-term hedging is when an organization expects to buy an asset in the future, but it aims to secure the price in the present [15].

Brown and Toft [7] stated that the choice of a hedging strategy got affect by the volatility of the market price and the quantity of the materials and the correlation between both variables. The size of the organization affects hedging decision [3]. In addition, market competition affects the hedging strategy especially when one organization's investment depends on its competitor's investment [21]. Bubere and Shihab [8] found that Swedish mining companies only hedge commodity prices when they encountered situations out of their control. On the other hand, the literature has shown contradictory results on the benefit of hedging in the raw materials industry. Jin and Jorion [18] found that hedging had no impact on the market value of oil and gas companies in the United States of America.

2.2 *Finance, Risk Management, and Hedging Theories*

This section explains the relevant theories for this research paper which are Modern Portfolio theory (MPT), Agent-Principal Theory (APT), and Efficient Market Hypothesis (EMH).

Modern Portfolio Theory (MPT) shows how investors can minimise the risk of a portfolio and how a portfolio can be designed to deliver the greatest return at any given level of risk or expected return. Most investors who choose a portfolio used MPT as a framework. The goal of this theory was to calculate the return that securities would provide in the future and accordingly decide on the selection of the portfolio [22]. The takeaway from this theory is that investors should alternate their portfolios across securities to minimize the overall risk. Nevertheless, one condition that diversification needs in order to occur is that the securities cannot be perfectly positively related because if they are, they both co-move and will have no diversification effect. This theory is relevant as it has affected investors' decision on the choice and management of their portfolio.

Agent-Principal Theory (APT) is more concern with the separation of ownership and control. Large firms are generally had two types of stakeholders, the shareholders who collectively owned the firm, and the upper management who managed and dominated the company for the owners. Often though, there is no strict difference between these two types since some managers would own company shares, and some shareholders might be part of the management team [25]. Nowadays, a lot of companies have multiple owners who might differ in their objectives and vision. The mixed ownership of such companies allows therefore separation of ownership and management. If there is no owner who is powerful enough to hold sole control, then there can be a benefit in engaging agents to work for all different owners so that all the objectives come into alignment. The agency relation is a legal relationship in which the agent or agents (management) has been conferred with a right to exercise control and management over the firm for the good of the principal or principals (owners) [17]. Additionally, the board of directors, who are owners' representatives, determine and enforce the rules and policies through internal controls of the company. External auditors are also employed by the board of directors to ensure everything is conducted according to current legislation and regulations [28].

Efficient Market Hypothesis (EMH) is another finance theory of significance that has influenced the finance field and was derived from the 'random walk hypotheses'. The 'random walk hypothesis' stated that prices ought to fluctuate in an unpredictable way and that no one should be able to forecast future prices and obtain extraordinary returns based on that [5]. This theory was further developed in an attempt to explain the correlation between stock prices, available information and the stock response to additional information. This would entail that prices should capture all of the data that investors have access to in an efficient market and that there cannot be a "beat the market" situation. Therefore, a firm can only get more return if it risks more [12]. This theory can be applied to the commodity markets to account for market efficiency, and its implications on firms. If, for instance, a market would be thought of as having a 'strong-form', then every one of the market players would be privy to all information, including insider information. This would make hedging speculation (suspecting "beat the market") impossible and companies should therefore not use this to hedge. The market efficiency is a very important consideration for companies in setting their hedging positions. Hence, the disparity of efficiency in commodity markets of minerals and metals can affect how individual companies decide to hedge, since all firms do not produce the same commodity [4].

3 Research Methodology

The purpose of this study is to investigate how Omani mining sector hedge and why they do so. The researcher follows a descriptive-deductive approach to theory (DDM). Descriptive-deductive research is a method of research that combines descriptive observations with the use of theories to interpret or comprehend contexts. It uses logical deduction to draw inferences from theories. The descriptive part makes sure the subject is properly described, and the deductive step links such findings to general theoretical models. This approach is particularly useful in the social sciences, law and humanities, where theory can be brought to bear on problems in practice [10].The researcher uses the descriptive-deductive method because it interweaves theory and practice, and delivers scholarly and practical results. Further, the approach helps test the validity and refinement of theory by applying it to real-world scenarios [23].

The literature has listed a number of reasons why firms would hedge. The study aims to highlight if the Omani mining companies are executing hedging strategies in accordance with these reasons or if there are other, more pressing reasons. The research design for this study is a qualitative case study design, using MDO as a case study. The empirical data collected through primary and secondary data form company documents and two semi-structured interviews. The two semi-structured interviews were conducted with the Chief Financial Officer (CFO), and Director of Treasury at MDO in order to get their expertise and opinions of hedging. The sampling population was selected because they have the most and relevant information and expertise that the researcher needs to analyse and answer the two research questions. The interviews allow the researcher to gain deep insights on how and why Omani mining sector hedge against risks. The outcomes of the study will extend the existing knowledge of hedging in mining industry by answering how and why the Omani mining sector hedge. The researcher chooses one case study of MDO as it is the first study in Oman to explore the hedging strategy in the mining sector. Therefore, qualitative case study was chosen to enable the researcher to obtain in depth knowledge and understanding of hedging strategies in Omani mining sector. The following research questions were designed to collect relevant data:

1. What hedging strategy do MDO have?
2. Which factors are important for MDO when deciding it‘s hedging strategy?

4 Findings

The findings addressing the research questions are as follows:

4.1 What Hedging Strategy Do MDO Have?

Based on the analysis of MDO's grey literature, the researcher found out that MDO is a state-owned company, and it is the largest mining group in the Sultanate of Oman. It has a healthy financial position with fixed deposits of about 21 million Omani Rials and has also cash and cash equivalent of about 3.3 million Oman Rials by end of 2023. In addition, MDO's net current assets is about 8.8 million Omani Rials by end of 2023 which was adequate to cover all company's operation expenses. MDO is owned by 76.8% to Oman Investment Authority. The researcher learned from the interviews that Oman Investment Authority set the directives and the policies to MDO which filters to the board of directors then to the executive management of the company.

The interviewees stated that MDO's financial risk management philosophy is to hedge the base metals and in particular to hedge copper price. The MDO's board has approved the hedging strategy for copper metals and the management still working on drafting the hedging policy which will be presented to the board for approval. MDO is intended to build its hedging book as they go especially when they see the market in their favour. The interviewees added that the draft hedging policy gives the board the authority to set the governance for setting the hedging ratio and gives the management the authority to select and recommend the best instruments to execute the hedging strategy. Regarding the financial risk evaluation and calculation process, the interviewees stated that the market highly integrated and the treasury team at MDO will monitor different indexes which may affect the hedging strategy such as United states of America Federal monetary reports, global supply and demand reports, Londen Merket Exchange (LME) and etc., and the main focus is to monitor the volatility chart (implied volatility) which gives indication about the status and the expecting changes of the commodity price. The risk calculations are done once a week.

The interviewees stated that the prioritized financial risks for MDO are as follow: (1) metal price risk, and (2) interest rate risk. They added that currency exchange risk is not critical as the payments and receivables will be in US dollars and there is a national hedge to protect that, and they believe that interest rate risk has reached its peak and is expected to decline in the coming period.

The CFO of MDO stated that the management believed that MDO is youth, and the management aims to protect the coming Mazoon copper project from any financial risk as it is the major project for MDO. It will help in protecting the company form exposing directly and fully to commodity price risk, while the disadvantage of this approch is the cost of hedging which entails paying premium cost to obtain this insurance. The Director of Treasury stated that hedging is a dynamic process, and the treasury team will monitor closely the market to ensure the relevance of the hedging strategy and recommends any changes to the board for approval through the board's risk committee.

The interviewees explained how MDO would hedge and stated that the duration of hedging is split between one, three, and five years. MDO will start with 5 years hedging plan for its copper project, starting from the production year in 2027 and ending in 2032. The hedging process is ongoing in which it is a rolling hedging mechanism. MDO will build its hedging portfolio depends on the market conditions. MDO might embark on its first hedge which is going to be 25% of ratio, then increase it to 50% hedging ratio if the market is in their favour. MDO aim to develop a balanced sort of hedging ration in which to hedge 50% of its copper production and exposed to the market by 50%. The management will choose and recommend to the board the hedging instruments among the four financial derivatives which are forwards, futures, swaps and options to trade risks of commodity price movements. Currently, MDO succeeded to hedge 25% of its copper production for 18 months starting from the production year which is 2027 by using a "put option" hedging instrument called "minimum price guarantee" in which the downside risk of the commodity price transferred to the seller. MDO hedging strategy will focus in using a combination of two financial derivatives which are "options" and "swaps". The interviewees stated that the two recommended instruments work for MDO based on the experience of similar mining companies in the industry and also these two instruments are relevant to MDO's needs. MDO will depend on internal and external expertise to execute the heading policy in which a shortlisted counter party are finalized to sign a formal bilateral agreements. Table 1 presents some of the MDO's hedging strategy features.

Example of the characteristics of MDO hedging strategy reported by the interviewees are as follows:

Table 1 Some of MDO's hedging strategy features

Feature	Description
MDO Financial Position	A healthy financial position
Financial risk management philosophy	To hedge copper price
Authority	• Board to approve the hedging strategy/policy and ratio • Risk Committee to monitor the mark to market (MTM) position • Management to execute the hedging policy and recommends the best hedging instruments
Financial risks Priority	(1) metal price risk (2) interest rate risk
Advantages of the risk strategy	Protect form exposing directly and fully to commodity price risk
Disadvantages of the risk strategy	The cost of hedging
Hedging Ratio	50%
Hedging Instruments	Combination of Options and Swaps
Hedging duration	A rolling hedge mechanism that starts with 5 years duration
Manging team of heading	Depends on internal and external expertise

> We feel that MDO is still youth and Mazoon copper project is the largest cooper project for MDO, and we need to protect the project from any financial risk. The intension is to build up our hedging portfolio as we go. We need to protect MDO's reputation in the market in order to secure our future investments (CFO)

> We excluded both futures and forwards hedging instruments because for futures instrument we will need a big team to manage it, and for forwards instrument it is operationally difficult to do as we need to do hedging for each production/shipment. We will build our hedging book as what we see in the market, for instance, now the market is in our favour, and we embarked on a first hedge which is 25% of ratio then when we see the market further in our favour, we will add another 25% of hedging ratio. (The Director of Treasury)

4.2 Which Factors Are Important for MDO When Deciding Its Hedging Strategy?

The interviewees stated that the main aim for hedging strategy at MDO is to stabilize the cash flow. They added that the copper price has encountered a fluctuation, and hedging is a risk management strategy to mitigate the risk in price fluctuation. In addition, the interviewees stated that 55% of worldwide copper production goes to China, and China's economy is slowing down. Therefore, hedging copper prices is like a protection from the uncertainty if the Chinese economy continue to slow down. Furthermore, it is a method to secure loans, as banks do not usually want to take price risk. Table 2 present some of MDO's factors when deciding its hedging strategy.

Example of the characteristics of the hedging factors reported by one of the interviewees are as follows:

> The main purpose of having hedging strategy is mainly to stabilize our cash flow. We are not doing any hedging for speculation, we are not doing any hedging to take view in the market, it is mainly to stabilize the cash flow. What I mean by stabilizing our cash flow is to mitigate the volatility from commodity. Now we are very exposed to base metals which is gold and copper, and these all LME priced metals, and as you know commodity is just like oil and gas, it can go up and it can go down, and we need to avoid this volatility. So, we are working on a policy to implement certain hedging instruments to stabilize the cash flow, instead of having 1-year good year and 1-year a bad year, which is exposing ourselves to the market. (The Director of Treasury)

Table 2 Some of MDO's factors when deciding its hedging strategy

Financial reasons	Market condition	Investment funding
• Stabilize the cash flow • To manage the volatility of the metal prices	A protection from the uncertainty if the Chinese economy continue to slow down	To secure loans for new investments as hedging the commodity is more appealing for lenders

5 Discussions

The findings indicated that MDO is currently working on hedging against future prices of the copper metal that they will sell in 2027. MDO's board has approved the hedging strategy, and the management are working on the policy draft which will be presented to the board for approval.

MDO depends on internal and external expertise to manage the hedging strategy by having some bilateral agreements with leading counterparties in this field. Regarding the authorities of hedging at MDO, the findings indicated that the decision making of hedging strategy at MDO is governed by the board and the management is responsible in executing the hedging policy. The board sets the hedging ratio while the management executes the policy. Risk Management Committee is responsible in monitoring the mark to market (MTM) which help in knowing the MDO's position in the hedging process. The management seems to have a significant influence on recommending the hedging ratio and heading instruments to the board for approval. Modern Portfolio Theory (MPT) highlighted that investors are obliged to secure their investments against risks through their own portfolio, and MDO's board, which is shareholders representative, seems to implement the same reasoning.

Regarding the reasons for hedging at MDO, the findings showed that the factors for heading at MDO are for financial, market, and investment reasons. According to Agent-Principal Theory (APT), giving the agent (management) control in an organization that has more than one owner could be beneficial in which the management could align the goals of shareholders into one. The literature showed that companies use hedging to decrease the anticipated financial distress costs [27]. Stulz [29] indicated that the market value of an organization increased by implementing hedging because it reduces the cost of fund raising. Thus, the fining of this study found out that the use of hedging does not seem to be a critical reason for raising fund for new investments as the company is in a good financial position, thus the interviewees highlighted that hedging the commodity is more appealing for lenders.

6 Conclusion

The fining of this study indicated that Omani mining sector hedge against commodity price risk to manage the volatility of the copper prices. This is in alignment with Modern Portfolio Theory (MPT) which indicate that investors have the right to select the most appropriate risk management approach that to be managed in their portfolio. In addition, the findings of this study indicated that hedging strategy decision influenced by both the board and the executive management in which board sets the hedging ratio while the management executes the policy. The factors for heading at MDO are for financial, market, and investment reasons, and highlighting that the main aim for hedging strategy at MDO is to stabilize the cash flow. The findings of this study extend the current literature about hedging by presenting new aspects of hedging in the context of Omani mining sector, which could motive other researcher to build on these findings. The main theoretical contribution of this study is the new knowledge of how and why Omani mining sector hedge as part of its risk management which did not exist before in literature. Furthermore, the finings can also provide practical insights to mining practitioners as well. Future quantitative research may study the impact of the hedging strategy on the financial sustainability of the mining industry. Future qualitative research may also involve all related shareholders and stakeholders to obtain more in-depth information about hedging.

References

1. Ahmed, Z., Gill, S.: Data-driven approaches in commodity price hedging. J. Financ. Risk Manag. **9**(3), 145–162 (2020)
2. Al-Muharrami, S., Matthews, K.: Risk management in GCC resource-based firms. Middle East. Econ. Stud. **18**(2), 201–218 (2021)
3. Anniyati, H., Hermanto, H., Hidayati, S.A.: The influence of firm size, financial distress, debt level, and managerial ownership on hedging decisions on manufacturing companies listed on the Indonesia Stock Exchange. Jim Unram-Master Manag. J. **9**(1), 28–41 (2020)
4. Blas, J.: Annual iron ore contract system collapses. *Financial Times* (2010, March 30)
5. Bodie, Z., Kane, A., Marcus, A.J. Investments, 9th edn. McGraw-Hill Education (2011)
6. Brown, G.W., Toft, K.B.: Technological advancements in risk management for mining firms. Glob. Financ. J. **35**(4), 467–489 (2020)
7. Brown, G.W., Toft, K.B.: How firms should hedge. Rev. Financ. Stud. **15**(4), 1283–1324 (2002)
8. Bubere, R., Shihab, M.: Hedging strategies of Swedish mining companies: a qualitative analysis of financial risk management. Stock. Sch. Econ. (2013)
9. Campello, M., Lin, C., Ma, Y., Zou, H.: The real and financial implications of corporate hedging. J. Financ. **66**(5), 1615–1647 (2011)
10. Creswell, J.W., Creswell, J.D.: Research Design: Qualitative, Quantitative, and Mixed Methods Approaches, 5th edn. SAGE Publications, Thousand Oaks, CA (2018)
11. Eiteman, D.K., Stonehill, A.I., Moffett, M.H.: Multinational Business Finance, 14th edn. Pearson Addison Wesley (2016)
12. Fama, E.F., Malkiel, B.G.: Efficient capital markets: a review of theory and empirical work. J. Financ. **25**(2), 383–417 (1970)
13. Hanafi, M.M.: Risk Management, 3rd edn. UPP STIM YKPN, Yogyakarta (2016)
14. Hull, J.C.: Options, Futures, and Other Derivatives, 10th edn. Pearson (2018)
15. Hull, J.C.: Options, Futures, and Other Derivatives, 7th edn. Prentice Hall (2009)
16. International Organization for Standardization: ISO 31000. Retrieved November 22, 2024 (2018) from https://www.iso.org/iso-31000-risk-management.html
17. Jensen, M.C., Meckling, W.H.: Theory of the firm: managerial behaviour, agency costs and ownership structure. J. Financ. Econ. **3**(4), 305–360 (1976)
18. Jin, Y., Jorion, P.: Firm value and hedging: evidence from US oil and gas producers. J. Financ. **61**(2), 893–919 (2006)
19. Johnson, L.L.: The theory of hedging and speculation in commodity futures. Rev. Econ. Stud. **27**(3), 139–151 (1960)
20. Karim, S., Majeed, A.: Adapting risk management practices in emerging markets. Econ. Horiz. **23**(2), 98–116 (2019)
21. Loss, F.: Optimal hedging strategies and interactions between firms. J. Econ. Manag. Strat. **21**(1), 79–129 (2012)
22. Markowitz, H.: Portfolio selection. J. Financ. **7**(1), 77–91 (1952)

23. Maxwell, J.A.: Qualitative Research Design: An Interactive Approach, 3rd edn. SAGE Publications, Thousand Oaks, CA (2013)
24. Minerals Development Oman.: MDO Audit Report and Consolidated Financial Statements (2023). Retrieved November 3, 2024, from https://www.mdo.om/publications.php
25. Ogden, S., Jenkinson, N., Mitchell, J.: The impact of corporate governance on financial performance in the public sector. Public Manag. Rev. **5**(3), 331–354 (2003)
26. Sianturi, C.N., Pangestuti, I.R.D.: Effect of liquidity, firm size, growth opportunity, financial distress, leveraging, and managerial ownership on hedging activities with derivative instruments. Diponegoro J. Manag. **4**(4), 1–13 (2015)
27. Smith, C.W., Stulz, R.M.: The determinants of firms' hedging policies. J. Financ. Quant. Anal. **20**(4), 391–405 (1985)
28. Solomon, J.: Corporate Governance and Accountability, 3rd edn. Wiley (2010)
29. Stulz, R.M.: Rethinking risk management. J. Appl. Corp. Financ. **9**(3), 8–25 (1996)
30. Yang, J., Sun, P., Wu, D.: Real-time analytics in commodity price hedging. Comput. Financ. J. **16**(2), 134–150 (2021)

A Connection Between the Sustainable Development Goals (SDGs) and the Value of Andragogy in Education and ELT

Rami Oweis, Safoura Navari, Salsabil El Kamili, and Ahmed Al Mata'ni

Abstract This study delves into how adult learning theory and practice, known as andragogy, can drive sustainable development in educational settings. Within the realm of adult education, the focus lies on how andragogy can contribute to accomplishing the United Nations' Sustainable Development Goals (SDGs). Specifically targeting SDG 4 (Quality Education), SDG 8 (Decent Work and Economic Growth), SDG 13 (Climate Action), and SDG 12 (Responsible Consumption and Production), this study seeks to explore the implementation of andragogical principles like experiential learning, self-directed learning, and practical application in real-world settings. Since motivation is an underlying force in learners' success, providing sufficient motivation to the learners so much so that they get more involved in various types of sustainable practices and efforts to obtain global sustainable competence is inevitable. This goal is to be achieved through concentration on the notion of andragogy, or adult education which paves the way for learners to become experts in environmental safekeeping, and to get better career opportunities which all can lead to being known as a lifelong learner in any field. One vital significance of maintaining sustainable development in education requires considering the approaches and underlying principles of andragogy in education in general and ELT in particular. The theoretical framework of this study aims to highlight the fact that the future of education can be more promising and sustainable as long as andragogy prerequisites and its pragmatic applications are put into practice in educational settings. To achieve these goals, this paper suggests that appropriate and pertinent frameworks and instructions should be deployed regarding andragogy with a reflection

R. Oweis · S. Navari (✉) · S. E. Kamili · A. Al Mata'ni
Modern College of Business and Science, Muscat, Oman
e-mail: Safoura.navarifarmad@mcbs.edu.om

R. Oweis
e-mail: rami.oweis@mcbs.edu.om

S. E. Kamili
e-mail: salsabil.elkamili@mcbs.edu.om

A. Al Mata'ni
e-mail: mubarak@mcbs.edu.om

B. Awwad (ed.), *AI and IoT: Driving Business Success and Sustainability in the Digital Age*, Studies in Systems, Decision and Control 590,
https://doi.org/10.1007/978-3-031-88874-8_85

on sustainable development. As the last remark, the necessity of encompassing the crucial impact of andragogy in the methods and techniques of English Language Teaching (ELT) as a tool to propagate and achieve sustainable development goals should not be neglected.

Keywords Adult education · Andragogy · Education · Education for sustainable development · ELT · Lifelong learning · SDGs · Sustainable development

1 Introduction

As the modern world revolves around complexity, so does adult education turn into a crucial concept undeniable for attention. As such, a vital task of educators is to engage in analyzing adult learning theories that would possibly contribute to equipping adult learners with skills and knowledge essential to participate actively in sustainable development, especially in the recently acclaimed realm of the Sustainable Development Goals (SDGs) outlined by the United Nations (UN). Given that the theory and practice of andragogy-adult learning- be implemented and cultivated by the requirements of such SDGs in education, this area can demonstrate vast progression in sustainably solving the problems of a global community. Education and environment-related SDGs such as Quality Education (SDG 4), Decent Work and Economic Growth (SDG 8), Climate Action (SDG 13), and Responsible Consumption and Production (SDG 12), can be flourished and materialized more appropriately if combined by the inclusion of andragogy guidelines to foster sustainable development, which is the focal point of this article. The common notions between andragogy and sustainable development goals have the immense capacity to create merits in educational perspectives, job appropriation, and decent life-education integration [1, 2].

Andragogy, with its focus on experiential learning, learner independence, and real-life problem-solving, serves as a useful method of preparing for the challenges to come [3]. What andragogy seeks is to adequately and practically equip adult learners to deal with the challenges of sustainability in life, grow a mindset to enable critical thinking, and implement all those andragogical guidelines to make a better and sustainably balanced future.

This research builds upon a theoretical framework that intertwines the fundamental concepts of andragogy with sustainable development goals. In this stream, andragogical insights of Malcolm Knowles can be interlaced with the rules of sustainability, offering an exact lens through which adult education can be assessed in the context of Sustainable Development Goals (SDGs). The mission can be accomplished by reinforcing the necessity of adult learning in encouraging sustainable behaviors, and emphasizing the revolutionary learning approaches such as self-directed, transformative, and experiential which are of such contribute to adult learners' career paths. It is mainly the personal, social, economic, and environmental achievements of adult learners that can be considered as the best tool to assess the benefits of such kind of education [4]. In this respect, the theoretical base proves to be very critical to understanding the role that adult education plays in the implementation of systemic change toward global sustainability. Thus, the need for adult education as the driving force to put the wheels of education into motion using global sustainability initiatives seems undeniable.

Fields like ELT aim, among other goals, to pave the way for international communication and collaboration. Similarly, experiential learning through andragogical principles entangled with sustainability issues would be possible for adult learners. Bearing in mind that the need for the inclusion of English as a lingua franca in environmental discourses between different nationalities has grown and educators will be obliged to continue the tradition of transforming traditional into modern education, hence the importance of emphasizing a focus on different kinds of education as a sustainable development initiative is not insignificant, including ELT [5].

The integration of sustainability into language learning has a positive impact on the improvement of English and intercultural understanding for a rise in global awareness competencies in dealing with questions of sustainability on a worldwide scale [6, 7]. It is worth noting that the integration of andragogy with ELT will eventually develop an integrated framework for the education of adults toward a sustainable and equitable future. It would be expected that such an improvement will inspire learners to initiate and engage in achieving the SDGs of the United Nations while reaching out for gains toward essential sustainable capabilities.

2 Literature Review

The research on andragogy has blossomed into a vast body of papers in the past couple of decades, and its links to education for sustainable development are still growing rapidly. This growth attests to the role of adult education in addressing the various challenges and threats to development as spelled out in the Sustainable Development Goals (SDGs) of the United Nations. With a philosophy that promotes adult learning, andragogy lays stress on the learner's right to a learner-centered approach respecting the person's experience, motivation, and preference for a self-directed learning model [4, 8].

Considering the territory of sustainability in education and owing to the fact that students are being encouraged to dive into the dynamics of honest-to-goodness, 'real-life issues', and real-world-applied learning, including the principles of andragogy in educational settings, seems promising. Andragogy has been investigated concerning possible roles it may play in increasing sustainable developmental behaviour in people, inhibiting factors that are critical for its integration into educational institutions, and economic contributions regarding possible SDGs such as Quality Education (SDG 4), Decent Work and Economic Growth (SDG 8), Responsible Consumption and Production (SDG 12), and Climate Action (SDG 13). This literature review is conducted for a general perspective on current research into the andragogy-sustainability nexus while considering the theoretical constructs applicable to the practical implementation of adult education [7, 9].

2.1 Andragogy's Function in Adult Education

The notion of andragogy as a theory on adult learning developed by Malcolm Knowles, emphasizes the ways adults differ from children in their learning. The andragogical principles through which adult educational methods take shape include the need to know; self-concept; experience; readiness to learn; orientation to learning; and motivation. Andragogy has been felt to lay a great deal of stress on adult learners and their past experiences, the demand of understanding the context for the subject, and the validation of the study concerning personal and career objectives. Distinct from the traditional pedagogical methods that consider the teacher-as-knowledge-source, this approach to learning focuses on learners. With andragogy, adults work with teachers in learning. Increasing evidence supports the view that with adult learners experiencing some freedom to make their own learning decisions, they will be increasingly more motivated to learn and thus more likely to succeed. Considering that andragogy emphasizes self-directed learning, wherein learners create their own learning goals, this learning, seemingly, aligns with improving adult persistence and lifelong education [10].

2.2 Sustainable Development and Andragogy

The UN looks at sustainable development as an integrated approach to the fulfillment of such basic social and economic demands with environmental concerns vital for humane lives in light of oncoming generations as well as the rapid global changes. In the present scenario, the contribution towards SDG, especially by the means of adult education among different subjects of SDGs, develops greater inclusive societies and opens paths of a more creative economy—[2, 6, 7]. Andragogy became the framework that best served to guide sustainability education toward active and self-directed learning and experiences.

As such, experiential learning, as one of the key concepts of andragogy, may assign pragmatic values to sustainable development education with engaging hands-on activities, such as conducting workshops on renewable energy, environmental evaluation, and grassroots projects on sustainability. These activities train the learners and empower them with practical skills for immediate problems like resource depletion, and climate change, which eminently is the essence of achieving SDG 12 Responsible Consumption and Production, and SDG 13 Climate Action. Andragogy puts much emphasis on the internalization of sustainable behaviors among adults with much intensity, hereby allowing the students to go ahead and practice their freshly acquired knowledge in their daily activities the way they should [4].

2.3 *Difficulties in Using Andragogy in Education for Sustainable Development*

Another option is a space for andragogy to engage in a vast range of sustainable developments; however, many barriers curb its functional possibilities. An important cause of concern is that qualified educators are few who are well-trained with methods of andragogy, especially in the area of education for sustainability [3, 10].

While there are ample opportunities for andragogy to advance sustainable development, several barriers hinder its extensive implementation. A key challenge lies in the limited availability of qualified educators trained in andragogical methods, particularly within the realm of sustainability education. The transition towards collaborative, learner-centric approaches could face obstacles due to the prevalent use of conventional, instructor-focused teaching methodologies, as indicated by research [10, 11].

2.4 *The Connection Between the SDGs and Andragogy*

As indicated in several studies, discussing the connection between SDGs and andragogy, and underlining the role of adult education in contributing to global sustainability goals are of immense significance. Actually, SDG 4 about Quality Education is very direct: there is no difference in whether continuous learning opportunities can be defined in equal terms for all people. It is, therefore, through andragogy that education can be developed in such a way as to meet the needs of adult students in their quest to implement sustainable practices in areas like business, agriculture, and energy. Learning via andragogy will question the reality of practical experience and application since it would arm adults with the proper armamentarium to become champions for sustainable programs [3, 12].

2.5 *English Language Teaching (ELT) in Sustainable Education*

As an immense body in education, the English Language Teaching (ELT) field is increasingly becoming prominent for the type of education to be delivered in a sustainable, integrated manner and to meet the needs of the entire globe to address sustainability issues. ELT is a core concern for education because it swathes students with language inputs that eventually help them address sustainability issues at a wider level. Access to English language empowering ways brings students to their doorstep and opens the doors to important information, dialogues with other cultures, and significant contributions to world sustainability initiatives [3, 7]. Sustainability discussions ranging broadly from climate change to sustainable development objectives will use English widely and therefore place ELT as an important mechanism for individuals to deal with such issues.

Analysis of the previous studies demonstrates quite evidently that merging ELT with sustainability content has twin benefits: improved language skills while bringing up social and environmental consciousness together. Such genuine materials regarding sustainability issues around social justice, renewable energy, and climate change have instructors produce student interest for motivating their learning in both the target language and subject-based material related to sustainability. Employing studying the English language, adult learners will get the necessary linguistic competencies for becoming involved in global forums as a means to advocate for the enforcement of policies on sustainability along with taking part in communal-based sustainability actions. On the other hand, ELT speaks of a communicative vein. It thus correlates well in principle with issues of experiential learning in adult education and bridges the gap between language learning and education for sustainability [5].

It should be noted that currently, there is a very strong linkage between ELT and the development of social responsibility along with critical engagement within the framework of sustainable education. Introduction of the issues of sustainability into the ELT curriculum will urge students for critical analysis of the various local and global problems of such a nature as inequality and poverty, and degradation of the environment. This in return makes them reflect on the contribution they are going to make for the sake of a sustainable future. Besides, this will widen their worldview. Such an opportunity allows adult learners to relate coursework with their life experiences and to provide it with social and individual conception [7, 11]. The principles of andragogy may allow for these issues and even in combination with the guidelines of ELT, they can raise learners' autonomy and aid them in their life journey.

By using the language learning process to engage students in real-life sustainability challenges, ELT provides responsibility to learners to allow them to effect change within their communities. In this regard, an important strategy for creating a citizenry that is informed, socially aware, and globally conscious-one that can contribute to the country's sustainable development since the important aspect of international standards of sustainability could be the integration of ELT and sustainability into adult education. This comprehensive adult learning approach therefore equips the learners with the linguistic skills, knowledge, and competencies to address the challenges of sustainability both locally and internationally.

3 Methodology

The current research is a qualitative study conducted on andragogy in its connection with sustainable development. It covers everything from academic books, papers, and articles that have been obtained from various sources such as government publications, international organizations, and scholarly journals. The primary intent of such a literature review has been to construct a conceptual understanding through the synthesis of findings of various studies on andragogy concerning its bearing on sustainable development in adult education.

In this regard, emphasis was given to those articles that focused on the integration of andragogical principles in sustainability education and those that propagate the combination of ELT and sustainable development practices. Synthesis of these materials provides a theoretical framework for integrating adult learning methodologies into sustainability education and determines the role of andragogy in response to global sustainability challenges. An andragogical perspective framework, concerning the ELT approach, brought the emphasis on experiential and self-directed learning that must be applied practically. While it encourages learners to engage with global issues that are crucial for sustainability more intensely, this mode of educational approach deepens skills in the language at the same time. The value of the integration of this andragogy with their teaching of the English language is that it may help learners apply theoretically learned knowledge to practical life.

4 Discussion

Andragogy has become a significant force in sustainable development education through emphasizing active participation and practical use due to the fundamental norms it upholds toward adult learning. The other key strength of andragogy is the emphasis on experiential learning, which is a foundational method in sustainability education. This mode of learning, contrary to classroom learning, guarantees instant exposure for adult learners to environmental, social, and economic problems. For instance, citizens may be directly involved in work with community-based sustainability projects on renewable energy, sustainable agriculture practices, or a recycling program. Such real-life engagement situations enhance the relevance and impact of education in helping to make complex ideas of sustainability plausible for students and enabling them also to apply these in day-to-day situations [1, 13].

Moreover, it emphasizes self-directed learning. This particular aspect needs constant addressing concerning dealing with complex matters about sustainable development. Adult learners constantly have to upgrade themselves regarding competencies and knowledge against newly emerging environmental and social challenges. By adopting the principles of andragogy, adults will be able to take responsibility for their progress concerning personal and professional development.

Although andragogy has a great potential to further develop teaching about sustainable development, several principal issues could be overcome in the real process of its implementation. One of the main problems relates to the fact that there is a serious shortage of qualified teachers who are experienced in the methods of andragogy in general, and particularly in the sustainability education context. Many educators in developing countries still rely on traditional teaching methods that are beyond the needs of sustainable development, such as passive learning through lectures and memorization, which are inadequate to solve complex, interdisciplinary aspects that require critical thinking, problem-solving, and collaboration [5]. Beyond this, economic barriers, lack of time, and insufficient technological resources are major obstacles for adult learners in pursuit of quality education, especially for disadvantaged groups. Overcoming these hurdles requires that educational institutions and policymakers emphasize investment in teachers' professional development and further assure equitable and affordable learning opportunities for all.

The next view regards the large andragogy challenges in education for sustainable development which deal with the ways to root both theoretical knowledge and practical skills into the curriculum. While andragogy calls for relevance and real-world application, most sustainability programs remain firmly rooted in academic knowledge without giving as much importance to hands-on learning [8]. In this regard, educators must develop curricula that will not only teach concepts of sustainability but also provide students with ample opportunities to apply the principles of sustainability in their daily lives and professional practices. Some of the strategies that can be adopted for engaging students in solving the challenges of sustainability include project-based learning, case studies, and fieldwork. The integration of theory with practical experience allows the adult learner to be better equipped to perform leadership duties concerning the sustainability process within both the communal and professional spectrums.

Also, the implementation of andragogy in sustainability education should consider cultural and geographical diversities. Andragogical methods may turn out to be effective due to the different challenges, resources, and cultural views of sustainability in different societies. For example, in some regions, a tradition of communal resource governance or agricultural practices may exist, hence offering a rich potential for the integration of sustainability into education.

In sum, it is not easy to put andragogy into practice, faced with challenges like institutional barriers and highly trained instructors. Most learning institutions are still anchored on traditional pedagogies that do not support critical thinking and active participation which is the essence of sustainability education. For instance, educational institutions should ensure access to programs by attracting diverse students, restructure their curricula to infuse the principles of sustainability, and offer training to teachers to professionalize them.

5 Conclusion

This study illustrates how andragogy acts as a sound basis for sustainable development through the education of adults, anchored in experiential learning, self-directed learning, and real-world learning. Andragogy nurtures critical thinking, problem-solving, and practical skills of adult learners to tackle such key sustainability challenges as economic development, climate change, and sustainable consumption. Despite existing implementation challenges, most especially those relating to institutional resistance and accessibility, andragogy has enormous potential for contributing to the attainment of the United Nations SDGs. This study investigates the potential of andragogy in terms of the SDGs through adult education, emphasizing how its principles address the needs of adult learners in the context of global issues. The review of literature provided evidence that andragogical methodologies, emphasizing experience and self-directed, learner-centered education, are particularly appropriate for developing the competencies so needed for sustainable development: the needed skills of critical thinking and problem-solving.

More significantly, when andragogy combines with sustainability education, a person can interact directly with a real-life scenario and use the gained knowledge effectively. It also contributes not only to the development of a person but also to the general goals of the UN regarding society as a whole. However, the distinct sociocultural and educational environment should be taken into consideration when considering the applicability of andragogical principles in the Arab region, specifically in Oman. Like other Arab countries, Oman is undergoing rapid demographic shifts, including having a sizeable and increasing adult population that is eager to pursue further education for employment and to contribute to the nation's development.

It will help the population acquire the skills related to emerging green enterprises and thus support the initiatives of sustainability. It will also fall in line with Oman's endeavoring towards a broader-based economy and sustainable development. Economic diversification of Oman with the conservation of the environment and involving the community falls in line with the principles of andragogy, which emphasizes practical and experiential learning opportunities.

Effective inclusivity in the andragogical programs will ensure that adult learners with disadvantaged or rural backgrounds participate massively in sustainability projects. For example, countries like Oman need to modify their traditional educational frameworks if they have to effectively accommodate even adult learners. While conventionally, regional educational institutions have moved in the direction of teacher-centered learning, there is an emerging awareness of the importance of learners' centricity and experiential learning methodologies, especially with the growing clamor for lifelong learning and practical skills.

Although andragogy holds a lot of promise in terms of enhancing sustainable development in Arabic countries, further research and increased funding will be necessary in order to develop its scope in those contexts. Research should cover the unique potentials and challenges of an andragogical approach in Arab countries, as well as define the best practices that are locally and culturally relevant for sustainability education. Moreover, cross-cultural research comparing and contrasting the impact of andragogy in different regions could provide a useful look at how principles of adult learning do or do not work in the frame of sustainability education across different cultures. In conducting such research, policymakers, and educators are better equipped to refine strategies in adult education that ensure sustainability education is both rooted in the local context and addresses global issues.

Policymakers should ensure that education becomes all-inclusive to widen participation in sustainability programs. This would involve the creation of inclusive education programs, taking into consideration the varied learning needs and styles of individuals, coupled with Oman's culture and values. In this regard, the shift by Oman towards a greener economy allows applying andragogy into the adult population for those skills needed in the new sustainable industries. Adult education in these countries will, therefore, be vital in the opening of better career opportunities once eco-friendly transitions are brought into effect in important areas like sustainable tourism, water conservation, and renewable energies. The andragogical method applied within such contexts may be serving effectively in developing students' practical problem-solving abilities related to complex environmental topics. By empowering such adults to participate actively, Oman can build a skilled workforce required to support the country on all levels of sustainability-albeit local or global.

Likewise, while adult learning programs are focused on reciprocity, local experience, and action learning, these programs have an imperative basis in the solution finding of key environmental and several other social problems affecting them at the community level. From this perspective, using andragogy presents promising opportunities that support sustainable economic development in the Arabic regional nations. This again mainly depends on its potential working relation within society or vicinity. Countries like Oman, coupled with other Arab nations, can forge ahead with the SDGs by responding to the specific needs, motives, and problems of adult learners in this region and integrating indigenous cultural values and traditions into sustainability education. There is a need for constant collaboration among researchers, educators, and policymakers on how best to embed andragogical principles into sustainability education. They thus facilitate the bringing up of a well-informed, resilient, and empowered adult population that ensures equality and sustainability for all concerned in the future.

6 Suggestions for Future Research

The long-term effects of andragogical methods on the adoption and promotion of sustainable practices by adult learners should also be researched further. There is a need for longitudinal studies focusing on adults who have participated in sustainability education programs to determine how knowledge gained is being applied in personal and professional lives. It would be even more practical to focus the research efforts on developing resources and strategies that can be used to embed andragogical principles into sustainability-focused curricula of educators. Investigation into possible benefits that digital learning platforms can bring in for andragogy-based sustainability education, especially in areas considered remote or with poor access to traditional classroom facilities, would also be of importance.

Therefore, it is concluded that future research efforts are highly needed from the perspective of andragogical model efficiency within sustainability education, especially at the Arab region level. Although andragogy is based on a sound theoretical foundation, a lot of comprehensive research is required to understand better how this applies across a wide set of cultures and geographical environments. For instance, the ability to understand how adult learners in Oman and other Arab countries engage with the subject matter of sustainability using andragogical methodologies can further pedagogical methods to ensure optimum effectiveness. Longitudinal studies that can measure the practical effects of using andragogy to attain selected SDGs are also needed. Longitudinal analysis would thus allow explanations to pinpoint persistent contributions of sustainability education to eventual behavior, professional development, and community involvement of mature adult graduates. The rigorous evidence could be used for the design and conceptualization of appropriate and culturally oriented programs.

Finally, research studies need to explore the potential of cross-cultural comparisons in assessing the effectiveness of andragogy in sustainability education across different geographical regions. Contrasting the learning experiences of adult learners in Oman with those in other Arab nations, or even globally, can provide valuable insights into the universal applicability of andragogical concepts across diverse cultural settings. By offering guidance on adapting these pedagogical approaches to the specific needs of various learner groups, such comparative studies can help identify optimal strategies for integrating andragogy into sustainability education and language teaching programs. This exploration has the potential to enhance the development of more effective sustainability education initiatives not only in the Arab world but also globally, by shedding light on how cultural factors influence the efficacy of andragogical techniques in adult learning contexts.

References

1. Knowles, M.: The Adult Learner: A Neglected Species. Gulf Publishing, Houston (1973)
2. United Nations: Transforming Our World: The 2030 Agenda for Sustainable Development. United Nations (2015)
3. Mezirow, J.: Learning as Transformation: Critical Perspectives on a Theory in Progress. Jossey-Bass, San Francisco (2000)
4. UNESCO: Adult Education for Sustainable Development: Global Perspectives and Trends. UNESCO, Paris (2020)
5. Fisher, A.: Experiential learning for climate action: empowering adults for a sustainable future. Environ. Educ. Res. **25**(1), 12–28 (2019)
6. Jackson, L., Brown, R.: Responsible consumption in adult education: a framework for sustainability. J. Environ. Educ. **49**(2), 133–149 (2018)

7. Benson, C.: Language education for sustainable development: the role of English language teaching (ELT) in promoting sustainability. Sustainability J. **11**(21), 6049 (2019)
8. Brookfield, S.D.: The power of Critical Theory: Liberating Adult Learning and Teaching. Jossey-Bass, San Francisco (2005)
9. Knowles, M.: Andragogy in Action: Applying Modern Principles of Adult Learning. Jossey-Bass, San Francisco (1980)
10. Thompson, L.M.: Barriers to implementing andragogy in adult education systems. J. Adult Contin. Educ. **25**(4), 240–255 (2018)
11. Dewey, J.: Experience and Education. Macmillan, New York (1938)
12. Hart, K., Uvin, P.: The role of adult education in fostering sustainable development. Adult Educ. Q. **67**(3), 225–240 (2017)
13. Rogers, A.: Teaching Adults. Open University Press, Buckingham (2002)
14. Fisher, A., Wells, K.: Sustainability in education: Challenges and solutions for adult learners. J. Sustain. Educ. **10**(3), 15–32 (2019)
15. Rajagopal, M.: Andragogy and its relevance in adult learning for sustainable development. J. Adult Educ. **44**(1), 50–67 (2016)
16. Holton, E.F., Swanson, R.A.: Foundations of Human Resource Development. Berrett-Koehler, San Francisco (2004)
17. Sifakis, N.: The role of English in promoting sustainable development: a global perspective. J. Engl. Acad. Purp. **28**, 30–43 (2017)

Control of Collaborative Robot Using a Hybrid Impedance and Admittance Control to Build Kinesthetic Learning Industrial Robot

Mostafa Gamal, Mohammed Aad Khudhair, Hothefa Shaker Jassim, and Husam A. Neamah

Abstract This work presents a hybrid control system that consists of Impedance and Admittance Control systems. This control system is developed to help in kinesthetic learning for the robot's control. The control system was simulated and tested, to ensure its credibility. A proof of concept of four degrees of freedom robotic arm was developed, to test the proposed control system and its functionality. By experimenting with the system and the robot; and analysing the resulting values. It was possible to conclude that the system is successful, as it remained within an acceptable error range. This was calculated while teaching the trajectory and while repeating it. The average error is less than 6% in the manipulation mood and less than 8.6% in the execution. The robot was easily manipulated, trained, and repeated the movement with high accuracy.

Keywords Robotics · Impedance control · Admittance control · Kinesthetic learning

M. Gamal · H. A. Neamah
Department of Electrical Engineering and Mechatronics, Faculty of Engineering, University of Debrecen, Debrecen, Hungary
e-mail: husam@eng.unideb.hu

M. A. Khudhair
Faculty of Informatics, University of Debrecen, Debrecen, Hungary

College of Engineering, National University of Science and Technology, Dhi Qar, Iraq

H. S. Jassim (✉)
Modern College of Business and Science, Muscat, Oman
e-mail: hothefa.shaker@mcbs.edu.om

H. A. Neamah
Department of Business Management, Al-Imam University College, Balad, Iraq

B. Awwad (ed.), *AI and IoT: Driving Business Success and Sustainability in the Digital Age*, Studies in Systems, Decision and Control 590,
https://doi.org/10.1007/978-3-031-88874-8_86

1 Introduction

During the last decades, Robots has become a huge contributor to human civilization and modern society [1–3]. With the revolution of the robot and computing technologies, the need for a modern way of controlling the robot systems became more of a necessity than at other times [4]. Among the highly used controlling method, Human–Robot teamwork had played a major role in improving the functionality of the robot. For that, the term co-pot was introduced in a 1997 US patent entitled “Co-bots” [5]. Advances in robotics include B-spline trajectory optimization for speed[6], Organic Human–Robot Interaction (O-HRI) [7], reinforcement learning for adaptability [8], advanced sensors for precision [9], and novel algorithms for efficiency [2, 10, 11]. Teaching a robot is known as Kinesthetic teaching, also referred to as leading-through programming. It is a user-friendly approach that does not require previous programming or robotics-related knowledge, it can be used to teach the robot a specific skill via direct physical interaction [12]. Kinesthetic teaching often uses backdrive motors with encoders to record joint movements, but this method is limited to small motors due to manual manipulation requirements. A preferred alternative involves active control with sensors placed on the robot to detect external forces, typically using force/torque sensors. Authors developed a hybrid system of Impedance-Admittance [13] Control, with the use of Electromyography [14], as the controlling signal was implemented. In another research [15], a they proposed controller for an upper limb exoskeleton used they used an admittance control system to do so, this system with implemented with a help of an additional PID control logic. The defining moment of these motors can be found in their small scale. In previous works [16] an approach to reach a kinesthetic control was developed, where they managed to teach the robot not only the torque but the trajectory position. Another implementation was on a humanoid robot [17]. The user might teach trajectories and keyframes in their work, resulting in specific positions. They propose that for certain and specific events, Instead of specifying the trajectory, the controller may merely wish to define the target point. On the other hand, the author of [18] used a different approach where they offer a control system classified as admittance, that determines the required joint velocities using the predicted torques in each joint as inputs. Their approach was to control the robot using a sensor-less approach. Their approach gives an estimation of the force projected on the robot. Multiple co-robots were produced to create that unique experience of the Human–robot interaction. For example, OUR-1 developed by Smokie Robotics [19] it was made possible to control the position of the working end using user control. Also, the Baxter [20] robot developed by Rethink Robotics in that robot the user can guide the robot into a specific path. In this work, the resulting value of moving the robot actuators, and the force are registered in the robot operating system to be executed later. A hybrid control system was developed for a 3D-printed, four-degree-of-freedom industrial robot. Denavit-Hartenberg parameters were calculated for kinematics, and a kinesthetic learning (KL) algorithm with force sensors enabled signal registration and replay. The robot operates in manipulation and execution modes using admittance

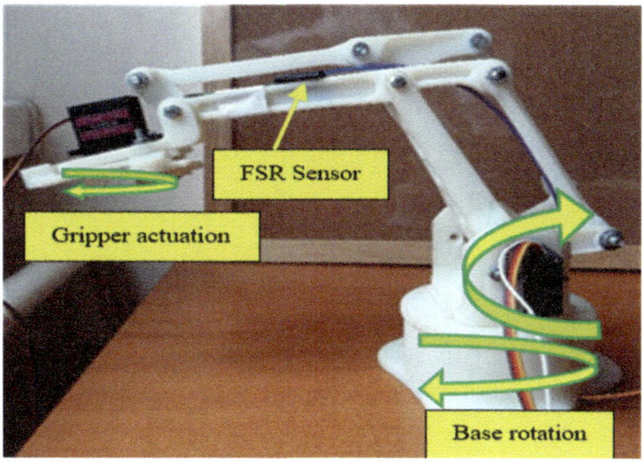

Fig. 1 The used 3D printed robot final assembly

and impedance control, with algorithms implemented in MATLAB and tested in Simulink. A GUI facilitates trajectory tracking and control, while Python visualizes saved paths. The system supports both software and physical interaction, ensuring accurate task execution even with external disturbances.

2 Design Specifications

The robotic arm is a simplified model to the industrial robot ABB IRB 460 [21]. The model is to be operated by an Arduino microcontroller [22]. It will be able to grasp, pick up, and shift various items based on their weight and shape. The manipulator, which is the gripper, is built to pick up geometric objects such as a box. The robot has three degrees of freedom in addition to one degree related to the gripper. The actuators used are micro servo motors controlling the movement. The 3d printed robot has a total of 19 parts individually. Figure 1 shows the final assembly of the robot.

3 Control System Implementation

Impedance and admittance work in the opposite way. While the first prevent the errors and noise that occurs to the system due to environmental changes. The second allows the interaction between the user and the robot. If we are looking at an n-DOF system, whose working end has a single DOF in which we have a mass m, and x will be the mass generalized inertia and displacement. We will assign a control force F and Fext are the control force and the outer force acting on the mass via the environment. Thus, the motion equation can be written as.

$$m\ddot{\chi} = F + F_{ext} \tag{1}$$

The object of both control systems Impedance and Admittance is aiming to design the control force F that from a trajectory X_0, it will create a relationship between both the external force *Fext* and the deviation $e = (X - X_0)$ Then a second order liner relationship can be driven as.

$$M_d\ddot{e} + D_d\dot{e} + K_d e = F_{ext} \tag{2}$$

where the positive constants M_d, D_d and K_d respectively represent the desired inertia, damping, and stiffness [23].

3.1 Admittance Control

The controller is designed to be a mechanical admittance, as the plant is position-controlled, and it has a mechanical impedance behaviour. The controller can be implemented using a PD controller with K_p and K_d as gains, the system dynamics equation where the first is the Position constant and second is the speed constant. Admittance control and be described as follows [24].

$$U = K_p(X_0 - X_d) + K_d\left(\dot{X}_0 - \dot{X}_d\right) \tag{3}$$

where U is the Input, X_d is the current position, X_0 is the reference position.

In our case, the force is from a human or humanoid source. The Admittance control is implemented by calculating the force the human applies to the joint. This measurement is possible by the use of force sensors on the robot body. Thus, that will allow the movement, the motor will change position according to the input. The controller will regulate the movement responding to the force.

3.2 Impedance Control

In contrast with the Admittance control, the impedance Control is a relationship between the external force interacting with the system and the position error. When a position is desired, and the process is interrupted. The force will be increased to correct the position error. This can be donated using the following equations.

$$F = K_p(x_d - x) - K_d\dot{x} \tag{4}$$

$$m\ddot{x} + k_d\dot{x} + k_d(x - x_d) = F_{ext} \tag{5}$$

$$M_d(\ddot{x}_d - \ddot{x}_0) + D_d(\dot{x}_d - \dot{x}_0) + k_d(x_d - x_0) = F_{ext} \tag{6}$$

Equation (6) is the final equation used for the impedance system.

where M_d, D_d and K_d are the desired impedance coefficients that result in adjusting the required position. M is the mass of the system, b is the damping, K is the rigidity, Xd is the current position, X0 is the reference position, and Fext is the external force.

4 Applying the Control Theory

In this work, a mutual system of Impedance and Admittance is introduced. The controlling Method is based on two main things.

Firstly, the trajectory measured values, which is given from the feedback of the servo motors as an angle. Secondly, the amount of force applied on the force sensors, for that another version of the previously mentioned equation is introduced.

4.1 Admittance Control

The Admittance Control calculate the required angle because of the environment interaction via the force sensors,

$$\theta_x = \theta_{x-1} + K_f\left(S_f - S_{f-1}\right) \tag{7}$$

where the θ_x is the desired angle, θ_{x-1} is the previously recorded angle, K_f is the Force constant, S_f is the sensed value, and S_{f-1} is the previous value. So, the required angle is calculated using the measured value from the force sensors.

4.2 Impedance Control

Impedance In similar yet the opposite, the Impedance Control calculate the change of the angle. This change in the angle occurs because of the change in the environment. This change in the environment can be measured via the force sensors. The controller then calculates the required correction angle and feed the motors with required changes. The definitive angle θ_d can be calculated as follows.

$$\theta_d = \theta_x + K_{cor} \times m \tag{8}$$

where θ_x is the originally required angle, K_{cor} is the correction constant responsible for determining the angle shift concerning the load.

5 The Hybrid Control System Algorithm Implementation

The System was developed using multiple platforms. The main system was developed on Jetson Nano with MATLAB for the calculation and Python for the 3d-Simulation. The interaction between the system and the robot sensors and actuators were done using Arduino. The date was sent and received via serial communication. Lastly, the testing of the Control system was done using Simulink simulation. In this work, a mutual system of Impedance and Admittance is introduced. the controlling Method is based on two main things. Firstly, the trajectory measured values, which is given from the feedback of the servo motors as an angle. Secondly, the amount of force applied on the force sensors, for that another version of the previously mentioned equation is introduced.

(1) *Force sensors calibration and force measure*

The FSR was connected to a basic voltage divider, and the resulting voltage indicating force was read using the embedded controller because of the circuit. To calibrate the sensor, various masses were applied to different parts of the sensor's surface.

(2) *Potentiometer Calibration and Position Measurement*

To reach an accurate position measurement a feedback signal is calibrated. By setting the motor values to zero-degree (home position), then increasing the steps gradually one step at a time. The value from the feedback signal is read and displayed along with the degrees passed. Till once it reaches the maximum degrees, in our case it was 180°, the inner potentiometer has a maximum value of 670 Ohm. Using Arduino, because of the potentiometer changing a corresponding voltage were measured. To convert the measured voltage into resistance reading the following equation is used;

$$\mathrm{Pr} = (\mathrm{Vs} * \mathrm{Vpin})/\mathrm{Mv}$$

where Pr is the required measured value, Vs is 5 V supplied by Arduino microcontroller. Vpin is proportional to the amount of voltage being applied to the pin. And Mv is the measured voltage. The result is shown in the following chart show that there is a positive relationship between the measured resistance and the angle of the motor.

(3) *The kinesthetic teaching algorithm Implementation.*

Because we can readily get each joint angle, as well as the location of the TCP, we will use this method. The way I implemented the Kinesthetic teaching in the robot is as follow. Firstly, When the robot is manipulating mood, the angles are sent to be processed and later saved, as the system continuously reads values from each of the four inner potentiometers of the motors. These values are saved in the form of angles. Secondly, Mapping these values into Degrees of movement. Thirdly, Sending the mapped values as an input value to the motors.

(4) *The Motion recording mechanism*

The code creates multiple arrays to save the current position of the servos. These positions are imported from the potentiometer real-time calibration loop. The values of the potentiometer are mapped and stored. Then the new position is compared to the old one to make sure that the motor moved and to save storing space.

(5) *The Controlling Assisting Algorithm.*

As explained before in the Admittance control section. Once an outer source of force is interacting with the robot, the loop responsible for force calibration measures the amount of force is interacting with the servo motor at that exact moment. Then as a result of that, and depending on the amount and direction of the applied force measured, it assumes where it should move. If the force is not applied anymore, the system corrects its position using the stored.

(6) Execution Algorithm

The function of this part of the code is to restore the recorded values from the arrays. After recalling the values, they got applied to the servo motors. In case there is any disturbance in the environment, for example, an outer force has interacted with the robot. The error is fixed using the Impedance Control, using the reference angle and force recorded by the system.

6 Tests and System Evaluation

Based on the equation mentioned before and the illustration of the implementation of the system. A MATLAB code was developed with the aim of providing a way of simulation to the control system, the result we can obtain from the MATLAB simulation is later are used to tune the control system The block diagram of admittance control is shown in Fig. 2.

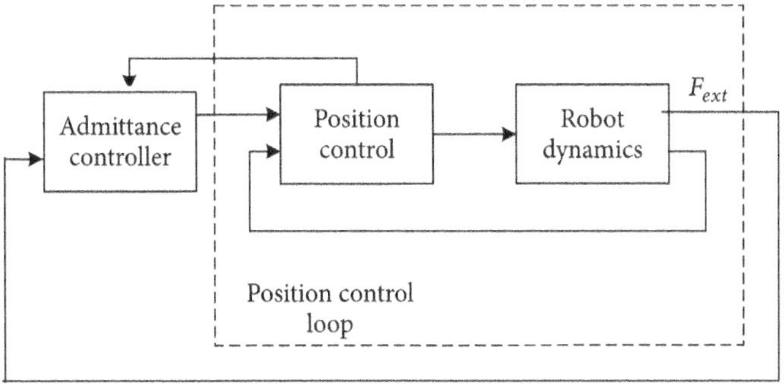

Fig. 2 Block diagram of the admittance control

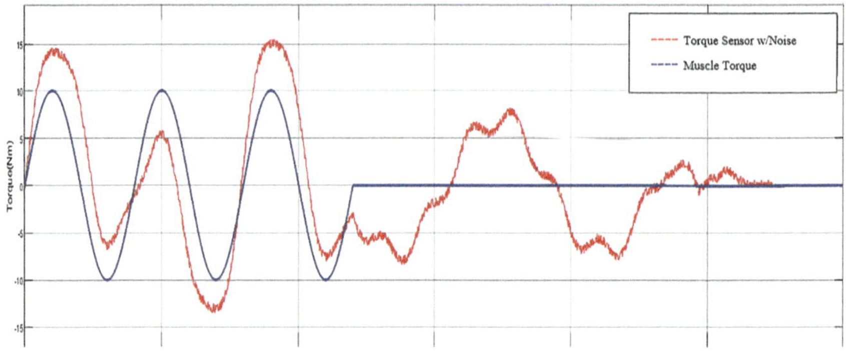

Fig. 3 The resultant Torque required to ensure the motion success

(1) Testing and simulation of the Admittance control

The MATLAB model initializes key control parameters, including Virtual Mass, Damper, Spring, and Environmental Stiffness, then processes the current angle θ and force signal to calculate the required angle and control signal. The system accurately predicts output angles, minimizing errors and achieving stability quickly, as demonstrated by torque and correction signal graphs, Fig. 3 shows the resultant torque required to ensure motion success.

As we can see in the above graph the system works on minimizing the error in the robot torque by measuring the human body torque and estimating the required force to do the task. The result is the torque sensor error and the difference between it and the required one. However, we can see a slight overshoot, in the end, this can be fixed by better tuning the system.

6.1 *Testing and Simulation of the Impedance Control*

The testing of this system works similarly to the Admittance, hence the measurement is needed to the system about the current position of the robot TCP. Also, whether or not the system is under human interaction. Based on that the system resulted from value change. Also, the same initial values are manually added to the simulation Virtual Mass, Virtual Damper, Virtual Spring, Environmental Stiffness, System Inertia, System Damper, and System spring.

Test 1: Testing Kinesthetic teaching function; Admittance control: The Kinesthetic teaching function was highly successful, with the robot accurately following the trained path. Using Admittance control, the TCP trajectory, located at the gripper's center, was recorded during manipulation mode and repeated ten times without external force. The process achieved a 7.694% average error, as shown in Fig. 4a, b.

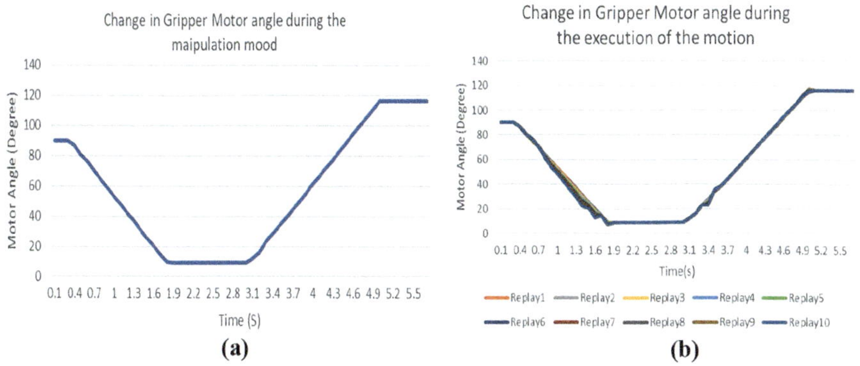

Fig. 4 **a** Recorded data for the motors, **b** execution gripper motor's angular velocity result

Test 2: In the following test, the robot was taught to draw a simple letter. For this example, the letter X was taught. In Fig. 5a, we can see the trajectory taught to the robot. In Fig. 6b, we can see the result of the executed function related to the previously saved trajectory. The gripper was adjusted to hold a pen and to follow path A, B, C, and D.

Test 3. Pick-and-place tasks involved two robots, with one transferring the test subject to the other, which then moved it to the desired position. In the second part of the test, an external force interrupted the robot's execution. Initially, the trajectory taught to the robot showed significant overshoot due to motor bouncing, as seen in Fig. 6a. The Impedance control calculated and reduced the error angle, and after multiple tests, tuning the correction *constant* (K_{cor}), improved system efficiency. The refined system achieved higher stability, with errors reduced to below 5.2%, as shown in Fig. 6b.

Test 4, Like test 2, the robot was taught to draw a simple latter. In this example, the letter X was taught. However, in this test, the robot was interrupted using external

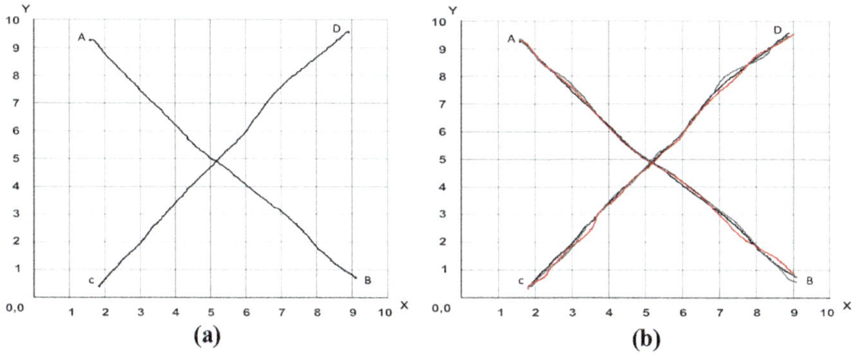

Fig. 5 **a** 2D graph of the desired path, **b** 2D graph of the executed paths

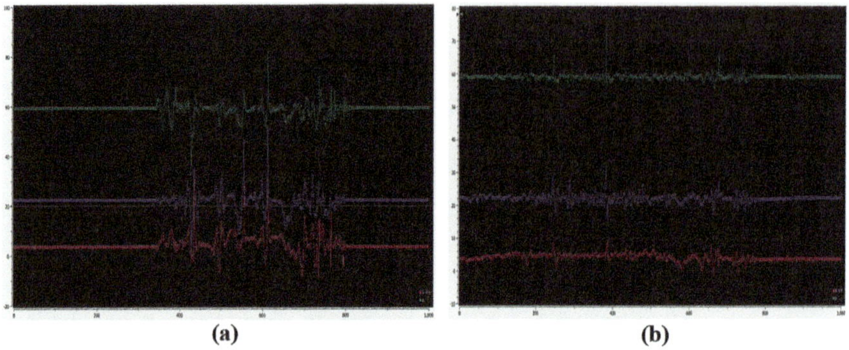

Fig. 6 **a** Error in X, Y, Z positions of the TCP. **b** The result step response after tuning the correction constant

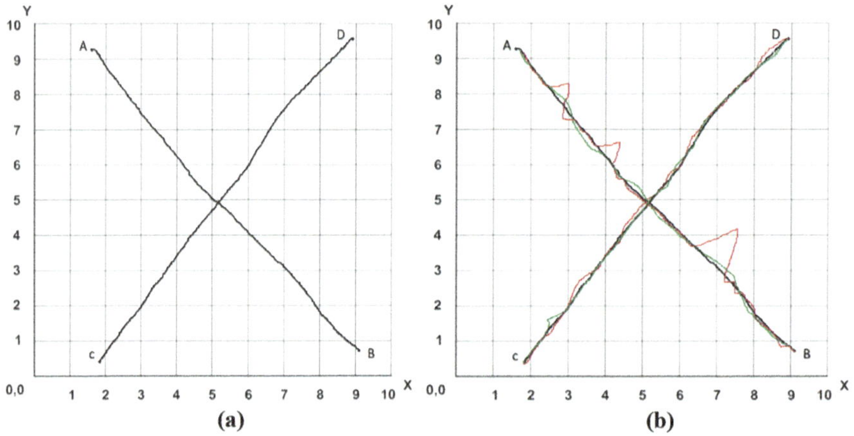

Fig. 7 **a** The gripper was adjusted to hold a pen, **b** 2D graph of the executed paths

force, during the execution of the required task. The desired and executed paths are shown in Fig. 7a, b.

7 Conclusion

For a real-time Kinesthetic teaching function robot, a hybrid control system is shown here. As a proof of concept, a four degree of freedom robot was used to test the suggested technology. We wanted to make it possible for people to learn and practice basic situations that may become more complex in the future. The work investigates whether the proposed systems can be applied to multiple robots working together with the human manipulator. Through the testing and evaluation, we can see the first experimental results are promising: almost all subjects could perform the tasks smoothly and accurately. The Human–robot interaction was easy and smooth as well. The result showed a low average error when the system is tuned properly. However, the average error increased when a higher load was added, it also increased when a very rough interruption error occurs. The implementation of the method is relatively easy and cheap to other methods. It can be used widely in many sectors like the tech industry, social robots, or even day to day tasks. As everyone can teach and interact with the robot. Future work could be training the robot using reinforcement learning, to ensure the stability of the system. Also, the system shall be tested on bigger industrial robots. Lastly, the control system lack stability when a high force is applied, better sensors and motors can be used to solve the problem.

References

1. Šabanović, S.: Robots in society, society in robots. Int. J. Soc. Robot. **2**(4), 439–450 (2010). https://doi.org/10.1007/s12369-010-0066-7
2. Neamah, H.A., Donát, E.A., Korondi, P.: Optimizing autonomous navigation in unknown environments: a novel trap avoiding vector field histogram algorithm VFH+T. Results Eng. In press, 1–29 (2024). https://doi.org/10.1016/j.rineng.2024.102625
3. Katona, K., Neamah, H.A., Korondi, P.: Obstacle avoidance and path planning methods for autonomous navigation of mobile robot. Sensors **24**(11), Art. no. 11 (2024). https://doi.org/10.3390/s24113573
4. Javaid, M., Haleem, A., Singh, R.P., Suman, R.: Substantial capabilities of robotics in enhancing industry 4.0 implementation. Cogn. Robot. **1**, 58–75 (2021). https://doi.org/10.1016/j.cogr.2021.06.001
5. US-5952796-A—Cobots I Unified Patents [Online]. Available: https://portal.unifiedpatents.com/patents/patent/US-5952796-A. Accessed: 27 Nov. 2024
6. Neamah, H.A., Butdee, R.: B-splined trajectory modified generation to maximize speed of the nonholonomic AMR robot. J. Mach. Eng. **24** (2024) [Online]. Available: http://jmacheng.not.pl/pdf-196037-116709?filename=116709.pdf. Accessed: 29 Nov. 2024
7. Keczan, L., et al.: Technical limitations of organic human-robot interaction (O-HRI) for mobile robots moving amongst humans. In: 2024 IEEE 21st International Power Electronics and Motion Control Conference (PEMC), pp. 1–6. IEEE (2024) [Online]. Available: https://ieeexplore.ieee.org/abstract/document/10726327/. Accessed: 16 Nov. 2024
8. Neamah, H.A., Mayorga Mayorga, O.A.: Optimized TD3 algorithm for robust autonomous navigation in crowded and dynamic human-interaction environments. Results Eng. **24**, 102874 (2024). https://doi.org/10.1016/j.rineng.2024.102874
9. Neamah, H.A., Ardi, D.A., Korondi, P.: Enhancing autonomous robot perception for precision positioning and localization. In: 2024 IEEE/SICE International Symposium on System Integration (SII), pp. 1058–1063 (2024). https://doi.org/10.1109/SII58957.2024.10417351
10. Neamah, H.A., Butdee, R.: Optimization modeling parameters for industrial AMR slippage using ANFIS system in dynamic environment. In: Advanced in Creative Technology-added Value Innovations in Engineering, Materials and Manufacturing, vol. 979; Janmanee, P., Chujuarjeen, S., Butdee, S., Srikhumsuk, P., Batako, A.D.L., Burduk, A., Xavior, M.A. (Eds.) Lecture Notes in Networks and Systems, vol. 979, pp. 214–223. Springer Nature Switzerland, Cham (2024). https://doi.org/10.1007/978-3-031-59164-8_18
11. Almusawi, H., Al-Jabali, M., Khaled, A., Korondi, P., Husi, G.: Self-driving robotic car utilizing image processing and machine learning. IOP Conf. Ser. Mater. Sci. Eng. **1256**, 012024 (2022). https://doi.org/10.1088/1757-899X/1256/1/012024

12. (PDF) Robot Programming by Demonstration. In: Research Gate. https://doi.org/10.1007/978-3-540-30301-5_60
13. da Silva, L.D.L., Pereira, T.F., Leithardt, V.R.Q., Seman, L.O., Zeferino, C.A.: Hybrid impedance-admittance control for upper limb exoskeleton using electromyography. Appl. Sci. **10**(20), Art. no. 20 (2020). https://doi.org/10.3390/app10207146
14. Neamah, H.A., Khudhair, M.A., Dhaiban, M.S.: Navigating attention-centric: a machine learning approach to EMG-based hand gesture recognition for interactive RC car. In: 2024 IEEE 21st International Power Electronics and Motion Control Conference (PEMC), pp. 1–6 (2024). https://doi.org/10.1109/PEMC61721.2024.10726353
15. Yu, W., Rosen, J., Li, X.: PID admittance control for an upper limb exoskeleton. In: Proceedings of the 2011 American Control Conference, pp. 1124–1129 (2011). https://doi.org/10.1109/ACC.2011.5991147
16. Paraschos, A., Rueckert, E., Peters, J., Neumann, G.: Probabilistic movement primitives under unknown system dynamics. Adv. Robot. **32**(6), 297–310 (2018). https://doi.org/10.1080/01691864.2018.1437674
17. Akgun, B., Cakmak, M., Yoo, J.W., Thomaz, A.L.: Trajectories and keyframes for kinesthetic teaching: a human-robot interaction perspective. In: Proceedings of the Seventh Annual ACM/IEEE International Conference on Human-Robot Interaction, in HRI '12, pp. 391–398. Association for Computing Machinery, New York, NY, USA (2012). https://doi.org/10.1145/2157689.2157815
18. Capurso, M., Ardakani, M.M.G., Johansson, R., Robertsson, A., Rocco, P.: Sensorless kinesthetic teaching of robotic manipulators assisted by observer-based force control. In: 2017 IEEE International Conference on Robotics and Automation (ICRA), pp. 945–950. IEEE (2017) [Online]. Available: https://ieeexplore.ieee.org/abstract/document/7989115/. Accessed: 28 Nov. 2024
19. Prestashop 1.5, RoboSklep. [Online]. Available: https://robosklep.com. Accessed: 28 Nov. 2024
20. "Baxter," ROBOTS: Your Guide to the World of Robotics [Online]. Available: https://robotsguide.com/robots/baxter. Accessed: 28 Nov. 2024
21. "IRB 460 | ABB Robotics," Robotics [Online]. Available: https://new.abb.com/products/robotics/robots/articulated-robots/irb-460. Accessed: 28 Nov. 2024
22. Barrett, S.F.: Arduino Microcontroller Processing for Everyone! In Synthesis Lectures on Digital Circuits & Systems. Springer International Publishing, Cham (2013). https://doi.org/10.1007/978-3-031-79864-1
23. (PDF) Robust hybrid impedance control of robot manipulators, ResearchGate, Oct. 2024. https://doi.org/10.1109/ROBOT.1991.131589
24. (PDF) Active Impedance Control of Bioinspired Motion Robotic Manipulators: An Overview, ResearchGate, Oct. 2024. https://doi.org/10.1155/2018/8203054

Mental Command-Driven Exoskeletons: Integrating EEG and ERP for Personalized Stroke Rehabilitation

Husam A. Neamah, Rayan Kazi, Suliman Abdulmalek, Hothefa Shaker Jassim, and Zaid Ameen Abduljabbar

Abstract Integrating conventional rehabilitation methods with robotics has led to devices that enhance recovery by supporting finger movement and promoting neural pathway formation. This project leverages EEG signals to directly control robotic appendages, eliminating the need for external control. By using EEG-detected brain signals to drive movement, individuals can see their fingers move in response to their thoughts, enhancing neuroplasticity and recovery. Unlike visual-only feedback from 3D or VR applications, this method provides real movement, similar to Mirror Therapy, but with greater impact. The device's control system processes EEG signals through an ERP pipeline, capturing motor actuation potentials.

Keywords Emotiv · EEGLAB · ERP · Rehabilitation · Arduino · Servo

H. A. Neamah · R. Kazi
Department of Electrical Engineering and Mechatronics, Faculty of Engineering, University of Debrecen, Debrecen, Hungary
e-mail: husam@eng.unideb.hu

H. A. Neamah
College of Engineering, National University of Science and Technology, Dhi Qar, Iraq

Department of Business Management, Al-Imam University College, Balad, Iraq

S. Abdulmalek
Faculty of Informatics, University of Debrecen, Debrecen, Hungary

H. S. Jassim (✉)
Modern College of Business and Science, Muscat, Oman
e-mail: hothefa.shaker@mcbs.edu.om

Z. A. Abduljabbar
Department of Computer Science, College of Education for Pure Sciences, University of Basrah, Basrah, Iraq

B. Awwad (ed.), *AI and IoT: Driving Business Success and Sustainability in the Digital Age*, Studies in Systems, Decision and Control 590,
https://doi.org/10.1007/978-3-031-88874-8_87

1 Introduction

The study of the brain has advanced neural science, aiding in diagnosing and treating disorders. EEG signals, despite their complexity, provide detailed insights into brain activity by recording micro-voltage fluctuations. Motor function impairments, often caused by strokes, damage neural pathways controlling hand movements. Through neuro-rehabilitation, the brain compensates by forming new pathways, typically aided by repetitive physiotherapy [1, 2].

EEG signals, detected by scalp electrodes, amplify potential differences generated during synaptic transmissions in the brain. When an action is intended, "action potentials" produce ionic currents at synapses, creating local field potentials detected by EEG sensors [3, 4]. These signals are classified into five rhythms: delta (0–3 Hz, deep sleep), theta (4–8 Hz, drowsiness/light sleep), alpha (8–13 Hz, relaxation/meditation), beta (13–30 Hz, alert/active states, in the frontal brain), and gamma (30–80 Hz, intense focus, linked to memory and cognitive tasks). Beta and gamma waves are particularly significant for experiments involving mental activity and cognitive skills [3, 5]. A standard EEG setup includes low-impedance electrodes for signal detection, amplifiers to magnify microvolt signals while filtering noise, analog-to-digital converters for data processing, and a computer to store the data [6]. Electrodes are labeled based on brain regions (FP: pre-frontal, F: frontal, C: central, T: temporal, P: posterior, O: occipital) and their placement (odd numbers: left side; even numbers: right).

Incoming signals consist of desired brain potentials, undesired potentials, line noise (50–60 Hz), electrode–skin interface signals, and general noise. Filters are applied to clean the data: high-pass filters (0.1–0.7 Hz) remove low-frequency bio-function noise,

The amplified EEG signal comprises five components: desired brain potentials, undesired potentials, line noise (50–60 Hz), electrode–skin interface artifacts, and general noise. Filtering is applied to clean the signal: high-pass filters (0.1–0.7 Hz) reduce low-frequency noise from bio-functions like breathing, while low-pass filters (below 40 Hz) remove high-frequency artifacts, preventing aliasing [7, 8].

Brain-Computer Interface (BCI) leverages EEG signals to perform tasks, offering transformative solutions for individuals with motor impairments and neurological disorders [9]. EEG signal variations, reflecting mental states and thought processes, are analyzed through a process involving four stages: pre-processing, feature extraction, post-processing, and result analysis. Pre-processing includes downsampling, artifact rejection, and feature scaling. Feature extraction is carried out in time, frequency, or time–frequency domains using methods like Fast Fourier Transform (FFT) and Wavelet Transform (WT), often utilizing band-pass filters [10]. Post-processing involves selecting and refining features for specific applications, typically through dimensionality reduction. BCI systems operate by associating user training signals with commands and using classifiers such as Neural Networks, SVM, kNN, or LDA to compare incoming signals and execute the appropriate actions [11]. Event-related potentials (ERPs) are brain micro-voltages generated in response to specific stimuli or events, derived from EEG signals. They are time-locked to motor, cognitive, or sensory activities, providing a noninvasive method to study neural responses during experiments [12]. Hand rehabilitation is challenging due to the complexity of its functions and the severity of impairments caused by neural or physical damage, such as stroke [13–16]. While simple movements may remain possible, the ability to perform essential daily tasks (e.g., gripping, writing) can be lost. Prompt rehabilitation is crucial to maintain neuronal connections with the upper extremities, although limited resources often prioritize lower extremities for mobility over hand recovery [17]. Neuro-rehabilitation focuses on re-establishing motor functions through repetitive training, leveraging neuroplasticity. This project leverages robotics and Brain-Computer Interfaces (BCI) to control a rehabilitation exoskeleton using EEG signals. Utilizing the Emotiv EPOC X headset and Node-RED software, the system trains both the brain and motor functions, facilitating recovery to a near-normal state (Fig. 1).

1.1 Related Works

Ang et al. [19] reviewed different strategies on the use of BCI technology to assist in physical practice in stroke patients as a method of neuro-rehabilitation. The basic strategy proposed was that of providing a physical movement for the affected limb using a robotic appendage in the form of a sensorimotor feedback, regardless of voluntary movement. Aguiar et al. [20] propose a BCI approach to controlling a robotic arm. Whereas this does not essentially relate to rehabilitation, it provides an insight into the working of the BCI and its applications for controlling actuators and the like. The goal of this project was to create an interface to control a robotic arm with an end effector to do some desired tasks. As shown in Fig. 2, the system processes EEG signals using STFT for event detection, evaluates event spaces, and discriminates motor actions based on thresholds. With the Emotiv EPOC X headset and Node-RED, it enables precise brain and motor function training for recovery [20].

They created their own algorithm to identify and classify these signals. A disadvantage of using such a system however was found to be that sudden motions of the head or eyes skewed results and outputs as the algorithm was not built to filter out unusual activity. Webb et al. [21], proposed a portable exoskeleton controlled using the Emotiv BCI. They employed an open loop system to receive input signals from the BCI system. These signals, which are based on the EEG signals processed by the BCI are sent to the microcontroller which is connected to the motor driver. This motor driver drives the connected motors which are present at the relevant joints. A potentiometer was integrated in order to limit the motion of the motor to the maximum positions to avoid undesirable positions. Figure 3, illustrates how EEG signals processed by BCI software control a microcontroller and motor via PWM and angular feedback, enabling precise rehabilitation exoskeleton movement [21]

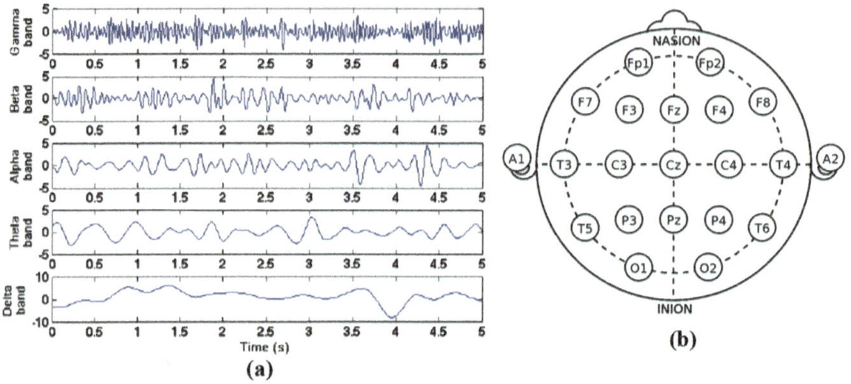

Fig. 1 **a** Various different frequency bands in an EEG Signal [18]. **b** 10–20 International Position of Electrodes [7]

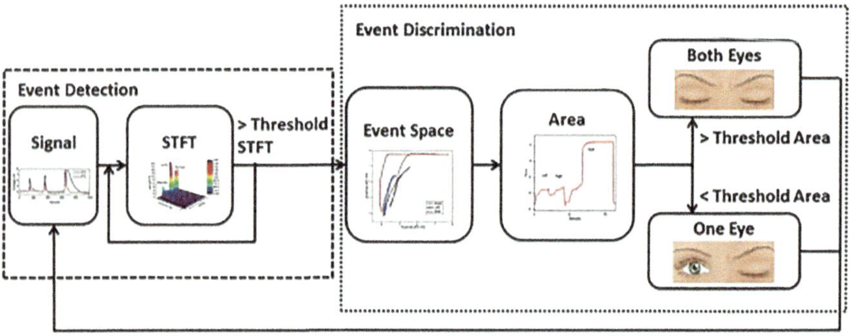

Fig. 2 Event classification algorithm for eye blink identification [20]

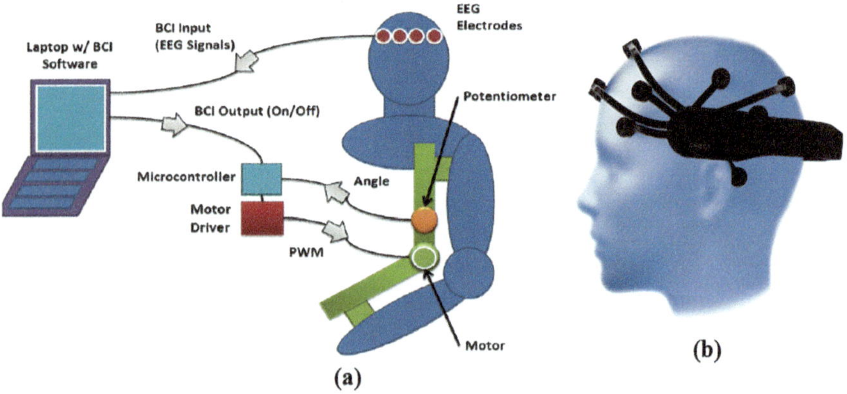

Fig. 3 **a** BCI-robotic rehabilitation system with potentiometer feedback for angle position [21], **b** Electrode placement on EPOC X

2 Project Development

The conception and development of this project was done in two stages, essentially, the planning—which involved the literature review and mapping out the steps necessary to achieve the goals of the project, and the execution of the methodology created. Figure 4, shows the methodology: preliminary study, headset setup, servo implementation and testing, EEG data analysis, and final assessment.

2.1 Emotiv Headset and Applications

A study by Das et al. [22] showed that the Emotiv headband, with more sensors than Neurosky, is better at measuring cognitive loads. To reduce computational complexity, only necessary electrodes can be processed. Ranky et al. [23] found the Emotiv EPOC's adaptability, portability, and ease of customization made it superior for robotic arm control compared to medical EEGs. The EPOC X, based on the 10–20 system, features nine arms with seven output and two reference electrodes, pre-processing raw EEG data with 2048 Hz sampling, dual notch filters (50/60 Hz), a 64 Hz low-pass filter, and downsampling to 128 or 256 Hz for clean signals. The Virtual Brainwear® system, part of Emotiv Launcher, emulates the headset for testing device states like battery and sensor quality. The Emotiv BCI links the headset with third-party systems, using proprietary algorithms for signal classification, baseline establishment, and features like Mental Commands and motion sensors. A virtual headset was configured with 100% battery and EEG quality and successfully tested. With Node-RED, the EMOTIV BCI enables drag-and-drop interface creation for brain-computer projects. BCI data is processed through Emotiv Toolbox nodes. A

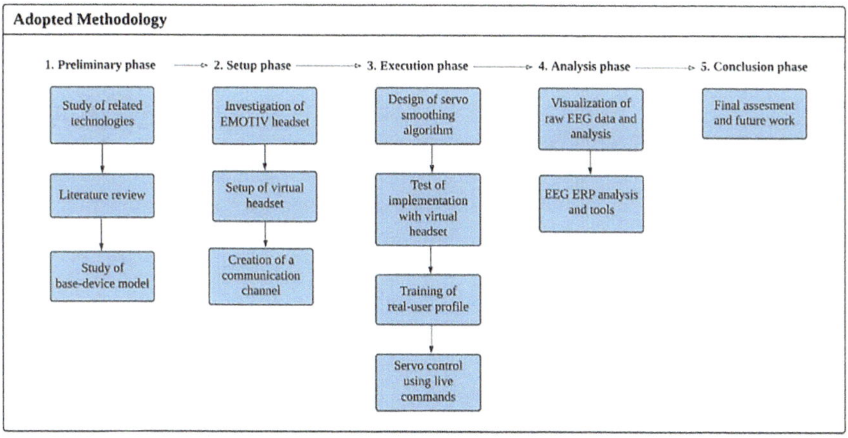

Fig. 4 Methodology adopted for the completion of the project

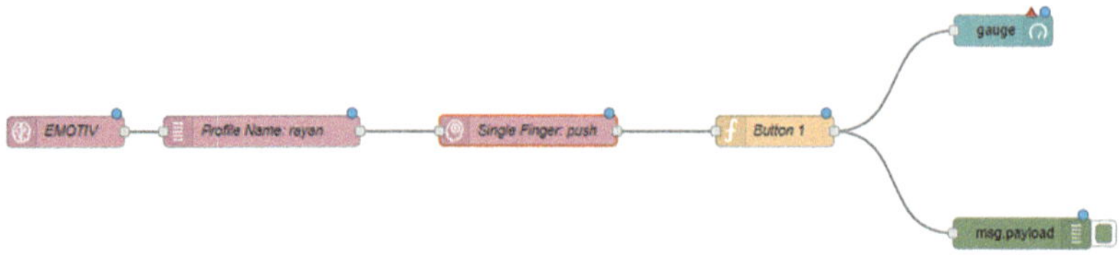

Fig. 5 Node-RED test flow for single button

test flow linked to the 'Trial 1' profile authenticated Mental Commands data, successfully triggering outputs when brain state intensity exceeded 50, ready for COM port integration. Figure 5, shows the Node-RED flow, where the Emotiv headset processes a finger push command, with outputs displayed on a gauge and payload message.

2.2 *Servo Algorithm on ATmega Controller*

An algorithm was developed to control servo movement using outputs from the Emotiv BCI application and establish communication with the microcontroller managing the servos. Initially, a simple algorithm was created to close the hand, with motors actuating upon receiving a trigger and returning to the original position when released. For testing, simple buttons were used as triggers for motor control. Figure 6, depicts the servo smoothing algorithm, where button actions adjust the servo position incrementally until the target position is reached, toggling between "open" and "closed" states.

The next step of the process was to replace the physical buttons with virtual ones controlled on the Node-RED platform. But for this a communication protocol was necessary. There are many ways to communicate between Node-RED and Arduino, however all of them are based on serial communication. Multiple methods using character recognition on the serial monitor and others using number triggers were attempted but all of these presented issues of latency, and did not resolve the issue. Ultimately, a simple function of bits and bytes was developed. Table 1, outlines the bitmask-to-motor mapping, where each button corresponds to a hexadecimal value controlling a specific motor, with the option to actuate all motors simultaneously.

Each button state was saved as a bit on a byte which contains 8 bits. This gives us the ability to control 8 buttons. Depending on which motor needed to be activated a different bit message was sent via the Serial node on Node-RED. Figure 7, illustrates the Node-RED flow, where the Emotiv headset processes mental commands ("push" and "pull") from a user profile. These commands control single or all joints of the system, with outputs directed to a payload and COM2 interface.

The received byte is deconstructed to identify motor numbers using bit masks, enabling motor control via "button" triggers. The Emotiv BCI was linked to an

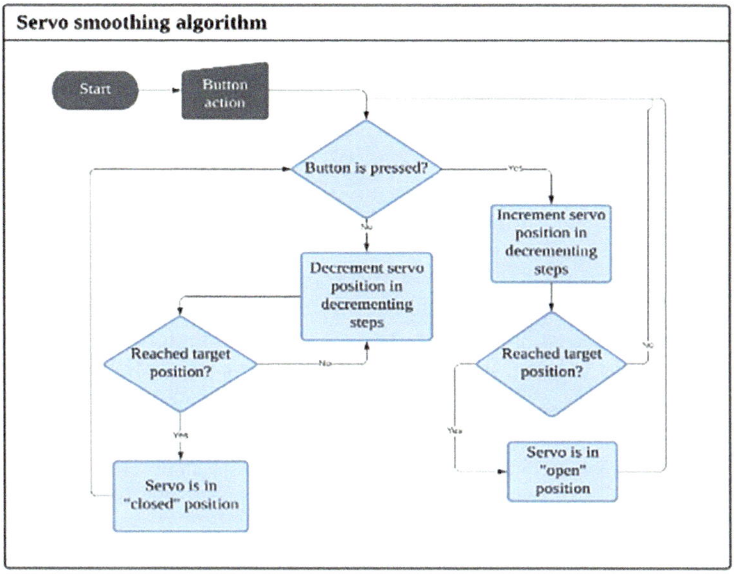

Fig. 6 Algorithm employed for smoothing movement of servo motor

Table 1 Bitmasks corresponding to button/motor numbers with hexadecimal values for actuation

Button–Bitmask	Hexadecimal	Motor
1	0X00000010	1
2	0X00000100	2
3	0X00001000	3
4	0X00010000	4
5	0X00100000	5
6	0X01000000	6
ALL BUTTONS	0X01111110	ALL MOTORS

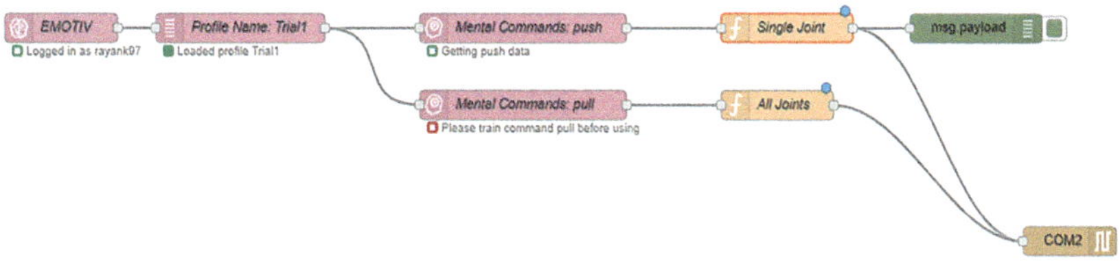

Fig. 7 Final flow setup on Node-RED for Communication with Arduino UNO

Arduino Uno, allowing smooth motor actuation with Mental Commands. After training the Virtual Brainwear headset, live tests with the EPOC X headset successfully actuated the device using EEG signals for push and pull commands.

3 Analysis and Results

Once the entire project was found to be working as designed and required, the process needed to be validated. To achieve this, it was required to analyze the raw EEG data coming from the headset, and analyzed on a processing platform. To obtain the raw EEG data from the headset, it was required to obtain a license for the EmotivPRO application, as this was the only way to gain access to the required datasets. Once the raw EEG data, preprocessed and containing the necessary additional information was imported, it was necessary to load it onto a platform for further processing. MATLAB, but more specifically, EEGLAB was chosen for this function.

To obtain the ERP data for a particular event, it is necessary to follow a set of steps which enable to filter the data correctly, remove any unwanted artifacts and find the required signal shape in an average of multiple trials. After importing the data and removing the DC offset by correcting the baseline, the next step is beginning the process of the pipeline. Figure 8, shows the ERP pipeline, starting with raw data import, followed by cleaning, artifact removal, filtering, and epoching. The process includes baseline removal, eliminating deviant epochs, and averaging trials for ERP analysis.

Even after cleaning and removing the artifacts with any methods, there may remain epochs specifically which have outlier values which will skew the results away from the general average. These must be removed by simply deleting the epochs. This however reduces the sample size that EEGLAB then has to work with, which is why

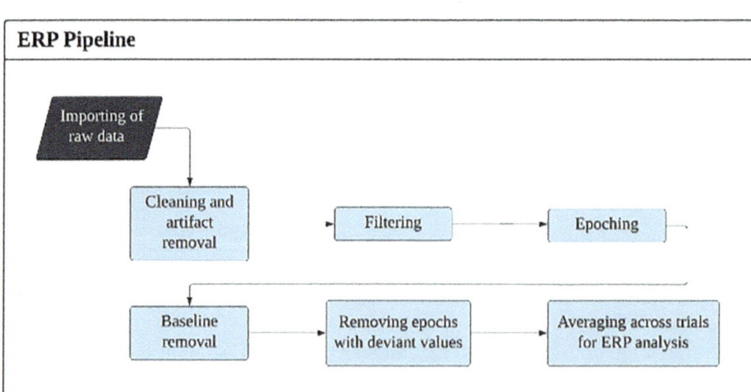

Fig.8 ERP Pipeline used to Obtain the Potential Related to the Motor Actuation

it is necessary to have a large number of trials, since it is highly likely that some epochs will be deleted, Fig. 9, shows Removing Epochs with Deviating Values.

After performing the steps of the pipeline, we arrive at the ERP results of the raw EEG data which can be seen below. From Fig. 10, we see that the highest potential –denoted by the dark blue line- is at approximately 14 μV with being from the 6th electrode component.

Figure 11a shows an ERP time–frequency analysis, with a heatmap representing signal power across frequencies and time, and the ERP waveform below indicating amplitude variations. The dashed line (time = 0 ms) marks event onset, with red and yellow indicating higher activity. Figure 11b highlights components 3, 6, and 10 with the highest intensity in the fronto-central and temporal regions, linked to motor activation. A refined ERP analysis revealed a microvolt spike at 300 ms, persisting until 450 ms, indicating the brain state triggering the push command.

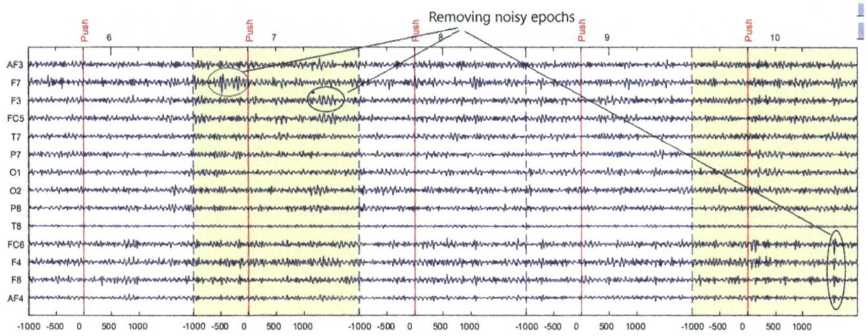

Fig. 9 Removing epochs with deviating values

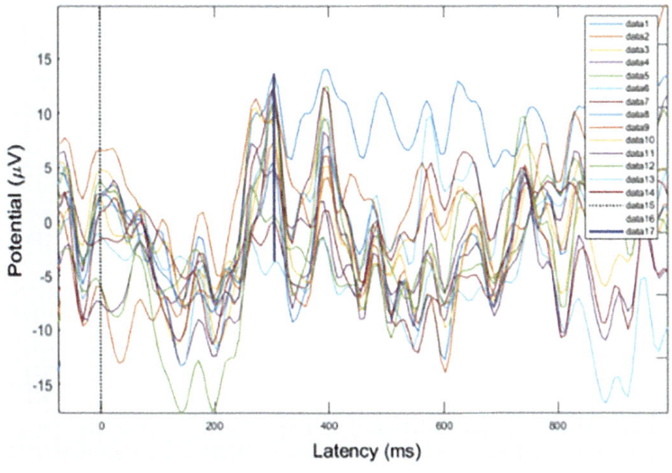

Fig.10 ERP Analysis of 14 Components of the Epochs

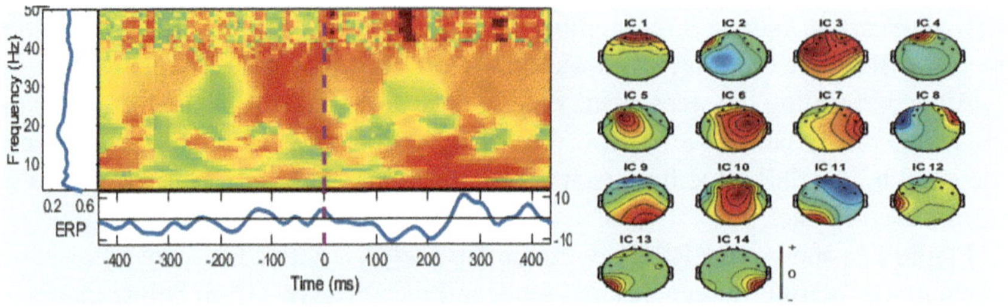

(a) (b)

Fig. 11 **a** Averaging dominant electrodes for ERP analysis. **b** 14 ERP components with varying degrees of intensity at different regions

4 Conclusion

This paper aimed to develop EEG-based control for a rehabilitation exoskeleton to assist stroke victims unable to benefit from traditional rehabilitation methods. The Emotiv EPOC X, with its Mental Commands feature, proved effective, while Virtual Brainwear was used to streamline testing. Node-RED connected the EEG device to the microcontroller controlling the exoskeleton motors. The system retained original functionalities, allowing easy integration of additional components. ERP analysis using EEGLAB validated brain activity during mental commands, highlighting subject-specific brain responses to the required imagery..

References

1. AbdulKareem, A.H., Adila, A.S., Husi, G.: Recent trends in robotic systems for upper-limb stroke recovery: A low-cost hand and wrist rehabilitation device. In: 2018 2nd International Symposium on Small-Scale Intelligent Manufacturing Systems (SIMS), pp. 1–6 (2018). https://doi.org/10.1109/SIMS.2018.8355302
2. Lee, H.: Approach to Understanding Brain Disorders and Improving Healthcare Outcomes, (n.d.). https://osf.io/hwzfy (Accessed December 1, 2024)
3. Guy-Evans, O.: What Happens at The Synapse? Simply Psychology (2021). https://www.simplypsychology.org/synapse.html.
4. Ray, A., Issa, N.P.: EEG Instrumentation and basics. In: Spanaki, M.V., Wasade V.S. (eds.), Understanding Epilepsy: A Study Guide for the Boards, Cambridge University Press, Cambridge, pp. 203–222 (2019). https://doi.org/10.1017/9781108754200.012.
5. Matsuo, M., Higuchi, T., Miyabara, H., Higashijima, M., Oshikawa, T., Nakamura, M., Yamaguchi, Y., Higashionna, T.: Assessing attentional task-related electroencephalogram signal variations by using mobile electroencephalogram technology: an experimental study. Medicine **102**, e35801 (2023). https://doi.org/10.1097/MD.0000000000035801
6. Teplan, M.: Fundamentals of EEG measurement. IEEE Meas. Sci. Rev. **2**, 1–11 (2002)
7. Matsuo, M., Higuchi, T., Abe, T., Ishibashi, T., Egashira, A., Kamashita, R.: Electroencephalogram signal correlations between default mode network and attentional functioning. J. Behav. Brain Sci. **14**, 119–134 (2024). https://doi.org/10.4236/jbbs.2024.144009

8. Zhang, G., Garrett, D.R., Luck, S.J.: Optimal filters for ERP research II: recommended settings for seven common ERP Components, (2023) 2023.06.13.544794. https://doi.org/10.1101/2023.06.13.544794
9. Sweeney, K.: Motion artifact processing techniques for physiological signals. RIAN. (2013). http://eprints.maynoothuniversity.ie/4466/
10. トマトン, Electrode locations of International 10–20 system for EEG (electroencephalography) recording, (2010). https://en.wikipedia.org/wiki/10%E2%80%9320_system_(EEG)#/media/File:21_electrodes_of_International_10-20_system_for_EEG.svg
11. Guy-Evans, O.: Frontal lobe function, location in brain and damage (2021). https://www.simplypsychology.org/frontal-lobe.html#:~:text=Damage%20to%20the%20Frontal%20Lobes,causing%20a%20variety%20of%20symptoms
12. 10-20 Sites and cortical functions, Brain Trainer (n.d.). https://brain-trainer.com/answers/about-the-brain/10-20-sites-cortical-functions/.
13. Almusawi, H., Husi, G.: Design and development of continuous passive motion (CPM) for fingers and wrist grounded-exoskeleton rehabilitation system. Appl. Sci. **11**, 815 (2021)
14. Talab, M.A., Péter, K., Almusawi, H.A.: Design of an individual fingers rehabilitation device. In: 2022 IEEE 20th International Power Electronics and Motion Control Conference (PEMC), pp. 734–739 (2022). https://doi.org/10.1109/PEMC51159.2022.9962909
15. Almusawi, H., Afghan, S.A., Géza, H.: Designing the mechanical parts of a low-cost hand rehabilitation CPM device for stroke patients. In: J. Machado, F. Soares, G. Veiga (Eds.), Innovation, engineering and entrepreneurship, Springer International Publishing, Cham, pp. 60–66 (2019). https://doi.org/10.1007/978-3-319-91334-6_9
16. Almusawi, H., Afghan, S.A., Geza, H.: Technological advancements in stroke hand rehabilitation devices: a review. In: International Mechatronical Student Micro-Conference, University of Óbuda, Budapest, pp. 80–90 (2016)
17. BLRI, Removal of Line Noise Component from EEG Signal (2016). https://blricrex.hypotheses.org/files/2016/09/LineNoise_Removel_MultiTaper.pdf
18. Abo-Zahhad, M., Ahmed, S.M., Abbas, S.N.: A new EEG acquisition protocol for biometric identification using eye blinking signals. IJISA **7**, 48–54 (2015). (Department of electrical and electronic engineering, Assiut University, Assiut, Egypt) https://doi.org/10.5815/ijisa.2015.06.05
19. Ang, K.K., Guan, C.: Brain-computer interface for neurorehabilitation of upper limb after stroke. Proc. IEEE **103**, 944–953 (2015). https://doi.org/10.1109/JPROC.2015.2415800
20. Aguiar, S., Yanez, W., Benitez, D.: Low complexity approach for controlling a robotic arm using the Emotiv EPOC headset. In: 2016 IEEE International Autumn Meeting on Power, Electronics and Computing (ROPEC), IEEE, Ixtapa, Zihuatanejo, Mexico, pp. 1–6 (2016). https://doi.org/10.1109/ROPEC.2016.7830526
21. Webb, J., Xiao, Z.G., Aschenbrenner, K.P., Herrnstadt, G., Menon, C. (2012) Towards a portable assistive arm exoskeleton for stroke patient rehabilitation controlled through a brain computer interface. In: 2012 4th IEEE RAS & EMBS International Conference on Biomedical Robotics and Biomechatronics (BioRob), IEEE, Rome, Italy, pp. 1299–1304. https://doi.org/10.1109/BioRob.2012.6290674
22. Das, R., Chatterjee, D., Das, D., Sinharay, A., Sinha, A.: Cognitive load measurement—a methodology to compare low cost commercial EEG devices. In: 2014 International Conference on Advances in Computing, Communications and Informatics (ICACCI), IEEE, Delhi, India, pp. 1188–1194 (2014). https://doi.org/10.1109/ICACCI.2014.6968528
23. Ranky, G.N., Adamovich, S.: Analysis of a commercial EEG device for the control of a robot arm. In: Proceedings of the 2010 IEEE 36th Annual Northeast Bioengineering Conference (NEBEC), IEEE, New York, NY, USA, pp. 1–2 (2010). https://doi.org/10.1109/NEBC.2010.5458188

Sustainable Development of Academia and Industry: The Role of Internships and the Eidaad Program

Beverly Baker-Kelly and Hamed Saif Al Hadhrami

Abstract Using the Eidaad Internship Program as a template, gives us a window towards understanding the role of sustainable development of academia and industry and the role of a well-structured program that calls for collaboration between industry and academia. Moreover, the Eidaad Program is aligned with Oman's Vision 2040. It is a win–win situation. Students can gain career clarity, critical skills for employment, and exposure to industry standards through internships. Oman's national workforce development is the fountainhead for addressing employability of graduates from our Omani educational institutions.

Keywords Internships · Academic learning · Industry demands · Eidaad Program

1 Introduction

In 1984 Kobe wrote about experiential learning by pegging experience as the important stepping-stone for learning and development. The focus was on bridging the gap between academic knowledge and the experience needed for workforce readiness.

This paper explores the transformative power of internships, with a particular focus on the Eidaad Program in Oman. It highlights how the program serves as a model for industry-academia collaboration, aligning academic outcomes with labor market needs. Here, the discussion includes an analysis of the program's impact, addressing both its successes and areas for improvement, and situates these findings within the broader context of workforce development in Oman and beyond [1–3]. Since the industrial revolution in the Western world, labor markets have been primed to look for work readiness through apprenticeships. This provides students with

B. Baker-Kelly (✉)
Modern College of Business and Science, Muscat, Oman
e-mail: bbakerkelly@gmail.com

H. S. Al Hadhrami
Petroleum Development Oman, Muscat, Oman

B. Awwad (ed.), *AI and IoT: Driving Business Success and Sustainability in the Digital Age*, Studies in Systems, Decision and Control 590,
https://doi.org/10.1007/978-3-031-88874-8_88

real-world skills essential for the transition between higher educational pursuits and industry related employment.

As a part of its Vision Oman 2040 espouses the need for strategic initiatives like the Eidaad Program to address the challenges of Omanisation in its workforce. This is an important goal for all sectors of the Omani government. The Omani leader clearly see the need for job placements for its Omani citizens.

2 Power of Internships

2.1 *The Transformative Power of Internships*

The transformative or catalytic power of internships is not a new idea. Four decades ago, ten years right after Omani educational institutes were established, using a learning theory model, educational scholars wrote about the importance of aligning college level curriculum with practical real-world job placement experiences. It was posited that students were better able to navigate complex, professional environments more effectively after having been engaged in a yearlong professional assignment without the rigors of time-consuming classes and assignments. This arrangement extended the students' college graduation by one year, but the benefits of the program come from pairing technical learning skills with the learning of soft skills, such as teamwork, leadership, and communication [3–5]. This allows the students to step into the shoes of real-time workers with a focus on the industry's role-defining priorities.

2.2 *Real-World Application of Knowledge*

Ten years ago, Jackson [4] found that employability skill development could be enhanced by exposure to formal in-class learning. Even special "knowledge-based industry classes" could be beneficial. Guest speakers from industry and others skilled in work-classroom alliances could be helpful. There were a few barriers to the adoption of transformative skill-based knowledge and skills designs.

The idea of being able to apply real world knowledge that is associated with academic learning helps to bridge the gap between theory and practice. Likewise, Klob paired experimental learning from hands-on experiences by reinforcing classroom learning and problem-solving approaches. Thirty years later, Hagler and Lee saw the need for enhancing the understanding of complex concepts and applying them in real world challenges on the job [2, 3].

2.3 Skill Development Across Disciplines

Both technical and soft skills can be acquired by participants in a work-related internship program with specific competencies related to industry that can be acquired alongside transferable skills like leadership and communication [4].

The Eidaad program was found to significantly improve students' time management and problem-solving skills. Project management skills made the participants more versatile and adaptable. The Eidaad program is based on a controlled environment where students can practice decision-making and critical thinking. These are skills that are essential for success across disciplines and related to educational exposure [1, 4, 6].

2.4 Career Exploration and Clarity

Through the Eidaad program, it became clear to the administrators of the program, the decision-makers in industry and the educators that students enhanced their ability to settle on specific careers trajectories based on their exposure to industry-based placements. The students had a clearer understanding of industry expectations and their work culture. This enabled them to make informed career choices as stated in Gault [7]. This was important because students identified strengths and weaknesses within themselves as areas for improvement. Jackson emphasized the alignment of student's academic and professional goals [4, 8].

2.5 Enhanced Employability

One of the most significant outcomes of internships is the transformative power of exposure to a particular industry and the specific placement and ultimate employability of the student-participants. Understandably employers often prefer candidates with internship experience because this demonstrates a readiness for the workplace. Smith [5] posited that building professional ethics, being able to adapt to the workplace, and job specific competencies make the internship participants more attractive to potential employers. Notably, the internship program has been instrumental in aligning graduates' skills with the demands of the labor market. This contributes to a smoother transition from student to employee [1, 3]. This also allows for the Omanisation goals of the government to be realized. The idea is to attract more Omanis into a variety of occupations.

2.6 Building Professional Networks

Students learn to understand the importance of establishing connections with varied industry professionals. This has become an invaluable resource from mentorship. Students should understand that access to long-term job opportunities comes from professional networks gleaned while being interns. [3]; and they learn to better interact with supervisors, colleagues, and other stakeholders, while developing relationships that can lead to extended professional growth and connections. The benefit of participating in the internship program can pave the way for meaningful industry collaborations and potential employment offers [1, 2] for both short-term and long-term needs of employees.

2.7 Confidence Building

The experience of having an industry-based internship helps students to become more resilient and prepared for future professional roles while enhancing students' overall confidence in their abilities. This firsthand job experience in professional settings helps to build confidence in handling adult workplace responsibilities. Constructive feedback from mentors helps those exposed to real-world challenges develop self-confidence and independence [7]. Interns can build confidence in their ability to work in professional situations by participating in work related internships.

2.8 Pathways to Employment

Pathways to full-time employment are goals of many graduating seniors. After a yearlong internship, students often feel that they may have a direct pathway to full-time employment. On the employer side, internships offer the employers low-risk situations to evaluate potential employees and then have a steadier basis for offering them permanent positions. Industry based internships can create a seamless transition from student to professional workplace employees [4].

3 Eidaad: A Model of Industry-Academia Collaboration

The Eidaad Program stands out as a pioneering initiative in Oman by addressing the persistent gap between Omanisation, academic education, and industry needs. By offering structured, immersive work placements, Eidaad fosters a dynamic partnership between academia and industry, empowering students with the skills and experiences necessary to thrive in competitive labor markets.

3.1 Comprehensive Coverage

Eidaad approaches the student participants in a holistic way which ensures a comprehensive coverage of professional, technical, and person development. Fortunately, the program is geared towards a wide range of academic disciplines, aligning student experiences with industry specific requirements. It caters to diverse industries, including manufacturing, technological industrial designs, and oil and gas. This ensures that students gain hands on knowledge in a real-world context and exposes them to areas of the industry that were unknown or relatively nonexistent before they entered the program. The breath of industry participating in the Eidaad Program ensures that students will encounter various workplace cultures and practices, enhancing their adaptability and intercultural competencies. Students cannot be exposed to all the industries that are available to the overall group applying for inclusion. Eventually, students can be prepared for not only the local job markets, but also for other inner cultural opportunities for job programs. This can encourage students to have a wider scope of readiness outside of the confines of professional opportunities in the work world. Note that this comprehensive coverage also extends to non-technical skills in problem-solving, teamwork and ethical decision-making. The graduates of the program become more well-rounded professionals [4].

3.2 Structured Internship Framework

Eidaad's structured intern program is in the process of being fine-tuned. Essentially it balances academic learning with practical applications. There exist other ad hoc internship models that simply place their eligible internship programs into miscellaneous internship programs. This contrasts with the Eidaad Program that incorporates (1) pre-placement training (2) clearly defined learning outcomes and, most importantly, (3) post placement evaluations. This specific organizational structure tries to ensure consistency in the quality of the internship experience across diverse industries, and most importantly, the development of the use of a standardized framework which allows for regular assessments of students, progress in technical skills, monitoring of the workplace, adaptability and soft skills evaluation and the inclusion in academic environments of clearly defined academic advisors, working with industry, which ensures, constructive feedback and continuing guidance for the interns. This structured approach minimizes variability in student outcomes and helps to minimize variability in student experiences. The goal is to guarantee that all participants meet a baseline level of professional readiness. This is not an easy standard to achieve. And the personnel needed to carry out these evaluations is daunting.

It is difficult to establish a standardized development framework, which does not interfere with the industry mentor's regular workplace requirements. This is a stumbling block in the system that requires good organizational techniques for the industry partners, continuance guidance and constructive feedback for both the industry mentor and the student participants. The goal is to minimize the variability in student experiences and guarantee that all participants, both interns and mentors, meet a baseline level of professional readiness. Notably, there is a need for improvement and tighter coordination between the Eidaad team and the industry partners.

3.3 Alignment with National Vision

It is essential to establish an alignment between the Eidaad program and Oman's National Vision 2040 which emphasizes economic diversification, human capital development, and enhanced employability among Omani youth. It cannot be overlooked that enhanced employability for Omani youth is one of the paramount goals of Oman's National Vision 2040. These decision makers want to put for the employment of Omanis as one of their most important national objectives in fostering a knowledge-based economy. Oman's National Vision 2040 includes the strategic importance of developing economic sectors piloted by Omani teams as a need in industry for enhancing the Omani economy. What is required is a partnership between professional industries and strategic planners in Oman, dealing with economic sectors and well trained newly recruited young Omanis who can carry out the vision of the industries.

3.4 Industry-Ready Graduates

As reiterated throughout this paper, the goal of the Eidaad program is to produce graduates who are ready to meet the demands of the workplace. A key administrator requirement of Eidaad is comprehensive feedback from students and supervisors about their work program and the values of Oman's strategic economic sectors. There is a partnership that should be emphasized and developed between the graduates of our higher educational institutions and the captains of industries, such as advanced manufacturing and logistics, renewable energy and supporting the nation's movement toward vocational efforts. There are key challenges outlined in Vision 2040, including Oman's high youth employment rate and the need for skilled local workforce to reduce dependency on expats. They need to offer long-term internships as part of academic training. This creates a seamless transition from education to employment fulfilling a critical national need, the challenges of creating a developmental approach to leadership skills and the advancement of technical competencies [4].

4 Balancing Benefits and Addressing Minor Concerns: The Success of the Eidaad Programme

Evaluation of the benefits and possible distractors of the Eidaad program are complicated. The Eidaad program has emerged as a highly attractive opportunity for Omani students. The demand for placement often surpasses the development of available seats in industry. This overwhelming interest reflects a five-year tug-of-way to obtain strong recognition among students of the programs, the necessary values in providing essential industry placement, the program is widely regarded as a fountainhead for bridging the gap between academic learning and professional readiness.

The end goal is to enhance students' employability and align this with the national workforce development goal. Even if the program enjoys popularity among a subset of students, there are significant reservations about committing to a yearlong internship based on cultural and personal concerns. A recent study in 2024 by Al Abri, explores the perception of engineering students at Sultan Qaboos University when they participated in the Eidaad program [1].

There was a recognition of its prior positive contribution to students' career clarity and professional development. However, the bottleneck develops when the participants have to make a very hard choice of leaving the university for an entire year, and becoming employed in an industry in which they know little about. The internship program emerges as a transformative initiative significantly enhancing the employability of many engineering students, however, there were other roadblocks that prohibited eligible SQU students in the engineering field from choosing to leave the university for a full year before they could continue with their professional studies [4].

4.1 Cultural Pressure to Graduate Ahead of the Scheduled Four-Year Curriculum

For Eidaad to become more successful program it is important to understand the hurdles in cultural expectations in Oman that compose challenges for some students. The educational the formal educational system for Oman started about 50 years ago. The values of the parents who encourage their children to continue in the higher educational system are rooted in the idea that they will get a job and that they will be employable at the end of their university program. The community, the extended family, and the parents want their children, relatives and acquaintances to be graduates. It is difficult for students to return home and tell their extended family or village that they are going to take off one full year of their educational pursuits to go into an internship with low pay, if any, and no guarantee of a job placement. This is difficult for the parents and the students to understand. They have been subtly taught to expect that the government will financially sponsor their college education. There is fear that they will not graduate if they stop for one year for an internship. The culture and the values of Omani people often hinder the signing on to the Eidaad program system.

4.2 Desire for Immediate Career and Personal Advancement

Students and their families must understand how delayed gratification can work for them and not against them if they choose the internships and opt for the long-term career benefits, by recognizing the benefits of internships as a pathway to enhance career perspective and skill development.

4.3 Fear of Delayed Graduation

As mentioned before, senior students who are asked to delay their graduation for a year and really delay their graduation for two years brings about a feeling of being left behind by peers graduating who will graduate within four years.

4.4 Positive Momentum and Recommendations

The Eidaad initiative has gained a foothold in the educational system in Oman. It is making a difference. Over 300 students are Eidaad graduates. But there may be more detractors than proponents, yet each year the Program becomes stronger and more comprehensive; it is an idea whose time has come. The floodgates of indispensable internships, allied to industry and job placement are at our door. No one can keep their finger in this dike. It is an educational tool that cannot be ignored. Those who do so will do so as at their own peril because the goals of the Omani administrators, are clear. They in who are deeply committed to Omanisation and to building a workforce that is both dedicated, Omani and skilled.

The recommendations promoting the Program include:

- **Cultural Awareness Campaigns**: Promote the program's alignment with societal and economic goals, emphasize the long-term benefits of internships to foster competitive graduates.
- **Seamless Academic Integration**: Ensure that the program is fully integrated into academic frameworks, allowing students to understand the need to prolong their studies to receive valuable training for a year.
- **Career Advancement Messaging**: Highlight success stories of program alumni to illustrate the career acceleration and opportunities provided by the internships.

5 Conclusion

Internships are transformative experiences without doubt that empower students to apply academic knowledge, develop essential skills, and explore career pathways. They play a pivotal role in bridging the education-employment gap, enhancing employability, and preparing students for the complexities of modern workplaces. Programs like Oman's Eidaad Program illustrate the potential of internships to contribute to national development goals by aligning education with market needs.

The Eidaad Program exemplifies a successful model of industry-academia collaboration, addressing critical gaps in employability and alignment with Oman's National Vision 2040. Through its comprehensive coverage, structured framework, and emphasis on producing industry-ready graduates, Eidaad sets a benchmark for experiential learning initiatives. Continued refinement, particularly in mentorship quality, task alignment, and collaboration between universities and host organizations, will further enhance its role as a cornerstone of Oman's human capital development strategy.

Despite its relatively high enrollment rates and widespread recognition, addressing the concerns of a subset of hesitant participants is mandatory. By tackling issues such as cultural pressures, the need for more academic integration and enhancing program design, Eidaad will continue to lead as a vital bridge between academia and industry. This will equip Omani students with the skills and experiences needed to excel in the modern workforce and support the nation's broader goals for economic diversification, Omanisation, and sustainable development.

References

1. Al-Abri, M., Denman, C., Al Alawi, M., Al Ajmi, M.: Enhancing employability through university-industry linkages: Omani engineering students' perspectives of the Eidaad internship programme. Human. Soc. Sci. Commun. **11**, 565 (2024) [Online]. Available: https://doi.org/10.1057/s41599-024-02779-y
2. Helyer, R., Lee, D.: The role of work experience in the transition from higher education to employment. High. Educ. Q. **68**(3), 242–257 (2014)
3. Kolb, D.A.: Experiential Learning: Experience as the Source of Learning and Development. Prentice-Hall, Englewood Cliffs, NJ (1984)
4. Jackson, D.: Employability skill development in work-integrated learning: barriers and best practice. Stud. High. Educ. **40**(2), 350–367 (2015)
5. Smith, C.D.: A framework for evaluating work-integrated learning curricula. J. Cooperat. Educat. Internships **42**(1), 6–15 (2008)
6. Mokhtar, M., Fauzi, F.: Enhancing employability through structured internships: insights from Malaysian universities. Adv. Bus. Res. Int. J. (2024)
7. Gault, J., Leach, E., Duey, M.: Effects of business internships on job marketability: the employers' perspective. Educ. Train. **60**(1), 44–56 (2018)
8. Oman Academic Accreditation Authority (OAAA): The State of Higher Education in Oman: Challenges and Opportunities. OAAA, Muscat (2018)

Artificial Intelligence and Economic Innovation: A Study of Oman's AI Potential

Mahmoud Abouraia, Ayman Bassam Nassoura, Hothefa Shaker, Fadi Abdelfattah, and Hisham Madi

Abstract This paper explores the issue of using Artificial Intelligence (AI) technology to boost economic development in Oman, with a view to the effects of incorporating AI technology in different segments of the economy. In AI development, the knowledge of AI in Oman's economic environment is essential for developing new solutions and achieving sustainable growth. Data was collected through a questionnaire answered by 145 professionals from different fields, such as education, healthcare, finance, and technology, to understand their awareness, adoption and perceived barriers to AI. The survey results indicate that though there is a considerable understanding of the benefits contributed by AI, the main issues, including infrastructure constraints, employee dearth, and ethical questions, still impede AI adoption. Governments and policymakers worldwide need to invest wisely in education, infrastructure, and the proper regulations to fully develop AI capabilities. This research helps identify challenges and opportunities in AI that can strengthen Oman's economy and encourage innovation. It also offers valuable advice for policymakers and business leaders.

Keywords Artificial intelligence (AI) · Economic innovation · AI adoption · Smart economy · Technology integration · Digital economy

1 Introduction

The place of Artificial Intelligence (AI) as an economic tool for change and development has grown more crucial while considering technological developments [1, 2]. Thus, countries implementing AI-related technologies within the national economy seek to unlock efficiency, productivity, and national competitiveness. In Oman, with

M. Abouraia · H. Shaker · F. Abdelfattah (✉) · H. Madi
Modern College of Business and Sciences, Muscat, Oman
e-mail: fadi.abdelfattah@mcbs.edu.om

A. B. Nassoura
City University Ajman, Ajman, UAE

B. Awwad (ed.), *AI and IoT: Driving Business Success and Sustainability in the Digital Age*, Studies in Systems, Decision and Control 590,
https://doi.org/10.1007/978-3-031-88874-8_89

Vision 2040 for the country's development, efforts are being made to move away from dependence on oil and gas, so AI technologies are integrated into education, health care, finance and government bodies, among others [3]. AI's importance and opportunities for economic growth can be mainly applied in Oman, encouraging efforts towards attaining a knowledge-based economy through actions such as the National Artificial Intelligence Strategy [1]. This vision focuses on utilizing Artificial Intelligence to improve everyday public services, optimize industries, and innovate new products across the private sector and public services.

The adoption of AI in Oman has economic implications in both a positive and negative sense. AI can contribute to economic growth by proactively helping organizations make better decisions, identifying cost-saving areas, and improving customer satisfaction [4, 5]. Furthermore, the growth of AI's productivity augmenting effect on industries has been established by literature as it has been found that AI technologies can cause productivity increases significantly in industries including manufacturing, logistics and financial services [6]. Nonetheless, many factors of concern to AI adoption remain, such as the skills required, the challenge of creating infrastructure for AI, and ethical issues [7]. Thus, the discovery of Oman's economy means that overcoming these challenges is crucial to achieving the best results with the help of artificial intelligence.

Moreover, the use of AI as a tool to create economic value in Oman is also highly dependent on the government's attempts to construct an innovative economy. Mature intelligent economies thus harness Digital technologies, especially AI, to foster sustainable economic growth and development [8]. Oman is gradually changing its economy for the digital sector by investing in artificial intelligence, the Internet of Things, and cloud technologies [3]. However, these technologies hold great promise, the actual application of which can only be achieved if the country deals with barriers that include infrastructure inadequacies, regulatory disincentives, and data privacy [9].

Consequently, we need to know how AI can potentially be used to foster Oman's economic development. Thus, the current extent of AI adoption is accompanied by emerging difficulties and prospects for Oman's economy. This work aims to study the prospects for using AI in Oman, how technology is progressing, and what measures need to be taken to prevent obstacles to AI development and implementation. This study adds to the extant literature on economic development via AI, particularly in the Sultanate of Oman, by stressing the potential of AI in this area.

2 Literature Review

The incorporation of AI into national economies has been widely discussed recently because of its applicability across industries. With the shift in the global economy to becoming more digital, a new frontier is set to be enabled by the use of AI technology. For instance, the applications of AI in creating new forms of economic value are emerging as a critical factor in the development plans of such countries concerning the transition towards non-COVID-19-related industries. Out of these, Oman, with its Vision 2040, is placing AI at the center of its plan to create a diversified and knowledge-based economy that best fits the Sultanate's current economic structure, as noted by [1].

2.1 AI's Role in Economic Growth and Innovation

Closely related, many academic works demonstrate that AI can positively impact economic development, mainly due to its capability to enhance innovation in various industries. [5, 8] maintains that Brynjolfsson and McAfee's view is primarily true concerning its claims about AI acting as an innovation tremor. This is especially significant because of the possibilities that the integration of AI fosters, especially for emergent economies such as Oman. Based on the work done by [8], AI technologies such as automation, analytics, and machine learning increase operational functions and promote innovations in numerous industries, including manufacturing, procurement, and financial sectors. The McKinsey Global Institute has also curated more data on how AI helps in the innovation of the economy, which is active in areas like retailing and financing where many uses of Artificial Intelligence, like automation of analyses of data, help in the engagement of customers, optimization of supply chain processes, and cutting costs. For instance, using financiers through artificial intelligence has led to better customer experience, investment a stronger economy through playing a significant role in eradicating fraud [10, 11]. Likewise, the contribution of AI in promoting automation in production must benefit industries, lowering operation costs and improving production, thus benefiting the economy of a nation, for instance, Oman.

2.2 AI Adoption in Oman and the Middle East

The strategies for adopting AI in Oman are contextualized in Vision 2040, the country's plan for economic diversification. In line with this vision, the Omani government considers AI and other inventive technologies as driving forces for economic growth. [3] highlights that the National Artificial Intelligence Strategy in Oman aims to build the base for the application of AI that ensures the improvement of governmental services, increasing effectiveness in public services, and developing innovative initiatives in the private sector. When implemented across various sectors such as healthcare, logistics and energy, the shocks of AI are believed to stimulate the Countries' economic transformation. However, with the adoption of AI, there are benefits in the economy that should be exploited effectively, although there are several issues in achieving maximum efficiency. As [1] also points out, the foundation of AI means that while a vast majority of organizations now use AI, the infrastructure that it requires—data centers and high-speed internet connections, for example—should increase at a similar pace. Also, a highly skilled workforce is essential because AI policies require deep domain knowledge and sectoral competencies in sub-areas of artificial intelligence, including machine learning, data analysis, and systems architecture [9]. However, like many other developing nations, one of the significant challenges in Oman is the insufficient skill quantity for AI and related areas for education and training [7].

2.3 Challenges of AI Integration

Oman and many Middle Eastern countries face a big problem: AI ethics. People worry about keeping their information private, safe, and fair as AI spreads. [7] It is crucial to create ethical rules to ensure AI is used correctly in essential areas like healthcare and money matters. In Oman, people might not want to use AI if their worries about data privacy and understanding how AI works are not addressed. Another issue is making rules. Many countries, including Oman, must develop good policies to control how AI is used and managed. The Organization for Economic Co-operation and Development [9] thinks governments should focus on making solid rules that allow new ideas and protect people. For Oman, this could mean making policies that help new ideas grow while also taking care of worries about keeping data private, stopping cyber-attacks, and ensuring fair AI decisions.

2.4 AI and the Future of Economic Diversification

In the long run, there is a possibility of reducing reliance on oil and achieving economic diversification in Oman by reducing costs or increasing new industries and developing old ones, according to Hassan et al. As noted in [4], AI can spur diversification by providing new areas of economic growth, including AI R&D, intelligent infrastructure, and self-driving cars. Further, as artificial intelligence technologies are becoming an integrated part of everyday business processes, the global competitiveness of Omani companies will also be enhanced due to more significant insights and improved decision-making. According to [3], AI can also significantly build a more sustainable and resilient economy by optimizing resource use and minimizing waste.

The necessity to optimize energy consumption, manage waste, and improve industries by making them sustainable in a better way makes AI one of the critical factors that will help Oman achieve its sustainability goals. The significance of AI in creating a more established, productive and sustainable future for a knowledge-based economy is pivotal as Oman approaches the 375 path. A literature review indicates that AI has great potential as an economic innovator for Oman. Nevertheless, much needs to be solved in the country for this potential to blossom, including infrastructure gaps, skilled labor and ethical concerns. Oman can also enable the transformative potential of AI for sustainable economic growth and diversification supported by developing the required technological, educational, and regulatory foundations. AI will shape the immediate future of Oman's cybersecurity strategy and become one of the most crucial components of Oman's economic strategy as its adaptation progresses.

3 Methodology

This study explores the relationship between artificial intelligence and economic innovation in the Sultanate of Oman using a quantitative approach. The objective of the study was to collect accurate and measurable data on the acceptance and perception of AI and the challenges faced by various economic sectors. The survey included 145 people from the technology, health, finance, and education sectors. The participants were carefully selected to reflect the major economic sectors of the Sultanate of Oman. To obtain data from experienced industry professionals, the selection process focused on those involved in technology, policy making, or decision-making responsibilities.

We developed the questionnaire after carefully reviewing relevant research and consulting with experts in the fields of AI and economic innovation. The questionnaire is divided into three main sections: Potential and Barriers, AI Awareness and Adoption, and Demographics. Demographics collected data on participants including industry, job title, and age. The AI Awareness and Adoption section explored how well people are aware of AI technology, whether they are currently using it in their respective businesses, and what benefits they perceive. Finally, the Opportunities and Barriers section addressed topics including poor infrastructure, shortage of skilled workers, moral dilemmas, and the expected financial impacts of AI adoption. Data was collected through online and in-person questionnaires, making participation simple and convenient. Respondents were more likely to give truthful and objective answers because they were assured of their privacy and anonymity. While simple statistics were used to describe trends in the data, more sophisticated techniques were used to identify relationships between various factors, such as the relationship between AI knowledge and applications in industry. This approach provided a clear and simple picture of the role of AI in Oman's economy and the obstacles preventing its widespread adoption.

The preferred strategies provided practical and fact-based insights into the current application of AI and its potential impact on Oman's economic expansion. Due to this study focused on quantitative facts rather than subjective judgments, the results are reliable and relevant to strategic planning and policy decisions.

4 Result

Table 1 shows that the sample consisted of 82 males and 63 females, with a majority falling in the 25–34 age group. Most respondents held a bachelor's degree, followed by those with master's or doctoral qualifications. The private sector contributed the largest proportion of participants, while a significant number had prior experience with AI, reflecting growing exposure to the technology in Oman.

Table 2 highlights that most respondents had moderate familiarity with AI, while only a more minor group demonstrated high familiarity. Nevertheless, a substantial majority regarded AI as essential for Oman's economic development, indicating widespread recognition of its potential to drive progress.

Table 1 Demographic profile of respondents

Demographic category	Classification	Number of respondents
Gender	Male	82
	Female	63
Age group	18–24	28
	25–34	57
	35–44	38
	45 and above	22
Educational level	High school or below	15
	Bachelor's degree	85
	Master's/PhD	45
Profession	Private sector	50
	Government sector	40
	Academia/students	30
	Entrepreneurs	25
Experience with AI	Yes	90
	No	55

Table 2 Familiarity with and perception of AI

Aspect	Familiarity level	Number of respondents
Familiarity with AI	High	40
	Moderate	75
	Low	30
Perception of AI's importance for economic development	Very Important	100
	Moderately important	35
	Not important	10

Table 3 Sectors that can benefit from AI

Sector	Mention frequency
Education	90
Healthcare	80
Finance	70
Logistics	65

As shown in Table 3, education and healthcare emerged as the top sectors respondents believed would benefit most from AI. Finance and logistics also ranked highly, demonstrating AI's relevance across various industries.

Table 4 shows that a majority of respondents considered AI's current utilization in Oman to be ineffective. Chatbots and virtual assistants were identified as the most prevalent applications, with limited adoption of advanced AI capabilities such as predictive analytics.

As reflected in Table 5, respondents viewed productivity enhancement and economic diversification as the most promising opportunities for AI in Oman. Job creation and public service improvements were also noted as significant benefits.

Table 6 highlights that lacking a skilled workforce was the most cited challenge to AI adoption, followed by high costs and infrastructure limitations. These findings underline the critical need for investment in education and infrastructure to overcome these barriers.

According to Table 7, the majority of respondents said Oman was only somewhat ready to implement AI. Nonetheless, a split view of the nation's preparation is shown by the equal percentage of respondents who indicated either high or low preparedness.

The majority of respondents had high confidence in AI's capacity to support Oman's Vision 2040 objectives, as seen in Table 8. Only a small percentage said AI

Table 4 Current utilization of AI in Oman

Aspect	Feedback level	Number of respondents
Effectiveness	Effective	50
	Not effective	95
Common applications	Chatbots/assistants	70
	Smart systems	50
	Predictive analytics	25

Table 5 Opportunities for AI in Oman

Opportunity	Mention frequency
Enhancing productivity	100
Economic diversification	95
Job creation	70
Improving public services	60

Table 6 Challenges in AI adoption

Challenge	Mention frequency
Lack of skilled workforce	85
High costs	70
Infrastructure limitations	60
Data issues	55

Table 7 Preparedness for AI Adoption

Preparedness level	Number of respondents
Well-prepared	30
Moderately prepared	85
Not prepared	30

Table 8 AI's contribution to vision 2040

Opinion	Number of respondents
Strong contribution	95
Moderate contribution	40
Limited contribution	10

could make a significant contribution, indicating a positive view of its potential for advancing national development.

Table 9 highlights the need for government intervention in the development of AI. respondents were a mighty advocate for funding research, promoting education, and establishing clear policies. To ensure successful implementation of AI.

Table 10 signifies the general optimism about the contribution of AI to economic growth. But privacy and ethical concerns have been major, underscoring how crucial it is to create a responsible AI framework to address these Hurdles.

Table 9 Role of government in AI development

Government role	Mention frequency
Funding research	80
Promoting education programs	75
Establishing AI policies	70

Table 10 Future perspectives and concerns

Perspective/concern	Mention frequency
Optimistic about AI	110
Ethical concerns	65
Privacy issues	55

5 Discussion

This paper delves how artificial intelligence (AI) and its potential for economic transformation in Oman are perceived and projected. The results show that the AI adoption landscape is dynamic, with new opportunities and challenges emerging regularly for a variety of industries and stakeholders. The respondents reflect a diverse background: younger professionals and people in the private sector are over-represented. This broad mix of backgrounds indicates the increased focus on AI as a general-purpose technology with the potential to change many sectors.

This suggests that a basic knowledge of AI is one of the prerequisites for promoting further adoption and embedding of these technologies into Oman's economy. Education and healthcare are two sectors that would be positioned to reap the most significant rewards from advances in AI. These sectors have the highest opportunity spaces for efficiency, accessibility, and quality of service improvement, aligned with national goals to raise public services and standard of living.

Nonetheless, the use of AI in Oman is still nascent, and there is not much assimilation in terms of practical implementations. The case calls for broader initiatives to enhance the scale and value of AI integration. AI provides enormous opportunities for Omanis to increase productivity and diversify economically. These opportunities are linked to national development strategies that prioritize innovation as one of the pillars of development; however, in order to fully benefit from AI technologies, there are a number of significant obstacles that must be addressed, including the lack of a suitable workforce, high execution costs, and infrastructure limitations. Sectors and stakeholders are not equally prepared for AI. However, they also see the symptoms of inadequate infrastructure and workforce preparation. Closing these gaps will make the shift to widespread AI usage easier.

People are hopeful that AI can help with national growth plans, make economies more diverse, and give countries an edge. The government plays a crucial role in helping AI grow. Those who answered the survey stress that we must put more money into research, education, and helpful policies to create a good setting for new AI ideas. Worries about ethics and privacy are also significant concerns, showing how important it is to use AI.

Being open and having clear rules about right and wrong is critical to keeping people's trust and making sure AI tech is used to help everyone. The research shows AI can improve Oman's economy, but it also points out the significant steps needed to get past current problems. If Oman puts money in the right places, makes good laws, and builds up its skills, it can become a top dog in using AI to spark new ideas in business.

6 Conclusion

The present research highlights how AI can be a crucial factor for economic innovation and diversification in Oman. The results revealed significant interest across sectors in adopting AI, with apparent potential excitement in education and healthcare for AI transformation. While excitement regarding the advantages of AI is high, the study shows the top issues that need to be solved before embracing its promise. These comprise the requirement for a skilled workforce, improved infrastructure and sound policies to back AI.

Organizations and stakeholders are at different reading states for AI, demonstrating excitement and caution. Even with the positive outlook of AI driving productivity and benefiting national aspirations, closing the training, resources, and ethical skills gap is critical for sustainable implementation. The government also plays a vital role by creating conducive environments, combining investment with education and regulation.

AI provides a one-off opportunity for Oman to leapfrog in new value-added industries and technology, making it a regional hub for innovation. To do this, Oman must ensure that it tackles all existing barriers while emphasizing ethical, agile and inclusive AI practices for the feature so that the nation can employ the transformative potential of AI to boost productivity, promote growth and uplift society. It is an approach that requires collaboration from government, industry, and academia to ensure that the rise of AI technology develops towards responsible deployment where all stakeholders find long-term value.

References

1. Al-Shanfari: National artificial intelligence strategy: implications for Oman's development. Oman Econom. Rev. **15**(2), 45–60 (2021)
2. Alexander, G.: Artificial intelligence: Catalyst or barrier on the path to sustainability? Technol. Forecast. Soc. Change **175** (2022)
3. Al-Farsi: Building a digital future: the role of AI in Oman's vision 2040. Middle East Technol. J. **18**(1), 22–34 (2021)
4. Hassan, et al.: AI as a catalyst for economic growth: a global perspective with focus on GCC economies. Int. J. Technol. Economy **9**(4), 90–101 (2022)
5. Abdelfattah, F., Al Halbusi, H., Al-Brwani, R., Dahleez, K. : Evaluating opportunities in e-entrepreneurship: the role of self-perceived creativity. In Mishrif, A. (ed.) Business Resilience and Market Adaptability: The Political Economy of the Middle East, Palgrave Macmillan, Singapore, pp. 45–60 (2024). https://doi.org/10.1007/978-981-97-2962-3_4
6. Brynjolfsson, McAfee.: The second Machine Age: Work, Progress, and Prosperity in a Time of Brilliant Technologies. Norton & Company (2017)
7. Binns: The ethical challenges of artificial intelligence in business and economy. J, Bus. Ethics **154**(3), 557–567 (2018)
8. Chui et al.: Artificial Intelligence: The Next Digital Frontier? McKinsey Global Institute (2018)
9. OECD: Artificial Intelligence in the Public Sector: Challenges and Opportunities for Innovation. Organization for Economic Co-operation and Development (2020)

10. Ministry of Economy: Oman Vision 2040: Economic Diversification and Technological Advancement. Government of Oman (2020)
11. Yogesh, et al.: Evolution of artificial intelligence research in technological forecasting and social change: research topics, trends, and future directions. Technol. Forecast. Soc. Change 192 (2023)
12. Dwivedi, et al.: Artificial Intelligence (AI): Multidisciplinary perspectives on emerging challenges, opportunities, and agenda for research, practice and policy. Int. J. Inf. Manage. **57** (2021)

Empowering SMEs in Jordan: Leveraging Artificial Intelligence for Digital Marketing Success

Mahmoud Abouraia, Ayman Bassam Nassoura, Hothefa Shaker, Fadi Abdelfattah, and Hisham Madi

Abstract This research aims to identify how AI could improve the concept of digital marketing to help Small and Medium-sized Enterprises (SMEs) in Jordan. Through surveying 125 Jordanian SMEs, the research paper discusses AI tool adoption, its efficacy, the difficulties faced, and the possibilities of future AI advancements. The study emphasized the fact that though the application of AI is touted as a means of enhancing customer interaction and promoting friendly marketing strategies, multiple challenges affect SMEs, exceptionally high implementation costs, severe scarcity of technical personnel, and data privacy issues; all of which need to be overcome for better business performance. Nevertheless, based on the responses, SMEs remain ready to adopt AI, meaning there is a good starting point for firms looking to go through the digital transformation process. Such a conclusion strengthens demand for specific development and educational programs focused on the SMEs' AI capabilities enhancement, presenting policy recommendations for those who want to support AI growth and development.

Keywords Artificial intelligence · Digital marketing · Small and medium-sized enterprises (SMEs) · Jordan · AI adoption challenges · Predictive marketing

M. Abouraia · H. Shaker · F. Abdelfattah (✉) · H. Madi
Modern College of Business and Science, Muscat, Oman
e-mail: fadi.abdelfattah@mcbs.edu.om

M. Abouraia
e-mail: Mahmoud.abouraia@mcbs.edu.om

H. Shaker
e-mail: hothefa.shaker@mcbs.edu.om

A. B. Nassoura
Department of MIS, City University Ajman, Ajman, UAE
e-mail: a.bassam@cu.ae.ae

B. Awwad (ed.), *AI and IoT: Driving Business Success and Sustainability in the Digital Age*, Studies in Systems, Decision and Control 590,
https://doi.org/10.1007/978-3-031-88874-8_90

1 Introduction

In the modern world of highly competitive business, AI presents digital marketing with a new round of tools that allows businesses to improve the advertising experience for customers, increase the speed and effectiveness of decision-making, and achieve greater efficiency when designing marketing initiatives. Machine learning, Natural Language Processing (NLP), and predictive analytics have made it easy for marketers to examine large datasets and forecast and target consumers with high precision [1]. It is essential for Small and Medium-sized Enterprises (SMEs) that, while they have a limited marketing budget, they may need new approaches to strengthen their competitive position. However, despite the broadly positive outlook, the integration of AI by SMEs is still not homogenous, primarily because of the scarcity of resources, knowledge and consciousness of its possible use [2, 3].

Digital marketing has advanced even more significantly with the birth of artificial intelligence, with AI tools that can create content, segment customers, or create smart advertisements [4]. Marketing automation studies prove that AI-based marketing tactics like recommendations, predictive models, and chatbots positively impact customers' experience and sales [5, 6]. The move towards digital channels is particularly crucial for SMEs in emerging markets such as Jordan, where pressures from a growing Private Economy and Commercial Activities economy have seen firms turn to technology to stay relevant. The SME sector in Jordan is essential to enhance employment opportunities, income generation, and innovation [7]. To these organizations, using artificial intelligence in marketing could offer a strategic means for growth, increasing customer interaction and productivity. However, this premise has not been realized due to studies showing that regarding AI uptake, Jordans SMEs are bound by issues concerning technical competence, getting data, and budget constraints [8].

This research will fill the abovementioned gap by analyzing how Jordanian SMEs use AI in their digital marketing methods and the strengths and weaknesses of AI utilization. Thus, by assessing the current levels of AI application, its efficacy, and perceived barriers, this study aims to identify perspectives that can help SMEs harness AI to enhance marketing outcomes. The work also looks at future possibilities for the creation of AI for SMEs in Jordan and gives a more positive view of how they can bypass the current limitations of utilizing AI to their advantage. This paper thus expands the current understanding of AI use in digital marketing to highlight recommendations for Jordan's SMEs through a survey of 125 firms across industries.

In the following sections of this paper, the literature on AI in digital marketing will be reviewed, a description of the method used for data collection will be provided, and a presentation and discussion of the survey findings to reveal the contemporary landscape and potential of SMEs in Jordan to leverage AI-integrated marketing. Having this understanding of the dimensions, Jordanian SMEs may not feel overwhelmed with trying to catch up with the shift necessary to compete in today's increasingly digital environment.

2 Literature Review

The integration of AI in the digital marketing domain has received significant focus over the past years as firms continue to employ AI solutions to promote value addition in customer experience, strategy formulation, and improvement of market decision-making. This section presents a literature review of the current state of AI in digital marketing, AI in marketing, AI deployment issues, and AI's prospects in digital marketing, mainly focusing on SMEs in emerging markets such as Jordan.

Another primary application of AI in the context of consumer behavior is the easy test. It is well known for its capacity to handle big data and mine them to help businesses better understand customer behavior, attitudes and buying habits [9]. Applying artificial intelligence in digital marketing ventures, such as machine learning, chatbots, predictive analytics, and automated content generation, will enhance clients' perceived value of the digital experience [10]. Today's artificial intelligence tools dominate marketing platforms since they enable targeted segments to be reached with customized messages and content based on their online activities [11].

In Jordan, therefore, digital marketing is crucial to SMEs, who are trying their best to compete in the market. An increasing number of Jordanian businesses, particularly SMEs that constitute most industries in the country, have yet to grasp the significance of DP, while AI has significant potential to improve brand awareness and customer engagement on such platforms [7]. However, many SMEs have yet to embrace the AI strategy, which usually engages in the most straightforward digital marketing trends combined with social media. The reasons AI is effective in digital marketing include empirical evidence showing how AI strengthens marketing impacts in certain areas. AI marketing applications like customer segmentation, content personalization, and automated lead generation are proven to optimize business operations, minimize cost, and enrich decision-making [12]. The first advantage is that AI offers marketers real-time data analysis to alter their programs and control marketing expenses following consumer trends.

Research has established that AI marketing strategies improve customer satisfaction and loyalty by pushing tailored marketing strategies in line with customer inclination [13]. For instance, the consumption by clients such as Amazon, Netflix, and other e-commerce platforms of synthetic intelligence-based recommendation systems have enhanced sales and client satisfaction due to the recommendations to them based on their past conduct [14]. The identified benefits of AI are precious for SMEs, which may have limited financial resources or the ability to invest large sums in traditional marketing advertising campaigns. Based on the above understanding, this paper regards AI as conducive to enabling SMEs to streamline specific segments of digital marketing and, more importantly, to lower operational costs while enhancing overall marketing responsiveness [15]. SMEs are also constantly adopting AI to leverage advanced tools for acquiring, engaging and retaining customers, putting them at par with large competitors.

However, integrating AI in the digital marketing environment has some emerging issues that can affect SMEs specifically and those from the Jordanian market in general. One of the challenges is that the employees in the organization do not possess adequate technical skills regarding artificial intelligence. Most small and medium business managers, together with their employees, may lack adequate skills to undertake the proper adoption and leveraging of AI technologies [2, 16]. Furthermore, there is frequently insufficient knowledge regarding the kinds of tools facilitated by AI and how they may be incorporated into marketing activities. As one of the critical disadvantages, another issue that can be pointed out is the cost of AI implementation. Despite the long-term benefits that may accrue, the initial outlay needed to develop AI technologies, purchase software, and train existing workers can be expensive for many SMEs, especially in territories undergoing the digital transition [8]. In addition, poor data quality may also be a problem that SMEs encounter in Jordan when training the AI models. AI systems demand massive datasets to function correctly, and insufficient or low-quality data is a barrier to AI in marketing [5, 17].

The two other critical factors that come into play when adopting AI include cultural and organizational factors. Skepticism concerning change and mistrust in automation systems are significant barriers to SMEs' adoption of AI-driven marketing strategies. There could also be issues associated with the appropriateness and ethics of AI applications in customer relations, including privacy issues and bias in the algorithms [16, 18]. This paper finds that advancements in AI for digital marketing seem promising, with a pivotal pointer for SMEs in emerging markets such as Jordan. The availability of AI technologies will increase the number of SMEs incorporating AI into their marketing strategies due to improved accessibility and cost of the technologies. AI's improvements, especially NLP, computer vision and machine learning algorithms, generate new possibilities for customer relations and data customization [13, 19].

Thus, the further advancement of truly usable AI tools for small and medium-sized enterprises that are both affordable and scalable is expected to enhance the companies' interest in applying those technologies in their operations. Also, the continuing advances in cloud and AI technologies will mean that investments in large AI systems are not as prohibitive to small businesses as once believed [11, 20]. With artificial intelligence in creating virtual assistants, chatbots, and predictive analysis, the marketing field will continue to expand and can provide better and faster customer experience. In Jordan, government and industry organizations have the potential to share some responsibilities for overall AI promotion by offering incentives, consultation and training. The increase in the adoption of digital solutions by SMEs in the country will also determine the expansion of the AI market and intensify the competition and innovation in the marketing field.

The literature suggests that while AI is seen as a significant enabler that could enhance digital marketing for SMEs, issues related to expertise, cost, and data accessibility persist. Still, as AI technologies become more developed and affordable, SMEs in Jordan and other new markets will likely receive enhanced opportunities to apply AI to improve marketing effectiveness. With the ever-increasing importance of digital marketing, the use of AI in helping businesses grow and engage with their customers cannot be overstated. AI, therefore, pointed out that adopting AI in marketing will be an essential value-add for Jordanian SMEs in a highly digitalized economy.

3 Methodology

The present study employed a quantitative method to examine AI's application in digital marketing in SMEs in Jordan. The first part of the study covered the survey, sample, data and analysis methods used to capture current usage of AI, challenges faced and the likelihood of future use of AI in marketing activities among SMEs. The survey was conducted with 15 questions that were built into the questionnaire as closed questions, aiming to possess more quantifiable results regarding the general utilization of AI in digital marketing as well as the perceived efficiency, impersonal difficulties, and the presumed tendency and forecasted likelihood of much more frequent use in the future. Adopted from the literature review, the questions were customized to Jordan's SMEs, with responses captured on a Likert scale of 5 addressing the extent and impact of AI applications.

A purposive sampling technique was employed to select SMEs from different industries in Jordan's economy. The respondents totaled 125, including those from retail, manufacturing, and service industries. Respondents were people involved in business at different managerial levels and who had a sufficient understanding of their firm's online marketing tactics. The participants' diversity enriched the study and provided broader and varied perspectives on the topic investigated across various types of businesses and contexts of functioning. Consequently, data collection took four weeks, and surveys were self-administered instruments, which offered a convenient way to access participants in Jordan. Because the survey was self-administered, respondents could complete it conveniently and quickly. Data collected was sorted for cleaning after coding and then analyzed to make it credible and reliable through statistical software. A quantitative approach of descriptive analysis of frequencies, means, and standard deviations gave an overview of the respondents' perception of AI in digital marketing.

Furthermore, inferential statistics like correlation and regression analyses found the possibility of a correlation between AI adoption and perceived business outcomes, which helped us understand how AI impacts marketing efficiency in the SME segment. To enhance the work's credibility, the survey instrument was pretested with a small sample of SME representatives to confirm the validity of the items used in the measurement instrument. A Cronbach alpha test was conducted to test the reliability of each sub-scale whereby the overall reliability coefficient was 0.809, which was more extraordinary than the pre-requisite 0.7, signifying the high reliability of the entire scale. This method provides an exhaustive and coherent strategy when studying the impact of artificial intelligence in digital marketing in small and medium enterprises in Jordan. It offers essential data on the state of current adoption, the problems faced, and SMEs' views on AI's capacities for advancing digital marketing activities in the future.

4 Result

The analysis in Table 1 shows that most respondents are decision-makers within their companies, including owners (40%) and marketing managers (28%). Most businesses have been operating between 4 and 10 years (36%) or 1–3 years (28%), indicating an established yet growing presence of SMEs in Jordan. Retail is the most represented industry (32%), followed by technology (20%) and hospitality (16%). Additionally, nearly half of the companies (40%) have fewer than ten employees, underscoring the small business scale of most respondents.

Table 2 reveals that chatbots are the most widely adopted AI tool (48%), with customer segmentation (40%) and predictive analytics (36%) also commonly used. These findings suggest that SMEs prioritize AI tools that enhance customer interaction and insights. A minority (12%) of respondents reported no AI adoption, indicating a lack of adoption among some companies, potentially due to resource constraints or limited knowledge of AI technologies.

According to the results, the majority of respondents believe that AI has "moderately improved" (40%) their marketing outcomes, while 24% see a "significant improvement." A smaller group (8%) reported no noticeable improvement, which may suggest that the effectiveness of AI can vary based on the specific applications used or the level of implementation within each SME. The results showed that high

Table 1 Demographic overview of respondents

Variable	Categories	Frequency	Percentage (%)
Role in company	Owner	50	40
	Marketing manager	35	28
	Marketing Specialist	30	24
	Other	10	8
Years in business	Less than 1 year	15	12
	1–3 years	35	28
	4–10 years	45	36
	More than 10 years	30	24
Industry	Retail	40	32
	Technology	25	20
	Hospitality	20	16
	Healthcare	15	12
	Education	10	8
	Other	15	12
Number of employees	Fewer than 10	50	40
	10–50	40	32
	51–100	20	16
	More than 100	15	12

Table 2 Adoption of AI technologies in digital marketing

AI technology implemented	Frequency	Percentage (%)
Chatbots for customer service	60	48
Predictive analytics for customer behavior	45	36
AI-powered content creation	35	28
Personalization of marketing content	40	32
Customer Segmentation Using Machine Learning	50	40
Voice search optimization	25	20
None	15	12

costs (52%) and lack of technical expertise (44%) are the main challenges faced by SMEs in adopting AI for digital marketing. Data privacy concerns (32%) and limited access to quality data (24%) are also significant obstacles, emphasizing the need for support in these areas to help SMEs overcome adoption barriers. In addition, the result indicates that 36% of respondents are "very likely" to increase their AI investment in the next two years, with an additional 28% "likely" to do so. This trend suggests that many SMEs are increasingly confident in AI's potential and are considering scaling their AI initiatives in digital marketing to improve competitive advantage and customer engagement.

5 Discussion

The collected data was further purified, coded, and analyzed using a statistical tool to measure their accuracy and reliability. A quantitative data analysis method used frequencies, means, and standard deviations to present the respondents' perceptions of using AI in digital marketing. Moreover, the survey gathered detailed data on the correlation between AI adoption and perceived business impacts; inferential statistics like correlation and regression analyses gave a deeper insight into the effects of AI on marketing efficiency within the SME context. The conclusion reached in this study offered insights into AI usage in digital marketing by Jordanian SMEs, concentrating on their adoption behavior, perceived advantages, constraints, and plans. The demographic profile shows that most respondents are critical employees in developed SMEs, mainly in the retail, technology, and hospitality industries. This demographic context is essential as such sectors have many customers and are suitable for using AI in marketing. SMEs employ various AI tools like chatbots, customer segmentation, and predictive analytics. Chatbots are relevant to world tendencies because chatbots provide customers with fast and personalized services that save time when human staff is scarce in small businesses. On the adoption of predictive analytics, there is a demonstration of understanding of customer insights well known to be critical in the marketing of products to customers and is well understood as a means of encouraging the customers to be loyal to the brands being sold in the market.

However, a significant percentage of the SME respondents who have not adopted AI could result from constraints like inadequate capital and technical know-how.

The perceived benefit of AI analysis shows that most SMEs benefit from its effect as it enhances the efficiency of marketing processes and improves customer relations. Most experience mild to considerable improvement, and a few report minimal improvement. It might be due to the difference in the depth of its implementation, the availability of resources, or the types of AI applications deployed. This research supports the idea that although AI has great potential, the optimal use of AI calls for the use of proper tools and enough training. It is noteworthy that transfer barriers in AI implementation have not diminished much for Jordanian SMEs and that cost and expertise create the main barriers. One of the challenges is high costs, which is still a significant factor for small businesses because of the lack of funding required to invest in new technologies. Also, the absence of internal competency referring to the application of AI tools reflects the probable skills shortage, which means that SMEs cannot maximize the utilization of these tools even if they procure or develop them. This leads to considerations of privacy and data quality as significant issues, highlighting the long-standing problem of poor data management and the lack of a comprehensive regulatory framework for business owners.

The view towards AI investment also suggests that the future is bright because most SMEs claim they will increase their investment in AI in the next few years. Such optimism means that, even in existing adversities, SMEs see the corresponding effectiveness of AI in advancing their marketing strategy. Further enhancements of technologies and greater availability of affordable and easy-to-use AI tools may also contribute to this process and aid SMEs in attaining edges and improving their interaction with clients. In conclusion, this study identifies the trend of using AI for the digital marketing of products in Jordanian SMEs with a focus on addressing challenges and opportunities to succeed in using AI for sustainable growth. The study provides evidence that enhancing the digital marketing effectiveness of SMEs in Jordan through AI is possible if they are provided with adequate support and resources. Such change might not only be beneficial for individual companies' results but could also foster the economies and SMEs' development in Jordan based on digitization.

6 Conclusion

This paper discussed the current state of artificial intelligence, AI trends, effectiveness and the significant challenges Jordan's SMEs face when adopting AI in digital marketing. The findings highlighted some promising potential and challenges AI presents to these enterprises. According to the study, SMEs are growing in adopting these tools, such as chatbots, customer segmentation, and predictive analytics. However, the most significant potential of AI in growing digital marketing is still unrealized because of threats, including high costs, lack of technical personnel, and privacy issues. However, research on the implementation of AI for SMEs shows that firms that have adopted AI are receiving improvements in overall results, especially in customer relations and marketing.

The findings underline conceivable gaps in AI solutions from the resource constraints of the SME organizations and the training programs to reduce the skills gap. Regarding the dynamic advancement in the technology sector, Jordanian SMEs seem to foresee more investment in AI in the future. Such a perspective for the future means there is an increased willingness within the SME sector to implement AI at large, consequently defining a strategy to amplify the competitiveness and stability of the sector. Therefore, to increase the use of AI for digital marketing by Jordanian SMEs, there is a need for cooperation between AI technology vendors, government, and industry organizations to address the challenges outlined above. Such mutual support would create a context in which AI will be a competitive solution, facilitating the development of reliable and sustainable growth directions for the Jordanian economy. The study thus offers a basis for appreciating the state of AI readiness among Jordanian SMEs and, therefore, offers a principled direction for positively influencing this sector's AI readiness and, by extension, its ability to engage in AI-based innovation.

References

1. Abdelfattah, F., Al Halbusi, H., Al-Brwani, R., Dahleez, K.: Evaluating opportunities in E-Entrepreneurship: the role of self-perceived creativity. In: Mishrif, A. (ed.) Business Resilience and Market Adaptability: The Political Economy of the Middle East, Palgrave Macmillan, Singapore, pp. 45–60 (2024). https://doi.org/10.1007/978-981-97-2962-3_4
2. Al-Natour, Gharib.: Digital marketing in jordan: challenges and opportunities for SMEs. Int. J. Bus. Manage. **16**(4), 25–38 (2021)
3. Chaffey: Digital Marketing: Strategy, Implementation, and Practice. Pearson (2020)
4. Chakraborty, et al.: Ethical implications of AI in marketing. J. Bus. Ethics **157**(3), 645–660 (2022)
5. Chung, et al.: The effectiveness of personalized marketing through AI in E-commerce platforms. J. Market. Sci. **40**(1), 68–82 (2022)
6. Gentsch.: AI in Marketing: How Artificial Intelligence is Changing the Marketing Landscape. Springer (2020)
7. Gupta, George: Barriers to artificial intelligence adoption in SMEs: the case of developing economies. Int. J. Entrepreneurial Behav. Res. **29**(2), 387–406 (2023)
8. Rahman, M.S., Bag, S., Abdelfattah, F., Rana, N.P.: The New wave of AI-powered luxury brands online shopping experience: the role of digital multisensory cues and customers' engagement. J. Retail. Consum. Serv. **72**, 103273 (2023). https://doi.org/10.1016/j.jretconser.2023.103273
9. Hussain, Mehmood:. Leveraging AI in small businesses: strategic insights and implementation. J. Small Bus. Manage. **60**(1), 112–128 (2022)
10. Infoway, S.: The rise of AI in digital marketing: how machine learning is transforming campaigns (2024, May 21). Retrieved from https://medium.com/@sakshiinfoway/the-rise-of-ai-in-digital-marketing-how-machine-learning-is-transforming-campaigns-c3db2c45de65
11. Madanchian, M.: The impact of artificial intelligence marketing on e-commerce sales. Systems **12**(10), 1–20 (2024)
12. Mikalef, et al.: Digital transformation through artificial intelligence: opportunities and challenges for SMEs. IseB **18**(2), 331–355 (2020)

13. Patel, et al.: The role of AI in digital marketing: a study on its impact in emerging markets. J. Digital Market. **8**(1), 13–28 (2022)
14. Abdelfattah, F., Salah, M., Dahleez, K., Darwazeh, R.: The future of competitive advantage in oman: integrating green product innovation, AI, and intellectual capital in business strategies. Int. J. Innovat. Stud. **8**(1), 635–655 (2024). https://doi.org/10.1016/j.ijis.2024.02.001
15. Pereira, et al.: Artificial intelligence and marketing: a comprehensive overview. Int. J. Market. Res. **42**(3), 117–134 (2020)
16. Qasem, et al.: Artificial intelligence adoption in SMEs in Jordan: challenges and perspectives. J. Small Bus. Enterp. Dev. **29**(4), 415–431 (2022)
17. Shankar, et al.: Artificial intelligence in marketing: understanding the benefits and implementation challenges. J. Bus. Res. **123**, 130–142 (2021)
18. Madi, H., Nabi, G.A., Abdelfattah, F., Madi, A.: Artificial intelligence and transforming digital marketing. In: Studies in systems, decision and control, Springer, Cham (2024).https://doi.org/10.1007/978-3-031-35828-9_35
19. Wang, et al.: The role of artificial intelligence in digital marketing: opportunities and impacts. J. Bus. Res. **139**, 83–92 (2021)
20. Zhang, et al.: AI tools and techniques in digital marketing: a review. Int. J. Market. Stud. **12**(2), 41–56 (2020)

Exploring the Impact and Adoption of Artificial Intelligence in Education: A Study of Jordan's Educational Landscape

Mahmoud Abouraia, Ayman Bassam Nassoura, Hothefa Shaker, Fadi Abdelfattah, and Hisham Madi

Abstract This research aims to determine the current use of artificial intelligence (AI) in Jordan's educational system, its implementation's effects, the benefits realized, impediments to its expansion, and prospects for the future. The target participants were students, educators, administrators, and policy-makers. A graded questionnaire was used to survey 245 participants. This study compares the quantitative data to assess the stakeholder's attitudes toward integrating AI into education. The findings show that the level of concern is high, and participants expressed their desire to know more about AI and how it can improve learning achievements and administration. Nevertheless, such restrictions and limitations as the lack of qualified specialists, inadequate infrastructure, organizational culture, and opposition to change proved critical for advanced AI implementation. Nevertheless, as highlighted in this study, Jordan could leverage investment in infrastructure, training, development, and policies to embrace AI as an enabler in the educational sector. At the end of this research, there is a recommendation on how to enhance awareness and ensure the availability of the appropriate resources to facilitate the integration of AI in educational institutions in Jordan.

Keywords Artificial intelligence (AI) · Education · AI adoption · Jordan · Educational technology · Policy development · Digital transformation

---

M. Abouraia · H. Shaker · F. Abdelfattah (✉) · H. Madi
Modern College of Business and Science, Muscat, Oman
e-mail: fadi.abdelfattah@mcbs.edu.om

M. Abouraia
e-mail: Mahmoud.abouraia@mcbs.edu.om

H. Shaker
e-mail: hothefa.shaker@mcbs.edu.om

A. B. Nassoura
Department of MIS, City University Ajman, Ajman, UAE
e-mail: a.bassam@cu.ae.ae

B. Awwad (ed.), *AI and IoT: Driving Business Success and Sustainability in the Digital Age*, Studies in Systems, Decision and Control 590,
https://doi.org/10.1007/978-3-031-88874-8_91

1 Introduction

The adoption of AI in education is gradually receiving increased attention globally as a new trend of making learning more effective, efficient, and enjoyable, besides advancing learning goals and objectives. Learning technologies, including machine learning, natural language processing, and intelligent tutoring systems, can enhance students' learning by delivering content and providing feedback to develop what suits them best [1]. The development of the application of AI in education systems and its contribution to enhancing education quality and access, as well as helping teachers and managers in decision-making, is appreciable in developed and developing countries [2]. However, for Jordan the adoption of AI in educational institutions has remained low despite the fact that its implementation is looming large. Jordan has grasped the economic role of AI innovation and is engaging in national actions on the use of AI in areas including schooling [3]. The Jordanian government has recorded massive success in the exploitation of technology for development The government of Jordan has thus devoted considerable effort to its vision of becoming a smart country. This supports the use of ICT in enhancing the delivery of public services and the emphasis include for example education. Yet, the potential of applying AI into the Jordanian educational system has the following concerns: ineffective physical learning environment, weak technical competency, and organizational resistance to change.

There is a vast body of literature around the world that addresses many aspects of AI in education: It ranges from the prospects for applying intelligent technologies for the teaching–learning processes to challenges of implementing it and the moral issues which arise when using AI in education. Research indicates that AI can assist in developing a plan that delivers differentiated instruction by showing teaching methods that adapt the pace and material of instruction to the learner's success rate [4]. In addition, AI can free teachers' time through shared activities, grading, and analyzing students' achievements. [5]. Furthermore, other ability-enhancing technologies like intelligent tutoring systems have revealed the potential for delivering timely feedback and skills practice in real-time, especially in mathematics and language learning, among others [6].

Another area that has attracted much attention from researchers in the past years is applying artificial intelligence in higher education institutions, where the application of AI in the delivery and management of education is considered in detail. In their study, Nassoura [7] Systematically reviews applied artificial intelligence applications in higher education institutions, elaborating on how different AI technologies have been employed to support learning achievement, amplify administrative proficiency, and create intelligent learning surroundings. The paper stated that the use of AI in improving the quality of higher education is emerging even more, beginning with the intelligent tutoring system, automated assessment for the student, and AI with result analysis for performance [7].

Nevertheless, issues arise from AI's application in education, as discussed below. Investigations show that there is a shortage of professional development for teachers, drawbacks regarding the presence of AI, their usage, and students' data protection that is crucial for the rights of learners [8]. There are also issues regarding the cost and availability of the technologies in question across many developing countries such as Jordan and the availability of these technologies in constrained educational settings [9]. Furthermore, the author also captured the cultural and social challenge of embracing AI-based education tools as another challenge by other scholars [10]. There are increased attempts to improve the quality of education, together with inclusive information and communication technologies, in the country towards achieving equitable access to learning resources [11]. In this relation, AI's potential advantages, including education provision, student performance, and education equality, are crucial policy, education, and research priorities in Jordan.

This research seeks to discover the state of AI integration in Jordan's education system and its prospects and constraints from the distinct stakeholder viewpoints of students, educators, guardians, administrators, and policy-makers. From these different viewpoints, this research aims to determine the AI expansion driver in Jordanian education while also ascertaining the practices required to mitigate the existing challenges. First, the study will expand the existing literature regarding the use of AI in education, especially in the Middle Eastern and North African (MENA) region; as such, it will offer recommendations that would help policy-makers and educators improve the utilization of AI technologies in the Jordanian educational system.

2 Literature Review

AI applicability and agent incorporation in education systems are among the most discussed topics among scholars and policy-makers, stressing how AI systems can improve learning-teaching paradigms. The application of AI in education can be summed up as personalization of learning and instruction, administrative optimization, and decision support. However, implementing AI in education systems, especially in Jordan's developing nations, is difficult. This literature review analyzes the conditions of utilizing AI for education, the current advantages and challenges, future motivation to integrate AI, and the strategies for impacting Jordan.

2.1 AI in Education: A Global Perspective

Around the world, AI has been applied in many facets of learning, including customized learning, grading, and feedback systems. Intelligent teaching and tutoring systems have been considered helpful in enhancing students' learning because they deliver content depending on their needs [11]. In their review of AI in learning [3] concluded that machine learning and predictive analysis can foster a personalized learning environment. The above-advanced technologies help track students' progress, adapt the courses being offered, and even provide feedback to students in real-time, increasing the ratio of students to courses offered and, in the process, boosting enrollment. In the same way, another study pointed out that AI in adaptive learning has more remarkable performance in mathematical courses because concepts can be mastered when presented in smaller segments depending on the learner's capability [12].

Apart from making learning personalized, AI in education can enhance efficiency and reduce the task volume that learners can perform. As a result, AI reduces the amount of time educators can spend on administrative tasks, for example, grading assignments or creating and redistributing the weekly/timetable [13]. For instance, makers of extensive online courses use auto-grading tools that enable the preparation of students' results and the acceptance of their assignments at a breakneck speed, thus saving the precious time of instructors. These AI systems can also take advice from student analysis data to look for struggling students and generate actions to gingerbread them; this leads to elevated academic accomplishment and consequently generative improved student accomplishment data [9]. In addition, the particularity of AI dealing with decisions in education management has been discussed. Education managers and policy-makers can use AI to gain insights from large datasets for policy-making, resource utilization, and curriculum planning [14]. They should help educational leaders by offering analysis of student conduct, learning processes, and institutional outcomes, giving a good foundation for decision-making and enhancing education quality and accessibility.

2.2 AI in Higher Education: Applications and Challenges

This Paper aims at identifying how AI has been implemented in the higher education institutions as a way of enhancing student performance as well as organizational efficiency. Since LMS can aid in improving students learning experience, AI has been adopted into the LMS. Luckin et al. [6] also point out that in practice AI presents spectacular real-time homework and review systems that can critically review student's work and provide them immediate feedback. It can be used to mark dissertation assignments, and even offer solutions to questions asked thus enhancing the quality and time taken to do the marking [13]. Peculiarly, higher education as a rule has not embraced AI as the end remedy. The problem is that there is no adequate organizational approach and the required qualifications in various organizations to implement specific AI systems [8]. Schools in poor countries like Jordan lack adequate funding and technology, and human capital, to create and sustain sophisticated AI applications [1]. Besides, there might be other reasons for this as arts-based AI tools may cause concerns among some faculty members related to their control over the process of education and humanities compared to the use of AI [12]. The second major issue is the ethical question or concern related to AI, which has solution in education system.

Algorithmic bias, data privacy, and the potential for discrimination within AI systems have also been raised by both academics and industry professionals [14]. For instance, when the AI system is using student data to come up with some of the decisions it will have the potential of being prejudiced by some agreed-on biases. Appropriateness: Yet another question regarding AI algorithm is its transparency particularly when it is used in high risk educational contexts as in admissions or grading [14]. Most universities should implement the application systems of artificial intelligence rightfully and appropriately while there needs to be adequate safeguard measures put in place in regard to students as the applications keep growing. Scholars and practitioners have concerns pertaining to data protection, privacy of the algorithms and discrimination [14]. For instance, if decision are made about students based on the data that has been collected, then it can be assumed that the decision that are taken may well be biased and unfair. Moreover, the amount of transparency when it comes to AI algorithms in important sections including admissions and grading is still alarming [15]. Given the fact that application of such technologies in institutions is rising at a very fast pace, it becomes mandatory that such institutions take measures that would ensure consumer rights are well protected, and that the students' welfare is well taken.

2.3 AI in Jordan's Education System: Opportunities and Barriers

The use of AI in education in Jordanian is relatively limited and the government as well as schools is only now beginning to cautiously take steps toward digital adaptation. In their study too, made in 2021, Al-Qudah et al. also pointed out that the application or adoption of AI results in the reform in the health sector of Jordanian Government Health Sectors to improving the education quality of education in Jordan. As a result, various national strategies for computer literacy for students and information technologies in classes in Jordan have been formulated. These are an advancement of the AI avails education and comprehensives and intention to introduce or implement AI in schools and college. However, there are some drawbacks that worth considering if AI has to be employed in the process of Jordan's education sector. These are the constraints: These are attributed to a poor development of infrastructure such as internet facilities and modern computing devices which are scarce and un-proportionately shared within learning institutions especially in rural areas [1]. Indeed, one of the biggest problems today is the scarcity of talented teachers that know how to use AI. Kassem and Nasser [5] provide evidence for the view that a large share of educators in Jordan remains either insufficiently informed about AI technologies or insufficiently prepared to implement them in learning.

In addition, there are external barriers considered to be obstructive to the education process and the integration of AI solutions including the prevailing mentality in the academic community, which might hold pessimistic attitudes towards AI using their hat as a teacher which may be replaced by AI in the future. Experts and tutors themselves remain fairly ignorant about the potential opportunities and risks of AI application in learning processes, and this makes it challenging to develop suitable policies and practices regarding AI's integration into the educational environment [2]. Nevertheless, there are several great opportunities to apply AI in Jordan's education system. It is also important in enhancing delivery of provision of education by providing students with better learning tools and delivering education online to those who are in hard to reach areas of the society. Also, learning analytics derived from AI can assist in the detection of learning patterns and hence offer the educators insights to the students that require more attention [3]. In addition, smart administrative services can help to minimize the administrative burden to improve the work of schools and other institutions [13]. This paper goes through various papers that analyze the incorporation of AI in education in general and a case of Jordan in particular.

3 Methodology

This study aims at assessing the current state of AI, advantages, disadvantages and future trends of AI in Jordanian education system. A quantitative research approach was adopted because the participants recruited included students, teachers, administrators and policymakers. The samples were collected using cross-sectional survey questionnaires which were distributed to participants in several education centers. This survey will therefore seek to capture the demographic attitude of respondent towards the adoption of AI in education and what perception and expectation they have about AI. The population targeted comprises all individuals involved in Jordan's educational process: students, teachers, principals, and other government officers associated with school and college. Using a stratified random sampling method the sample size of 245 participants has been taken from schools universities and vocational training centers. The sample was subdivided according to occupation of respondent as student or teacher, in a school for general, vocational, or higher learning. This approach ensured that all the diversified view of the educational stakeholders were given an equal chance.

This data required for this study was gathered through a structured closed ended pre tested questionnaire. The survey includes questions connected to main concepts of AI in learning, including participants' awareness of AI, current use of AI in the learning context and its advantages and/ or disadvantages and further potential of AI. While developing the questionnaire both closed-ended and Likert-scale questions were incorporated to know the perception of the respondents toward AI in education. This was intersubjectively established about the validity and reliability of the instrument among a panel of experts. The survey was then conducted electronically to the selected sample via email and their social media handles on 20/09/2024. The respondents were given the option of two weeks to fill out the questionnaire, and reminder emails were sent to them. As for the student participants,

245 completed responses were received, making the response rate about 80%. The information obtained from the questionnaires was analyzed using statistical software (SPSS v. 15). Frequency, percentage, and mean scores were used to analyze the respondents' demographic data and attitudes to AI in education. Chi-square analysis was employed on the respondents to assess the extent of association between various factors, such as the one that focused on the respondents' experience in educational setup and the other one that concerned the level of awareness of AI. The answers were analyzed and categorized to identify AI-related advantages and disadvantages, and the results were described qualitatively.

4 Result

Table 1 shows that students form the largest group of respondents, making up 40% of the sample, followed by teachers/instructors at 29.4%. This indicates significant representation from those directly involved in the learning process. Regarding the type of institution, schools and universities dominate, with 41.6% and 39.2% of respondents, respectively. This highlights the relevance of AI in traditional educational settings.

Table 2 illustrates respondents' familiarity with AI in education. Most respondents (41.6%) are somewhat familiar, while 27.8% are familiar. Only 8.2% report being very familiar, indicating room for increased awareness and training to better understand AI applications in education.

Table 3 indicates that 53.9% of respondents confirm using AI technologies in their institutions, signifying that over half of the sample has had exposure to AI in education. However, 32.7% reported no usage, and 13.5% were unsure, emphasizing a knowledge gap that needed to be addressed.

Table 1 Demographic profile of respondents

Category	Frequency	Percentage (%)
Role in education		
• Student	98	40
• Teacher/instructor	72	29.4
• Administrator	38	15.5
• Policy-maker	15	6.1
• Others	22	9
Type of institution		
• School	102	41.6
• University	96	39.2
• Vocational training	32	13.1
• Others	15	6.1

Table 2 Familiarity with AI in education

Level of familiarity	Frequency	Percentage (%)
Not familiar	55	22.4
Somewhat familiar	102	41.6
Familiar	68	27.8
Very familiar	20	8.2

Table 3 Current use of AI technologies

AI usage in education	Frequency	Percentage (%)
Yes	132	53.9
No	80	32.7
Not Sure	33	13.5

In addition, the results revealed that 65.3% of respondents see improved personalization as the key benefit of AI in education. Reduced teacher workload and enhanced decision-making rank highly, highlighting AI's dual impact on teaching efficacy and administrative efficiency. The results highlight funding as the most significant challenge, cited by 69.4% of respondents. Limited technical expertise and infrastructure gaps also pose considerable barriers. Although cited less frequently, ethical concerns remain essential in discussions about responsible AI deployment. The results also showed that nearly half of the respondents (44.9%) anticipate widespread adoption of AI within 3–5 years, reflecting optimism for its integration in the medium term. However, 18.4% foresee a longer timeline, pointing to potential barriers that must be addressed.

5 Discussion

Based on the data derived from this study, it was evident that AI is becoming increasingly significant in Jordan's educational industry. Even so, the measures of experience with AI, its utilization, and its anticipation of its application remain diverse. The results show that most participants seem quite informed about AI technologies but still demonstrate limited knowledge of how AI could be best applied to teaching practices. This is primarily why education needs to be continued and developed, as well as for educators to enhance their knowledge and pass it on to students. AI has shown a considerable possibility of improving the quality of education in aspects such as the classroom approach, the reduction of bureaucratic tasks, and the ability to aid in critical decision-making. This one where people believe that integration of AI will enhance the personalization of learning is by far the most outstanding benefit. Mentor also implies that artificial intelligence can effectively address learner needs, which is quite important, given that schools have various students' needs. This aligns with the trends implemented in other parts of the world where the incorporation of

AI in enhanced learning environments is enhanced to provide students with selective learning procedures and prompt feedback.

Thus, we can conclude that the application of artificial intelligence in education has all sorts of benefits, but the challenges that must be overcome before implementing it cannot be underestimated. Lack of funds, small groups of specialized persons in Jordan, and minimum formal structure are significant challenges for AI utilization in Iraqi institutions. These challenges can be considered global trends in the application of AI in weak-resource environments. The challenges will be surmountable through concerted efforts by government agencies, educational institutions, and private sector partners to attract the right funding and implement the proper infrastructure to support AI systems. Another challenge discovered from the research involves resistance to change; hence, any switch in organizational culture is likely to meet much resistance, especially in places like schools where the leadership employs conventional ways of executing their responsibilities. AI implementation in such contexts may also encounter cultural and organizational resistance that hinders integration. To deal with this challenge, it will be advisable for stakeholders to embark on a culture change focused on the potential of AI to augment the teaching–learning process rather than supplanting it. The second set of enablers classified by the evaluated sources is based on the ethical considerations of adopting AI, which are cited less often. With the advancement in AI technologies, it is essential to guarantee that the adoption of AI is conducted in a privacy-friendly, fair, and transparent manner. Going forward, the emphasis should be on ethical principles and guidelines for using AI in learning processes to help prevent threats such as bias in algorithms and uneven adoption of AI technologies. In the future, the study proves that there is hope for the uptake of AI in education since most people believe that it will soon be prevalent in the next few years. This expectation is based on the understanding of AI's ability to improve the quality and effectiveness of learning.

Nevertheless, everything is not dark, as a brighter future depends on overcoming the abovementioned challenges, mainly funding, expertise, and infrastructure. AI always discloses endless prospects to innovate education systems in Jordan; however, the process must overcome more barriers. Jordan has a golden opportunity to use AI technologies to enhance performance in education by increasing training for teachers who can guide students endowed with such modern technologies to face the future world dominated by information technology.

6 Conclusion

This paper discusses the emergence of hope in applying AI to enhance learning systems in Jordan. Some of the issues highlighted by the literature review must be resolved before AI can be integrated into the system. The potential of AI to facilitate differentiated instruction and minimize administrative workloads while improving the quality of decision-making processes brought much interest to the subject. There is a need to foster AI in education as several barriers limit its adoption, including funding issues, lack of technical ability, infrastructure deficiencies, and negative enemas to change. The results also address the calls for enhancing awareness and expertise in machine learning technologies among teachers, learners, and officials. Given the growth of AI and its application in different areas, it is crucial to set norms and standards regarding intelligence applications. To guard against such risks as algorithmic bias and student/visitor privacy risks, these frameworks shall be adopted to ensure all stakeholders' fair utilization of AI in the education sector. However, there is a great deal of optimism regarding the future implementation of AI in educational organizations in Jordan. As the analysis has shown, innovation in education due to the power of AI may drastically change the learning process, supplying it with the elements of personalization, efficiency, and interest. Hence, tripartite stakeholders, including policy-makers and educational leaders, among others, should develop ways of creating enablers that facilitate the integration of AI into the educational systems to develop the future of education that involves technology in shaping the next generation of learners.

References

1. Al-Khatib: Artificial intelligence and its role in shaping the future of education in Jordan. J. Educ. Technol. Dev. **5**(2), 45–60 (2020)
2. Al-Qudah, et al.: Digital transformation in Jordan's education system: challenges and opportunities. Int. J. Educ. Technol. **12**(3), 203–220 (2021)
3. Baker, et al.: Machine learning for education: personalized learning pathways. J. Learn. Anal. 6(1), 72–85 (2019)
4. Chen, et al.: Adaptive learning and AI: enhancing student success in higher education. Educ. Technol. Soc. **23**(4), 42–55 (2020)
5. Kassem, Nasser: Barriers to AI integration in Jordanian educational institutions. J. Educ. Adm. Policy **39**(2), 58–70 (2021)
6. Luckin, et al.: Intelligence unleashed: an argument for AI in education. Pearson Educ. (2016)
7. Nassoura: Applied artificial intelligence applications in higher education institutions: a systematic review. Webology **19**(3), 1168–1183 (2022)
8. Sharma, Yadav: AI and its adoption in developing countries' education systems: challenges and opportunities. Int. J. Educ. Dev. **42**(2), 122–134 (2020)
9. Siemens: Learning analytics: the emergence of a discipline. Am. Behav. Sci. **57**(10), 1361–1374 (2013)
10. UNDP.: Jordan digital transformation strategy 2020–2025. United Nations Development Programme (2020)
11. VanLehn: The relative effectiveness of human tutoring, intelligent tutoring systems, and other tutoring systems. Educ. Psychol. **46**(4), 197–221 (2011)
12. Williams, Taylor: AI in education: cultural resistance and the future of learning. J. Educ. Innovat. **33**(1), 30–44 (2021)
13. Chassignol, et al.: AI for teachers: enhancing efficiency and effectiveness in educational settings. Educ. Tech. Res. Dev. **68**(4), 1715–1734 (2020)
14. Zawacki-Richter, et al.: Systematic review of AI applications in education: a framework for research and development. J. Educ. Res. **11**(3), 122–141 (2019)
15. Binns: Ethical issues in AI applications in education: a critical review. AI and Ethics **4**(3), 345–361 (2018)

Exploring Opportunities and Challenges of AI-Driven Marketing in the UAE: A Survey of Industry Perspectives

Mahmoud Abouraia, Ayman Bassam Nassoura, Hothefa Shaker, Fadi Abdelfattah, and Hisham Madi

Abstract This study examines the adoption of AI in marketing in the UAE, focusing on current applications, challenges, and future trends. A survey of 125 professionals across industries, including IT, retail, banking, healthcare, and real estate, provided quantitative insights into the role of AI in marketing. Data analysis using descriptive and objective methods revealed the application of AI in customer targeting, content personalization, and performance evaluation. The study highlights critical benefits such as improved customer personalization, enhanced decision-making, and increased efficiency. However, challenges include high implementation costs, data privacy concerns, and a shortage of skilled professionals. Despite these barriers, respondents anticipate significant growth in AI-based marketing investments, particularly in personalized communications and predictive analytics. This research contributes to understanding the transformative impact of AI on marketing in the UAE and provides practical insights for organizations aiming to integrate AI technologies into their strategies effectively.

Keywords Artificial intelligence (AI) · Marketing strategies · AI adoption · Predictive analytics · Marketing challenges · Customer personalization

---

M. Abouraia · H. Shaker · F. Abdelfattah (✉) · H. Madi
Modern College of Business and Science, Muscat, Oman
e-mail: fadi.abdelfattah@mcbs.edu.om

M. Abouraia
e-mail: Mahmoud.abouraia@mcbs.edu.om

H. Shaker
e-mail: hothefa.shaker@mcbs.edu.om

A. B. Nassoura
Department of MIS, City University Ajman, Ajman, UAE
e-mail: a.bassam@cu.ae.ae

B. Awwad (ed.), *AI and IoT: Driving Business Success and Sustainability in the Digital Age*, Studies in Systems, Decision and Control 590,
https://doi.org/10.1007/978-3-031-88874-8_92

1 Introduction

Machine learning affects multiple sectors globally, and marketing is not exempted; here, AI improves operations, decision-making, and customer experience [1–4]. As a rich environment for implementing technologies, including AI, in the country's business processes, the UAE is suitable for integrating AI and technological innovations in commercial use [5]. It is only in recent years that AI has become popular within marketing practices; customers within different sectors of the economy seek to incorporate the use of tools driven by AI while enhancing the process of customer engagement, market targeting, and the overall outcomes of marketing practices [6, 7].

Marketing practitioners in the UAE use three AI technologies: predictive analytics, machine learning, and natural language processing to personalize the clients, reduce human interference, and manage masses of data [7–9]. This paper has explored how AI revolutionizes marketing strategies by automating decision-making, predicting consumer behaviors, and delivering more personalized content in an ever-changing marketplace to ensure businesses are well-positioned [10]. Nevertheless, like every technology, AI generates numerous possibilities that are holistic, integrated, superior, or quicker; AI has remarkable hurdles concerning availability, cost, justification, protection of data and skilled human force required for AI [11].

This paper explores the opportunities, challenges, and future expectations of AI in marketing in the UAE. By surveying a diverse group of marketing professionals, business executives, data scientists, and AI specialists, the research investigates how AI is adopted in marketing, its key benefits to businesses, and the barriers that hinder its broader implementation. The study also aims to examine the future trends and investment outlook for AI in marketing in the UAE, providing insights into how businesses in the region plan to harness the power of AI to drive growth and innovation.

The scholars of this paper hope to discuss the potential implementations, potential disadvantages, and prospects related to AI marketing in the UAE. Thus, through the research conducted with 120 marketing practitioners, business leaders, Data Scientists, and AI professionals, the study explores the current state of AI utilization in marketing, the benefits that adopting AI provides to a business and the challenges that are seen to prevent its mass application. The study also seeks to explore the future trend and investment plan of AI in marketing in the UAE about the steps organizations in the UAE are willing to take to embrace AI's power to enable enthusiasts to unleash the potential of AI in delivering value in marketing.

1.1 Literature Review

Marketing automation is an advanced field rapidly growing due to the application of artificial intelligence (AI) in marketing. In this literature review, the main areas of AI in marketing are covered, which will allow an understanding of the current application of artificial intelligence, its potential, and problems at the moment, as well as the prospects for its further development according to the questions stated in the research questionnaire. Based on the findings, the review supplements the contextually relevant literature to externally validate the research and extend the analysis to discuss AI's implications for marketing—theoretically and as a burgeoning reality within the UAE.

1.2 Current Use of AI in Marketing

The capability of AI propels the marketing application of this concept to streamline processes, customize customers' experience, and increase the capacity for analyzing big data. Digital marketing tools such as predictive analytics, machine learning, NLP, and chatbots are used consistently and more frequently by marketing professionals across industries [10]. Jarek et al. [7] found that the roles of AI tools in marketing are customer targeting, content personalization, and social media analysis. Other authors have also pointed out the use of AI, particularly in customer segmentation, since it helps marketers develop improved customer personas based on behavior, preferences, and buying habits [12].

Nevertheless, recent studies show that many organizations in the UAE are not interested in AI adoption in marketing [13]. The most frequent AI application areas are customer targeting, content personalization, and marketing automation, though organizations' AI usage is frequently constrained by resources, human capital, and technological support [14, 15]. Moreover, companies have yet to note the ways, means, and ways of implementing or making AI tools compatible with the extant marketing strategies to offer value.

1.3 Opportunities in AI Marketing

AI holds much promise regarding the interventions that it can help bring into marketing. In turn, one benefit is the possibility of developing aspects of customer personalization. AI enables marketers to offer individualized customer experiences based on each customer's past behavior and anticipated future behavior [15, 16]. With the help of machine learning, it is possible to Make real-time content suggestions, products, and advertisements according to the consumer's previous engagement with a particular brand or company. Such a focused strategy increases customer interaction, time, and utilization, improving conversion rates and customer loyalty [10].

Another significant advantage is the enhanced decision-making brought about by AI through statistical results. With AI and machine learning algorithms, AI can take enormous amounts of consumer data and understand specific patterns that would otherwise remain unnoticed. This assists marketers in creating better marketing methods, increasing the effectiveness of executing methods, and better anticipating future market shifts [12]. Also, real-time management of customer relationships is achievable through AI since real-time customer engagement is doable [14]. Traditionally, industries like retail, healthcare, and finance in the UAE have been quick to identify that leveraging AI would provide businesses with a competitive edge by offering customized marketing solutions and services.

1.4 Challenges in AI Marketing

Nevertheless, multiple issues prevent AI from becoming more popular in marketing. These challenges include High capital investment and recurrent costs, privacy and data security, and lack of competent professionals [5, 11]. The deployment of AI systems is typically associated with certain initial and recurrent costs. Hence, whether SMEs in the UAE are willing to bear such costs or will they resist such efforts and argue that the returns are not immediate [13]. Another challenge continues to emerge from fears of data privacy and security since most AI systems base their procedure on the customer data they gather. According to [13], data protection regulations are still developing in the UAE, but businesses remain ambiguous concerning data privacy laws concerning data collection, storage, and usage.

Moreover, challenges like unfair algorithms or consequences of excessive use of automation must be solved to develop the ethical use of AI [11]. However, lacking skilled human personnel in AI and data science has been identified as one of the most significant barriers to AI in marketing. In [7] view, companies must pursue training and development activities in the growing AI field to possess the skills to harness AI. It has become more evident in the UAE, where the adoption and implementation of AI are on the rise, but talent is still abysmal.

1.5 Future Trends in AI Marketing

AI will continue to revolutionize marketing practices in the UAE in the future. The most positive tendency is the growth in applying new personal and targeted advertising methods. With the help of advanced AI algorithms, companies are refining their ability to assign personalized experiences to customers. As artificial intelligence customer insights continue to develop, marketing will have even more accurate ways to predict customer behavior and tailor marketing [10]. Marketing automation using artificial intelligence will also rise due to the growing number of businesses looking to cut operations and reduce automation. The trends in OM that will help organizations adapt to the marketing situation include the use of predictive and prescriptive analytics so that firms can estimate the needs of consumers and make necessary changes ahead of time [12]. Further, combining AR and VR with AI captures the essence of customer desires in today's economy, where more industries are searching for ways to provide unique and superior value experiences [14].

Public demand in real-time data analytics and processing coupled with instantaneous response responsiveness will enhance and promote an increased loyalty customer business interaction. AI's place in marketing efficiency is expected to grow more prominently as more organizations adopt AI systems to keep up with their rivals [15, 16]. AI offers many benefits to marketing practices in the UAE, including better decisions and personalization for efficiency and competitive advantage. However, several barriers must be met for AI to achieve its maximum capability, including excessive costs, privacy issues, and the number of people required to be employed. The prospects for Artificial Intelligence in marketing in UAE seem strong, as such directions as improved personalization, automation, and real-time customer interaction are already on the way to becoming the next breakthrough with AI within the industry.

2 Methodology

The current research employs a quantitative research paradigm and includes a survey questionnaire as the primary data collection technique to evaluate the extent of AI adoption, prospects, and issues of AI integration in marketing in the UAE. Hence, the research objective is to understand how AI performs in marketing and within diverse industries through the views of marketing practitioners, business managers, data scientists, AI experts, and academics. The study adopted a cross-sectional survey designed to collect data at a one-time point for professionals working in the UAE. A standard questionnaire was designed to get demographic data and ideas about AI use in marketing. The survey covers five key areas: personal information of the respondents, role, experience, and industry of operation, the current application of AI in marketing, evaluation of currently used AI technologies, marketing activities supported through AI, and its perceived level of influence on marketing performance; perceived opportunities in AI marketing for the UAE, identification of significant opportunities AI present for marketing in the UAE; perceived challenges in AI marketing with emphasis on cost, data privacy, and skills; and future perspectives on AI marketing. The target population for the study was employees involved in marketing, AI, and related industries in the UAE, as well as the technology, retail, finance, health, and real estate sectors. To make the study statistically relevant and to contain data collection efforts, a sample of 125 respondents was used. Respondents were selected purposively using online networks and professional connections and snowball sampling. It also allowed us to reach a diverse group of respondents with different amounts of experience and knowledge. Sampling and data collection were done by administering an online questionnaire through digital media, including professional sites like LinkedIn, relevant forum groups and email invitations, conducted in the UAE. The survey was live for four weeks, and while some questions were closed, others were of the Likert-scale type that ensures the collection of measurable data. In addition to closed questions, more broad questions were asked to get more insights into the current opportunities and threats posed by

AI in the marketing context in the UAE. The demographics and AI usage profiles gathered from 125 valid surveys were described using frequency statistics. The likeability indices of items with Likert-scale results have been analyzed through metrics of central tendency, mean and median to understand the consensus about AI in the marketing industry. To get the frequency ranking of the multiple-choice questions, a frequency analysis was conducted on the first set of questions to determine the most common AI technologies and the marketing activities supported by AI. The qualitative responses, the sectors with a good potential for adopting AI in their marketing functions and implementation challenges, were conducted using the theme analysis method. Some core topics include personalization, automation, cost, data privacy, and skill development. All procedures complied with the necessary ethical standards to protect participants' identities. Participants were advised of the nature of the research and told that their involvement was voluntary and that their identities would remain unknown. Written informed consent was sought before the survey, and the respondent's right to withdraw from the survey was granted without prejudice. The information was collected and processed with proper storage and utilized only with the intent of the research. However, the convenience sampling technique restricts the broader application of the results regarding generality.

Further, the survey responses are self-measures, which will give a bias; for instance, people may exaggerate the use of converting AI or underemphasize the problems met. Subsequent work might overcome these limitations by using random sampling and including open-ended interviews to understand the use of AI in marketing better. The approach used in this research gives a solid foundation to identify the strengths, weaknesses, and future trends for AI use in marketing in the UAE. Due to adopting an explorative research method that applies to a quantitative method, the research draws findings from numerous professionals from various fields, providing the study with valuable data about AI in the marketing segment in the region.

3 Result

As shown in Table 1, the majority of respondents are marketing managers (28%), followed closely by business executives (24%), data scientists, or AI specialists (20%). This diversity reflects a broad perspective on AI in marketing from both strategic and technical roles.

However, most respondents work in the Technology industry (24%), followed by Retail (20%) and Finance (16%). These sectors are known for early AI adoption, indicating an alignment with the survey's focus on AI-driven marketing. Table 2 shows that most respondents have over five years of experience (36%) or three to five years (32%). Participants' experience level suggests they have a solid foundation for evaluating AI's impact on marketing.

Table 1 Respondents' role in organization

Role	Frequency	Percentage (%)
Marketing manager	35	28
Business executive	30	24
Data scientist/AI specialist	25	20
Academic/researcher	15	12
Other	20	16
Total	125	100

Table 2 Years of experience in marketing or AI roles

Experience level	Frequency	Percentage (%)
Less than 1 year	10	8
1–3 years	30	24
3–5 years	40	32
More than 5 years	45	36
Total	125	100

68% of respondents currently use AI in their marketing activities, showing a substantial adoption rate. This finding suggests that AI is widely regarded as a valuable tool for marketing within the UAE.

The result illustrates that Machine Learning (60 respondents) and Customer Segmentation (55 respondents) are the most used AI technologies in marketing, highlighting the focus on predictive capabilities and personalized customer interactions. The result showed that customer targeting, segmentation (65) and content personalization (60) are the primary marketing activities supported by AI, reflecting a trend toward more tailored and data-driven marketing strategies.

The results also indicate that enhanced customer personalization (55 respondents) and improved decision-making (50 respondents) are the top perceived opportunities of AI in marketing, emphasizing the role of AI in creating more customized and data-informed strategies. The results showed the main challenges of high implementation and maintenance costs (60) and data privacy/security concerns (55). This points to potential financial and regulatory barriers that organizations in the UAE face.

On one hand, the results indicate that 40% of respondents are "very likely" to increase their AI investment within the next five years, suggesting a solid commitment to leveraging AI in marketing. On the other hand, the results highlight that advanced personalization and targeting (65) and marketing automation (55) are anticipated to be the most impactful AI trends, reflecting the increasing demand for tailored, efficient customer engagement.

4 Discussion

The results of this study provide a nuanced understanding of the present state and future direction of AI in marketing for different industries within the UAE. It is, therefore, crucial to understand the opportunities AI provides for the marketing function and the challenges experienced by marketing practitioners to help the organization that seeks to implement AI-led marketing strategies. Based on survey results, UAE marketing professionals are keenly interested in implementing AI to improve customers' persona and decision-making processes and reduce paperwork. More profoundly, there was the recognition of the increased potential of customer personalization, which dovetails with the current global pivot to personalizing customer relations. Due to the UAE's elevated levels of digital adoption, it has become possible for brands to adopt AI personalization techniques to ensure that consumers are engaged at the right time with what they want to see. This is especially useful when the consumer touch point and experience quality determine the customer devotion and market outcome, as in the retail and finance industries. Enhanced decision-making by embracing data is another opportunity, highlighted next [16]. These AI technologies, like predictive analytics and machine learning, help marketers make sense of the millions of data points favorable in the UAE, where data is plentiful. Using these tools, their marketer can have better solutions within lesser amounts of time, hence driving the strategic direction of the marketer and the operation of the marketer. Also, AI enhances the overall effectiveness of operations and streamlines repetitive tasks, allowing marketers to focus on more enriching, functional components during marketing campaigns. As the authors have argued in the marketing discipline, harnessing the power of AI means different things to different organizations, as various challenges affect its adoption in the UAE. That is why the significant challenges include high implementation and maintenance costs, data privacy, and a lack of skilled workforce in artificial intelligence. These are especially important given the high implementation and maintenance costs in small to mid-sized businesses; thus, they might be unable to invest in such high-tech AI technologies. These financial barriers highlight the increasing importance of affordable and affordable AI solutions that might appeal to various organizations.

Considering the UAE authorities' impressive concentration on compliance and data privacy, it is unsurprising that marketers are hesitant to deal with client data using AI. This concern poses the need for standard data governance policies and safe means of handling consumer data to comply with the regulations. Marketers in the UAE must be sensitive to data privacy as artificial intelligence processes more data in the future.

Future expectations of the participants revealed that respondents expressed a positive outlook about the role of artificial intelligence in UAE marketing, and more than two-thirds indicated their plans to enhance their AI spending over the next five years. Also, the interest in augmented and virtual reality applications in marketing supports the overall trend of the UAE as one of the representatives of the technology-savvy country with a receptive customer base to experiential marketing strategies.

Combining AI-augmented reality (AR) and virtual reality (VR) could provide brands with next-level ways to set themselves apart, potentially especially in industries like retail clothing and real estate, where the customer experience is paramount.

5 Conclusion

This paper has offered a state-of-the-art understanding of the current adoption level, potential uses, benefits, and potential drawbacks of AI in marketing and the overall future expectations of the UAE Market. The study shows that AI is evolving into a core competency skill set for marketers and their organizations to deliver improved value to their clientele through personalization, analytics, and effectiveness. These advantages benefit the UAE market, especially concerning customer relations, product and service differentiation, and efficient operations. The study established that self-educating systems, computer-aided statistics, patterns, and natural language breaks are current technologies that have broad reception across different commercial segments with a core concentration on enhancing advertising success. One of the main advantages for marketers in the UAE market was the possibility of providing customers with customized experiences and targeting customers more accurately. The reason for the continuous growth in AI companies' demand for marketing solutions is that organizations continue to show a positive outlook on the future AI trends and development in the next several years for them to invest in. However, the study identified the following hurdles that have to be bridged to realize effective AI deployment. Gremlins, such as high costs for full-scale implementation, data privacy issues, and availability of qualified labor to actualize the change, are significant demerits among organizations, tiny ones. Such challenges must be addressed equally effectively, including enhanced and affordable solutions, robust data management structures, and skill enhancement activities.

References

1. Dennis, M., Sam, J.: How artificial intelligence is transforming marketing. [Online]. Available: https://owdt.com/article/how-artificial-intelligence-is-transforming-marketing/. [Accessed: 2 Dec 2024]
2. Peter, S.: How AI is transforming marketing [Online]. Available: https://marketinghire.com/career-advice/how-ai-is-transforming-marketing/. [Accessed: 2 Dec. 2024]
3. SAP: What is AI in retail? Oct. 10, 2024. [Online]. Available: https://www.sap.com/resources/ai-in-retail/ [Accessed: 2 Dec 2024]
4. Sodiq, O., Ola, A., Okoro, M.: The role of AI in marketing personalization: a theoretical exploration of consumer engagement strategies. Int. J. Manage. Entrepreneurship Res. **6**(3), 936–949 (2024)
5. U.S.-U.A.E. Business Council, The UAE's big bet on artificial intelligence, U.S.-U.A.E. Business Council, 2024. [Online]. Available: https://usuaebusiness.org/wp-content/uploads/2024/02/SectorUpdate_AIReport_Web.pdf. [Accessed: 2 Dec 2024]

6. Haleem, M.K., Suman, S.G.: Artificial intelligence (AI) applications for marketing: a literature-based study. Int. J. Intell. Netw. **3**, 119–132 (2022)
7. Jarek, J., Mazurek, G.: The role of artificial intelligence in marketing. J. Mark. Trends **4**(2), 35–43 (2020)
8. Vision Tech Solutions: How will AI impact digital marketing of the future in the UAE [Online]. Available: https://www.visiontechsolutions.ae/. [Accessed: 2 Dec 2024]
9. Abdelfattah, F., Salah, M., Dahleez, K., Darwazeh, R.: The future of competitive advantage in Oman: integrating green product innovation, AI, and intellectual capital in business strategies. Int. J. Innovat. Stud. **6**(2), 1–12 (2024). https://doi.org/10.1016/j.ijis.2024.02.001
10. Chaffey, D.: Digital Marketing: Strategy, Implementation, and Practice, Pearson Education (2019)
11. Brynjolfsson, E., McAfee, A.: The Second Machine Age: Work, Progress, and Prosperity in a Time of Brilliant Technologies. Norton and Company (2017)
12. Davenport, T.H., Ronanki, R.: Artificial intelligence for the real world. Harv. Bus. Rev. **96**(1), 108–116 (2018)
13. Alashoor, T.: AI adoption in the UAE: opportunities and challenges. Int. J. Mark. **12**(3), 45–59 (2021)
14. Loureiro, S.M., Guerreiro, M., Tussyadiah, A.: Exploring the impact of artificial intelligence on marketing in the tourism sector. J. Travel Res. **60**(5), 1181–1197 (2021)
15. Wedel, M., Kannan, P.: Marketing Analytics: A Practitioner's Guide to Marketing Analytics and Research Methods, Springer (2016)
16. Al Halbusi, H., Al-Sulaiti, K., Abdelfattah, F., Ahmad, A.B., Hassan, S.: Understanding consumers' adoption of e-pharmacy in Qatar: applying the unified theory of acceptance and use of technology. J. Sci. Technol. Policy Manage., vol. ahead-of-print (2024). https://doi.org/10.1108/JSTPM-03-2023-0042
17. Dahleez, K., Abdelfattah, F., Al Halbusi, H., Salah, M., Madi, H.: An investigation of government involvement and green information systems in driving management practices. In: Lecture Notes in Networks and Systems, vol. 1081, Springer, Cham, pp. 310–320 (2024). https://doi.org/10.1007/978-3-031-67437-2_31

Performance Evaluation of Islamic Banks in Oman: An Analysis Based on Islamic Banking Assets

Dhanuskodi Rengasamy, Mohammad Abdul Matin Chowdhury, Mohammad Belayet Hossain, Ravi Thirumalaisamy, Velmurugan Ramasamy, Sudarvel Jayaraj, Mahmoud Abouraia, and Hisham Madi

Abstract The country's economic development heavily relies on the banking sector's progress, with banks serving as growth drivers. Islamic banking, one of the fastest-growing financial sectors, follows Shariah principles, focusing on profit-sharing and prohibiting interest. Bank Nizwa and Alizz Islamic Bank are fully operational Islamic banks in Oman. This study analyzes their performance based on Islamic assets, classifying assets into traditional and Islamic for comparison. Using secondary data from bank annual reports, the study covers 2014–2023 for Bank Nizwa and 2015–2023 for Alizz Islamic Bank. Tools like CAGR and trend analysis show that both banks demonstrate positive, progressive growth.

D. Rengasamy · M. A. M. Chowdhury · M. B. Hossain · H. Madi
Department of Accounting and Finance, Faculty of Business, Curtin University, Miri, Malaysia
e-mail: dhanuskodi@curtin.edu.my

M. A. M. Chowdhury
e-mail: martinchy@curtin.edu.my

M. B. Hossain
e-mail: belayet.hossain@curtin.edu.my

R. Thirumalaisamy · M. Abouraia
Department of Accounting, Finance and Economics, Faculty of Business and Economics, Modern College of Business and Science, Muscat, Sultanate of Oman
e-mail: hisham.madi@mcbs.edu.om

R. Thirumalaisamy
e-mail: ravi@mcbs.edu.om

M. Abouraia
e-mail: Mahmoud.Abouraia@mcbs.edu.om

V. Ramasamy · S. Jayaraj
Karpagam Academy of Higher Education, Coimbatore, Tamil Nadu, India
e-mail: drvelsngm@gmail.com

S. Jayaraj
e-mail: j.sudarvel@gmail.com

B. Awwad (ed.), *AI and IoT: Driving Business Success and Sustainability in the Digital Age*, Studies in Systems, Decision and Control 590,
https://doi.org/10.1007/978-3-031-88874-8_93

Keywords Islamic banking assets · Performance evaluation · Islamic banks in Oman · Growth rate analysis · Compound annual growth rate (CAGR)

1 Introduction

Banks play a crucial role in the economic development of every nation. They control over a large part of the supply of money in circulation. Economic development is a dynamic and continuous process where the banks are the primary stimulus to the economic progress of a country, and commercial banks perform various key economic functions [6]. A study conducted by Dominic et al. [8] mentioned that the banking industry is a key contributor to inclusive growth and economic development, and it serves as the foundation for capital allocation and financial intermediation within economies. In general, economic development is based on the mobilization of resources, investment, and the operational efficiency of the various segments of the economy, banks have been consciously used as an engine of development. The banking sector provides various services, including loan provisions and deposit acceptance. The existing literature demonstrates that the banking sector moderates economic growth and industrial development [7, 13]. The banking sector mainly connects the gap between the sector having funds and the business's capital requirement [15]. According to a Statista research development publication (published January 31, 2024, cited 3 October 2024), the value of bank assets worldwide has increased gradually for 2022 and 2023 to 182.9 trillion US dollars and 183 trillion US dollars, respectively.

Islamic banking is one of the important subsectors of the financial industry, and it is the fastest-growing sector as well. Financial operations that comply with Shariah (Islamic law) are called Islamic banking, Islamic finance, or Shariah-compliant finance. The fundamental principles of Islamic banking are sharing profits and losses and prohibiting lenders and investors from collecting and paying interest. An article from the Standard Chartered Banks website describes Islamic banks as a method of carrying out banking operations by the Shariah principles, in which forbidden activities, including interest (Riba), gharar (uncertainty), and engaging in prohibited businesses such as gambling or alcohol [18].

Most of the studies having a similar concept of the origin of modern Islamic banks indicated that the first modern Islamic banking was officially established in the year 1963, named Mit Ghamr Savings Bank in Egypt, and its operations were based on Shariyah principles and norms [5, 9, 14] . Following the first Islamic bank, several countries' governments, including Islamic countries, initiated the establishment of this type of bank in their countries. In this way various Islamic banks were established, for example, Islamic Development Bank (IDB) in Jeddah (1975), Dubai Islamic Bank in Dubai (1975), Bahrain Islamic Bank in Bahrain (1979), Abu Dhabi Islamic Bank in Abu Dhabi (1997), Islamic Bank Bangladesh Limited, Bangladesh (1983), Bank Islam Malaysia Berhad, Malaysia (1983), Al Meezan Investment Bank, Pakistan (2002), and Bank Nizwa in Oman (2013).

2 Review of Literature

Bank performance measures a bank's effectiveness and efficiency in achieving its objectives, directly impacting its overall stability. It involves evaluating how well a bank utilizes its resources to meet goals, with various indicators showing its current status and success in achieving intended outcomes. Numerous studies focus on both traditional and Islamic banks' performance. For instance, Asutay and Ubaidillah [3] analyze Islamic banks' performance through intellectual capital, while Al-Sharif [1] examine the impact of macroeconomic variables such as economic growth and inflation. Other studies explore performance from a social perspective [11], or measure factors like non-performing financing [17], profitability [10], fintech innovations [12], and financing products [2, 4]. This study differs by focusing specifically on Islamic assets as a performance indicator for Islamic banks in Oman.

The research gap lies in the lack of studies specifically analyzing the performance of Islamic banks in Oman based on Islamic assets. While existing research covers various performance factors, none focuses on how Sharia-compliant assets (e.g., Wakala, Ijarah) influence Omani Islamic banks' financial stability and operational effectiveness.

3 Significance of the Study

The study aims to evaluate the growth and performance of Islamic assets in Oman's Islamic banks by analyzing growth percentages and compound annual growth rates (CAGR). Islamic banks in Oman have expanded significantly, yet there is limited research on the growth patterns of their key assets. By examining the trend and growth rates of Islamic assets, this study seeks to provide insights into these banks' financial performance and stability. Identifying areas with fluctuating or stable growth can highlight the potential for improvement and strategic investment, enhancing the overall contribution of Islamic banking to Oman's financial sector.

4 Objectives of the Study

In accordance with the statement of the problem, the following objectives have been framed:

1. To study the growth percentage and compound annual growth rate (CAGR) of Islamic assets in Islamic banks in Oman
2. To measure the trend and growth of Islamic assets in Islamic banks in Oman.

5 Research Methodology

The data required for the study is secondary. Secondary data have been collected from the respective bank's annual reports. The survey method was adopted to select Islamic banks in Oman. The study period ranges between 2014 and 2023 for Bank Nizwa and 9 years from 2015 to 2023 for Alizz Islamic bank. The collected data has been analyzed and presented in two aspects: (i) growth percentage and CAGR of Islamic assets, and (ii) growth rate and trend of Islamic assets in Islamic banks in Oman for the study period.

6 Licensed Banks in the Sultanate of Oman

According to the Central Bank of Oman's annual report (2023), the licensed banks in the Sultanate of Oman consist of four major categories: local banks, foreign banks, specialized banks, and Islamic banks and windows. Table 1 provides details of the banks, their year of establishment, and their operating offices.

7 Islamic Banks in the Sultanate of Oman

Bank Nizwa and Alizz Islamic Bank are Oman's two full-fledged Islamic banks, evaluated here based on their Islamic assets. The Central Bank introduced the Islamic Banking Regulatory Framework (IBRF) on December 18, 2012, allowing fully operational Islamic banks and Islamic windows within traditional banks. IBRF ensures Sharia compliance through comprehensive regulatory requirements, including Sharia governance frameworks and licensing guidelines, all overseen by the Central Bank to support Islamic banking operations in Oman.

7.1 Bank Nizwa

Bank Nizwa, Oman's first Islamic bank, provides Sharia-compliant products and services. Licensed by the Central Bank of Oman under Royal Decree No. 114/2000, it began operations on December 23, 2012. By 2023, the bank expanded to 22 branches across Oman, offering services like deposit acceptance and financing through Islamic instruments such as Wakala Investments, Musharaka, Ijara, and Wakala Bil Istethmar. From 2014 to 2023, Bank Nizwa's total asset value grew by 84%.

Table 1 Licensed banks in the Sultanate of Oman

S. No.	Name of the bank	Year of establishment	Operating offices
Local banks			
1	National Bank of Oman	1973	61
2	Oman Arab Bank	1973	50
3	HSBC Bank Oman	1975	0
4	Bank Muscat	1981	146
5	Bank Dhofar	1990	87
6	Bank Sohar	2007	77
7	Al Ahli Bank	1997	24
Foreign banks			
1	Standard Chartered Bank	1968	1
2	Habib Bank Ltd.	1972	0
3	Bank Melli Iran	1974	1
4	First Abu Dhabi Bank	1976	7
5	Bank Saderat Iran	1976	2
6	Bank of Baroda	1976	1
7	State Bank of India	2004	1
8	Bank of Beirut	2006	4
9	Qatar National Bank	2007	5
Specialized banks			
1	Oman Housing Bank	1977	9
2	Oman Development Bank	1977	19
Islamic banks			
1	Bank Nizwa	2012	21
2	Al Izz Islamic Bank	2013	14
Islamic window			
1	BM-Meethaq	2013	27
2	NBO-Muzn	2013	6
3	BD-Misara	2013	22
4	AHB-Al hilal	2013	22
5	BS-Sohar Islamic	2013	10
	Total		617

Source Central Bank of Oman, Annual Report 2023

7.2 *Alizz Islamic Bank*

Alizz Islamic Bank, a fully operational Islamic bank in Oman, follows Sharia principles and was established under Royal Decree No. 69/2012. Licensed by the Central Bank on September 5, 2013, it began operations on September 30, 2013. On June 30, 2020, Oman Arab Bank SAOC acquired Alizz Bank, making it a subsidiary and delisting it from the Muscat Securities Market. By 2023, Alizz Islamic Bank expanded to 14 branches, with total assets growing by 90% since 2014.

8 Significance of the Study

Bank performance analysis generally involves examining financial activities and evaluating overall business health [16]. To assess their stability, Yilmaz [19] analyzed Greek banks' financial performance using key financial ratios and statement components, such as assets, capital, ATMs, and branch numbers. Many studies apply traditional financial ratios—profitability, liquidity, etc.—to gauge bank performance. However, this study focuses specifically on Islamic banks in Oman and assesses their Islamic assets. The findings offer insights into the financial health of Oman's Islamic banking sector, underscoring stability and performance based on Islamic assets.

9 Islamic Assets in the Statement of Financial Position

The financial statements of the two Islamic banks in Oman consist of assets classified into traditional and Islamic assets. Traditional assets include cash, bank balances with the Central Bank of Oman, receivables, financial assets, property, and equipment. Islamic assets consist of Interbank Wakala Investments, Ijarah assets, Ijara Muntahia Bittanleek, Wakala Bil Istithmar, Murabaha receivables, and Diminishing Musharaka. This study examines the performance of Islamic banks through their Islamic assets, making it essential to understand each asset type's role in bank performance.

9.1 *Inter-Bank Wakala Investments*

Wakala is an Islamic contract where the customer (Muwakkil or principal) provides capital, and the bank (Wakil or agent) invests it in Sharia-compliant ventures for returns. The customer's contribution is treated as a bank liability. However, when the bank further invests this capital with other local or foreign banks, it becomes an asset, classified as inter-bank Wakala investments. This structure allows Islamic banks to generate returns while adhering to Sharia principles.

9.2 *Investment in Ijarah Asset*

Ijarah is a contract between a lessor and lessee in which the lessor leases a tangible asset to the lessee for a specified period and rental price. This arrangement allows the lessee to use the asset without ownership. Essentially, it's an agreement to rent an asset to benefit from its use. Islamic banks in Oman utilize this model by investing in Ijarah assets, aligning with Sharia principles while generating returns through leasing.

9.3 *Ijara Muntahia Bittanleek*

This financing mode, used by Islamic financial institutions, involves a contract where banks fund property or equipment with agreed-upon rental installments, including profit and principal. The profit rate may be fixed or variable, as specified in the agreement. Ownership transfers to the client upon lease expiration and final payment.

9.4 *Wakala Bil Istithmar*

This asset is created through an agreement where the bank (principal or Muwakki) appoints the client as its agent (wakil) to manage investments in Sharia-compliant activities. The goal is to generate returns from business operations, with the investment amount and any profit paid upon maturity, aligned with expected profit rates.

9.5 *Murabaha Receivables*

The bank financing the money for buying the commodity, which represents the Murabaha contract, later resells this commodity to the customers through purchase orders. The consumer repays the sale price (cost plus profit margin) in one lump sum or in installments over the predetermined time frame.

9.6 Diminishing Musharakah

This partnership form involves a bank and client jointly contributing capital to acquire an asset, with both parties owning the capital. The client gradually acquires the bank's share of the asset until full ownership. The client may use the bank's portion in exchange for covering the takaful and maintenance costs related to the bank's share.

10 Findings

10.1 Growth Percentage and CAGR

Table 2 shows the growth percentage and CAGR of Islamic assets of the bank Nizwa, and Table 3 provides the same information for Alizz Islamic Bank.

The analysis indicates that Inter-Bank Wakala Investments in Oman's Islamic banks experience year-to-year fluctuations due to the bank's wakala investments in other banks. These investments fall into three categories: investments in local banks using local currency, investments in foreign banks using foreign currency for terms under one year, and those over one year. This volatility results in an inconsistent growth percentage. Notably 2015, substantial changes occurred as local bank investments and long-term foreign investments were withdrawn, leading to an overall negative compound annual growth rate (CAGR) of 13%. In contrast, Murabaha Receivables show steady growth, except in 2019 and 2023, achieving a CAGR of 59%. Other assets, Ijara Muntahia Bittamleek and Wakala Bil Istithmar have CAGRs of 16% and 40%, respectively. Overall, the growth of Islamic assets in Oman's Islamic

Table 2 Growth percentage and CAGR of Islamic assets of Bank Nizwa (Amount in RO)

Year	Bank Nizwa							
	Inter-bank Wakala Investments		Investment in Ijarah Asset		Ijara Muntahia Bittamleek		Wakala Bil Istithmar	
	Amount	Growth %	Amount	Growth %	Amount	Growth %	Amount	Growth %
2014	55,710,675		7,353,240		59,339,054		17,671,500	
2015	9,625,000	– 82.72	0	– 100.00	155,580,391	162.19	37,647,530	113.04
2016	41,397,905	330.11	799,441		234,053,765	50.44	62,850,817	66.95
2017	16,397,550	– 60.39	25,726,666	3118.08	282,109,561	20.53	91,310,805	45.28
2018	0	– 100.00	59,880,031	132.75	316,397,423	12.15	137,923,549	51.05
2019	454,888		116,658,786	94.82	318,133,875	0.55	183,671,834	33.17
2020	0	– 100.00	176,652,000	51.43	303,905,000	– 4.47	298,692,000	62.62
2021	26,841,000		260,170,000	47.28	291,610,000	– 4.05	339,422,000	13.64
2022	29,053,000	8.24	364,836,000	40.23	250,548,000	– 14.08	339,446,000	0.01
2023	15,315,000	– 47.29	484,986,000	32.93	228,309,000	– 8.88	371,924,000	9.57
CAGR		– 13		59		16		40

Table 3 Growth percentage and CAGR of Islamic assets of Alizz Islamic Bank (Amount in RO)

Year	Alizz Islamic Bank							
	Wakala Bil Istithmar		Murabaha Receivables		Ijarah Assets-Ijarah Muntahia Bittamleek		Diminishing Musharaka	
	Amount	Growth %	Amount	Growth %	Amount	Growth %	Amount	Growth %
2015	25,312,392		42,989,409		130,090,007		1,050,407	
2016	55,288,924	118.43	60,592,103	40.95	184,410,642	41.76	12,503,859	1090.38
2017	100,305,502	81.42	83,183,614	37.28	234,964,599	27.41	38,752,764	209.93
2018	111,542,706	11.20	104,243,923	25.32	264,997,859	12.78	71,528,324	84.58
2019	105,291,139	− 5.60	95,973,518	− 7.93	252,211,378	− 4.83	118,519,582	65.70
2020	123,129,167	16.94	96,067,530	0.10	251,728,615	− 0.19	235,675,948	98.85
2021	147,507,999	19.80	105,211,172	9.52	226,561,121	− 10.00	322,494,125	36.84
2022	179,478,969	21.67	123,976,679	17.84	192,127,921	− 15.20	411,937,356	27.73
2023	185,466,328	3.34	115,835,609	− 6.57	152,209,181	− 20.78	518,818,395	25.95
CAGR		28.27		13.19		1.98		117.12

banks is satisfactory, with positive CAGRs for all assets except Inter-Bank Wakala Investments.

The results in Table 3 reveal that Wakala Bil Istithmar assets show positive growth throughout the study period, except in 2019, with a positive CAGR. Murabaha Receivables and Ijarah Muntahia Bittamleek display volatile growth but maintain a positive CAGR. Diminishing Musharaka records consistent growth with a notably high CAGR of 117.12%. Thus, hypotheses one and two are accepted, confirming that growth prospects for these assets are satisfactory and the overall CAGR is positive.

10.2 Trend and Growth Rate of Islamic Assets in Bank Nizwa

An attempt is made to find the trend and growth rate of Islamic assets in Bank Nizwa, presented in Table 4.

The analysis shows that Inter-Bank Wakala Investments' growth rate is negative (− 2,115,680.982). The trend value is 19479501.8. In the case of investments in Ijarah Asset, Ijara Muntahia Bittamleek, and Wakala Bil Istithmar, the growth rate is positive, and the trend value is also progressive. Overall, the trend, the growth rate, and the trend value are progressive.

Table 4 Trend and growth rate of Islamic assets in Bank Nizwa (Amount in RO)

Year	Inter-bank Wakala Investments	x	Trend
2014	55,710,675	− 4.5	29,000,066.5
2015	9,625,000	− 3.5	26,884,385.5
2016	41,397,905	− 2.5	24,768,704.5
2017	16,397,550	− 1.5	22,653,023.5
2018	0	− 0.5	20,537,342.5
2019	454,888	0.5	18,421,661.5
2020	0	1.5	16,305,980.5
2021	26,841,000	2.5	14,190,299.5
2022	29,053,000	3.5	12,074,618.5
2023	15,315,000	4.5	9,958,937.5
Growth Rate b = Σxy/Σx2 = − 2,115,680.982			
Trend = a + bx (where a = Σy/N = 19,479,501.8)			
Year	Investment in Ijarah Asset	x	Trend
2014	7,353,240	− 4.5	− 86,447,085.22
2015	0	− 3.5	− 33,968,573.75
2016	799,441	− 2.5	18,509,937.73
2017	25,726,666	− 1.5	70,988,449.2
2018	59,880,031	− 0.5	123,466,960.7
2019	116,658,786	0.5	175,945,472.1
2020	176,652,000	1.5	228,423,983.6
2021	260,170,000	2.5	280,902,495.1
2022	364,836,000	3.5	333,381,006.5
2023	484,986,000	4.5	385,859,518
Growth Rate b = Σxy/Σx2 = 52,478,511.47			
Trend = a + bx (where a = Σy/N = 149,706,216.4)			
Year	Ijara Muntahia Bittamleek	x	Trend
2014	59,339,054	− 4.5	174,714,787.2
2015	155,580,391	− 3.5	190,111,191.6
2016	234,053,765	− 2.5	205,507,596
2017	282,109,561	− 1.5	220,904,000.3
2018	316,397,423	− 0.5	236,300,404.7
2019	318,133,875	0.5	251,696,809.1
2020	303,905,000	1.5	267,093,213.5
2021	291,610,000	2.5	282,489,617.8
2022	250,548,000	3.5	297,886,022.2
2023	228,309,000	4.5	313,282,426.6

(continued)

Table 4 (continued)

Year	Ijara Muntahia Bittamleek	x	Trend
Growth Rate b = Σxy/Σx2 = 15,396,404.37			
Trend = a + bx (where a = Σy/N = 243,998,606.9)			
Year	Wakala Bil Istithmar	x	Trend
2014	17,671,500	− 4.5	− 12,442,439.48
2015	37,647,530	− 3.5	32,112,770.08
2016	62,850,817	− 2.5	76,667,979.63
2017	91,310,805	− 1.5	121,223,189.2
2018	137,923,549	− 0.5	165,778,398.7
2019	183,671,834	0.5	210,333,608.3
2020	298,692,000	1.5	254,888,817.8
2021	339,422,000	2.5	299,444,027.4
2022	339,446,000	3.5	343,999,236.9
2023	371,924,000	4.5	388,554,446.5
Growth Rate b = Σxy/Σx2 = 44,555,209.55			
Trend = a + bx (where a = Σy/N = 188,056,003.5)			

10.3 Trend and Growth Rate of Islamic Assets in Alizz Islamic Bank

According to Table 5, the growth rate of all four assets is positive, and the trend value is progressive.

11 Conclusion

This study evaluates the performance of Oman's two Islamic banks—Bank Nizwa and Alizz Islamic Bank—through analysis of their Islamic banking assets over 10 years and 9 years, respectively. Using secondary data from annual reports, the study assesses key assets. For Bank Nizwa, Interbank Wakala Investments display high fluctuation, with negative growth rates in 2015, 2017, 2018, 2020, and 2023, resulting in a negative compound annual growth rate (CAGR). Ijarah assets show positive growth except in 2015, with a positive CAGR of 59%. Ijara Muntahia Bittamleek records positive growth in the initial five years, turning negative in subsequent years but achieving a positive CAGR of 16%. Wakala Bil Istithmar maintains positive growth throughout, with a CAGR of 40%. Although these assets have generally positive CAGRs, improvement in growth rates would enhance performance. For Alizz Islamic Bank, Wakala Bil Istithmar maintains positive growth except in 2019, with a CAGR of 28.27%. Murabaha Receivables show a positive CAGR (18.19%) despite

Table 5 Trend and growth rate of Islamic assets in Alizz Islamic Bank (Amount in RO)

Year	Wakala Bil Istithmar	x	Trend
2016	55,288,924	− 3.5	66,299,547.2
2017	100,305,502	− 2.5	83,357,202.8
2018	111,542,706	− 1.5	100,414,858.4
2019	105,291,139	− 0.5	117,472,514
2020	123,129,167	0.5	134,530,169.6
2021	147,507,999	1.5	151,587,825.2
2022	179,478,969	2.5	168,645,480.8
2023	185,466,328	3.5	185,703,136.4
Growth Rate b = Σxy/Σx2 = 17,057,655.6			
Trend = a + bx (where a = Σy/N = 126,001,341.8)			
Year	Murabaha Receivables	x	Trend
2016	60,592,103	− 3.5	70,399,450.75
2017	83,183,614	− 2.5	77,466,898.68
2018	104,243,923	− 1.5	84,534,346.61
2019	95,973,518	− 0.5	91,601,794.54
2020	96,067,530	0.5	98,669,242.46
2021	105,211,172	1.5	105,736,690.4
2022	123,976,679	2.5	112,804,138.3
2023	115,835,609	3.5	119,871,586.3
Growth Rate b = Σxy/Σx2 = 7,067,447.929			
Trend = a + bx (where a = Σy/N = 98,135,518.5)			
Year	Ijarah Assets-Ijarah Muntahia Bittamleek	x	Trend
2016	184,410,642	− 3.5	243,042,522.6
2017	234,964,599	− 2.5	236,430,777.4
2018	264,997,859	− 1.5	229,819,032.3
2019	252,211,378	− 0.5	223,207,287.1
2020	251,728,615	0.5	216,595,541.9
2021	226,561,121	1.5	209,983,796.7
2022	192,127,921	2.5	203,372,051.6
2023	152,209,181	3.5	196,760,306.4
Growth Rate b = Σxy/Σx2 = − 6,611,745.167			
Trend = a + bx (where a = Σy/N = 219,901,414.5)			
Year	Diminishing Musharaka	x	Trend
2016	12,503,859	− 3.5	− 45,395,309.29
2017	38,752,764	− 2.5	29,368,720.25
2018	71,528,324	− 1.5	104,132,749.8

(continued)

Table 5 (continued)

Year	Diminishing Musharaka	x	Trend
2019	118,519,582	− 0.5	178,896,779.3
2020	235,675,948	0.5	253,660,808.9
2021	322,494,125	1.5	328,424,838.4
2022	411,937,356	2.5	403,188,868
2023	518,818,395	3.5	477,952,897.5
Growth Rate b = Σxy/Σx2 = 74,764,029.54			
Trend = a + bx (where a = Σy/N = 216,278,794.1)			

negative growth in 2019 and 2023. Ijarah assets record positive growth in the first three years while Diminishing Musharaka maintains consistent positive growth with a high CAGR of 117.12%. Both banks could focus on enhancing growth rates for select assets to strengthen their performance further.

References

1. Al Sharif, B.M.M.: The impact of macroeconomic variables on the performance of Islamic banks: an empirical study. Int. J. Prof. Bus. Rev. **8**(4), 22 (2023)
2. Aress, A.H., Abdul, F.: Islamic banking products and financial performance of banks in Nairobi County, Kenya. Int. Acad. J. Econ. Fin. **3**(9), 374–394 (2023)
3. Asutay, M., Ubaidillah: Examining the impact of intellectual capital performance on financial performance in Islamic banks. J. Knowl. Econ. **15**(1), 1231–1263 (2024)
4. Awwad, B.S., Razia, B.S., Razia, A.S.: Challenges and obstacles to issuing Islamic Sukuk in Palestine: an analytical study of Islamic banks operating in Palestine. Compet. Rev. Int. Bus. J. **34**(1), 163–178 (2024)
5. Bilal, Z.O., Durrah, O.M., Atiya, T.M.: Comparative study on the performance of Islamic banks and conventional banks: evidence from Oman. Int. J. Econ. Fin. Issues **6**(4), 1835–1841 (2016)
6. Carney, M.: What are banks really for? Methods (2024)
7. Da Rin, M., Hellmann, T.: Banks as catalysts for industrialization. J. Fin. Intermed. **11**(4), 366–397 (2002)
8. Dominic, J., Joseph, A., Sisodia, G.: The role of the banking sector in economic resurgence and resilience: evidence from Pacific Island countries. Glob. J. Flex. Syst. Manag. **24**(Suppl 1), 9–30 (2023)
9. Faizulayev, A.: Comparative analysis between Islamic banking and conventional banking firms in terms of profitability, 2006–2009. Ph.D. diss., Eastern Mediterranean University (EMU) (2011)
10. Halim, N.A., Mazlan, N.H.: Profitability performance of full-fledged Islamic banks and economic growth in Malaysia: a panel data approach. J. Numer. Optim. Technol. Manage. **1**(1), 29–41 (2023)
11. Meskovic, A., Kozarevic, E., Avdukic, A.: The influence of national and individual Islamic governance on Islamic banks' social performance. J. Islamic Account. Bus. Res. **15**(6), 911–941 (2024)
12. Rahman, A.A.A., Ur Rahman, H., Meero, A., Amin, A.R.: Fintech innovations and Islamic banking performance: post-pandemic challenges and opportunities. Banks Bank Syst. **18**(4), 281 (2023)

13. Rioja, F., Valev, N.: Does one size fit all? A re-examination of the finance and growth relationship. J. Dev. Econ. **74**(2), 429–447 (2004)
14. Riyas, K., Ahmed, M.I., John, S.: A study on the challenges and prospects of Islamic banking in Sultanate of Oman. Int. J. Bus. Adm. Res. Rev. **2**(10), 279–282 (2015)
15. Saunders, A., Cornett, M.M.: Financial Institutions Management: A Risk Management Approach, 7th edn. McGraw Hill (2011)
16. Sharma, S.: Comparative financial performance analysis of neo banks and conventional banks in India (2019–2023): an empirical study. Techno Rev. J. Technol. Manage. **4**(2), 01–11 (2024)
17. Sjarief, L., Ghoni, M.A., Taufiq Affandi, M.T.: The role of financial performance on the profitability of Indonesian Islamic banks. Jurnal Ekonomi & Keuangan Islam, 277–285 (2023)
18. Standard chartered website (cited on 24 October 2024). https://www.sc.com/ae/stories/basics-of-banking/all-you-need-to-know/#:~:text=Islamic%20banking%20is%20a%20system,alcohol%2C%20gambling)%2C%20etc
19. Yilmaz, N.: Financial performance analysis of Greek banks: the SECA method. In: Pioneer and Contemporary Studies in Social, Human and Administrative Sciences, p. 267 (2023)

Impact of Macroeconomic Factors on Non-performing Assets of Indian Scheduled Commercial Banks

Sumi Prasad, Ravi Thirumalaisamy, Velmurugan Ramasamy, Sudarvel Jayaraj, Mahmoud Abouraia, and Dhanuskodi Rengasamy

Abstract It is essential to understand the impact of macroeconomic factors on non-performing assets (NPAs) on the performance of commercial banks, a key aspect of financial stability in a country. With this primary aim, the study was undertaken to analyze the impact of fiscal and monetary tools, along with domestic macroeconomic factors, on the Non-Performing Assets of Indian scheduled commercial public sector banks. It relies on secondary data sourced from the official websites of the Reserve Bank of India, the World Bank, and the International Monetary Fund, covering the financial period from April 2008 to March 2022, encompassing 14 years of analysis. The collected data have been analyzed by employing correlation and stepwise regression. The results indicate monetary tools, fiscal tools, and domestic macro-economic factors such as industry growth, exports, imports, government tax, statutory liquidity ratio, inflation, and government expenses were found to be the most significant variables, contributing 71.20% to the variation in non-performing assets (NPAs). This high contribution underscores the importance of these factors in understanding and managing NPAs. The feasible suggestions are provided for improving financial

S. Prasad (✉)
Economics and Business Studies Department, Mazoon College, Muscat, Sultanate of Oman
e-mail: sumidinu@gmail.com

R. Thirumalaisamy · M. Abouraia
Department of Accounting, Finance and Economics, Faculty of Business and Economics, Modern College of Business and Science, Muscat, Sultanate of Oman
e-mail: ravi@mcbs.edu.om

M. Abouraia
e-mail: Mahmoud.Abouraia@mcbs.edu.om

V. Ramasamy · S. Jayaraj
Karpagam Academy of Higher Education, Coimbatore, Tamil Nadu, India
e-mail: drvelsngm@gmail.com

S. Jayaraj
e-mail: j.sudarvel@gmail.com

D. Rengasamy
Department of Accounting and Finance, Curtin University, Miri, Malaysia
e-mail: dhanuskodi@curtin.edu.my

B. Awwad (ed.), *AI and IoT: Driving Business Success and Sustainability in the Digital Age*, Studies in Systems, Decision and Control 590,
https://doi.org/10.1007/978-3-031-88874-8_94

stability and reducing surging NPAs in the banking industry and for the Central Bank of India.

Keywords Macroeconomic factors · Non-performing assets (NPA) · Statutory liquidity ratio · Inflation · Reserve Bank of India · GDP

1 Introduction

The macroeconomic factors and their impact on non-performing assets (NPAs) of Indian scheduled commercial banks are the most critical areas of analysis. The NPAs affect the banking sector's financial health and the economy in a broader term. The NPAs, representing unpaid loans that are close to default, can be significantly influenced by factors such as GDP growth, inflation, and interest rates [1]. The economic downturn often leads to an increase in NPAs as the loan repayment by the borrowers slows down. Particularly in the most vulnerable sectors, such as infrastructure and real estate [2]. The increase in inflation rates reduces purchasing power, making it difficult for individuals and businesses to service their debt, leading to an increase in NPA levels [3]. Interest rates also play a significant role, as the increase in borrowing costs may lead to an increased rate of default, and augment NPA levels [4]. In addition, volatility in exchange rate impacts businesses with foreign debt exposure and augments NPA levels, particularly for export-oriented businesses [3]. Understanding these macroeconomic dynamics is crucial for policymakers and banks to develop effective risk mitigation strategies and stabilize the financial sector.

2 Review of Literature

The research by Mujibar and Debasish [5] examined 27 Indian public sector banks for the period 2000–2015, using panel data regression, finding that priority sector lending, operational inefficiencies, business per employee, real interest rate, and inflation rate positively influence the NPAs. Conversely, non-interest income, net interest margin, credit orientation, high capital adequacy, financial intermediation, and GDP growth negatively influence the NPAs. Hersugondo et al. [6] assessed that the public and private commercial banks' performances are negatively impacted by the increase in NPA, as reflected in lower net interest margins, return on assets, and return on equity. The Reserve Bank of India (RBI) introduced several measures to control the increasing NPA rates. Vivek [2, 7] highlighted that the Prompt Corrective Action (PCA) framework, revised in 2017, placed restrictions on banks with low capital adequacy ratios and high NPAs, lending limits, and expanding their branches. Additionally, the Strategic Debt Restructuring (SDR) scheme of 2015 permitted banks to convert the defaulted debt of borrowers into equity. The SDR scheme also replaced the defaulting promoters to recoup debts. The Sustainable Structuring of

Stressed Assets (S4A) was another initiative that enabled banks to restructure 50% of bad loans to be converted into equity, which are recovered as the company values improved. The establishment of the Central Repository of Information on Large Credits (CRILC) in 2014 and the Asset Quality Review (2015–16) were aimed at high transparency in loan reporting and detecting early signs of default.

Supriya [8] discussed the decline in lending due to increased NPAs, which in turn, affects economic growth, particularly for priority sectors and budding entrepreneurs. The overall economic stability was impacted by lower asset returns, reduced profits, asset-liability mismatches, and lower capital adequacy caused by high NPAs. According to Abhishek [9], industrial credit growth declined from 6.1% in 2019 to 0.8% in 2020 due to limited credit availability. The industries such as chemicals, infrastructure, and jewellery suffered. Kaul (2020) noted that the government's easy money policies from 2009 to 2011 led to high inflation and economic slowdown, limited banks' ability to lend as they faced high interest rates. Industries with foreign trade exposure affected by Currency depreciation, often due to economic crises, further increased NPA.

The Corporate Debt Restructuring (CDR) Framework in 2001, an additional regulatory framework established by RBI, allowed banks to restructure viable borrower debts without legal action. The SARFAESI Act (2002) allows banks to recover secured debts without any court intervention. The creation of Asset Reconstruction Companies (ARC) in 2003 helped banks offload NPAs from their balance sheets by selling bad loans at a discount, facilitating fresh lending. The Financial Stability Report (2021) projected that Gross NPAs (GNPAs) could rise under stress scenarios, with banks maintaining sufficient capital due to improved Capital Risk-Weighted Assets Ratio (CRAR) and a high provision coverage ratio. The slow recovery post-pandemic, supply disruptions, inflation, and economic shifts continue to challenge the banking sector and the Indian economy.

3 Research Gap

The existing literature primarily addresses macroeconomic factors affecting NPAs and regulatory measures like PCA and SDR. However, there is a research gap across public and private sector banks in examining the long-term effectiveness of these factors. Subsequently, limited studies focus on how priority lending in specific subsectors contributes to NPAs. Future research could explore differences in NPA management strategies between public and private banks, sector-specific NPA drivers, and regulatory impacts over time.

4 Statement of the Problem

The economic growth is affected by the increasing non-performing assets (NPAs) of Indian scheduled commercial banks, as it poses a significant threat to financial stability by affecting banks' profitability (Rajeev and Mahesh 2010; Dash and Kabra 2010). Macroeconomic factors have complex and varying impacts on NPA levels, yet the mechanisms remain inadequately understood, especially factors such as inflation, interest rates, and GDP growth, along with sector-specific lending practices (Ranjan and Dhal 2003; Misra and Dhal 2010). This study aims to identify and analyze the influence of macroeconomic factors on NPAs within Indian banks, highlighting the existing gaps in regulatory measures and management strategies. Addressing this issue is crucial for strengthening the banking sector and fostering a resilient financial environment in India (Ghosh 2015; Mehta and Bhavani 2017).

5 Objective of the Study

To analyze the impact of domestic macroeconomic factors and fiscal and monetary tools on Non-Performing Assets of Indian scheduled commercial public sector banks.

6 Research Methodology

The study depends on secondary data from the official websites of the Reserve Bank of India, the World Bank, and the International Monetary Fund. The Indian public sector scheduled commercial banks are focused in this study for a period covering April 2008 to March 2022, encompassing analysis for 14 years. Following the merger of weaker banks with stronger counterparts during the 2019–20 period, the existing are 12 state-owned banks, reduced from 27 banks in 2017. In this study, all 12 public sector scheduled commercial banks are included. The correlation and step-wise regressions are employed to analyze the collected data.

7 Significance of the Study

The significance of the study is to understand the impact of macroeconomic factors on non-performing assets (NPAs) of Indian scheduled commercial banks, a key aspect of financial stability. To strengthen banking resilience, improve lending practices, and reduce NPAs, the study has identified variables like inflation, interest rates,

GDP growth, and other variables that affect NPAs, which provides insights for policymakers and banks. It is essential to address these factors for a stable financial environment, sustainable growth, and increased investor confidence in India.

8 Limitations of the Study

This study is confined to the period between 2008 and 2022, during which extreme events such as the pandemic could have influenced the variables, potentially skewing the findings related to the Asset Quality Review. Additionally, the researcher utilized data from merged entities, as individual bank data prior to the mergers was unavailable. Another limitation is the reliance on secondary data; despite sourcing it from official published sources, there is a possibility that the data for this period may suffer from nondisclosures.

9 Explanation for Fiscal and Monetary Tools

9.1 *Fiscal Tools*

9.1.1 Government Taxes

Lower taxes, an expansionary policy, increase disposable income for individuals and businesses, boosting economic activity and borrowers' repayment capacities, thus reducing non-performing assets (NPAs). Conversely, higher taxes and a contractionary policy limit disposable income, potentially harming borrowers' financial health and increasing NPAs. This taxation-NPA relationship underscores the significant impact of fiscal policy on banking sector stability.

9.1.2 Government Expenditure

As an expansionary fiscal policy, increased government spending boosts employment and household income, enhances borrowers' repayment capacity, and reduces gross non-performing assets (NPAs). Conversely, contractionary policies can reduce income levels, increasing NPAs.

9.2 Monetary Tools

9.2.1 Bank Rate

The lending rate for short- and medium-term financing is the bank rate, impacting private sector borrowing. Borrowers' debt burden increases with the increased interest rates, leading to defaults and subsequently increasing non-performing assets (NPAs) in banks [4].

9.2.2 Repurchase Rate

The interest rate at which banks borrow from the Reserve Bank of India is the Repurchase rate. Contractionary tools are used to curb inflation and discourage excessive borrowing, leading to an increase in the Repo rate. This, in turn, raises the cost of capital, reducing investments and potentially lowering household incomes. It also increases loan repayment difficulties and, thus, non-performing assets (NPAs) [10, 11]. The borrower defaults due to elevated rates and worsening NPAs.

9.2.3 Cash Reserve Ratio

Banks must keep a proportion of deposits as liquid cash with the Reserve Bank of India. For a contractionary monetary policy, CRR is increased to reduce liquidity, which can lead to refinancing challenges and higher non-performing assets [1, 4].

9.2.4 Statutory Liquidity Ratio

The percentage of total deposits that banks must hold as liquid assets. Similar to the Cash Reserve Ratio (CRR), an SLR increase reduces liquidity, potentially escalating the gross non-performing assets [1, 4].

9.2.5 Broad Money Growth

Broad money growth increases the money supply, lowering interest rates and encouraging investment. This boost in household income and business profits strengthens borrowers' ability to meet obligations, reducing non-performing assets (NPAs) [1].

9.3 *Domestic Macro Economic Factors*

9.3.1 Gross Domestic Investment

An increase in Gross Domestic Investments (GDI) generates employment opportunities, enhancing household income and overall economic resilience. With higher household income, consumer demand rises, which, in turn, leads to increased profits for businesses. Improved economic conditions and profitability contribute to a reduction in gross non-performing assets, as credit risk diminishes.

9.3.2 Unemployment

High unemployment decreases individual income levels, signaling reduced capital investments and lower household earnings [12]. This drop in disposable income dampens demand, further impacting investments. Consequently, income instability grows, creating revenue uncertainties for companies and leading to potential loan defaults.

9.3.3 Inflation

Inflation is a sustained rise in price levels. Moderate inflation within a central bank's target typically does not harm economic stability, but unchecked inflation can be risky [10, 13]. High inflation affects non-performing assets (NPAs) in various ways. It reduces the real value of money, making loan repayments easier if wages rise [5], thereby reducing NPAs. Currency depreciation also makes exports more competitive, potentially boosting GDP and lowering NPAs, but it depends on export elasticity [12]. However, inflation can increase production costs and reduce export competitiveness, particularly if inflation disparities widen. Central banks may raise interest rates to control inflation, suppressing demand and economic growth, which can raise NPAs as defaults increase [3]. Some studies suggest a negative inflation-NPA relationship, while others, like Prasanna et al. [11], find it insignificant with no direct causation.

9.3.4 Exports

Exports of goods and services boost GDP by contributing valuable market services to the global economy. A trade surplus, where exports exceed imports, strengthens economic growth and can reduce banks' non-performing assets (NPAs) through improved national income [3].

9.3.5 Imports

The value of foreign products consumed domestically is the value of Imports. Imports lower consumer prices, which, on the other hand, challenges domestic industries, reducing their revenues and debt repayment capacity. When imports exceed exports, a trade deficit lowers the GDP, impacting national income and increasing non-performing assets (NPAs) in the banking sector [3].

9.3.6 Industries, Agriculture and Services

Growth in the manufacturing, agriculture, and services sectors contributes to GDP expansion, signaling the increased national income. The debt repayment capacity of the borrowers reduces banks' non-performing assets (NPAs) [3, 4, 5, 10, 14, 15]. However, some studies suggest a positive GDP-NPA correlation [12].

9.3.7 Credit Growth in the Private Sector

Credit growth in the private sector indicates an increase in Investments from the private sector. Capex increases during the growth phase of an economy and results in higher employment, higher corporate profits, and higher household income, reducing the gross non-performing assets. On the other hand, procyclical lending and an overheated economy might lead to defaults and result in surging gross non-performing assets [3, 14].

9.4 Analysis and Interpretation

The **correlation analysis** is used to examine the relationship between the dependent variable NPA to total advances and the independent variables such as domestic macro economic factors,monetary and fiscal tools. (Table 1).

Out of 16 macroeconomic and policy variables analyzed, 14 significantly impacted Non-Performing Assets (NPA). Key variables, significant at the 1% level, include Broad Money Growth, Cash Reserve Ratio, Government Expenses, Government Tax, Export Growth, Service Value Added, Inflation (CPI), Gross Domestic Investment, Unemployment, Industry Growth, Import Growth, Statutory Liquidity Ratio, Bank Rate, and Repo Rate.

Broad Money Growth is negatively correlated with NPA, accounting for 28.7% of its variation. Similarly, a decrease in the Cash Reserve Ratio, Government Expenses, Export Growth, and Industry Growth correlates with an increase in NPAs, explaining 35.9%, 18.6%, 43.6%, and 59.5% of the variation, respectively. Inflation (CPI) and Gross Domestic Investment also show negative correlations with NPA, explaining 51.3% and 51.6% of NPA variation, respectively. Conversely, Government Tax,

Table 1 Correlation analysis of selected domestic macro economic, fiscal, and monetary factors on the NPAs of all banks

Variables	r	r^2
Gross domestic investment	− 0.718**	0.516
Inflation	− 0.716**	0.513
Unemployment	0.235**	0.055
Export	− 0.660**	0.436
Import	− 0.673**	0.453
Industrial growth	− 0.772**	0.595
Agricultural growth	− 0.113	0.013
Service sector growth	0.741**	0.550
Private sector credit growth	− 0.001	0.000
Government tax	0.312**	0.097
Government expenses	− 0.431**	0.186
Bank rate	− 0.220**	0.048
REPO rate	− 0.334**	0.112
Cash reserve ratio	− 0.599**	0.359
Statutory liquidity ratio	− 0.791**	0.626
Broad money growth	− 0.536**	0.287

Source Computed data

Service Value Added, and Unemployment positively correlate with NPA, suggesting that increases in these variables contribute to higher NPA levels, with government tax alone explaining 9.7% of the variation.

Other notable influences include Import Growth, the Statutory Liquidity Ratio, the Bank Rate, and the Repo Rate, all of which are negatively correlated with NPA and explain between 4.8% and 62.6% of the variation in NPA. These findings underscore the complex dynamics between domestic macroeconomic factors and fiscal and monetary policies in influencing NPA levels, emphasizing the need for careful policy management to mitigate NPA risks.

1. **Gross Domestic Investment (GDI)**: As Barro (1991) argues, higher GDI is linked to economic growth through capital accumulation and productivity improvements. By increasing income and employment, higher GDI enhances borrowers' repayment capabilities, stabilizing the banking sector and reducing NPAs. Barro's research highlights the positive effects of investment on financial stability.
2. **Inflation (Consumer Price Index)**: Mild inflation encourages spending, as Keynes (1936) suggested in The General Theory of Employment, Interest, and Money, leading to higher revenues for businesses and stronger debt-repayment capacities. Phillips (1958) further illustrates the relationship between inflation and employment, noting that controlled inflation can support economic activity, reducing the risk of loan defaults.

3. **Export Growth**: Export growth strengthens the economy by increasing foreign revenue, which improves the financial position of export-driven firms, as Krugman (1979) explains in his theory of international trade. According to Krugman, firms with a diversified revenue stream are less likely to default on loans due to a larger export market.
4. **Import Growth**: Grossman and Helpman (1991) argue that increased imports can signal consumer and capital goods demand, fostering productivity. Increased productivity allows businesses to generate revenue and better service debt, thereby reducing NPAs. This aligns with Helpman and Krugman's (1985) views on the relationship between trade and growth.
5. **Government Expenses**: Keynes (1936) argued that government spending can create a stable economy, stimulating aggregate demand and employment. Increased public spending, especially on infrastructure, leads to job creation and boosts economic stability, reducing the likelihood of loan defaults as borrowers become more financially secure.
6. **Bank Rate**: Mishkin (2007) explains that the central bank's bank rate affects the cost of borrowing for commercial banks, influencing lending practices. When bank rates are high, banks adopt a more cautious lending approach, which, in turn, reduces the chances of defaults and NPAs. Mishkin's work in The Economics of Money, Banking, and Financial Markets discusses interest rates' influence on financial stability.
7. **Repo Rate**: A higher repo rate encourages banks to lend selectively, mitigating default risk. Their work demonstrates that the monetary policy instrument's repo rate affects the financial sector's health. Thus, repo rate directly impacts liquidity in the banking system, as noted by Bernanke and Blinder (1992).
8. **Cash Reserve Ratio (CRR)**: Promoting financial stability, as highlighted by Diamond and Rajan (2001), a higher CRR limits the funds banks can lend in their research on liquidity and banking. The potential for over-leveraging is reduced with the CRR restrictions, thus lowering NPA risk. Diamond and Rajan emphasize that The banks can protect from excessive exposure to risk by adhering to the reserve requirements.
9. **Statutory Liquidity Ratio (SLR)**: According to Allen and Gale (2000), higher SLR mandates that banks invest in safe assets to curb risky lending. Their research on financial systems and liquidity risk demonstrates that banks with ensured regulatory liquidity ratios like SLR maintain a conservative lending portfolio, reducing exposure to NPAs.
10. **Broad Money Growth**: A balanced broad money growth coupled with liquidity, without creating excessive inflationary pressure, supports economic activity. This aligns with Friedman's (1982) work, which explains how controlled money supply growth can positively impact financial stability by reducing the risk of defaults, provided inflation is kept in check.
11. **Unemployment**: As unemployment increases, household income and consumer spending power decrease, leading to a higher likelihood of loan defaults, as Minsky argued (1986). Minsky's Financial Instability Hypothesis explains the increase in NPA due to the rise in unemployment.

12. **Service Sector Growth**: In economies lacking a strong industrial base, growth in the service sector may signal economic imbalances, as noted by Baumol (1967). This growth often involves low-productivity jobs that may lead to income volatility and credit defaults. Baumol's theory on "unbalanced growth" suggests that economies are more susceptible to financial instability when the service sector expands without parallel industrial growth.
13. **Government Tax**: As taxes increase, disposable income for businesses and individuals decreases, limiting their ability to service debt. Barro (1974) observed this effect, suggesting that higher taxes reduce cash flows, making it harder for borrowers to meet loan obligations. Barro's work on government fiscal policy explains how higher taxes, especially in times of economic downturn, can exacerbate NPAs by straining borrowers' financial resources.

A Stepwise regression test is employed to find variables that prominently influence the piling up of NPA.

The stepwise regression results, as shown in Table 2, indicate that Industrial Growth, Export, Import, Government Tax, Statutory Liquidity Ratio, Inflation, and Government Expenses significantly contribute to Non-Performing Assets (NPA), collectively explaining 71.2% of the variation in NPA to Total Advances.

10 Suggestions

- The Reserve Bank of India should enforce periodic Asset Quality Review with fair disclosures so that subsequent policy actions can arrest NPAs and minimize bank capital erosion.
- Stable government and predictable policies of the government enable companies to make better investment decisions, minimizing the risk of NPAs. Therefore, the government should be cautious in its policy decisions, as their effect on stakeholders is substantially high.
- The government should resolve the delay in the lawsuit related to the loan repayments. Fast-tracking the judicial process through exclusive tribunals, increasing the workforce, and leveraging technology can yield better results.
- The government and the Reserve Bank of India should work towards growing and developing the debt market in India. A well-developed debt market can eliminate the issue of liquidity and enable companies to access capital from the debt market, decreasing their dependency on the banks.
- The culture of the political parties announcing loan waivers in their election manifesto and the subsequent implementation of the same discourages repayment and encourages non-repayment. Governments should be cautious about the same and identify better social programs.
- Providing immunity against unwarranted lawsuits could make this reform more effective.

Table 2 Stepwise regression analysis of selected domestic macro economic, fiscal, and monetary factors on the NPAs of selected banks

Step	Constant	Industrial growth	Export	Import	Government tax	Statutory liquidity ratio	Inflation	Government expenses	R^2
1	0.727	− 0.023							0.539
2	0.735	− 0.018	− 0.008						0.585
3	0.946	− 0.026	− 0.023	0.014					0.627
4	0.660	− 0.023	− 0.019	0.010	0.020				0.647
5	0.592	− 0.022	− 0.018	0.010	0.027	− 0.007			0.660
6	0.551	− 0.023	− 0.016	0.009	0.030	− 0.014	0.001		0.688
7	0.611	− 0.033	− 0.022	0.015	0.025	− 0.018	0.002	0.016	0.712

Source Computed data

- IBC, with a time-bound act, would be instrumental in recovering bad debt, along with the advancement of technology and progressive amendments over a period to be on par with global standards. Further relaxation and regressive changes in the reforms would lead to the act's ineffectiveness.
- Addressing the principal-agent issues such as poor corporate governance, too many bosses, absence of individual directors, and too much interference by the promoters in allocating loans resulted in inefficiency, demotivation, confusion, and resistance from the top managers of the Public Sector Banks.

The advancement of technology in banks and financial institutions brings enormous accountability, proactive resolution in place by monitoring the borrowers, efficient recovery of stressed assets at the right time with cost efficiency.

11 Conclusion of the Study

The study concluded that out of the three hypotheses tested with selected variables for 12 banks over 13 years, monetary tools, fiscal tools, and domestic macro-economic factors such as industry growth, exports, imports, government tax, statutory liquidity ratio, inflation, and government expenses were found to be the most significant variables, contributing 71.20% to the variation in non-performing assets (NPAs). It was observed that these macroeconomic factors, influenced by fiscal and monetary tools, contributed to the increase in NPAs. Contrary to expectations, NPAs rose with decreased Cash Reserve Ratio (CRR) and Statutory Liquidity Ratio (SLR). This could be attributed to three factors: firstly, the decreasing CRR and SLR during the study period reflected tight liquidity conditions, leading to refinancing problems and NPAs. Secondly, the slowdown in the Indian economy during the study period contributed to NPAs, as loans made during economic upswings turned problematic in downturns. Thirdly, the Reserve Bank of India's Asset Quality Review led to increased NPA disclosures. Increased government spending improved employment and household income, enhancing borrowers' repayment ability and reducing NPAs. Furthermore, a decline in inflation, industry growth, and trade imbalances with greater imports than exports contributed to the increase in NPAs. Exports, however, had a favorable effect on reducing NPAs.

12 Scope of the Study

The researcher explores/investigates the areas in detail that were not explored and had limited study by past research. Further research can be done on public sector scheduled commercial banks for a different period. The study can also be extended to private sector banks and other financial institutions. Further research can be done with a sophisticated tool such as an Artificial Neural Network.

References

1. Reserve Bank of India: Report Committee on Review of workings of Asset Reconstruction Companies. Reserve Bank of India (2021)
2. Vivek, K.: Baby Step Subbarao. In: Kaul, V. (ed.) Bad Money: Inside the NPA Mess and How It Threatens the Indian Banking System, pp. 65–78. HarperCollins, Noida (2020)
3. Bayar, Yilmaz: Macroeconomics institutional and bank-specific determinants of non-performing loans in emerging market economies: a dynamic panel regression analysis. J. Central Bank. Theory Pract. **8**(3) (2019)
4. Kadanda, D., Raj, K.: Non-performing assets (NPA) and its determinants: a study of Indian public sector banks. J. Soc. Econ. Dev. **20**(2), 193–212 (2018)
5. Mujibar Rahaman, S.K., Debasish, S.: Identifying key drivers of non-performing assets in Indian public sector banks. IIM Kozhikode Soc. Manage. Rev. (2021)
6. Hersugondo, Anjani, N., Pamungkas, Imang, D.: Non-performing assets, capital adequacy and insolvency risk on bank performance: a case study in Indonesia. J. Asian Fin. Econ. Bus. **8**(3), 319–329 (2021)
7. Vivek, K.: No borrowing, no lending, no problem. In: Vivek, K. (ed.) Bad Money: Inside the NPA Mess and How it Threatens the Banking System, 1st edn., pp. 180–189. HarperCollins (2020)
8. Supriya, R.: Non-performing Assets and Its Effects on the Indian Banking Sector. Tygar Law Corporate (2020)
9. Abhishek, S.: NPA and its effect on banks profitability. The Times of India (2020)
10. Saunders, A., Strock, E., Traylos, N.G.: Ownership structure, deregulation, and bank risk taking. J. Financ. **45**(2), 643–654 (1990)
11. Prasanna, K., Thenmozhi, Rana, N.: Determinants of non-performing advances in Indian banking system. Banks Bank Syst. **9**(2), 65–77 (2014)
12. Gokul Kumar, S., Jayanthi, M., Prasanth, A.: A study on the impact of institutional specific and macroeconomic indicators on the non-performing assets of new private sector banks in India. Int. J. Latest Technol. Eng. Manage. Appl. Sci. **6**(7), 125–132 (2017)
13. Memdani, L.: Macroeconomics and bank specific determinants of non-performing loans (NPL) in the Indian banking sector. Stud. Bus. Econ. **12**(2), 125–135 (2017)
14. Changjun, Z., Probir, K., Niluthpaul, S.: Industry-specific and macroeconomic determinants of non-performing loans: a comparative analysis of ARDL and VECM. Sustain. J. **12**(1), 325–330 (2019)
15. Petr, J., Thomas, R.: What are the key determinants of non-performing loans in CESEE? IMF Working Papers. iv (2014)

Leveraging ICT for Academic Excellence: Insights from Indian Higher Education Institutions

Niyaz Panakaje, S. M. Riha Parvin, K. Madhura, Shafeena Jabbar, Mahammad Shahid, Niha Sheikh, Afsana Mehar, and K. M. Abdul Rasheed

Abstract This study seeks to elucidate on the adoption of ICT in higher education and its effectiveness, aiming to provide informative perspectives into how ICT directs the educational landscape by modulating student performance and placement opportunities in higher education institutions along with discerning the challenges and opportunities associated with it. This study employs a quantitative and descriptive research design, adopting a structured questionnaire. The study revealed students' positive perceptions towards the importance of ICT contribute to its adoption, thereby enhancing student performance, placement, and opportunities. Nevertheless, challenges in ICT adoption can impede student performance. By combining the Technology Acceptance Model and Technology Enhanced Learning Theory, this research has added theoretical depth to the body of existing research.

Keywords Information and communication technology (ICT) · Student performance · Higher educational institutions · Effectiveness · Placement opportunities · Challenges

1 Introduction

ICT was initially used in academic environments during the 1980s, computers were not often utilized within these contexts. However, ICT precipitated a revolutionary transformation in higher education during the era of economic liberalization in the 1990s and took a significant turning point by the founding of National Mission on Education through Information and Communication Technology (NMEICT) in 2009 [1, 2]. The expansion of online learning environments and the 2020 National Education Policy (NEP) in higher education have highlighted the critical role played

N. Panakaje (✉) · S. M. Riha Parvin · K. Madhura · S. Jabbar · M. Shahid · N. Sheikh · A. Mehar · K. M. Abdul Rasheed
Yenepoya (Deemed to be University), Mangaluru, Karnataka, India
e-mail: niyaz0191@gmail.com

B. Awwad (ed.), *AI and IoT: Driving Business Success and Sustainability in the Digital Age*, Studies in Systems, Decision and Control 590,
https://doi.org/10.1007/978-3-031-88874-8_95

by technology in higher education in the twenty-first century. The NEP acknowledged the necessity for technology-driven education models, demonstrating a commitment to modernizing and adapting the higher education system to contemporary demands [3] and also to meet the needs of today's learners by the integration of ICT in higher education [4].

The Government of India reports 993 universities, 39,931 colleges, and 10,725 independent institutions [5] for the 2 million student population in independent schools, 7 million in universities, and 26 million in other colleges [6]. However, integrating ICT in this sector faces infrastructural and socioeconomic hurdles, unlike wealthier nations [7]. Despite these challenges, India can reskill from these countries to effectively incorporate ICT, for enhancing information sharing and educational quality [8]. A growing concern revolves around how ICT influences pedagogy and learning outcomes. Studies indicate that interactive, technology-driven teaching methods have significantly replaced traditional lecture-based approaches [9] into a more engaging educational environment [10].

This study evaluates whether ICT enhances student performance in higher education by examining its effect on engagement, information acquisition, critical thinking, problem-solving, and academic achievement [10, 11]. It investigates the integration of ICT and its role in empowering students, enhancing academic outcomes, and preparing them for a knowledge-driven society [12]. The research emphasizes the opportunities and challenges posed by technology in education, highlighting its potential to transform higher education. It builds upon previous research that has primarily focused on the theoretical aspects of ICT's impact on student performance.

A notable research gap exists in studying the specific impact of information and communication technology (ICT) on student performance as well as placement opportunities in Indian higher education. The proposed research aims to fill this void by understanding how ICT initiatives can optimize student performance, benefiting educational stakeholders in India and fostering societal advancements aligned with technological changes.

2 Related Works

The literature reviews highlight the importance of ICT, providing detailed justifications for its fundamental concepts. This section also includes comparison between previous studies related to present research (see Table 1).

2.1 Information and Communication Technology (ICT)

The integration of ICT in education has revolutionized the pedagogical strategies and methods of learning [13]. Moreover, it also increases the access to educational materials and creates opportunities for global connectedness. This research

Table 1 Comparison table

Aspects	Prior literature	Current research
Focus	It focuses on online learning, virtual classrooms, Technology Acceptance Model (TAM) by stressing the usage of Digital Information Access, Exchange, and Modification (AIMS)	It concentrated on different characteristics of features of ICT in higher education and its effect on placement opportunities and student performance
Objectives	Asses the satisfaction of the students for the usage of ICT in the classrooms and also focused on problem solving skill development digital classrooms, through mobile learning, and also emphasizing the individual skill development	Role of ICT in higher education, its effect on student placement, performance along with the hindrance due to challenges
Novelty	It is emphasis the importance of mobile learning, digital classrooms, TAM, role of teachers, and contributing to innovation and robustness in the field	Diverse perspective of ICT in Higher education sector
Hypothesis	General hypothesis on effect of ICT on students' performance	It specifies the personal profile influence on ICT usage, adoption ICT on students' perspective, its impact on their performance, challenges facing in the usage of ICT and its impact on placement opportunities, and their performance (See Fig. 1)

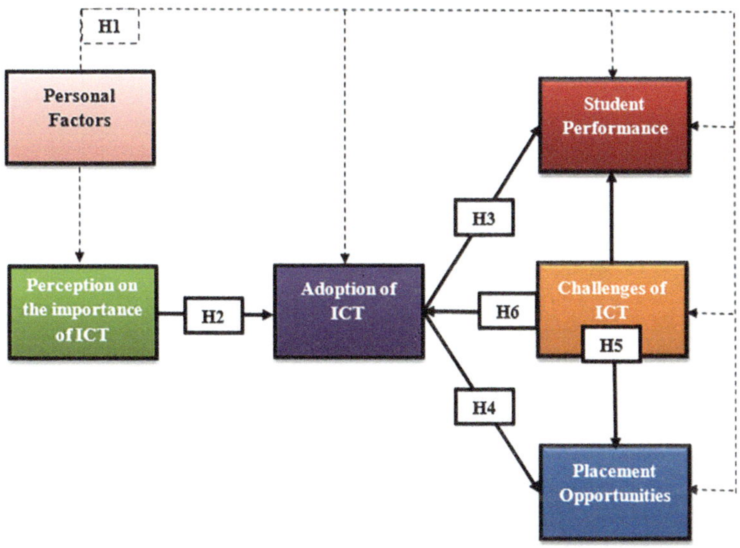

Fig. 1 Conceptual model. Source Developed by Researcher

particularly explores the adoption of Digital Information Access, Exchange, and Modification (AIMS) within the context of ICT. Additionally, the study employs Technology Acceptance Model (TAM), as suggested by Al-Rahmi et al. [14] and it focuses on evaluating student satisfaction with ICT in online learning environments. Recent studies highlighted that, the students' previous technology experiences and digital literacy influence their ICT usage [15]. Furthermore, the student's personal factors such as age, gender residential status, type of college, stream, and specialization significantly influences student's engagement on ICT [16]. Hence, it can be hypothesized that;

> ***H1***: *Usage of Information and Communication Technology in education is significantly determined by their personal and educational profile.*

2.1.1 Importance and Adoption/Usage Level of ICT in Higher Education

The adaptation of ICT like SWAYAM, NPTEL, JMOOC, Coursera and many more in higher education helps in promoting digital literacy, access to educational resources, enabling interactive and collaborative learning experiences among the students. The adoption of mobile learning is influenced by three primary factors: the IT framework, awareness, and support from senior management. These aspects encompass elements such as digital archives, flipped classrooms, and online repositories for teaching and content of the subject [17]. Authors have discussed resistance of the users in the usage of knowledge management systems (KMS) in higher education [18]. Mumtaz et al. [19] and Tennakoon et al. [20] contribute new perspectives on recent ICT usage patterns among college students and the impact of the COVID-19 pandemic on e-learning, respectively. Hence, it can be hypothesized that;

> ***H2***: *Students' Perceptions of ICT's importance will positively contribute to adoption*

2.1.2 Effectiveness of ICT on Student's Performance

Effectiveness of ICT on student's performance refers to the what extent the student's usage of ICT tools and resources significantly help in achieving their learning outcomes [21]. Studies on ICT's impact on student performance include varied approaches. Comi et al. [22] explore how instructors' use of ICT affects student outcomes. Chakraborty et al. [23] highlights how ICT contribute in augmenting accessibility by utilizing distance learning modes, knowledge connectivity. Moreover, a study showed that ICT have a positive impact on the development of student's transversal skills and literacy [24]. Fannakhosrow et al. [25] compare ICT with conventional instructional techniques, emphasizing ICT's crucial role in enhancing academic enthusiasm, contentment, and motivation, thereby leading to elevated academic achievements. These studies collectively advance the understanding of ICT's benefits in education. Hence, it can be hypothesized that;

H3: *Adoption of ICT in higher education significantly impacts students' Performance*

2.1.3 Placement Opportunities

This research collectively emphasizes the importance of ICT in placement opportunities and career building [26]. The integration of ICT produces high-quality e-content, transforms recruitment processes, expands job markets, provides on-demand content, and new avenues for career advancement. It also circulates information on vacancies through different online platforms, irrespective of geographical boundaries [27]. Iqbal et al. [28] explore the connection between curriculum implementation, ICT proficiency, adaptability, flexibility, decision making, self-discipline, self-management, and entrepreneurial skills. Thus, involving in ICT-mediated learning gives students the confidence that they will get the appropriate job suitable to their capabilities. Hence, it can be hypothesized that;

H4: *Adoption of ICT in higher education significantly impacts students' Placement Opportunities*

2.1.4 Challenges and Opportunities ICT

Several studies offer unique insights into the benefits and challenges of ICT across different sectors. The holistic viewpoint presented by several studies. In which, Selmi's [29] analysis examines how digitization transform education, advancing learning, customized delivery of relevant education material to each individual learner. Shukla's [30] discusses the challenges of ICT adoption in education, like poor internet connectivity, inadequate number of ICT tools, less training, lack of procurement funds, and implementation cost. In addition to addressing integration issues, Kaur's [31] explores the adaptability of ICT-based training, underscoring its advantages over traditional methods and providing insights on pedagogical strategies. The challenges can be addressed through establishment of cross sector collaboration [32] with government, businesses to overcome the gap in digital access by supplying cost-effective access and materials [33]. Hence, it can be hypothesized that;

H5: *Challenges in the usage of ICT hinder placement opportunities and other opportunities for using ICT*
H6: *Challenges in the adoption of ICT usage will hinder the influence of ICT adoption on placement opportunities and student performance*

2.2 Conceptual Model

3 Research Methods and Procedure

The study employs quantitative and descriptive research design methodologies. These methods are suitable for gathering numerical data to analyze ICT effectiveness in higher education, as demonstrated in studies on blended learning [34] and virtual learning [35]. The whole research process is demonstrated in Fig. 2.

3.1 Data Source

This research investigates the role of ICT in higher education using primary and secondary data. Secondary sources initially provided conceptual clarity and identified gaps. Primary data, collected through a standardized survey.

3.2 Construction of Research Instrument

The questionnaire was developed from prior research, observations, interviews, and expert opinions. The use of ICT in education is connected to a number of theories, including the technology enhanced learning theory and the information processing theory [36]. Apart from the theories, Measurement variables were sourced as follows;

Perception on the Importance of ICT. 11 items from Oliver [4]; Babu and Sridevi [37]; Saadé and Bahli [38]; Fathi Vajargah and Azadmanesh [39].

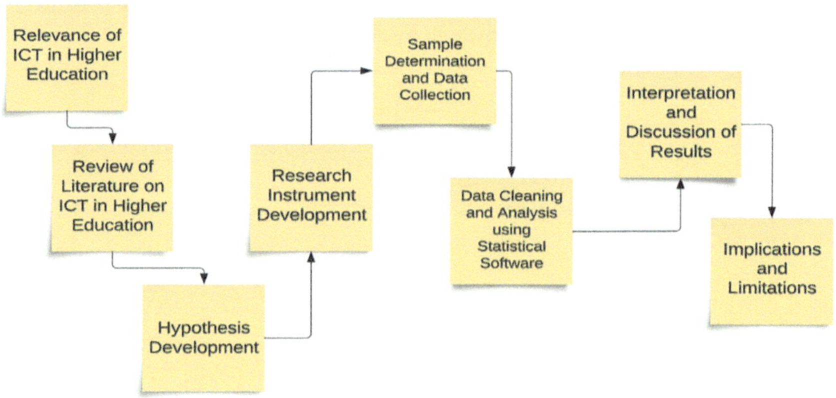

Fig. 2 Research process. *Source* Developed by researcher

Adoption of ICT. 12 items from Ganzert et al. [40]; Ibrahim et al. [41]; He and Li [42]; Abdullah et al. [18]; Mumtaz et al. [19] and Tennakoon et al. [20].

Student Performance. 11 items from Comi et al. [22]; Chakraborty et al. [23]; Asrizal and Utami [24]; Germo [43]; Fannakhosrow et al.

Placement Opportunities. 16 items from Phutela and Dwivedi [44]; Sepulveda-Escobar and Morrison [45]; Ajjawi et al. (2020); Wang et al. (2021); Iqbal et al. (2022) [28].

Challenges and Opportunities. 19 items from Alenezi et al. [26]; Shukla [30]; Selmi [29]; and Kaur [31].

The above measures taken from previous studies have been slightly altered to match the context of the study.

3.3 Content Validity, Pilot Study, Reliability and Validity Test

The questionnaire's content validity was ensured based on feedback from 5 research scholars and 2 education faculties. A pilot study with 50 students (25 M. Com and 25 MBA) preluded the main survey, facilitating a reliability and validity test. Through Confirmatory Factor Analysis the items with good factor loadings are retained (Above 0.5). The Cronbach's Alpha and composite reliability for all variables exceeded 0.7, confirming reliability. Convergent validity was established as the Average Variance Extracted (AVE) exceeded 0.5, and discriminant validity was confirmed as for each pair of constructs, the square roots of AVE were higher than the correlation coefficients as recommended by Fornell and Larcker [46] (see Table 2).

Table 2 Results of reliability and construct Validity

Variable	Cronbach Alpha	Composite Reliability	AVE	MSV	ASV	IICT	AICT	SP	PO	CO
IICT	0.915	0.862	0.530	0.524	0.500	0.737	-	–	–	–
AICT	0.907	0.868	0.541	0.529	0.515	0.733	0.744	–	–	–
SP	0.886	0.862	0.585	0.555	0.540	0.654	0.728	0.774	–	–
PO	0.980	0.950	0.625	0.575	0.554	0.726	0.743	0.759	0.745	–
CO	0.895	0.883	0.625	0.596	0.569	0.663	0.719	0.732	0.720	0.801

Note AVE Average variance extracted; *MSV* maximum shared variance; *ASV* average shared variance; *IICT* Importance of ICT; *AICT* adoption of ICT; *SP* student performance; *PO* placement opportunities; *CO* challenges and opportunities

3.3.1 .

3.4 Population and Sample

The research targeted 960 MBA students and 431 M. Com students from 10 colleges in the district who has high ICT usage in the curriculum. Using the Taro Yamane formula, 489 responses were collected, including 282 MBA students and 207 M. Com students. The sampling was conducted through proportionate stratified procedures, validating a representation of ICT usage in higher education within the district.

3.5 Statistical Techniques Used

The study employed SPSS 26 and SPSS AMOS 23 for comprehensive data analysis, including Percentage Analysis, Descriptive Statistics, Independent t-test, Regression Analysis, and Structural Equation Modeling (SEM), Regression and Multicollinearity tests were also conducted.

4 Results

4.1 Sample Profile

Out of 489 respondents, majority are under 24 years old (90.2%) and pursuing MBA. The gender distribution is nearly even, with 45.4% male and 54.6% female students, and a small proportion (8%) are married. The respondents are predominantly from rural areas (67.7%), 64% of the respondents are MBA students, with 43% specializing in Accounting and Finance. Students are mainly from Government (37.4%), Private (27.61%), Autonomous (19.63%), and Aided (15.34%) colleges.

4.2 ICT Usage

Among 489 students, 83.84% primarily use computers and mobile phones for ICT learning. For presentations, computers and LCD projectors are preferred, while a smaller group use scanners (13.5%), pen drives (11.45%), tablets (6.75%), and other devices like printers, cameras, and televisions. Internet use is widespread, with 95% accessing it daily. Microsoft Office tools are universally used: Word (daily-26%, weekly-34%), Excel (daily-27%, weekly- 33%), and PowerPoint (daily-23%, weekly- 39%). SPSS is used by 26.6%, while 84.45% do not use it. Email usage is nearly universal at 99%, MOOCs engagement is 38%, Wikipedia 54%, YouTube 78%, and online thesis repositories 4%.

The frequency and percentage of ICT tools used in the institute. Of the 322 respondents, 65.85% use ICT tools for accessing online materials. Digital libraries are frequently used by 59.10%, offering varied academic resources, while 13.70% use digital repositories for course materials like lecture notes. The flipped classroom approach is adopted by 11.45%, involving pre-class video lectures. A substantial 74.8% engage in multimedia learning via audio–video equipment. Moreover, 85.48% utilize projectors for lectures and presentations, and 93.25% use online tools like PowerPoint or Google Slides for presentations.

4.3 Perception on the Adoption and Importance of ICT

In a study on ICT adoption, students displayed positive perception of ICT adoption was indicated in a mean of 4.07 ± 1.00 and overall perception on the importance of ICT was positive with a mean of $4.022 \pm 0.0.860$.

4.4 Perception About ICT

The students confirmed ICT's role in enhancing literacy with a mean of 4.0777 and it positively impacts student achievement with a mean of 3.94 ± 0.86 and placement opportunities with mean of 3.91 ± 0.84. Challenges like electricity issues with a mean of 3.9141 and institutional support with a mean of 3.4560 were noted. Opportunities in customized education delivery with a mean of 4.1309 and monitoring student progress were also recognized, representing an overall positive perception on opportunities while using ICT with a mean of 3.94835 ± 0.811223.

4.5 Hypothesis Testing

H1: Usage of Information and Communication Technology in education is significantly determined by their personal profile. Usage of Information and Communication Technology in education is significantly determined by their gender ($t = -2.573$, $P = 0.010$) and the importance of ICT ($t = -4.256$, $P = 0.000$), with males having higher levels of ICT utilization perception. Additionally, was insignificant in terms of age, marital status and religion ($p > 0.05$). Therefore, H1 is rejected.

In order to test H2, H3, H4 and H5 regression analysis was used and demonstrated in the Table 3.

Table 3 Hypotheses testing on ICT in higher education

Hypothesis	Model	R	R2	F	β	t-value	Sig	Inference
H2	Students' perception → adoption	0.620	0.384	303.772	0.620	17.429	0.000	Accepted
H3	Adoption of ICT → students' performance	0.551[a]	0.303	212.202	0.551	14.567	0.000	Accepted
H4	Adoption of ICT → students' placement opportunities	0.487[a]	0.237	151.186	0.487	12.296	0.000	Accepted
H5.1	Challenges → placement opportunities	– 0.293	0.086	45.674	– 0.293	– 6.758	0.000	Accepted
H5.2	Challenges → other opportunities	– 0.254	0.065	33.705	– 0.254	– 5.806	0.000	Accepted

Source Primary Data

H2: Students' Perceptions of ICT's importance will positively contribute to adoption. Students' perception of ICT's importance significantly influences its adoption (R = 0.620), explaining a variance of 38.4% (r square). Coefficient results (Table 1) prove this, with a highly significant contribution ($P < 0.001^{**}$). The standardized beta coefficient reveals a 62% increase in ICT adoption for every unit increase in perception (β = 0.620). Hence, students' perception positively drives ICT adoption (see Table 3).

H3: Adoption of ICT in higher education significantly impacts students' Performance; H4: Adoption of ICT in higher education significantly impacts students' Placement Opportunities. ICT adoption is strongly correlated with the students' performance (R = 0.551) and students' placement opportunities (R = 0.487), explaining a variance of 30.3% (r square) and 23.7% (r square) respectively. Coefficient findings in (Table 3) proves this, highlighting a highly significant impact ($P < 0.001^{**}$). The standardized beta coefficient indicates a 55.1% performance increase and 48.7% increase in placement opportunities per unit increase in ICT adoption. Thus, ICT adoption significantly influences higher education students' performance and placement opportunities (see Table 3).

H5: Challenges in the usage of ICT hinder placement opportunities and other opportunities for using ICT. Challenges in ICT usage are negatively correlated with students' placement opportunities (R = – 0.293) and other opportunities of ICT usage (R = – 0.254), explaining a variance of 8.6% (r square) and 6.5% (r square) respectively. Coefficient results confirm this, highlighting a significant contribution ($P < 0.001^{**}$). The standardized beta coefficient indicates a 29.3% decrease in placement opportunities and 25.4% decrease in ICT usage opportunities with per unit increase in challenges and in ICT usage. Thus, challenges in ICT usage hinder placement opportunities and other opportunities in higher education.

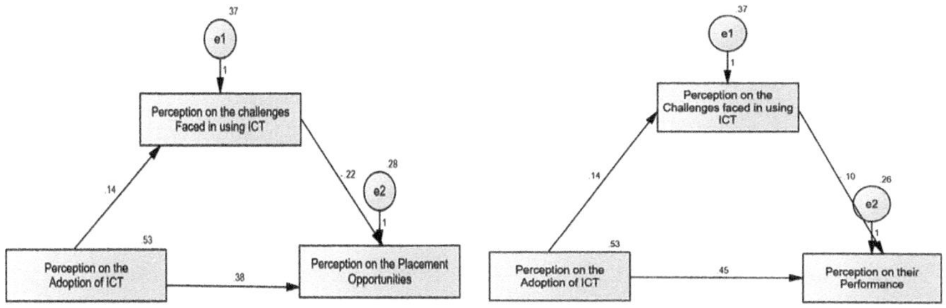

Fig. 3 Challenges in the adoption of ICT usage as a mediator between ICT adoption on placement opportunities and student performance. Source AMOS 23

H6: Challenges in the adoption of ICT usage will hinder the influence of ICT adoption on placement opportunities and student performance. Structural equation modeling (SEM) was employed in the study to evaluate the mediating effect of ICT usage issues on the relationship between ICT adoption and student performance and placement opportunities (see Fig. 3). Model significance was confirmed by SEM results (see Table 3), revealing that while ICT adoption significantly impacts placement opportunities (E = 0.383, $P < 0.000$**) and student performance (E = 0.452, $P < 0.000$**), it slightly affects ICT usage challenges (E = 0.142, $P < 0.001$**). The mediation analysis showed that challenges in ICT usage significantly impede both placement opportunities (E = − 0.219, $P < 0.000$**) and student performance (E = − 0.096, $P = 0.012$). The overall effect of ICT adoption on placement opportunities was estimated at 0.354 and on student performance at 0.438, both highly significant. However, challenges in ICT usage do not mediate the relationship between ICT adoption and placement opportunities or performance, as indicated by negligible and adverse indirect effects. The overall effect was mostly influenced by the direct effects of ICT adoption on placement opportunities (Estimate = 0.383, $P = 0.000$) and student performance (Estimate = 0.452, $P = 0.000$), rather than the total effect.

5 Discussion

Online learning platforms have made it possible for universities to offer extensive online courses and degree programs in the modern higher education landscape, which is characterized by globalization, technological improvements, and changing student demographics [47]. This research found that students frequently utilize the internet for educational purposes and have a positive attitude towards the accessibility of online learning. In line with this, previous studies such as Zhao et al. [48], Kozlova and Pikhart [49] and Patra et al. [50] have also argued that ICT-based learning is often perceived as favorable by college students, that can improve learning, creativity, and student engagement [51]. Nevertheless, due to high dropout rates and low engagement, participation in MOOCs and flipped classrooms is noticeably low [52]. The study shows that Gender stands out as a significant influence, with men demonstrating higher participation which is further found true [53, 54]. The study by Youssef et al. [55] highlights the role of ICT in raising student achievement and also explores the negative effects of ICT usage barriers on student performance and employment opportunities. But the current study gets into the depth by further delving into the issues, demonstrating that barriers for ICT usage not only obstruct placement and opportunities but also have a detrimental impact on student performance. However, obstacles such as insufficient internet access and the intricacy of ICT instruments hinder these prospective advantages. This nuanced viewpoint deepens our knowledge of the importance of ICT in education.

6 Conclusion

The revolutionary potential of information and communication technology (ICT) in global industries—particularly education—is examined in this study. As Traditional classroom settings are being replaced by online and virtual ones, the present study particularly assesses the influence of student's perception regarding the importance of ICT on their ICT adoption and its relevance in contributing to their performance and future opportunities. As per the results, students' perception positively drives ICT adoption which again enhances students' performance and placement opportunities. The first portion of this result aligns with Technology Acceptance Model and further the contribution of ICT adoption on student outcome and future opportunities supporting the theory of technology enhanced learning. However, study also discovered obstacles along with the opportunities, such as insufficient internet access, the intricacy of ICT instruments, lack of initiatives by Institutions and low motivation by the students with respect to ICT adoption need to be further addressed. Nevertheless, in conclusion, study reports that by giving students the access to a wealth of materials, ICT adoption in education improves student performance, assuring students' competitiveness in the digital era and creating worldwide opportunities.

7 Practical Implications

By combining the Technology Acceptance Model, which focuses on how students perceive the importance of ICT, with the Technology Enhanced Learning theory, which suggests integrating technology for improved student performance and opportunities, this research adds theoretical depth to the body of existing research. Moreover, study also argued that ICT adoption issues impede student performance and employment opportunities, highlighting the necessity of infrastructure development, cybersecurity training, and digital literacy. Hence, present study recommends that in order to enhance student results and assist the development of policy frameworks that support them, educational institutions should freely provide a variety of ICT tools, providing provide a road map for academicians to promote cooperation with industry professionals for real-world ICT use, as well as complete educator training programs that incorporate MOOCs, research and publications, educational technology conferences and events, collaboration, peer learning and Flipped Classroom techniques. Improving academic and job preparedness require equal access to technology and knowledge of its importance.

8 Limitations and Future Research

By investigating a wide range of tools, the study could gain a more diverse and nuanced understanding of how students utilize technology for learning purposes. Additionally, recognizing the study's findings of limited usage of MOOCs and the Flipped Classroom could lay the groundwork for further research.

References

1. Abdullah, T.: Information and communication technology in higher education in India: prospects and challenges with special reference to NME-ICT Project. Prof. Mohammad Iqbal Mattoo, p. 153 (2019)
2. Avgerou, C.: The link between ICT and economic growth in the discourse of development. In: Organizational Information Systems in the Context of Globalization, pp. 373–386. Springer, Boston (2003). https://doi.org/10.1007/978-0-387-35695-2_23
3. Wankhade, R.S.: Higher education and NEP-2020. Int. J. Res. Social Sci. Inform. Stud. **8**(1), 51–56 (2021)
4. Oliver, R.: The role of ICT in higher education for the 21st century: ICT as a change agent for education. Retrieved April, Vol. 14, p. 2007 (2002)
5. Jena, P.K.: Impact of Covid-19 on higher education in India. Int. J. Adv. Educ. Res. (IJAER) **5** (2020)
6. Dutta, A.: Impact of digital social media on Indian higher education: alternative approaches of online learning during COVID-19 pandemic crisis. Int. J. Sci. Res. Publ. **10**(5), 604–611 (2020)
7. Youssef, B. A., Dahmani, M.: The impact of ICT on student performance in higher education: direct effects, indirect effects and organisational change, HAL (2008)
8. Wulantari, N.P., Rachman, A., Sari, M.N., Uktolseja, L.J. and Rofi'i, A.: The role of gamification in english language teaching: a literature review. J. Educ. **6**(1), 2847–2856 (2023)
9. Webb, M.E.: Affordances of ICT in science learning: implications for an integrated pedagogy. Int. J. Sci. Educ. **27**(6), 705–735 (2005)
10. Andyani, H., Setyosari, P., Wiyono, B., Djatmika, E.: Does technological pedagogical content knowledge impact on the use of ICT in pedagogy? Int. J. Emerg. Technol. Learn. (IJET) **15**(3), 126–139 (2020)
11. Orr, D., Luebcke, M., Schmidt, J.P., Ebner, M., Wannemacher, K., Ebner, M., Dohmen, D.: Higher education landscape 2030. Springer International Publishing, Cham (2020). https://doi.org/10.1007/978-3-030-44897-4
12. Shukri, A., Nordin, L., Salleh, F.I.M., Raidzwan, S.N.M., Ahmad, R.: UniKL students' perception on synchronous learning using ICT as learning tools to learn english. J. Crit Rev. **7**(8), 793–796 (2020)

13. Chatterjee, P., Gantait, A., Swamy, G.A., George, B.: Information and communication technologies in education: a framework for transforming the Indian education system through smart learning, pp. 283–301 (2023). https://doi.org/10.1007/978-3-031-24101-7_16
14. Al-Rahmi, W.M., Alzahrani, A.I., Yahaya, N., Alalwan, N., Kamin, Y.B.: Digital communication: information and communication technology (ICT) usage for education sustainability. Sustainability **12**(12), 5052 (2020). https://doi.org/10.3390/su12125052
15. Nguyen, L.A.T., Habók, A.: Digital literacy of EFL students: an empirical study in Vietnamese Universities. Libri **72**(1), 53–66 (2022). https://doi.org/10.1515/libri-2020-0165
16. Brata, W.W.W., Padang, R.Y., Suriani, C., Prasetya, E., Pratiwi, N.: Student's digital literacy based on students' interest in digital technology, internet costs, gender, and learning outcomes. Int J Emerg Technol Learn (IJET) **17**(03), 138–151 (2022). https://doi.org/10.3991/ijet.v17i03.27151
17. Almaiah, M.A., Ayouni, S., Hajjej, F., Lutfi, A., Almomani, O., Awad, A.B.: Smart mobile learning success model for higher educational institutions in the context of the COVID-19 pandemic. Electronics **11**(8), 1278 (2022). https://doi.org/10.3390/electronics11081278
18. Abdullah, M.R., Ahmad, K., Elias, N.: Knowledge management system implementation and the performance of higher education institutions in the developing countries: a conceptual framework. J. Theor. Appl. Inf. Technol. **100**(24), 4783–4798 (2022)
19. Mumtaz, L., Ali, M., Fatima, A., Shaikh, F.: Information communication technology (ICT) among university students in Pakistan. Pak. J. Human. Social Sci. **11**(1), 132–143 (2023). https://doi.org/10.52131/pjhss.2023.1101.0335
20. Tennakoon, H., Hansen, J.M., Saridakis, G., Samaratunga, M., Hansen, J.W.: Drivers and barriers of social sustainable development and growth of online higher education: the roles of perceived ease of use and perceived usefulness. Sustainability **15**(10), 8319 (2023). https://doi.org/10.3390/su15108319
21. Almulla, M.A., Al-Rahmi, W.M.: Integrated social cognitive theory with learning input factors: the effects of problem-solving skills and critical thinking skills on learning performance sustainability. Sustainability **15**(5), 3978 (2023). https://doi.org/10.3390/su15053978
22. Comi, S.L., Argentin, G., Gui, M., Origo, F., Pagani, L.: Is it the way they use it? Teachers, ICT and student achievement. Econ. Educ. Rev. **56**, 24–39 (2017). https://doi.org/10.1016/j.econedurev.2016.11.007
23. Chakraborty, D., Dhara, S.K., Santra, A.: Effectiveness of ICT in strengthening the process of higher education system in India. Amity J Manage Res **3**(1), 40–53 (2018)
24. Asrizal, A., Utami, A.W.: Effectiveness of mechanical wave learning material based on ICT integrated CTL to improve students learning outcomes. Jurnal Penelitian Pendidikan IPA **7**(4), 632–641 (2021). https://doi.org/10.29303/jppipa.v7i4.837
25. Fannakhosrow, M., Nourabadi, S., Ngoc Huy, D.T., Dinh Trung, N., Tashtoush, M.A.: A comparative study of information and communication technology (ICT)-based and conventional methods of instruction on learners' academic enthusiasm for L2 learning. Educ. Res. Int. **2022**, 1–8 (2022). https://doi.org/10.1155/2022/5478088
26. Alenezi, M., Wardat, S., Akour, M.: The need of integrating digital education in higher education: challenges and opportunities. Sustainability **15**(6), 4782 (2023). https://doi.org/10.3390/su15064782
27. Ashour, S.: How COVID-19 is reshaping the role and modes of higher education whilst moving towards a knowledge society: the case of the UAE. Open Learn. J. Open Distance e-Learning **39**(1), 52–67 (2024). https://doi.org/10.1080/02680513.2021.1930526
28. Iqbal, J., Yi, X., Ashraf, M.A., Chen, R., Ning, J., Perveen, S., Imran, Z.: How curriculum delivery translates into entrepreneurial skills: the mediating role of knowledge of information and communication technology. PLoS ONE **17**(5), e0265880 (2022). https://doi.org/10.1371/journal.pone.0265880
29. Selmi, S.A.: Use of ICT to promote literacy in Sub-Saharan Africa. Am. J. Multidiscip. Res. Innov. **2**(1), 49–64 (2023). https://doi.org/10.54536/ajmri.v2i1.1182
30. Shukla, D.R.S.S.: Significance of digital technologies and future of education: trends, challenges, and opportunities. SEAS Trans. 1(1) (2023)

31. Kaur, K.: Teaching and learning with ICT tools: issues and challenges. Int. J. Cybern. Inform. **12**(3), 15–22 (2023). https://doi.org/10.5121/ijci.2023.120302
32. Devineni, S.K.: Bridging the data divide: secure sharing and governance for effective cross-agency and industry collaboration. J. Artif. Intell. Cloud Comput. 1–7 (2023). https://doi.org/10.47363/JAICC/2023(2)188
33. Ali, K.S., Faroque, A.R.: Addressing the complexity of the digital divide and the role of government in addressing it: role of government in bridging the digital divide. In: Fostering Sustainable Businesses in Emerging Economies, pp. 127–145. Emerald Publishing Limited (2023). https://doi.org/10.1108/978-1-80455-640-520231009
34. Alajab Ismail, A.M.: The Effect of adopting blended learning approach on students' achievement and satisfaction with learning in a post graduate course in educational statistics. Arab Gulf J. Sci. Res. 1–18 (2019).https://doi.org/10.51758/AGJSR-01-2019-0002
35. Al Shaibani, T.A., Naguib, Y.M., Abdul Razzak, R.L., Ali, F.A.: Impact of covid19- pandemic on students' performance in a problem-based learning system: comparative study between face-to-face and virtual learning. Arab Gulf J. Sci. Res 75–85 (2020). https://doi.org/10.51758/AGJSR-02-2020-0005
36. Puhakka, S.A.: Incorporating cultural and linguistic diversity in information and communication technology in education in Kenya. In: Culturally Responsive Education, pp. 83–94. Routledge (2017)
37. Babu, G.S., Sridevi, K.: Importance of E-learning in higher education: a study. Int J Res Culture Soc **2**(5), 84–88 (2018)
38. Saadé, R., Bahli, B.: The impact of cognitive absorption on perceived usefulness and perceived ease of use in on-line learning: an extension of the technology acceptance model. Inform. Manage. **42**(2), 317–327 (2005). https://doi.org/10.1016/j.im.2003.12.013
39. Fathi Vajargah, K., Azadmanesh, N.: The feasibility of ICT application in curriculum development in higher education. Res. Plan. Higher Educ. Inst. Res. Plan. High. Educ. **12**(4), 49–70 (2023)
40. Ganzert, M., Huber, S., Kaya, M., Sepin, S., Melzer, P.: Adoption, usage, and pedagogy of E-learning tools in university teaching (19) (2017)
41. Ibrahim, A.M., Hassan, M.S.H., Gujbawu, M., Khoshrouzadeh, J.: Influence of ICT adoption on tertiary education: investigating ICT usage in e-learning context among Nigerian students. J. Educ. Practice **9**(16), 21–34 (2018)
42. He, L., Li, C.: Students' adoption of ICT tools for learning english based on unified theory of acceptance and use of technology. Asian J. Educ. Social Stud. **44**(3), 26–38 (2023). https://doi.org/10.9734/ajess/2023/v44i3963
43. Germo, R.R.: Blended learning approach in improving student's academic performance in information communication, and technology (ICT). TransNav, The Int. J. Marine Navig. Safety Sea Transport. **16**(2), 251–256 (2022). https://doi.org/10.12716/1001.16.02.07
44. Phutela, N., Dwivedi, S.: A qualitative study of students' perspective on e-learning adoption in India. J. Appl. Res. High. Educ. **12**(4), 545–559 (2020). https://doi.org/10.1108/JARHE-02-2019-0041
45. Sepulveda-Escobar, P., Morrison, A.: Online teaching placement during the COVID-19 pandemic in Chile: challenges and opportunities. Eur. J. Teach. Educ. **43**(4), 587–607 (2020). https://doi.org/10.1080/02619768.2020.1820981
46. Fornell, C., Larcker, D.F.: Evaluating structural equation models with unobservable variables and measurement error. J. Market. Res. **18**(1), 39–50 (1981)
47. Dieguez, T., Loureiro, P., Ferreira, I.:, Entrepreneurship and leadership in higher education to develop the needed 21st century skills. In: ECMLG 2021 17th European Conference on Management, Leadership and Governance, Academic Conferences limited, p. 143 (2021)
48. Zhao, Y., Sánchez Gómez, M.C., Pinto Llorente, A.M., Zhao, L.: Digital competence in higher education: students' perception and personal factors. Sustainability **13**(21), 12184 (2021). https://doi.org/10.3390/su132112184
49. Kozlova, D., Pikhart, M.: The use of ICT in higher education from the perspective of the university students. Procedia Comput Sci **192**, 2309–2317 (2021). https://doi.org/10.1016/j.procs.2021.08.221

50. Patra, G., Datta, S., Mukherjee, R.: Students' perception of ICT use in higher secondary school students: an exploratory factor analysis approach. In: Digital Technologies for Smart Business, Economics and Education: Towards a Promising Future, pp. 243–259. Springer, Berlin (2023)
51. Basri, M., Paramma, M.A.: EFL students' perspective on the usefulness of ICT based learning in Indonesian higher education. ELT Worldwide State Univ Makassar **6**(2), 105–120 (2019)
52. Klemke, R., Eradze, M., Antonaci, A.: The flipped MOOC: using gamification and learning analytics in MOOC design—a conceptual approach. Educ. Sci. **8**(1), 25 (2018). https://doi.org/10.3390/educsci8010025
53. García, F.A.M., Salguero, F.L., Fernández-Sánchez, M.R., Carrión del Campo, J.L.: Disconnected university students?' Knowledge and use of digital technologies among university students. J. Educ. Online ERIC **19**(2), n2 (2022)
54. Maiti, A, Jana, S.K.: ICT Use in higher education in Eastern States of India: an analysis. In: Developing Countries and Technology Inclusion in the 21st Century Information Society, pp. 230–245. IGI Global (2021)
55. Ben Youssef, A., Dahmani, M., Ragni, L.: ICT use, digital skills and students' academic performance: exploring the digital divide. Information **13**(3), 129 (2022). https://doi.org/10.3390/info13030129

Control and Data Processing of Creep Test Machine in Compliance with ISO 9967

Hanna Madbak, Husam A. Neamah, Hothefa Shaker Jassim, and Zaid Ameen Abduljabbar

Abstract The Mechanical Engineering department carried out an industry-based project on the long-term deformation of thermoplastic pipes under constant stress. A 42-day study was carried out to assess the mechanical stability and performance of the pipes, following ISO 9967 requirements. These included repeatedly measuring at specific intervals and processing and analyzing the acquired data for the ongoing project which needed to be accurate and suitably time-efficient. To meet these demands, a complete diagnostic system was created, including the multiple elements needed to automate the measurement process. Accurate data acquisition, minimal manual intervention, and adherence to international standards were achieved through a similar combination of handheld wireless systems, which allowed for a detailed understanding of the deformation characteristics of thermoplastic materials undergoing prolonged loading conditions.

Keywords Creep · Complementary metal-oxide semiconductor (CMOS) · Thermoplastic-pipe

H. Madbak · H. A. Neamah
Department of Electrical Engineering and Mechatronics, Faculty of Engineering, University of Debrecen, Debrecen, Hungary
e-mail: husam@eng.unideb.hu

H. A. Neamah
College of Engineering, National University of Science and Technology, Dhi Qar, Iraq

Department of Business Management, Al-Imam University College, Balad, Iraq

H. S. Jassim (✉)
Faculty of Informatics, University of Debrecen, Debrecen, Hungary
e-mail: hothefa.shaker@mcbs.edu.om

Z. A. Abduljabbar
Modern College of Business and Science, Muscat, Oman

Department of Computer Science, College of Education for Pure Sciences, University of Basrah, Basrah, Iraq

B. Awwad (ed.), *AI and IoT: Driving Business Success and Sustainability in the Digital Age*, Studies in Systems, Decision and Control 590,
https://doi.org/10.1007/978-3-031-88874-8_96

1 Introduction

Creep refers to the phenomenon where a material gradually deforms or changes shape when exposed to a constant load or stress [1]. The deformation that will be done is inelastic deformation, meaning permanent deformation [2], which is usually seen around the 2-year mark in thermoplastic pipes [3]. When pipes are installed in soil, deflection often increases within two years, influenced by soil and installation factors [4]. The two-year creep ratio aids in long-term calculations as pipes endure consistent pressure [5]. Creep tests (42 days) assess performance under load, revealing creep behavior and modulus [3]. Flexible pipes, like plastic or GRP, can deflect over 2% without damage [6]. Their design focuses on minimizing deflection to ensure durability [7]. Unlike rigid pipes, flexible ones depend on soil for load distribution, with deflection reflecting changes in vertical diameter [8]. Proper installation per ASTM D 2321 prevents excessive deflection [9]. Creep tests are vital for assessing plastic materials' suitability [10]. This project focuses on thermoplastic pipes such as CPVC, HDPE, PB-1, PE, PP, and PVC [11]. Additionally, the creep testing potential applications can inspire developments in robotics and mechatronic systems, particularly in areas requiring precise material deformation analysis, structural durability assessment, and adaptive system responses in dynamic environments such as [12–16]. This specific project involves testing thermoplastic tubing, where a constant force will be applied to measure displacement using a specific sensor, in this case, we are using the KEYENCE-GT2-70 digital contact sensor, which is widely used in material testing [17]. The creep testing process was automated using a Raspberry Pi, displacement sensor, data amplifier, and a cloud-based system. This standard offers comprehensive guidelines for determining specific measurement intervals throughout the 42-day testing period, as well as the mathematical calculations necessary to ascertain the creep ratio accurately. In addition to developing a customized User Interface (UI) for the machine, this initiative aims to enhance user-friendliness and facilitate clear differentiation between the two primary operational modes, per the ISO 9967 standard and a fully dynamic variant. Users can seamlessly commence the process by providing essential information, including the starting date and their email address.

2 Description of Technical Solution

The KEYENCE-GT2-70 digital contact sensor. uses a high-precision linear scale to detect the displacement of an object and convert it into a digital signal that can be read through the GT2 Series including USB-type devices for easy connection to PCs at inspection stations and other locations [18]. There are several advantages of using a contact displacement sensor over a non-contact sensor in certain applications. For one, contact sensors are more precise and repeatable because the displacement of the target object is measured directly, whereas the signals of non-contact sensors could be affected by ambient factors like temperature, humidity, and surface conditions.

Second, contact sensors are more stable than non-contact sensors, as they are not influenced by external factors such as vibration or air movements that may lead non-contact sensors to give misleading readings. Third, contact sensors can be used in more areas, like rough surfaces, non-trans parent materials, and objects with complex shapes or textures, while non-contact sensors cannot be applied in such cases or may require special adjustments. Finally, contact sensors are often cheaper than non-contact sensors, which means that in many applications, they are a more affordable option [18]. This gives this creep test machine a big advantage over others that use non-contact sensors. The newly developed HL LEDs offer remarkably powerful and consistent point-source illumination, boasting nine times the intensity of standard LEDs. This ensures high luminance (HL) levels. The high-resolution CMOS imaging element exhibits exceptional sensitivity, receiving LED light passing through the absolute value glass scale and delivering output signals with double the resolution of conventional models [19]. This guarantees precise displacement measurements regardless of the object's surface color or texture. The I-Processor, a specialized integrated circuit, employs a unique algorithm for swift, high-resolution computation of CMOS sensor output signals [17].

2.1 Data Amplifier

The KEYENCE The DL-EN1 amplifier is a versatile data amplifier that can be linked with various sensors and can be set to control all of them in the same way. This function provides flexible setups, so the sensor's model is not so important but all of them can be controlled through the amplifier. Furthermore, the amplifier is a high-precision device, and it shows a fast response time which is the main reason why it is an excellent choice for use with various types of sensors such as the KEYENCE-GT2-70 digital contact sensor. Overall, the KEYENCE DL-EN1 amplifier lets us amplify sensor data (the one that is needed for measurement and analysis) with the best accuracy and thus is a reliable and effective tool in the process. A TCP/IP connection can be a green choice since it uses less electricity and allows for faster, more reliable performance. Nonetheless, the gains in energy efficiency that are associated with the TCP/IP connection may be spoiled by the fact that many PCs are turned on and are therefore keeping the network alive. Utilization of a Raspberry Pi is the solution here and it can be the beginning of a change that will in turn bring us a more economical and environmentally friendly reading process as well [6]. Specifically, in Fig. 1, we can spot how the displacement sensor is going to be put relatively close to the thermoplastic pipe. A spindle is used in a specific sensor to have the mechanism to conduct contact with a measured target object.

The flow chart in Fig. 2 outlines the system's operation. It begins by prompting the user to input details such as the starting date, time, hour, and minute, along with their email address and desired Excel file name. The user then selects the mode of operation: either using the preset ISO standard measurement times or inputting their own timings. When it's time for reading, the system sends a signal to the sensor to

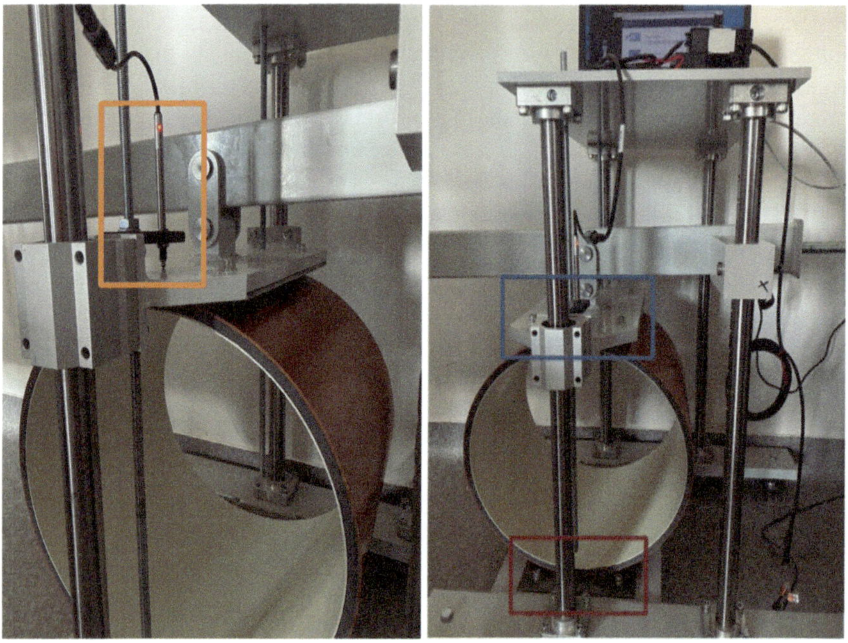

Fig. 1 KEYENCE DL-EN1 data amplifier, with TCP/IC connection, and shows how everything comes together: Plates on the creep test machine

take a measurement, which is then processed and written in a more organized format before being saved in the designated Excel file with the corresponding date and time. Finally, a copy of the file is sent to the user for their records.

In addition to that for safety and accurate readings, a warning system was implemented that alerts the user via email when the measured distance surpasses a specific value (minimum of 0.5 [mm] and maximum of 12.5 [mm], indicating that the sensor has been displaced and needs to be repositioned. This feature serves as a preventive measure to ensure accurate and reliable data collection.

2.2 User Interface

The bespoke user interface designed for this machine facilitates intuitive interaction and comprehension of its operation. The illustration below provides a visual representation of the primary page, or the operational interface, of the system. In Fig. 3a, the uppermost section of the UI is dedicated to essential information input. Enclosed within a section labeled 'Information,' users are prompted to provide their email address for communication with the machine. Additionally, users are required to specify the desired filename for the Excel document that will be generated, containing

Fig. 2 Flow chart of the operation of the creep test machine

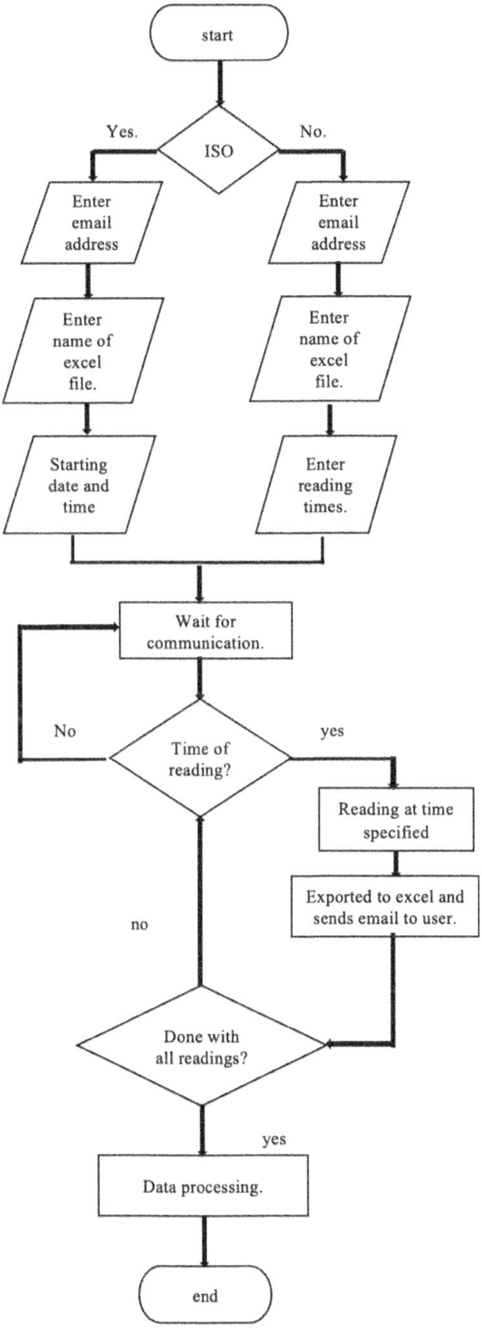

updated values and data, to be subsequently sent to the user via email after each reading. The section labeled 'Starting Date' encompasses a functional calendar interface, allowing users to input specific details such as the year, month, day, hour, and minute.

Figure 3b, illustrates an error warning triggered when a date in the past is entered. This caution is issued because setting a prior day or hour would lead to operational issues within the system, making it infeasible to initiate from a previous point in time. At the bottom of the UI, we can see a box titled 'reading times', here the user can decide if they want to use the standardized time by the ISO 9967 standard or fill in when they want the measurements to be taken as described in Fig. 4a.

The second page of the UI, as depicted in Figs. 4b and 5b, provides a comprehensive overview after the input of all requisite data. Notably, it presents a deflection and

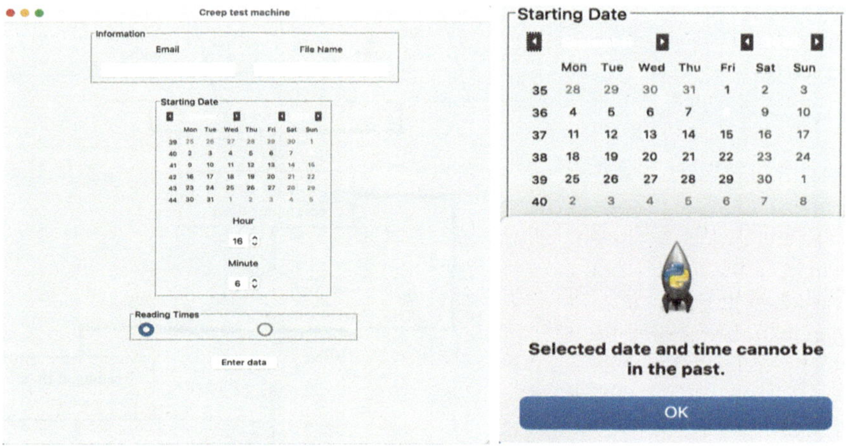

(a) (b)

Fig. 3 **a** The main page of UI. **b** Shows an error from entering the wrong time

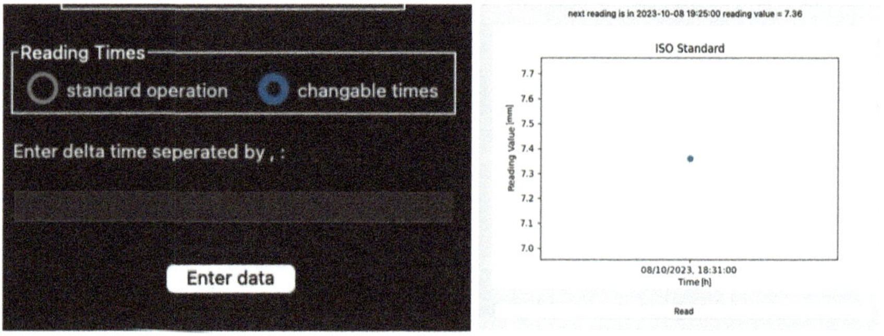

(a) (b)

Fig. 4 **a** Standard and dynamic modes. **b** Deflection and time graph

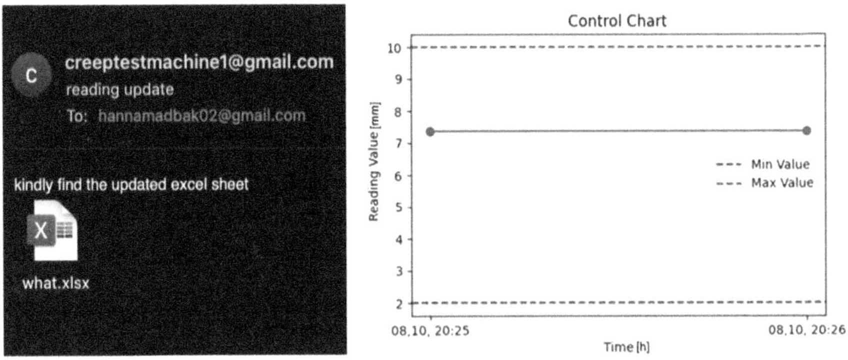

(a) (b)

Fig. 5 **a** Email after reading. **b** Control graph

time graph that serves as a visual aid for understanding their relationship. The blue dot at the center signifies the initial reading. At the top of the page, there is an informative indicator that apprises the user of the timing for the next reading see Fig. 5. This timing can vary depending on whether the program is executed in compliance with ISO 9967 standards or using the dynamic option. Furthermore, the page offers a 'Read' button, as shown at the bottom of Fig. 5a, enabling the user to promptly acquire a real-time reading. After every reading, the value, as well as the time it was taken in is then sent to the designated Excel file, which is then sent to the user via email, being straight to the point and having the subject of 'reading update' as demonstrated in Fig. 5. Another notable development includes the capability to send emails to the machine, thereby verifying its online status and operational integrity. Commands like "next reading?" allow users to receive responses via email. Additionally, issuing the command "update" prompts the system to send an email containing the latest Excel file as an attachment. The second graph which is shown in the second or 'operating' page of the user interface is a control graph. As shown in Fig. 5b.

This graph has been generated to address a critical concern: if the deflection exceeds a certain threshold, there is a risk of the sensor dislodging, thereby compromising the accuracy of the measurement. To mitigate this, a warning system has been implemented. If the deflection surpasses this threshold, an alert email will be promptly dispatched to the user. This allows them to promptly inspect the machine and verify the integrity of all its components.

2.3 *Equations*

One of the test reports will be used as an example for the data processing, pipe A has an inner diameter of 299.218 [mm] and has a starting measured value (MV) of 12.33

[mm]. This is one of four measurements of different cuts of the same thermoplastic pipe to calculate the average two-year deflection as well as the average creep ratio.

In Table 1, the recorded displacement values at various intervals during the 42-day period are presented. These intervals were determined based on the ISO 9967 standard.

The provided plot visually represents the relationship between the hours, as specified by the ISO standard, and the corresponding measured displacement values. It offers a comprehensive graphical depiction of this relationship.an inverse relationship can be seen between displacement and time. Figure 6. Represent the graphical results of MV.

The calculated pre-load in Newtons F_0 can be calculated using Eq. 1 [2] (ISO9967).

$$F_0 = 0.00025\, d_n \cdot L \tag{1}$$

Table 1 Test pipe values

Time (t_i) [h]	Measured value piece A [mm]	Measured value piece B [mm]	Measured value piece C [mm]	Measured value piece D [mm]
0.1	7.479	7.7205	8.2665	7.4325
1	7.4215	7.452	8.19	7.342
4	7.3645	7.386	8.1255	7.2685
24	7.254	7.2545	8.023	7.1315
168	7.034	7.057	7.85	6.9005
336	6.934	6.94	7.729	6.76
504	6.849	6.8932	7.672	6.663
600	6.8305	6.837	7.6405	6.567
696	6.778	6.8012	7.6185	6.5595
840	6.753	6.7605	7.5775	6.5275
1008	6.727	6.711	7.5345	6.503

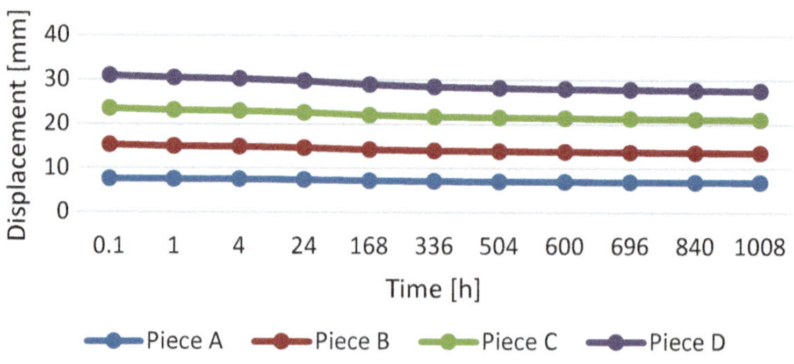

Fig. 6 Measured values and time plot

where:

F_0 calculated pre-load [N], d_n is the nominal outside diameter of the pipe. [mm] and *L* is the calculated average length of the pipe [mm].

Therefore:

$$F_0 = 0.00025 * 398.21 * 1600 = 159.284\ [\mathrm{N}] \tag{2}$$

For each of the 4 test pieces we must plot the deflection in meters against the logarithm of time in hours in a single logarithmic co-ordinate system then determine by linear regression the equation of the straight line.

$$\mathrm{Yt} = B + Mlogt \tag{3}$$

through all eleven points and through the last ten points, through the last nine points, …, and through the last five points, where the constants B and M and the coefficient of correlation, R, are determined using Eqs. (3) to Eq. (5), i.e., the method of least squares [2] (ISO9967).

$$M = \frac{N \sum x_i * y_i - \sum y_i \sum x_i}{N \sum x_i^2 - (\sum x_i)^2} \tag{4}$$

$$B = \frac{\sum y_i - M \sum x_i}{N} \tag{5}$$

$$R = \left[\frac{M(N \sum x_i y_i - \sum x_i \sum y_i)}{N \sum y_i^2 - (\sum y_i)^2}\right]^{1/2} \tag{6}$$

$$yi = y0 - measured\ value \tag{7}$$

the deflection y_i is calculated by using Eq. 7 which is by subtracting the Starting measured value (y_0) which is 12.33 and the measured value at the specific times t_i.

Where:

B: is the theoretical deflection, in millimeters. At t = 1 h, M: is the gradient coefficient, N: is the number of points on the deflection curve used for the linear regression, R: is the coefficient of correlation (if R has a value between 0.99 and 1 it is assumed that the plotted points lie on a straight line), Y_2: is the extrapolated deflection in millimeters after 2 years (17,520 h), t_i: is the time for point i, $X_i := \log(t_i)$ and Yi: is the measured total deflection at time t_i.

The 2-year deflection can be calculated using Eq. 8.

$$y2 = B + M * \log 17520 \tag{8}$$

Now having the deflection, we can determine the creep ratio using Eq. 9.

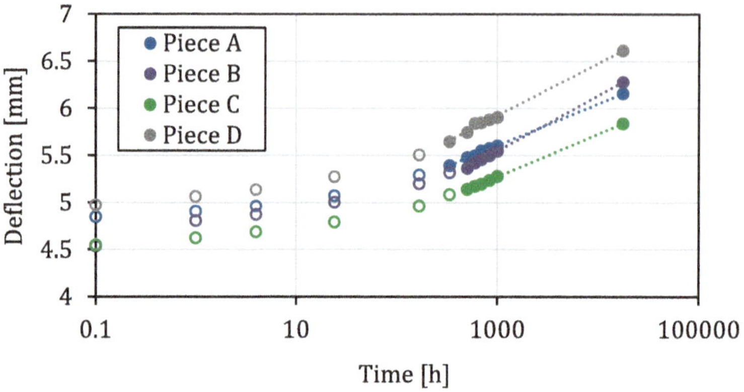

Fig. 7 Deflection and logarithmic time graph

$$Y_a = \frac{y_2\left(0.0186 + 0.025\frac{y_0}{d_i}\right)}{y_0\left(0.0186 + 0.025\frac{y_2}{d_i}\right)}$$

$$Y_a = \frac{6.158501323\left(0.0186 + 0.025\frac{4.851}{299.218}\right)}{4.851\left(0.0186 + 0.025\frac{6.158501323}{299.218}\right)} = 1.262276715 \tag{9}$$

where:

Y_0 initial deflection [mm] and
d_i pipe inner diameter [mm].

With

Piece B having a creep ratio of 1.3727732.
Piece C having a creep ratio of 1.276089283.
Piece D having a creep ratio of 1.320399297.
Average creep ratio of 1.307884624.

Finally, the last step is to determine the logarithmic time and deflection plot. Figure 7 represent the graphical results of Logarithmic Time graph.

3 Conclusion

This project automated the creep testing process for thermoplastic pipes having a sustained load over a 42-day period, aligning with the ISO 9967 standard. Precise measurements were achieved through the integration of the KEYENCE digital contact sensor, Raspberry Pi, data amplifier, and a cloud-based interface. The utilization of a user-friendly interface streamlined operation, enabling email notifications for safety and data accuracy. The showcased results underscore the system's efficiency and reliability, as the test pipe demonstrated deformation. In this example we are unable to accept nor reject this pipe as each use needs a specific creep ratio, but here this shows that we have a rigid pipe that is less prone to deformation because the creep value is close to 1.

References

1. Rosti, J., Koivisto, J., Laurson, L., Alava, M.J.: Fluctuations and scaling in creep deformation. Phys. Rev. Lett. **105**, 100601 (2010). https://doi.org/10.1103/PhysRevLett.105.100601
2. Blum, W., Eisenlohr, P., Breutinger, F.: Understanding creep—a review. Metall. Mater. Trans. A **33**, 291–303 (2002). https://doi.org/10.1007/s11661-002-0090-9
3. ISO 9967:2016(en), Thermoplastics pipes Determination of creep ratio (n.d.). https://www.iso.org/obp/ui/#iso:std:iso:9967:ed-3:v1:en. Accessed 6 Dec 2024
4. Lyu, C., Liu, J., Ren, Y., Liang, C., Liao, Y.: Study on very long-term creep tests and nonlinear creep-damage constitutive model of salt rock. Int. J. Rock Mech. Min. Sci. **146**, 104873 (2021). https://doi.org/10.1016/j.ijrmms.2021.104873
5. Le GEMMA: modes de marches et d'arrêts, GRAFCET de coordination des tâches, conception des systèmes automatisés de production sûrs : réglementation et normes de sécurité, BAC, STI, BTS, DUT, IUP, écoles d'ingénieurs - Simon Moreno - Librairie Mollat Bordeaux, n.d. https://www.mollat.com/livres/1516769/simon-moreno-le-gemma-modes-de-marches-et-d-arrets-grafcet-de-coordination-des-taches-conception-des-systemes-automatises-de-production-surs-reglementation-et-normes-de-securite-bac-sti-bts-dut-iup. Accessed 6 Dec 2024
6. Sargand, S.M., Hazen, G.A., White, K., Moran, A.: Time-dependent deflection of thermoplastic pipes under deep burial. Transp. Res. Rec. **1770**, 236–242 (2001). https://doi.org/10.3141/1770-30
7. Picinbono, B.: On deflection as a performance criterion in detection. IEEE Trans. Aerosp. Electron. Syst. **31**, 1072–1081 (1995). https://doi.org/10.1109/7.395235
8. Hiyoshi, N., Itoh, T., Sakane, M., Tsurui, T., Tsurui, M., Hisaka, C.: Development of miniature cruciform specimen and testing machine for multiaxial creep investigation. Theoret. Appl. Fract. Mech. **108**, 102582 (2020). https://doi.org/10.1016/j.tafmec.2020.102582
9. Standard Practice for Underground Installation of Thermoplastic Pipe for Sewers and Other Gravity-Flow Applications (n.d.). https://www.astm.org/d2321-20.html. Accessed 6 Dec 2024
10. Soderberg, C.R.: The interpretation of creep tests for machine design. Trans. Am. Soc. Mech. Eng. **58**, 733–743 (2023). https://doi.org/10.1115/1.4020382
11. Yu, K., Morozov, E.V., Ashraf, M.A., Shankar, K.: A review of the design and analysis of reinforced thermoplastic pipes for offshore applications. J. Reinf. Plast. Compos. **36**, 1514–1530 (2017). https://doi.org/10.1177/0731684417713666
12. Neamah, H.A., Butdee, R.: B-splined trajectory modified generation to maximize speed of the nonholonomic AMR robot. J. Mach. Eng. **24** (2024). http://jmacheng.not.pl/pdf-196037-116709?filename=116709.pdf. Accessed 29 Nov 2024
13. Neamah, H.A., Khudhair, M.A., Dhaiban, M.S.: Navigating attention-centric: a machine learning approach to EMG-based hand gesture recognition for interactive RC car. In: 2024 IEEE 21st International Power Electronics and Motion Control Conference (PEMC), pp. 1–6 (2024). https://doi.org/10.1109/PEMC61721.2024.10726353
14. Neamah, H.A., Mayorga Mayorga, O.A.: Optimized TD3 algorithm for robust autonomous navigation in crowded and dynamic human-interaction environments. Results Eng. **24**, 102874 ((2024)). https://doi.org/10.1016/j.rineng.2024.102874
15. Neamah, H.A., Donát, E.A., Korondi, P.: Optimizing autonomous navigation in unknown environments: a novel trap avoiding vector field histogram algorithm VFH+T. Results Eng. In press, 1–29 (2024). https://doi.org/10.1016/j.rineng.2024.102625

16. Neamah, H.A., Butdee, R.: Optimization modeling parameters for industrial AMR slippage using ANFIS system in dynamic environment. In: Janmanee, P., Chujuarjeen, S., Butdee, S., Srikhumsuk, P., Batako, A.D.L., Burduk, A., Xavior, M.A. (eds.) Advanced in Creative Technology—Added Value Innovations in Engineering, Materials and Manufacturing, pp. 214–223. Springer Nature Switzerland, Cham (2024). https://doi.org/10.1007/978-3-031-59164-8_18
17. Downloads: High-Accuracy Digital Contact Sensor—GT2 series | KEYENCE America (n.d.). https://www.keyence.com/products/measure/contact-distance-lvdt/gt2/downloads/. Accessed 6 Dec 2024
18. Thathachary, S.V., George, B., Kumar, V.J.: Digital converter for a contactless displacement sensor. IEEE Trans. Instrum. Meas. **64**, 2155–2164 (2015). https://doi.org/10.1109/TIM.2015.2390957
19. Hosticka, B.J.: CMOS sensor systems. Sens. Actuators, A **66**, 335–341 (1998). https://doi.org/10.1016/S0924-4247(98)00008-9

The Role of Digital Leadership in Achieving Digital Transformation: A Qualitative Study on Government Institutions in the Sultanate of Oman

Hamda Ali Sorour Al Shamsi and Muhammad Al-Jeddawi

Abstract The research aims to study digital leadership in achieving digital transformation within government institutions in the country, by understanding how digital leadership affects digital transformation, and how to overcome obstacles that may face government institutions. The qualitative approach was used here, through semi-organized interviews with a group of officials, the sample was selected based on their experience in managing digital transformation, and leadership, in their institutions, and the research found that digital leadership has a role in the success of digital change within institutions, and the existence of challenges facing digital leadership, by rejecting change, lack of training, accordingly, the research recommended the need to provide specialized training programs to enable administrators to acquire digital skills effectively, promote digital culture within government institutions, and encourage Employees to interact and adopt digital practices, in conclusion, the study confirms that digital management contributes to the success of digital transformation in institutions in the country.

Keywords Digital leadership · Digital transformation · Government institutions · Digital innovation

1 Research Introduction

In the modern era, the use of digital transformation has become an essential part of innovation in government and private institutions, aimed at improving efficiency, reducing materiality, facilitating access to services, to adopt these technologies and implement successful digital transformation projects.

H. A. S. Al Shamsi (✉)
Al-Buraimi Governorate Office, Al-Buraimi, Oman
e-mail: Hamdaalisoroor@gmail.com

M. Al-Jeddawi
Modern College of Business and Science (MCBS), Muscat, Oman
e-mail: mohamed.elgeddawy@mcbs.edu.om

B. Awwad (ed.), *AI and IoT: Driving Business Success and Sustainability in the Digital Age*, Studies in Systems, Decision and Control 590,
https://doi.org/10.1007/978-3-031-88874-8_97

In the Sultanate of Oman, the Omani government has begun to take serious steps towards achieving digital transformation, launching several initiatives and programs aimed at digitizing government services and improving their delivery to citizens and residents. Studies have shown that digital leadership is a key element in the success of these initiatives, through its ability to make strategic decisions, catalyze change, and ensure that the necessary resources are available to support this transformation. Annual [1].

1. **Research problem**: The research problem is that with many challenges in organizations, resistance to change from some employees, lack of trained cadres, digital leadership is an essential part of the success of digital transformation, as it plays a major role in determining strategies, change, and motivation.
2. **The importance of the study**: From being focused on digital management within government organizations in the country, one of the priorities in Oman Vision 2040, the study helps to understand how digital leadership can be a key driver in improving government performance, and enhancing efficiency, as the results of the study can provide practical insights for officials and decision-makers, to achieve effective results in digital transformation programs.
3. **Objectives of the study**:
    (a) Study the importance of digital leadership in achieving digital transformation within government organizations in the country.
    (b) Analyze obstacles within government agencies during the implementation of digital programs.
    (c) Compare digital leadership plans, and how they affect the success of digital transformation.
    (d) Provide recommendations to improve the capacity of government departments to effectively manage digital transformation.
4. **Study Methodology**: The qualitative approach in order to deepen the understanding of the context that will affect the digital transformation process in the local environment through the collection of rich and detailed data.

2 Literature Review

Through the development of government services and the adoption of modern technologies to improve the administrative performance of citizens, digital leadership is a key factor in accelerating and directing this digital transformation, as these processes require leadership with an understanding of digital transformation in government institutions.

- Al-Batly [2]: Transformation faces significant challenges related to organizational structure, as the ability of leadership to manage change and use technology strategically is a decisive factor in the success or failure of digital transformation.

- Al-Badri study [3]: clarified the improvement of government performance, and explained that digital management contributes to accelerating the provision of electronic services and improving efficiency in dealing with citizens. The following research has shown the importance of digital management in the success of digital transformation:
- Al-kindi study [4]: It dealt with achieving permanent leadership among school principals in the Sultanate in the country, the study showed that digital management contributes to trust between teachers and directing them to adopt digital systems.

1. **Challenges facing digital transformation in government institutions:** Government institutions face several challenges, the most prominent of which is resistance to change, lack of digital skills, and lack of a clear strategy for implementation, as indicated by the study of Al-Batli [2] that digital transformation in government institutions faces great challenges related to the organizational structure, as the ability of leadership to manage change and use technology strategically is a decisive factor in the success or failure of digital transformation.
2. **Gaps addressed by the current research:** Lack of in-depth studies on digital governance in Oman: Research generally focuses on other countries such as China or Western countries, so the need for in-depth study of state institutions. The cultures of countries differ by transformation, so this topic needs studies in the Omani context. Through this literature review, it has become clear that there is an urgent need to conduct more studies and provide practical recommendations for government leaders to develop appropriate digital transformation strategies.

3 Theoretical Framework

This study has the conceptual basis of digital leadership and its role in achieving digital transformation within organizations. Based on previous literature, this framework provides an analysis of the conceptual and applied elements that form the basis of digital transformation.

1. **The concept of digital leadership**: Digital leadership refers to the ability of leaders to guide teams and implement strategies that support digital transformation using modern tools and technologies. According to Al-Azmi [5] study, digital leadership is a key driver of organizational innovation, as it focuses on enhancing digital competencies and improving processes using advanced technologies.
2. **Digital transformation**: its concept and elements: Digital transformation is a comprehensive redesign of institutional practices using modern technology with the aim of improving performance and increasing efficiency. Elements of digital transformation include: Technology: The use of digital systems to accelerate processes.

3. **Organizational change**: Develop a more flexible and responsive organizational structure. Human Resources: Raising the efficiency of employees through continuous training [3].
4. **The relationship between digital leadership and digital transformation**: The literature suggests a close relationship between digital leadership and digital transformation. Al-kindi [4] study showed that digital leadership fosters a culture of innovation and change, helping organizations adapt to the demands of the digital age.
5. **Digital leadership in promoting organizational innovation Digital leadership**: is a key factor in fostering innovation. According to Ben Jumaa [6], leaders with strategic vision and technology capabilities contribute significantly to the implementation of innovative digital initiatives, increasing the competitiveness of organizations.
6. **Challenges facing digital leadership**: The main challenges hindering digital leadership include: Technical challenges: lack of technical competencies and lack of technological resources. Organizational challenges: resistance to change and lack of a clear strategy. Leadership challenges: Poor digital leadership skills and inability to manage teams in digital environments [7].
7. **Digital leadership strategies to achieve digital transformation**:

   (a) Develop continuous training programs to enhance digital skills.
   (b) Investing in modern technologies to enhance institutional efficiency.
   (c) Building teams capable of innovating and adapting to changes.

8. **Recommendations for future research**: Comparative studies between Oman and other countries in the region or globally are recommended to examine how digital leadership strategies differ and their impact on transformation.

   (a) Study the impact of the societal cultural challenge on the application of digital leadership in more detail, focusing on the factors that may hinder changing the organizational mindset towards digital transformation.
   (b) Conducting future studies focusing: on young and emerging leaders within government institutions.
   (c) Study the impact of artificial intelligence technologies in accelerating the transformation of government institutions.
   (d) Quantitative studies to study the relationship between the level of digital leadership and the achievement of digital transformation goals.' [4].

4 Conclusion

This study summarizes that digital leadership has a critical role in achieving digital transformation in Omani government institutions. Through the results of the study, it is clear that effective leadership contributes to defining strategic vision, managing change, and promoting digital innovation. To achieve digital transformation in government institutions in the Sultanate of Oman, through several axes: Digital leadership has helped define clear plans for digital transformation, within institutions. Leaders in government organizations have adopted plans for successful change management and employee engagement. Digital management has contributed to strengthening the culture of innovation within government institutions. Through the research, it was found that digital leadership positively affects the outcomes of digital transformation in government institutions, such as improving the quality of government services, increasing operational efficiency, and enhancing citizen participation in digital services.

References

1. Annual Report: Ministry of Transport, Communications and Information Technology, Sultanate of Oman. Strategic Directions for Digital Transformation in Oman (2022)
2. Al-Batli, A.: I Challenged the Application of Digital Leadership from the Point of View of Employees, a Field Study. King Saud University, Saudi Arabia (2023)
3. Al-Badri, K.H.: Digital transformation in Omani government institutions: challenges and opportunities. J. Manage. Econ. Digi. Leaders. (2022)
4. Al-kindi, L.: The Role of Digital Technology in Achieving Sustainable Leadership Among School Principals in the Sultanate of Oman. The digital repository of Zawiya University (2021)
5. Al-Azmi, H.: The reality of the application of digital leadership in the Kuwaiti secondary stage and its relationship to the level of job satisfaction of teachers. J. Faculty Educ. Mansoura Univ. (2022)
6. Bin Juma'a, N.A.: The Reality of Academic Leaders' Practice of Digital Transformation Skills at the Saudi University. Faculty of Education, Mansoura University (2024)
7. Shuaili, S.: The level of availability of digital leadership skills among administrative leaders in the Directorate General of Education in the Governorate of Al Dakhiliya in the Sultanate of Oman in the light of some contemporary models. Link for Arab educators (2023)

MANET Hybrid Routing Protocols: Definition, Advantages and Limitations

Hothefa Shaker Jassim, Zeyad T. Sharef, Muna Mohammed Al Mukhaini, and Seemaa Abbas Jawdat

Abstract Several categories of Ad Hoc Mobile network protocols can be classified as proactive, reactive, and hybrid routing protocols. The best features of proactive and reactive routing protocols are gathered in hybrid routing protocols. This research paper discusses four different hybrid routing protocols in mobile ad hoc networks and the mechanism they follow based on the reactive and proactive routing protocols to establish the best route between the communicated nodes which are: CEDAR, ZRP, ZHLS, and SHARP. Also, it presents the advantages and the limitations of each protocol and the reason for them. Additionally, it summarizes some of the comparison criteria between them which are: Network Overhead, Packet Loss, Latency, Link Quality, and Network type.

Keywords Hybrid routing protocols · MANET · Protocols · Reactive · Proactive protocols

1 Introduction

Mobile Ad Hoc Network (MANET) is a temporary wireless network that uses wireless technology media to connect the sender and receiver without a pre-build infrastructure. The nodes communicate directly with each other without a core or centralized intermediate between them. In MANETs, all nodes are in an equivalent scale where all of them act as self-coordinators in a constant network topology change where they can behave as a client, server, or router. Nodes in MANET can join and

H. S. Jassim (✉) · M. M. Al Mukhaini
Modern College of Business and Science, Muscat, Oman
e-mail: hothefa.shaker@mcbs.edu.om

M. M. Al Mukhaini
e-mail: muna.almukhaini@mcbs.edu.om

Z. T. Sharef · S. A. Jawdat
College of Electronics Engineering, Ninevah University, Mosul, Iraq
e-mail: zeyad.tariq@unoninevah.edu.iq

B. Awwad (ed.), *AI and IoT: Driving Business Success and Sustainability in the Digital Age*, Studies in Systems, Decision and Control 590,
https://doi.org/10.1007/978-3-031-88874-8_98

leave the network at any time, and they are fully free to move and change their locations. Thus, the wireless links are continuously originated or destroyed. Because of the highly vital domain and other characteristics of MANET, a routing protocol is extremely needed to manage the information exchanged between the source and destination nodes.

Proactive, Reactive and Hybrid protocols are types of mobile ad hoc routing protocols that play a role in managing communications and controlling the information interchanged between the ad hoc network nodes as well as finding the route of the communication.

Proactive routing protocol—also known as Table Driven—allows all the ad hoc devices to store a routing table that contains all route information to all possible destinations and whenever the network topology changes, the table keeps on updating its route records and modifying the path details. On the other hand, Reactive routing protocol—also called On-Demand Routing Protocol- is a real-time path discovery to the destination whenever the connection is needed. The Reactive protocol has two major components which are route discovery: to get the path, and route maintenance, which maintains the communication failure. A combination of both previous protocols generates a Hybrid Routing Protocol which uses the route table mechanism of the Proactive routing protocol and the discovery maintenance mechanism of the Reactive routing protocol. The hybrid routing protocol uses the characteristics of the Proactive and the Reactive protocol to get less overhead (Proactive Routing Protocol) of the network and get better latency (Reactive Routing Protocol) when sending information over the network. Generally, there are several hybrid routing protocols: Core Extraction Distributed Ad Hoc Routing protocol (CEDAR), Zone Routing Protocol (ZRP), Zone-based Hierarchical Link and State Routing Protocol (ZHLS), and Sharp Hybrid Adaptive Routing protocol (SHARP). Despite the mixture of the two protocols (Proactive and Reactive) to get better routing performance, hybrid routing protocols have some limitations and issues. This research paper discusses four hybrid routing protocols and their advantages and limitations [1–4].

2 Hybrid Routing Protocols

The hybrid routing protocols in MANET combine the best features of proactive and reactive routing protocols, as shown in Fig. 1. Those features are exploited in different mechanisms to accomplish better routing in the network. CEDAR, ZRP, ZHLS, and SHARP are some of the hybrid routing protocols that follow different routing methods.

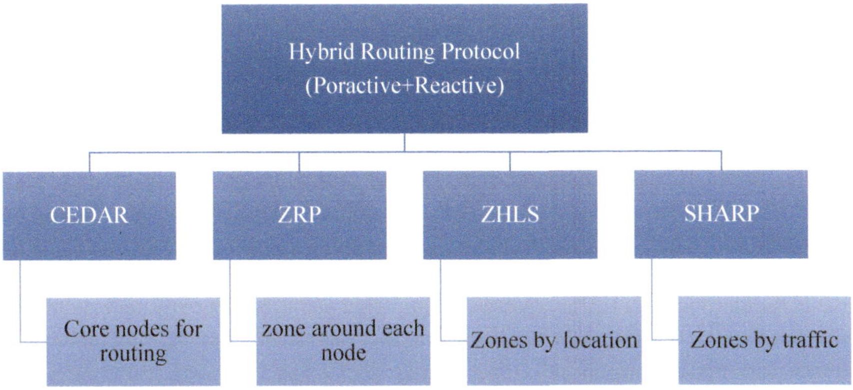

Fig. 1 Classification of hybrid routing protocols

2.1 Core Extraction Distributed Ad Hoc Routing Protocol (CEDAR)

CEDAR from its name (Core), institutes a network centralization for efficient packet delivery. In CEDAR, the institution of the route uses the Reactive routing protocol properties that are performed by the core nodes. The protocol has three components that are segregated into three phases which are:

- ***Core Extraction***: In this phase, several nodes are dynamically nominated to form the network cores
- ***Link State Propagation***: This phase covers more information about bandwidth. The details of stable and available high-bandwidth connections are broadcast to the core nodes in the network, while information about unsettled and low-bandwidth links is not propagated but kept locally.
- ***Route Computation***: the core route is used to build the path between the sender and the receiver.

In this protocol, in every three hops there is at least one selected core node and the route between the core nodes is known as virtual path. Usually, the transmission of the packets is performed by the core nodes in a unicast form. For more efficient packet transmission, every core node must know about its core neighbor nodes and if the neighbor core node is pulled away, all its previous attached nodes start looking for a new core node to establish the virtual path. Also, in this protocol, each node in the network selects a core node as its dominator and its maximum distance is one hop count.

To establish the communication between the source node and destination node, a core route should be found between them and that path is the most stable and has a high bandwidth. First, the sender checks the destination whether it is in the core node's topology table or not. If it is in the core topology table, a route will be established directly between them, and they start the communication but if it is not in the core node records, the sender sends a broadcast route request message to all the core nodes of the network. If the destination is attached to one of the core nodes, a replay message is sent to the source core node to establish the core path but if not, the core node forwards the request to its core node neighbor. When a reply message arrives at the source's core node, the core node evaluates the route quality in terms of the bandwidth and the link stability. If the link quality fits the core node, the connection starts. However, if the link is not as needed, the core node discards the route and sends a request to the other core nods again till it gets the best route quality. Figure 2 shows the source node (1), Distention node (15), Core nodes of the network (3, 5, 11, 12, 13), Network links (between the nodes), and Virtual links (between the Core nodes). While the dominator of node 1 is core node 5, node 12 is donated by node 15.

The main advantages of this protocol are the full utilization of the core nodes to fulfill the path establishment and the quality of the link to ensure smooth and efficient packet delivery. The full utilization of the core node leads to reduced overheads. In addition, in case the link between the core nodes is broken, the protocol is very flexible to restore the connectivity in a short time and efficient way. On the other hand, the main drawback of the protocol is all the communication depends on the core nodes as well as the route establishment and the link computation for checking the link quality. Besides, as the nodes in MANET are free to move, the core nodes may move as well and that will affect the performance of the protocol where the movement leads to assigning new core nodes, calculating the path quality again whenever are

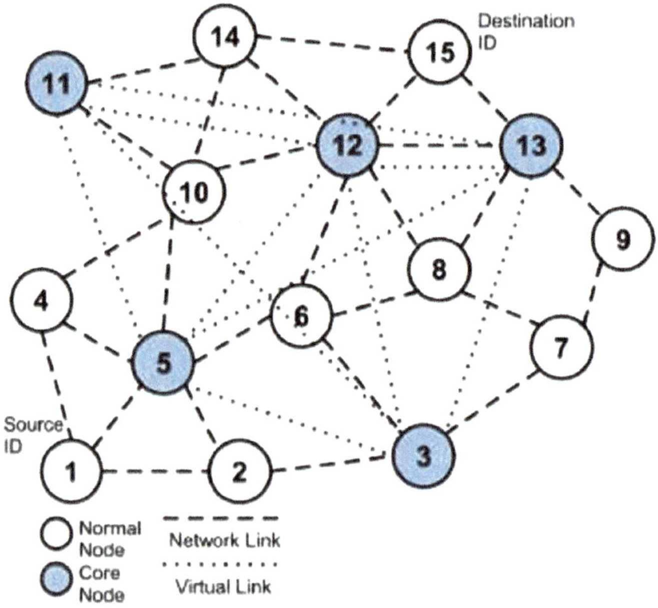

Fig. 2 Connection between source (1) and distention (15) in CEDAR protocol

the core nodes moved. Additionally, the resource consumption of the core nodes is high [5–7].

2.2 *Zone Routing Protocol (ZRP)*

Zone Routing Protocol is a type of hybrid routing protocol that is created to speed up the delivery of information and reduce the processing of the overheads. It was the initial hybrid protocol that used the combination of best features in both proactive and reactive routing protocols by reducing the control overheads which is the feature of the proactive routing protocol and decreasing the latency which is generated by the path discovery in the reactive routing protocol. From its name (ZONE), ZRP allows the nodes of the network to segregate the network into multiple zones by setting a zone around every device containing the nodes *and* number of neighbors. The node that creates the zone considers the next node as a hop distance. It is formulated by two protocols based on the proactive and reactive protocols which are: Intra Zone Routing protocol (IARP) and Inter Zone Routing Protocol (IERP). If the destination is inside the zone, ZRP uses the proactive protocol (IARP) that stores the routing table to deliver the message directly to the distention, but if the receiver is extended outside the zone and located in another zone, IERP takes part to deliver the packet by discovering the route to reach to the destination. Whenever the distention is outside the zone, Bordercast Resolution Protocol (BRP) is used. BRP is a packet delivery service which responsible for sending a broadcast message called Route Request to all the nodes located at the zone's borders (peripheral nodes) and if the destination is in the routing zone, a Route Reply message is sent to the source node, but if not, the Route Request will be forwarded to the node's zone peripherals till the distention found. Figure 3 illustrates how IERP and IARP work:

- IARP: as mentioned previously, this protocol of ZRP is used for route discovery between the source and the destination that are both located in the same zone is shows that source *H* wants to communicate with destination *J* where IARP calculates the best route between the two nodes.
- IERP: whenever the source node and the destination are in different zones, IERP is used to discover the path by exploiting the border nodes of the sender zone. As an example, the sender *S* would reach the receiver *X*, S sends a broadcast request using IERP to all the nodes placed at the border of the zone (Gray nodes: *H*, *G*, *J*, *I*) and the border nodes start to search on their routing table to find whether the destination *X* in reachable to them or not. In Fig. 3, X is recorded in *I*'s routing table, so *I* send a reply to *S* informing them that the destination is reachable to establish the connection [8, 9].

In ZRP, the main strength of the protocol is that it is efficient for large network sizes and the information is updated. Moreover, because of using both proactive and reactive protocol schemes, ZRP has low overheads inside and outside the zones

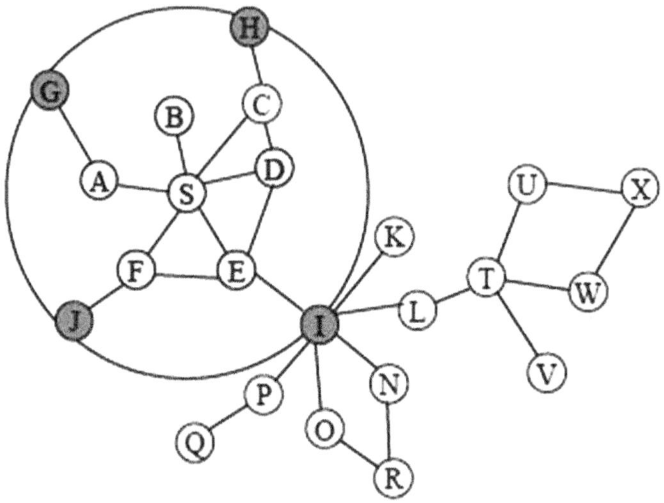

Fig. 3 IARP and IERP of ZRP

by periodic flooding and more effective latency inside the zone. Additionally, ZRP reduces bandwidth waste as compared with the reactive scheme where the mechanism of Route Request (RREQ) flooding is used. In contrast, the drawback of the ZRP is that, the route zone overlap. Since each node in the network has its zone, the overlaps of the zones are very heavy, and due to the forwarding of the RREQ that the border nodes sent, the message may reach the nodes many times without a valid control. Also, the same node may again forward the RREQ message and that increases the complexity of the network. Besides, the route to the destination may not be optimal. In addition, the ZRP consumes the memory of the nodes so the node should have a high memory specification to run ZRP [3, 4, 10–12].

2.3 Zone-Based Hierarchical Link State (ZHLS) Routing Protocol

ZHLS is a hierarchical hybrid routing protocol that utilizes information on the geographical location of the nodes and splits them into non-overlapping zones. There are two main approaches that ZHLS is based on, which are node ID and zone ID which they both assigned for each node where they are used to compute the graphical location of the nodes. By implementing ZHLS, each node knows the connectivity of the nodes within its zone and that is known as the Low Level or Node Level. In the High level or Zone level, the node knows the zone connectivity within the entire network topology. In the same way, all nodes contain two routing tables named as intra-zone routing table (contains information on the node level) and Interzone routing table (includes information on the zone level). Because of that, there are two link state updates (Link state Packet) generated to update the path which are: node level updates—LSP—where each node broadcasts it periodically to all the nodes in its zone and zone level updates—LSP—which is performed by the gateway nodes to update the route between the zones. When the source wishes to send to the distention, it checks its intra-zone table and if the destination is in the source's zone, the communication is started directly because the route details are already recorded. Otherwise, the source sends a location request containing the source and destination IDs through the gateway nodes to all zones. If the destination is found, the gateway of the destination node sends a location reply that includes the destination zone ID to the source. After knowing the destination information, the communication is established based on the source inter-zone routing table. Figure 3 shows an example of how ZHSL split the zones without overlaps. Tables 1 and 2 illustrates the node-level updates and zone-level updates of Fig. 4 network.

The main strength of this protocol against ZRP is that it splits the overlaps of the zones. Also, the information in zone-level topology is spread to all nodes so that no cluster head, thus a single point of failure is avoided, and the overhead of the network is reduced. Besides, because of using the node ID and zone ID of the destination node for routing, this protocol is more flexible and efficient for the dynamic network especially if the destination node does not move from its zone where no need to search for the destination node. However, producing additional traffic is the main disadvantage of ZHLS and that is due to the maintenance of the zone-level topology [5, 7, 11, 13, 14].

Table 1 Node level updates (LSP)

Source	Node LSP
6	8, 15, 16, G
8	6, 12, 16, B, D
12	8, 16, F
15	6, 16, G
16	6, 8, 12, 15

Table 2 Zone level updates (LSP)

Source	Zone LSP
A	B, D, F, G
B	A, C
C	B, G
D	A, B, E
E	D, F
F	A, E, G
G	A, C, F

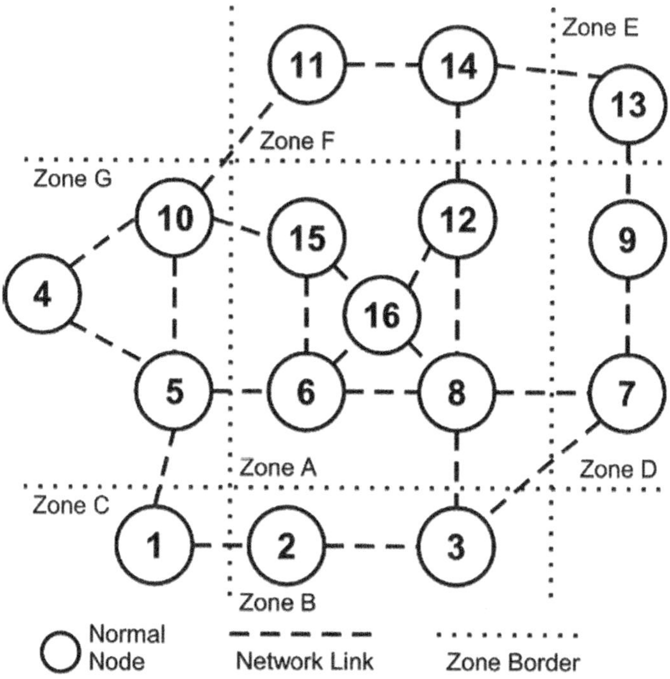

Fig. 4 ZHLS network topology

2.4 Sharp Hybrid Adaptive Routing Based Protocol (SHARP)

SHARP is a hybrid routing protocol that figured out the poise degree between a proactive routing protocol and a reactive routing protocol by adapting the point in which the path details spread proactively against the point in which the route information is required to be discovered reactively. It takes advantage of the proactive routing algorithm which has two powerful features which are efficiency and analytical tractability. Also, it uses any reactive routing mechanism that their cost can be analytically characterized. While the application that uses traditional routing protocols has no control over the performance, SHARP allows the application to keep track of different metrics to achieve the increment of the overhead for proactive information propagation and the decrement of the latency and loss amount. Hence, each node that uses SHARP can independently follow the application performance metrics.

SHARP defines the route details proactively by setting a proactive zone around some nodes. Those zone nodes specify a zone radius to define the nodes within the proactive zone where the nodes within the same or less than the zone radius are members of the proactive zone whereas nodes outside the zone use a reactive routing protocol to establish the route. For the node destination that receives the date from multiple nodes, SHARP originates a proactive zone. When the packets reach a node

Table 3 Analysis of hybrid routing protocols

	Network overhead	Packet loss	Latency	Link Quality	Network type	
					MANET	VANET
CEDAR	Low	Low	Low	High	Applicable	Not used
ZIP	Low	Low	Low	Medium	Applicable	Not applicable for large network
ZHLS	Low	Low	Low	Medium	Applicable	Not Sufficient for a highly dynamic network
SHARP	Depending on the radius of the zone	Depending on the radius of the zone	Low	High	Applicable	Not used

located in the proactive zone, the zone works as a collector where it collects the packets and forwards them efficiently to the destination. This protocol reduces the cost of the route maintenance of the routes that reach the destination in the proactive zones among all the source nodes that communicate with the destination.

The main strength and weakness of the SHARP protocol is based on the zone radius that Sharp creates where it acts as a virtual knob. If the zone radius of the proactive zone increases, it will cover more nodes in the zone so that the loss of the packets and the delay decrease. However, that will cost a large overhead due to the route maintenance of the large zone. In contrast, when the radius of the zone decreases, Sharp decreases the overheads since a small number of nodes are in the zone. Thus, a small number of nodes need to proactively be updated, but that may cause more delay and a high amount of packet loss as compared to the high radius.

3 Findings and Discussion

Generally, hybrid routing protocols follow different methods to establish better routing and more efficient network performance. Table 3 illustrates a comparison between the hybrid routing protocols in terms of Network Overhead, Packet Loss, Latency, Link Quality, and Network type.

4 Conclusion

To conclude, the best features of reactive proactive routing protocols are combined to become hybrid routing protocols such as CEDAR, ZRP, ZHLS, and SHARP. Some researchers believe that the Fisheye state Protocol (FSR), the Temporally Ordered

Routing Algorithm (TORA), and the Optimized link state routing protocol (OLSR) are hybrid routing protocols. They are not hybrid routing protocols, but hybrid routing protocols use some of the mechanism and algorithm features of FSR, TORA, and OLSR to operate the hybrid protocols. FSR and OLSR are proactive routing protocols and TORA is one of the members of the link reversal family. Based on the needs of the network, different hybrid routing protocols are used.

References

1. Cañas, D., et al.: Hybrid ACO routing protocol for mobile ad hoc networks. Int. J. Distrib. Sensor Netw. (2013)
2. Abdulleh, M., et al.: Comparative study of proactive, reactive, and geographical MANET routing protocols. Commun. Netw. **7**(02), 125–137 (2015)
3. Hryschuk, I., et al.: Analysis of routing protocol characteristics in an ad-hoc network. Inf. Telecommun. Sci. **1**, 12–17 (2024)
4. Dubey, S.: Analysis of performance matrices of OLSR, AOMDV & ZRP protocols in the VANET scenario. Int. J. Eng. Sci. Res. Technol. (2014)
5. Onuora, A., et al.: An adaptive hybrid routing protocol for efficient data transfer and delay control in mobile ad hoc network. Int. J. Eng. Trends Technol. **71**(5), 251–258 (2023)
6. Al Hilali, et al.: Study of geographical and energy-aware MANET routing protocols. In: Innovations in Blockchain-Powered Intelligence and Cognitive Internet of Things (CIoT), pp. 135–228. IGI Global Scientific Publishing (2025)
7. Shaker, H., et al.: Vehicle to infrastructure routing protocols: AI and finance perspectives. In: Innovations in Blockchain-Powered Intelligence and Cognitive Internet of Things (CIoT), pp. 341–388. IGI Global Scientific Publishing (2025)
8. Sathish, S., Thangavel, K., Boopathi, S.: Comparative analysis of DSR, FSR and ZRP routing protocols in MANET. In: International Conference on Information and Network Technology (2011)
9. Chandra, A., Thakur, S.: Qualitative Analysis of hybrid routing protocols against network layer attacks in MANET. Int. J. Comput. Sci. Mobile Comput. (2015)
10. Haglan, H.M.: Determining an optimum zone radius for zone routing protocol (ZRP) based on node mobility (2021)
11. Al-Rabayah, M.: Performance evaluation of hybrid routing protocols for multimedia transmission in vehicular ad hoc networks. Int. J. Appl. Eng. Res. **11**. ISSN 0973-4562 (2013)
12. Deshpande, P.M., et al.: Survey on various routing protocols for effective bandwidth optimization in MANET. Int. J. Sci. Res. Educ. (2015)
13. Raheja, K., Maakar, S.K.: A survey on different hybrid routing protocols of MANET. Int. J. Comput. Sci. Inf. Technol. (IJCSIT) (2014)
14. Nawneet Raj et al. (2015). International Journal of Computer Science and Mobile Computing: *Qualitative Based Study of Hybrid Routing Protocols in MANET*.

GPSR Compliance

The European Union's (EU) General Product Safety Regulation (GPSR) is a set of rules that requires consumer products to be safe and our obligations to ensure this.

If you have any concerns about our products, you can contact us on ProductSafety@springernature.com

In case Publisher is established outside the EU, the EU authorized representative is:

Springer Nature Customer Service Center GmbH
Europaplatz 3
69115 Heidelberg, Germany

Batch number: 10755461

Printed by Printforce, the Netherlands